# Switchmode RF and Microwave Power Amplifiers

# Switchmode RF and Microwave Power Amplifiers

## Second Edition

**Andrei Grebennikov**

**Nathan O. Sokal**

**Marc J. Franco**

AMSTERDAM • BOSTON • HEIDELBERG • LONDON
NEW YORK • OXFORD • PARIS • SAN DIEGO
SAN FRANCISCO • SINGAPORE • SYDNEY • TOKYO
Academic Press is an imprint of Elsevier

Academic Press is an imprint of Elsevier
The Boulevard, Langford Lane, Kidlington, Oxford OX5 1GB, UK
225 Wyman Street, Waltham, MA 02451, USA

First edition 2007
Second edition 2012

**Notice**

No responsibility is assumed by the publisher for any injury and/or damage to persons or property
as a matter of products liability, negligence or otherwise, or from any use or operation of any
methods, products, instructions or ideas contained in the material herein. Because of rapid advances
in the medical sciences, in particular, independent verification of diagnoses and drug dosages
should be made

**British Library Cataloguing-in-Publication Data**
A catalogue record for this book is available from the British Library

**Library of Congress Cataloging-in-Publication Data**
A catalog record for this book is available from the Library of Congress

ISBN: 978-0-12-415907-5

For information on all Academic Press publications
visit our web site at books.elsevier.com

Typeset by MPS Limited, Chennai, India
www.adi-mps.com

Printed and bound in Great Britain

12 13 14 15 16   10 9 8 7 6 5 4 3 2 1

# Contents

About the Authors ................................................................................xi
Foreword ...........................................................................................xiii
Preface ...............................................................................................xv
Acknowledgments ............................................................................xxi

**CHAPTER 1  Power Amplifier Design Principles** ...........................1
    **1.1**  Spectral-domain analysis...................................................1
    **1.2**  Basic classes of operation: A, AB, B, and C ..................7
    **1.3**  Load line and output impedance......................................13
    **1.4**  Classes of operation based upon a finite number of harmonics ..17
    **1.5**  Active device models ......................................................20
        1.5.1  LDMOSFETs ......................................................20
        1.5.2  GaAs MESFETs and GaN HEMTs............................24
        1.5.3  Low- and high-voltage HBTs.................................29
    **1.6**  High-frequency conduction angle....................................32
    **1.7**  Nonlinear effect of collector capacitance .......................38
    **1.8**  Push−pull power amplifiers ...........................................42
    **1.9**  Power gain and impedance matching ..............................47
    **1.10** Load−pull characterization.............................................52
    **1.11** Amplifier stability .........................................................54
    **1.12** Parametric oscillations ..................................................62
    **1.13** Bias circuits ..................................................................67
    **1.14** Distortion fundamentals .................................................72
        1.14.1 Linearity.............................................................72
        1.14.2 Time variance ....................................................73
        1.14.3 Memory...............................................................73
        1.14.4 Distortion of electrical signals .............................73
        1.14.5 Types of distortion..............................................74
        1.14.6 Nonlinear distortion analysis for sinusoidal
             signals − measures of nonlinear distortion ....................75
    References........................................................................78

**CHAPTER 2  Class-D Power Amplifiers** ......................................83
    **2.1**  Switchmode power amplifiers with resistive load .......83
    **2.2**  Complementary voltage-switching configuration...........92
    **2.3**  Transformer-coupled voltage-switching configuration ...............97
    **2.4**  Transformer-coupled current-switching configuration................99
    **2.5**  Symmetrical current-switching configuration ............103
    **2.6**  Voltage-switching configuration with reactive load ...............107

**2.7**  Drive and transition time ....................................................111
**2.8**  Practical Class-D power amplifier implementation .................118
**2.9**  Class D for digital pulse-modulation transmitters....................123
References .............................................................................127

**CHAPTER 3  Class-F Power Amplifiers**................................................**129**
**3.1**  Biharmonic and polyharmonic operation modes........................129
**3.2**  Idealized Class-F mode ..........................................................139
**3.3**  Class-F with maximally flat waveforms...................................143
**3.4**  Class-F with quarterwave transmission line .............................151
**3.5**  Effect of saturation resistance and shunt capacitance...............157
**3.6**  Load networks with lumped elements .......................................162
**3.7**  Load networks with transmission lines......................................169
**3.8**  LDMOSFET power amplifier design examples.........................176
**3.9**  Broadband capability of Class-F power amplifiers....................181
**3.10** Practical Class-F power amplifiers and applications ...............183
References .............................................................................190

**CHAPTER 4  Inverse Class-F**................................................................**195**
**4.1**  Biharmonic and polyharmonic operation modes........................195
**4.2**  Idealized inverse Class-F mode ..............................................202
**4.3**  Inverse Class-F with quarterwave transmission line .................205
**4.4**  Load networks with lumped elements .......................................208
**4.5**  Load networks with transmission lines......................................212
**4.6**  LDMOSFET power amplifier design examples.........................222
**4.7**  Examples of practical implementation .....................................226
**4.8**  Inverse Class-F GaN HEMT power amplifiers for
        WCDMA systems................................................................231
        References .............................................................................242

**CHAPTER 5  Class-E with Shunt Capacitance** ...............................**245**
**5.1**  Effect of a detuned resonant circuit........................................245
**5.2**  Load network with shunt capacitor and series filter .................250
**5.3**  Matching with a standard load................................................256
**5.4**  Effect of saturation resistance ...............................................260
**5.5**  Driving signal and finite switching time .................................263
**5.6**  Effect of nonlinear shunt capacitance......................................270
**5.7**  Optimum, nominal, and off-nominal Class-E operation ...........272
**5.8**  Push–pull operation mode.....................................................277
**5.9**  Load networks with transmission lines.....................................281

**5.10** Practical Class-E power amplifiers and applications ................291
References ...........................................................................................300

**CHAPTER 6    Class-E with Finite DC-Feed Inductance** .................305
**6.1** Class-E with one capacitor and one inductor ............................305
**6.2** Generalized Class-E load network with finite DC-Feed
inductance .................................................................................313
**6.3** Subharmonic Class-E ..............................................................320
**6.4** Parallel-circuit Class-E............................................................324
**6.5** Even-harmonic Class-E ...........................................................330
**6.6** Effect of bondwire inductance .................................................332
**6.7** Load network with transmission lines ......................................333
**6.8** Operation beyond maximum Class-E frequency.........................340
**6.9** Power gain.............................................................................345
**6.10** CMOS Class-E power amplifiers...............................................348
References ...........................................................................................354

**CHAPTER 7    Class-E with Quarterwave Transmission Line** ............357
**7.1** Load network with parallel quarterwave line............................357
**7.2** Optimum load-network parameters .........................................364
**7.3** Load network with zero series reactance .................................367
**7.4** Matching circuit with lumped elements ...................................372
**7.5** Matching circuit with transmission lines..................................373
**7.6** Load network with series quarterwave line and
shunt filter................................................................................376
**7.7** Design example: 10-W, 2.14-GHz Class-E GaN
HEMT power amplifier with parallel quarterwave
transmission line.......................................................................378
References ...........................................................................................385

**CHAPTER 8    Broadband Class-E**................................................**387**
**8.1** Reactance compensation technique .........................................387
8.1.1 Load networks with lumped elements ............................388
8.1.2 Load networks with transmission lines...........................394
**8.2** Broadband Class-E with shunt capacitance...............................400
**8.3** Broadband parallel-circuit Class-E .........................................409
**8.4** High-power RF Class-E power amplifiers .................................416
**8.5** Microwave monolithic Class-E power amplifiers.......................419
**8.6** CMOS Class-E power amplifiers................................................424
References ...........................................................................................426

**CHAPTER 9    Alternative and Mixed-Mode High-Efficiency Power Amplifiers** ..................................................**429**

**9.1**    Class-DE power amplifier..........................................430
**9.2**    Class-FE power amplifiers.........................................444
**9.3**    Class-E/F power amplifiers........................................462
    9.3.1  Symmetrical push−pull configurations...................465
    9.3.2  Single-ended Class-E/F$_3$ mode........................471
**9.4**    Biharmonic Class-E$_M$ power amplifier.....................488
**9.5**    Inverse Class-E power amplifiers ..............................495
**9.6**    Harmonic tuning using load-pull techniques................503
**9.7**    Chireix outphasing power amplifiers...........................512
References .................................................................524

**CHAPTER 10    High-Efficiency Doherty Power Amplifiers**..................**529**

**10.1**    Historical aspects and conventional Doherty architecture.......529
**10.2**    Carrier and peaking amplifiers with harmonic control...........540
**10.3**    Balanced, push−pull, and dual Doherty amplifiers.................543
**10.4**    Asymmetric Doherty amplifiers .....................................546
**10.5**    Multistage Doherty amplifiers .......................................550
**10.6**    Inverted Doherty amplifiers ..........................................556
**10.7**    Integration ..................................................................559
**10.8**    Digitally driven Doherty amplifier .................................562
**10.9**    Multiband and broadband capability ..............................564
References.................................................................568

**CHAPTER 11    Predistortion Linearization Techniques**......................**575**

**11.1**    Modeling of RF power amplifiers with memory ..................576
**11.2**    Predistortion linearization ...........................................582
    11.2.1  Introduction ....................................................582
    11.2.2  Memoryless predistorter for octave-bandwidth amplifiers........................................................584
    11.2.3  Predistorter with memory for octave-bandwidth amplifiers........................................................589
    11.2.4  Postdistortion.................................................590
**11.3**    Analog predistortion implementation .............................591
    11.3.1  Introduction ....................................................591
    11.3.2  Reflective predistorters ....................................591
    11.3.3  Transmissive predistorters.................................593
**11.4**    Digital predistortion implementation .............................598
    11.4.1  Introduction ...................................................598

11.4.2 Principles of memoryless digital predistortion .............598

11.4.3 Digital predistortion adaptation .................................601

11.4.4 Digital predistorter performance ................................603

References .................................................................................604

**CHAPTER 12  Computer-Aided Design of Switchmode Power Amplifiers** ..........................................................**607**

12.1  HB-PLUS program for half-bridge and full-bridge direct-coupled voltage-switching Class-D and Class-DE circuits ......608

12.1.1 Program capabilities ...................................................608

12.1.2 Circuit topologies .......................................................609

12.1.3 Class-D *versus* Class-DE ...........................................611

12.2  HEPA-PLUS CAD program for Class-E ...............................613

12.2.1 Program capabilities ...................................................613

12.2.2 Steady-state periodic response .....................................614

12.2.3 Transient response .......................................................614

12.2.4 Circuit topology ..........................................................614

12.2.5 Optimization ...............................................................615

12.3  Effect of Class-E load-network parameter variations ..............616

12.4  HB-PLUS CAD examples for Class-D and Class-DE .............619

12.4.1 Class-D with hard switching ........................................620

12.4.2 Class-DE with soft switching .......................................623

12.5  HEPA-PLUS CAD example for Class-E ................................626

12.5.1 Evaluate a candidate transistor ...................................626

12.5.2 Use the automatic preliminary design module to obtain a nominal-waveform Class-E design ................627

12.5.3 Simulate the nominal-waveforms circuit ......................629

12.5.4 RF output spectrum .....................................................629

12.5.5 Optimize the design, using the nominal-waveforms design as a starting-point ...........................................631

12.5.6 Use the SWEEP function .............................................635

12.6  Class-E power amplifier design using SPICE .........................638

12.7  ADS circuit simulator and its applicability to switchmode Class-E ................................................................................644

12.8  ADS CAD design example: high-efficiency two-stage 1.75-GHz MMIC HBT power amplifier .................................649

References .................................................................................668

Index .................................................................................................669

# About the Authors

**Dr Andrei Grebennikov** is a Senior Member of the IEEE and a Member of Editorial Board of the *International Journal of RF and Microwave Computer-Aided Engineering*. He received his Dipl. Ing. degree in radio electronics from the Moscow Institute of Physics and Technology and PhD degree in radio engineering from the Moscow Technical University of Communications and Informatics in 1980 and 1991, respectively. He has gained long-term academic and industrial experience working with the Moscow Technical University of Communications and Informatics, Russia, Institute of Microelectronics, Singapore, M/A-COM, Ireland, Infineon Technologies, Germany/Austria, and Bell Labs, Alcatel-Lucent, Ireland, as an engineer, researcher, lecturer, and educator. He lectured as a Guest Professor in the University of Linz, Austria, and presented short courses and tutorials as an Invited Speaker at the International Microwave Symposium, European and Asia-Pacific Microwave Conferences, Institute of Microelectronics, Singapore, and Motorola Design Centre, Malaysia. He is an author or co-author of more than 80 technical papers, five books, and 15 European and US patents.

**Nathan O. Sokal** was elected a Fellow of the IEEE in 1989, for his contributions to the technology of high-efficiency switchmode power conversion and switchmode RF power amplification. In 2007, he received the Microwave Pioneer award from the IEEE Microwave Theory and Techniques Society, in recognition of a major, lasting, contribution − development of the Class-E RF power amplifier. In 2011, he was awarded an honorary doctorate from the Polytechnic University of Madrid, Spain, for developing the high-efficiency switchmode Class-E RF power amplifier. In 1965, he founded Design Automation, Inc., a consulting company doing electronics design review, product design, and solving 'unsolvable' problems for equipment-manufacturing clients. Much of that work has been on high-efficiency switchmode nonlinear and linear RF power amplifiers at frequencies up to 2.5 GHz, and switchmode dc−dc power converters. He holds eight patents in power electronics, and is the author or co-author of two books and approximately 130 technical papers, mostly on high-efficiency generation of RF power and dc power. During 1950−1965, he held engineering and supervisory positions for design, manufacture, and applications of analog and digital equipment. He received B.S. and M.S. degrees in Electrical Engineering from the Massachusetts Institute of Technology, Cambridge, Massachusetts, in 1950. He is a Technical Adviser to the American Radio Relay League, on RF power amplifiers and dc power supplies, and a member of the Electromagnetics Society, Eta Kappa Nu, and Sigma Xi honorary professional societies.

**Marc J. Franco** holds a PhD degree in electrical engineering from Drexel University, Philadelphia. He is currently with RFMD, Technology Platforms, Component Advanced Development, Greensboro, North Carolina, USA, where he is involved with the design of advanced RF integrated circuits and integrated front-end modules. He was previously with Linearizer Technology, Inc. Hamilton, New Jersey, where he led the development of advanced RF products for commercial, military, and space applications. Dr Franco is a regular reviewer for the Radio & Wireless Symposium, the European Microwave Conference, and the MTT International Microwave Symposium. He is a member of the MTT-17 HF-VHF-UHF Technology Technical Coordination Committee and has co-chaired the IEEE Topical Conference on Power Amplifiers for Radio and Wireless Applications. He is a Senior Member of the IEEE. His current research interests include high-efficiency RF power amplifiers, nonlinear distortion correction, and electromagnetic analysis of structures.

# Foreword

Among the many power amplifier books on my shelf, not a single one contains more than a chapter or two on switchmode RFPAs. This book fills that gap, and does it well. The authors have contributed to this field for decades and across continents. They have published hundreds of papers in the PA field, hold dozens of patents, are active in current technical conferences, and have led industry projects in advanced RF product development for commercial, military and space applications. It is therefore not surprising that the book presents a comprehensive treatment of power amplifiers, with detailed theoretical analysis as well as practical insights which will be useful for anyone who aspires to become an expert in this field.

A thorough introduction to RF and microwave power amplifiers and fundamental classes of operation is complemented with a useful discussion of currently common device technologies with the fundamentals of their physical principles of operation and circuit modeling. The discussion of parasitic reactances and their effects on the load line gives insight into this important high-frequency effect that many texts do not mention. The next few chapters present a detailed treatment of Class D amplifiers and harmonically-terminated Class F and $F^{-1}$ topologies, including their lumped element lower frequency and transmission-line high frequency implementations. About a third of the book is devoted to Class-E power amplifiers, and is the most thorough treatment of this topic that I am aware of. Various circuit topologies, including broadband designs, are discussed theoretically, and implementations for a range of power levels and frequencies with devices technologies ranging from CMOS to GaN are presented in a systematic way. Important practical PA design components, such as bias circuits and various implementations of matching networks, are presented in detail throughout the text. The last few chapters cover topics that are relevant to the main theme of the book and are needed to complete the understanding of high-efficiency power amplifiers: mixed classes of operation; Doherty topologies; linearization techniques; and CAD for the highly nonlinear PA modes of operation using various specialized and commercial circuit simulators.

This book can be used at various levels and will be a useful aid to practicing engineers who need to design and implement efficient transmitters, graduate students who wish to understand the fundamental theoretical background of high-efficiency operation at high frequencies, and microwave engineers who are interested in the state of the art in high-efficiency power amplifiers.

**Zoya Popovic**
IEEE Fellow
Distinguished Professor and Hudson Moore Jr. Chair
University of Colorado, Boulder

# Preface

The main objective of this book is to present all relevant information required to design high efficiency RF and microwave power amplifiers, including well-known and novel theoretical approaches and practical design techniques. Regardless of the different operation classes like Class D, Class E, or Class F and their combination, an efficiency improvement is achieved by providing the non-linear operation conditions when an active device can subsequently operate in pinch-off, active and saturation regions resulting in nonsinusoidal collector current and voltage waveforms, for example, symmetrical for Class-F and asymmetrical for Class-E modes. As a result, the power amplifiers operated in Class F can be analyzed explicitly in the frequency domain when the harmonic load impedances are optimized by short-circuit and open-short terminations in order to control the voltage and current waveforms at the collector to obtain maximum efficiency. However, the power amplifier operated in Class E can be fully analyzed analytically in the time domain when an efficiency improvement is achieved by realizing the ideal on-to-off active device operation in only pinch-off and voltage-saturation modes only so that high voltage and high current at the collector do not occur at the same time. Unlike the single-ended power amplifiers operated in Class-F or Class-E modes, a Class-D power amplifier represents a switchmode power amplifier using two switching active devices driven on and off that one of the switches is turned on when the other is turned off, and *vice versa*. Since the transmitting signal is characterized by a high peak-to-average power ratio in modern wireless communication systems with increased bandwidth and high data rate when improving efficiency and linearity simultaneously have become critical, several analog and digital techniques are described to minimize the distortion of high-efficiency RF and microwave power amplifier.

Generally, this book is intended for and can be recommended to practicing RF circuit designers and engineers as an anthology of a wide family of high efficiency RF and microwave power amplifiers based on both well-known and novel switchmode operation conditions with detailed description of their operational principles and applications and clear practical demonstration of theoretical results. To bridge the theoretical idealized results with real practical implementation, the theory is supported by design examples, in which the optimum design approaches combine effectively analytical calculations and simulation, resulting in practical schematics of high efficiency power amplifier circuits using different types of field-effect or bipolar transistors.

The introductory Chapter 1 describes the basic principles of power amplifiers design procedures. Based on the spectral-domain analysis, the concept of a conduction angle is introduced with simple and clear analyses of the basic Classes A,

AB, B, and C of the power-amplifier operation. Nonlinear models are given for MOSFET, MESFET, HEMT, and bipolar devices (including HBTs), which have very good prospects for power amplifiers using modern microwave monolithic integrated circuits. The effect of the device input parameters on the conduction angle at high frequencies is explained. The concept and design of push-pull amplifiers using balanced transistors are presented. The possibility of the maximum power gain for a stable power amplifier is discussed and analytically derived. The parasitic parametric effect due to the nonlinear collector capacitance and measures for its cancellation in practical power amplifier are discussed. Finally, the basics of the load-pull characterization and distortion fundamentals are presented.

In Chapter 2, the voltage-switching and current-switching configurations of Class-D power amplifiers are presented, the increased efficiency of which is a result of operating the active devices as switches. The basic switchmode power amplifiers with resistive load of different configurations as well as the current-switching and voltage-switching configurations based on complementary and transformer-coupled topologies are analyzed. The effects of the transistor saturation resistance, rectangular and sinusoidal driving signals, nonzero switching transition times, and parasitic shunt capacitance and series inductance are demonstrated. Practical design examples of voltage-switching and current-switching Class-D power amplifiers that are intended to operate at high frequencies and microwaves are described. Finally, a high-frequency Class-S amplification architecture employing a switchmode Class-D power amplifier to efficiently amplify digitally modulated transmitting signal is shown and discussed.

Highly efficient operation of the power amplifier can be obtained by applying biharmonic or polyharmonic modes when an additional single-resonant or multi-resonant circuit tuned to the odd harmonics of the fundamental frequency is added to the load network. An infinite number of odd-harmonic resonators results in an idealized Class-F mode with a square voltage waveform and a half-sinusoidal current waveform at the device output terminal providing ideally 100% collector (or drain) efficiency. Chapter 3 describes the different Class-F techniques using lumped and transmission-line elements including a quarterwave transmission line. The effects of the transistor saturation resistance and parasitic shunt capacitance are demonstrated. Design examples and practical RF and microwave Class-F power amplifiers are given and discussed.

An inverse Class-F mode can be obtained by using a single-resonant or a multi-resonant circuit tuned to the even harmonics of the fundamental frequency added to the load network. An infinite number of even-harmonic resonators results in an idealized inverse Class-F mode with a half-sinusoidal voltage waveform and a square current waveform at the device output terminal. Chapter 4 describes the different inverse Class-F techniques using lumped and transmission-line elements including a quarterwave transmission line. Design examples and practical RF and microwave inverse Class-F power amplifiers are presented.

The switchmode Class-E tuned power amplifiers with a shunt capacitance have found widespread application due to their design simplicity and high efficiency operation. In a Class-E power amplifier, the transistor operates as an on/off switch and the shapes and relative timing of the current and voltage waveforms are such that high current and high voltage do not occur simultaneously. That minimizes the power dissipation and maximizes the power-amplifier efficiency. Chapter 5 presents the historical aspect and modern trends of Class-E power amplifier design. Different circuit configurations and load-network techniques using the push-pull mode, with lumped and transmission-line elements are analyzed. The effects of the device saturation resistance, finite switching times, and nonlinear shunt capacitance are described. Practical RF and microwave Class-E power amplifiers and their applications are given and discussed.

In Chapter 6, we discuss and analyze the switchmode second-order Class-E configurations with one capacitor and one inductor and generalized load network including the finite dc-feed inductance, shunt capacitance and series reactance. The results of the Fourier analysis and derivation of the equations for the idealized operation of the circuit are presented. Effects of the device output bondwire inductance on the optimum circuit parameters are demonstrated. The possibilities to realize a Class-E approximation with transmission lines and broadband Class-E load networks are shown and discussed. The operating power gain of a parallel-circuit Class-E power amplifier is evaluated and compared with the operating power gain of a conventional Class-B power amplifier. Circuit design examples and practical implementations of CMOS Class-E power amplifiers are also given.

Chapter 7 presents the results of exact time-domain analysis of the switchmode tuned Class-E power amplifiers with a quarterwave transmission line. The load-network parameters are derived analytically. The idealized collector voltage and current waveforms demonstrate the possibility of 100% efficiency. We consider load-network implementation, including output matching circuit at RF and microwave frequencies, using lumped and transmission-line elements. Accurate derivation of the matching circuit parameters is provided. Switchmode Class-E power amplifiers with a quarterwave transmission line offer a new challenge for RF and microwave power amplification, providing high efficiency and harmonic suppression.

In Chapter 8, the reactance compensation technique applied to the circuits with lumped elements and transmission lines to provide a broadband operation of Class-E power amplifiers is introduced and analyzed. This technique can be directly used to design a parallel-circuit Class-E power amplifier because its load network configuration has an exactly the same structure with shunt and series resonant circuits. Different circuit configurations and Class-E load-network techniques corresponding to broadband high-power RF power amplifiers operating in VHF and UHF bands, microwave monolithic integrated circuits of power amplifiers, and CMOS power amplifiers operating in a wide frequency range with high efficiency are given and described.

Chapter 9 presents alternative and mixed-mode configurations of high-efficiency power amplifiers. The Class-DE power amplifier is the combination of a voltage-switching Class-D mode and Class-E switching conditions, thus extending the switching Class-D operation to higher frequencies. The switchmode Class-E/F power amplifier can provide lower peak voltage on the switch, while providing the Class-E zero voltage and zero voltage-derivative switching conditions required to eliminate capacitor-discharge power dissipation. This is achieved by harmonic tuning using resonant circuits tuned to selected harmonic frequencies, realizing Class-F mode. Also described is the biharmonic Class-$E_M$ mode, which can reduce the switching power dissipation of a Class-E circuit operating at higher frequencies, at which the turn-off switching time occupies a larger fraction of the RF period. The requirements of jumpless voltage and current waveforms and sinusoidal load waveform, with nonzero output power delivered to the load, can be provided by using nonlinear reactive elements in the load network to convert fundamental-frequency power to a desired harmonic frequency or by injecting the harmonic-frequency power into the load network from an external source. The inverse Class-E power amplifier is the dual of the classical Class-E power amplifier with a shunt capacitor, in which the load-network inductor and capacitor replace each other. However, this dual circuit is limited to low operating frequencies or low output powers because the transistor must discharge its output capacitance from the peak voltage to zero, in every RF cycle. As a result, the transistor must be sufficiently small to reduce the capacitor-discharge power dissipation to an acceptable level. Harmonic-control techniques for designing microwave power amplifiers are given, with a description of a systematic procedure of multiharmonic load-pull simulation using the harmonic balance method and active load-pull measurement system. Finally, outphasing modulation systems are considered where a variable envelope output is created by the sum of two constant-envelope signals with varying phases which therefore can be amplified by highly efficient power amplifier operated in a switching Class-D, Class-E, or Class-F operation mode, or their combinations.

Chapter 10 describes the historical aspect of the Doherty approach to the power amplifier design and modern trends in Doherty amplifier design techniques using multistage and asymmetric multi-way architectures. To increase efficiency over the power-backoff range, the switchmode Class-E, conventional Class-F, or inverse Class-F operation mode by controlling the second and third harmonics can be used in the load network. The Doherty amplifier with a series connected load, inverted, balanced, push-pull, and dual Doherty architectures are also described and discussed. Finally, examples of the lumped Doherty amplifier implemented in monolithic microwave integrated circuits, digitally driven Doherty technique, and broadband capability of the two-stage Doherty amplifier are given.

Several techniques can be used to minimize the distortion of an RF power amplifier, since the RF power amplifier remains the major contributor of distortion in the communication channel. Among the simplest is to provide an operation of the power amplifier within its linear region. The predistortion is the most popular due to its ease of use and possibility to be implemented utilizing analog or digital techniques. Chapter 11 treats the power amplifier from a system perspective, introducing a simple black-box model. It continues describing the fundamentals of predistortion linearization, including the sensitivity of the method to various parameters. Finally, it presents practical predistorters implemented using analog and digital techniques.

Nonlinear circuit simulation in the frequency and time domains is a very important tool for analysis, design and optimization of high-efficiency switchmode power amplifiers of Classes D, DE and E. The advantages are the significantly reduced development time and final product cost, better understanding of the circuit behavior and faster obtaining of the optimum design. It is especially important at very high frequencies, including microwaves, and for MMIC development, where the transistor and circuit parasitics can significantly affect the overall power amplifier performance. Therefore, it is very important to incorporate into the simulator as accurate a transistor model as possible to approximate correctly the device behavior, at the fundamental frequency, and at the second and higher harmonics of the operating frequency. Chapter 12 focuses on five CAD programs for analyzing the time-domain and frequency-domain behaviors of the switchmode high-efficiency power amplifiers in frequency ranges from high frequencies to microwaves: HB-PLUS and SPICE CAD tools for Class-D and Class-DE circuits, and HEPA-PLUS, SPICE and ADS CAD tools for Class-E circuits.

# Acknowledgments

Dr. Marc J. Franco would like to express his gratitude to his long time friend Sergio R. Caprile for his valuable advice and help in improving the quality of this work.

Nathan O. Sokal would like to express sincere appreciation to Prof. Alan D. Sokal of New York University, U.S.A. and University College of London, U.K., for his joint conception of the Class-E power amplifier and his mathematical analysis of the circuit operation, and his reviews and suggestions on many of Nathan Sokal's subsequent published papers.

Finally, Dr. Andrei Grebennikov especially wishes to thank his wife, Galina Grebennikova, for performing important numerical calculations and computer artwork design, and for her constant encouragement, inspiration, support, and assistance.

# Power Amplifier Design Principles

## INTRODUCTION

This introductory chapter presents the basic principles for understanding the power amplifiers design procedure in principle. Based on the spectral-domain analysis, the concept of a conduction angle is introduced, by which the basic Classes A, AB, B, and C of the power-amplifier operation are analyzed and illustrated in a simple and clear form. The frequency-domain analysis is less ambiguous because a relatively complex circuit often can be reduced to one or more sets of immittances at each harmonic component. Classes of operation based upon a finite number of harmonics are discussed and described. The different nonlinear models for various types of MOSFET, MESFET, HEMT, and BJT devices including HBTs, which are very prospective for modern microwave monolithic integrated circuits of power amplifiers, are given. The effect of the input device parameters on the conduction angle at high frequencies is explained. The design and concept of push−pull amplifiers using balanced transistors are presented. The possibility of the maximum power gain for a stable power amplifier is discussed and analytically derived. The device bias conditions and required bias circuits depend on the classes of operations and type of the active device. The parasitic parametric effect due to the nonlinear collector capacitance and measures for its cancellation in practical power amplifiers are discussed. In addition, the basics of the load−pull characterization and distortion fundamentals are presented.

## 1.1 Spectral-domain analysis

The best way to understand the electrical behavior of a power amplifier and the fastest way to calculate its basic electrical characteristics such as output power,

power gain, efficiency, stability, or harmonic suppression is to use a spectral-domain analysis. Generally, such an analysis is based on the determination of the output response of the nonlinear active device when applying the multiharmonic signal to its input port, which analytically can be written as

$$i(t) = f[v(t)] \tag{1.1}$$

where $i(t)$ is the output current, $v(t)$ is the input voltage, and $f(v)$ is the nonlinear transfer function of the device. Unlike the spectral-domain analysis, time-domain analysis establishes the relationships between voltage and current in each circuit element in the time domain when a system of equations is obtained applying Kirchhoff's law to the circuit to be analyzed. Generally, such a system will be composed of nonlinear integro-differential equations in a nonlinear circuit. The solution to this system can be found by applying the numerical-integration methods.

The voltage $v(t)$ in the frequency domain generally represents the multiple-frequency signal at the device input which is written as

$$v(t) = V_0 + \sum_{k=1}^{N} V_k \cos(\omega_k t + \phi_k) \tag{1.2}$$

where $V_0$ is the constant voltage, $V_k$ is the voltage amplitude, $\phi_k$ is the phase of the $k$-order harmonic component $\omega_k$, $k = 1, 2, \ldots, N$, and $N$ is the number of harmonics.

The spectral-domain analysis, based on substituting Eq. (1.2) into Eq. (1.1) for a particular nonlinear transfer function of the active device, determines the output spectrum as a sum of the fundamental-frequency and higher-order harmonic components, the amplitudes and phases of which will determine the output signal spectrum. Generally, it is a complicated procedure that requires a harmonic-balance technique to numerically calculate an accurate nonlinear circuit response. However, the solution can be found analytically in a simple way when it is necessary to only estimate the basic performance of a power amplifier in terms of the output power and efficiency. In this case, a technique based on a piecewise-linear approximation of the device transfer function can provide a clear insight to the basic behavior of a power amplifier and its operation modes. It can also serve as a good starting point for a final computer-aided design and optimization procedure.

The piecewise-linear approximation of the active device current−voltage transfer characteristic is a result of replacing the actual nonlinear dependence $i = f(v_{in})$, where $v_{in}$ is the voltage applied to the device input, by an approximated one that consists of the straight lines tangent to the actual dependence at the specified points. Such a piecewise-linear approximation for the case of two straight lines is shown in Fig. 1.1($a$).

The output-current waveforms for the actual current−voltage dependence (dashed curve) and its piecewise-linear approximation by two straight lines (solid curve) are plotted in Fig. 1.1($b$). Under large-signal operation mode, the waveforms corresponding to these two dependences are practically the same for the

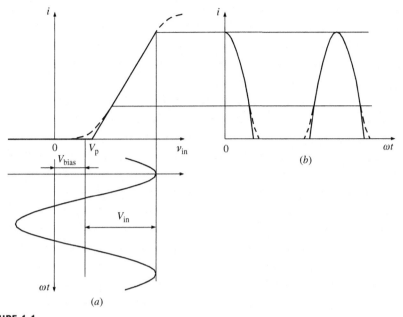

**FIGURE 1.1**

Piecewise-linear approximation technique.

most part, with negligible deviation for small values of the output current close to the pinch-off region of the device operation and significant deviation close to the saturation region of the device operation. However, the latter case results in a significant nonlinear distortion and is used only for high-efficiency operation modes when the active period of the device operation is minimized. Hence, at least two first output-current components, dc and fundamental, can be calculated through a Fourier-series expansion with a sufficient accuracy. Therefore, such a piecewise-linear approximation with two straight lines can be effective for a quick estimate of the output power and efficiency of the linear power amplifier.

The piecewise-linear active device current–voltage characteristic is defined as

$$i = \begin{cases} 0 & v_{in} \leq V_p \\ g_m(v_{in} - V_p) & v_{in} \geq V_p \end{cases} \tag{1.3}$$

where $g_m$ is the device transconductance and $V_p$ is the pinch-off voltage.

Let us assume the input signal to be in a cosine form,

$$v_{in} = V_{bias} + V_{in} \cos \omega t \tag{1.4}$$

where $V_{bias}$ is the input dc bias voltage.

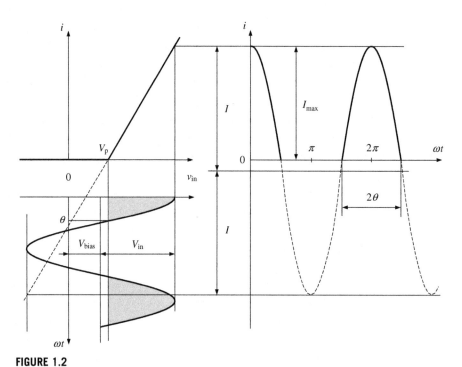

**FIGURE 1.2**

Schematic definition of a conduction angle.

At the point on the plot when the voltage $v_{in}(\omega t)$ becomes equal to a pinch-off voltage $V_p$ and where $\omega t = \theta$, the output current $i(\theta)$ takes a zero value. At this moment,

$$V_p = V_{bias} + V_{in} \cos \theta \qquad (1.5)$$

and the angle $\theta$ can be calculated from

$$\cos \theta = -\frac{V_{bias} - V_p}{V_{in}}. \qquad (1.6)$$

As a result, the output current represents a periodic pulsed waveform described by the cosinusoidal pulses with maximum amplitude $I_{max}$ and width $2\theta$ as

$$i = \begin{cases} I_q + I \cos \omega t & -\theta \leq \omega t < \theta \\ 0 & \theta \leq \omega t < 2\pi - \theta \end{cases} \qquad (1.7)$$

where the conduction angle $2\theta$ indicates the part of the RF current cycle, during which a device conduction occurs, as shown in Fig. 1.2. When the output current $i(\omega t)$ takes a zero value, one can write

$$i = I_q + I \cos \theta = 0. \qquad (1.8)$$

Taking into account that $I = g_m V_{in}$ for a piecewise-linear approximation, Eq. (1.7) can be rewritten for $i > 0$ by

$$i = g_m V_{in}(\cos \omega t - \cos \theta). \tag{1.9}$$

When $\omega t = 0$, then $i = I_{max}$ and

$$I_{max} = I(1 - \cos \theta). \tag{1.10}$$

The Fourier-series expansion of the even function when $i(\omega t) = i(-\omega t)$ contains only even components of this function and can be written as

$$i(\omega t) = I_0 + I_1 \cos \omega t + I_2 \cos 2\omega t + \cdots + I_n \cos n\omega t \tag{1.11}$$

where the dc, fundamental-frequency, and $n$th harmonic components are calculated by

$$I_0 = \frac{1}{2\pi} \int_{-\theta}^{\theta} g_m V_{in}(\cos \omega t - \cos \theta) d\omega t = I\gamma_0(\theta) \tag{1.12}$$

$$I_1 = \frac{1}{\pi} \int_{-\theta}^{\theta} g_m V_{in}(\cos \omega t - \cos \theta)\cos \omega t \, d\omega t = I\gamma_1(\theta) \tag{1.13}$$

$$I_n = \frac{1}{\pi} \int_{-\theta}^{\theta} g_m V_{in}(\cos \omega t - \cos \theta)\cos(n\omega t) d\omega t = I\gamma_n(\theta) \tag{1.14}$$

where $\gamma_n(\theta)$ are called the coefficients of expansion of the output-current cosine waveform or the current coefficients [1, 2]. They can be analytically defined as

$$\gamma_0(\theta) = \frac{1}{\pi}(\sin \theta - \theta \cos \theta) \tag{1.15}$$

$$\gamma_1(\theta) = \frac{1}{\pi}\left(\theta - \frac{\sin 2\theta}{2}\right) \tag{1.16}$$

$$\gamma_n(\theta) = \frac{1}{\pi}\left[\frac{\sin(n-1)\theta}{n(n-1)} - \frac{\sin(n+1)\theta}{n(n+1)}\right] \tag{1.17}$$

where $n = 2, 3, \ldots$.

The dependences of $\gamma_n(\theta)$ for the dc, fundamental-frequency, second-, and higher-order current components are shown in Fig. 1.3. The maximum value of $\gamma_n(\theta)$ is achieved when $\theta = 180°/n$. Special case is $\theta = 90°$, when odd current coefficients are equal to zero, i.e. $\gamma_3(\theta) = \gamma_5(\theta) = \cdots = 0$. The ratio between the fundamental-frequency and dc components $\gamma_1(\theta)/\gamma_0(\theta)$ varies from 1 to 2 for any values of the conduction angle, with a minimum value of 1 for $\theta = 180°$ and a

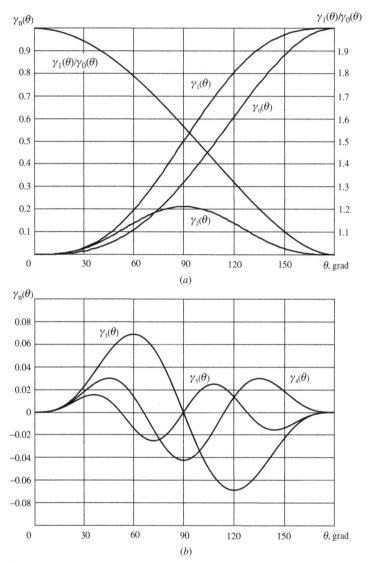

**FIGURE 1.3**

Dependences of $\gamma_n(\theta)$ for dc, fundamental, and higher-order current components.

maximum value of 2 for $\theta = 0°$, as shown in Fig. 1.3($a$). Besides, it is necessary to pay attention to the fact that the current coefficient $\gamma_3(\theta)$ becomes negative within the interval of $90° < \theta < 180°$, as shown in Fig. 1.3($b$). This implies the proper phase changes of the third current harmonic component when its values are negative. Consequently, if the harmonic components, for which $\gamma_n(\theta) > 0$, achieve positive maximum values at the time moments corresponding to the middle points of the current waveform, the harmonic components, for which $\gamma_n(\theta) < 0$, can achieve negative maximum values at these time moments too. As a result, a combination of different harmonic components with proper loading will result in flattening of the current or voltage waveforms, thus improving efficiency of the power amplifier. The amplitude of the corresponding current harmonic component can be obtained by

$$I_n = \gamma_n(\theta)g_m V_{in} = \gamma_n(\theta)I. \tag{1.18}$$

In some cases, it is necessary for an active device to provide a constant value of $I_{max}$ at any values of $\theta$ that require an appropriate variation of the input voltage amplitude $V_{in}$. In this case, it is more convenient to use the coefficients $\alpha_n$ defined as a ratio of the $n$th current harmonic amplitude $I_n$ to the maximum current waveform amplitude $I_{max}$,

$$\alpha_n = \frac{I_n}{I_{max}}. \tag{1.19}$$

From Eqs (1.10), (1.18), and (1.19), it follows that

$$\alpha_n = \frac{\gamma_n(\theta)}{1 - \cos\theta} \tag{1.20}$$

and maximum value of $\alpha_n(\theta)$ is achieved when $\theta = 120°/n$.

## 1.2 Basic classes of operation: A, AB, B, and C

As established at the end of 1910s, the amplifier efficiency may reach quite high values when suitable adjustments of the grid and anode voltages are made [3]. With resistive load, the anode current is in phase with the grid voltage, whereas it leads with the capacitive load and it lags with the inductive load. On the assumption that the anode current and anode voltage both have sinusoidal variations, the maximum possible output of the amplifying device would be just a half the dc supply power, resulting in an anode efficiency of 50%. However, by using a pulsed-shaped anode current, it is possible to achieve anode efficiency considerably in excess of 50%, potentially as high as 90%, by choosing the proper operation conditions. By applying the proper negative bias voltage to the grid terminal to provide the pulsed anode current of different width with the angle $\theta$, the anode current becomes equal to zero, where the double angle $2\theta$ represents a conduction angle of the amplifying device [4]. In this case, a theoretical anode efficiency approaches 100% when the conduction angle, within which the anode current flows, reduces to zero starting from 50%, which corresponds to the conduction angle of 360° or 100% duty ratio.

Generally, power amplifiers can be classified in three classes according to their mode of operation: *linear mode* when its operation is confined to the substantially linear portion of the active device characteristic curve; *critical mode* when the anode current ceases to flow, but operation extends beyond the linear portion up to the saturation and cutoff regions; and *nonlinear mode* when the anode current ceases to flow during a portion of each cycle, with a duration that depends on the grid bias [5]. When high efficiency is required, power amplifiers of the third class are employed since the presence of harmonics contributes to the attainment of high efficiencies. In order to suppress harmonics of the fundamental frequency to deliver a sinusoidal signal to the load, a parallel resonant circuit can be used in the load network which bypasses harmonics through a low-impedance path and, by virtue of its resonance to the fundamental, receives energy at that frequency. At the very beginning of 1930s, power amplifiers operating in the first two classes with 100% duty ratio were called the Class-A power amplifiers, whereas the power amplifiers operating in the third class with 50% duty ratio were assigned to Class-B power amplifiers [6].

To analytically determine the operation classes of the power amplifier, consider a simple resistive stage shown in Fig. 1.4, where $L_{ch}$ is the ideal choke inductor with zero series resistance and infinite reactance at the operating

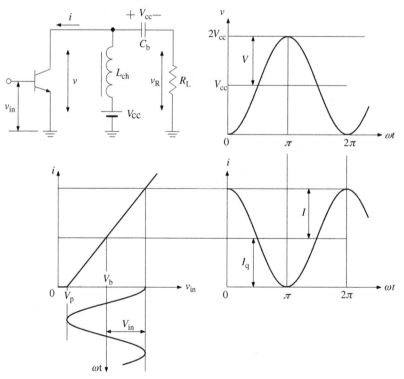

**FIGURE 1.4**

Voltage and current waveforms in Class-A operation.

frequency, $C_b$ is the dc-blocking capacitor with infinite value having zero reactance at the operating frequency, and $R_L$ is the load resistor. The dc supply voltage $V_{cc}$ is applied to both plates of the dc-blocking capacitor, being constant during the entire signal period. The active device behaves as an ideal voltage- or current-controlled current source having zero saturation resistance.

For an input cosine voltage given by Eq. (1.4), the operating point must be fixed at the middle point of the linear part of the device transfer characteristic with $V_{in} \leq V_{bias} - V_p$, where $V_p$ is the device pinch-off voltage. Usually, to simplify an analysis of the power-amplifier operation, the device transfer characteristic is represented by a piecewise-linear approximation. As a result, the output current is cosinusoidal,

$$i = I_q + I\cos\omega t \tag{1.21}$$

with the quiescent current $I_q$ greater or equal to the collector current amplitude $I$. In this case, the output collector current contains only two components — dc and cosine — and the averaged current magnitude is equal to a quiescent current $I_q$.

The output voltage $v$ across the device collector represents a sum of the dc supply voltage $V_{cc}$ and cosine voltage $v_R$ across the load resistor $R_L$. Consequently, the greater the output current $i$, the greater the voltage $v_R$ across the load resistor $R_L$ and the smaller the output voltage $v$. Thus, for a purely real load impedance when $Z_L = R_L$, the collector voltage $v$ is shifted by 180° relative to the input voltage $v_{in}$ and can be written as

$$v = V_{cc} + V\cos(\omega t + 180°) = V_{cc} - V\cos\omega t \tag{1.22}$$

where $V$ is the output voltage amplitude.

Substituting Eq. (1.21) into Eq. (1.22) yields

$$v = V_{cc} - (i - I_q)R_L \tag{1.23}$$

where $R_L = V/I$, and Eq. (1.23) can be rewritten as

$$i = \left(I_q + \frac{V_{cc}}{R_L}\right) - \frac{v}{R_L} \tag{1.24}$$

which determines a linear dependence of the collector current *versus* collector voltage. Such a combination of the cosine collector voltage and current waveforms is known as a Class-A operation mode. In practice, because of the device nonlinearities, it is necessary to connect a parallel $LC$ circuit with resonant frequency equal to the operating frequency to suppress any possible harmonic components.

Circuit theory prescribes that the collector efficiency $\eta$ can be written as

$$\eta = \frac{P}{P_0} = \frac{1}{2}\frac{I}{I_q}\frac{V}{V_{cc}} = \frac{1}{2}\frac{I}{I_q}\xi \tag{1.25}$$

where

$$P_0 = I_q V_{cc} \tag{1.26}$$

is the dc output power,

$$P = \frac{IV}{2} \tag{1.27}$$

is the power delivered to the load resistance $R_L$ at the fundamental frequency $f_0$, and

$$\xi = \frac{V}{V_{cc}} \tag{1.28}$$

is the collector voltage peak factor.

Then, by assuming the ideal conditions of zero saturation voltage when $\xi = 1$ and maximum output-current amplitude when $I/I_q = 1$, from Eq. (1.25) it follows that the maximum collector efficiency in a Class-A operation mode is equal to

$$\eta = 50\%. \tag{1.29}$$

However, as it also follows from Eq. (1.25), increasing the value of $I/I_q$ can further increase the collector efficiency. This leads to a step-by-step nonlinear transformation of the current cosine waveform to its pulsed waveform when the amplitude of the collector current exceeds zero value during only a part of the entire signal period. In this case, an active device is operated in the active region followed by the operation in the pinch-off region when the collector current is zero, as shown in Fig. 1.5. As a result, the frequency spectrum at the device output will generally contain the second-, third-, and higher-order harmonics of the fundamental frequency. However, due to the high quality factor of the parallel resonant $LC$ circuit, only the fundamental-frequency signal is flowing into the load, while the short-circuit conditions are fulfilled for higher-order harmonic components. Therefore, ideally the collector voltage represents a purely sinusoidal waveform with the voltage amplitude $V \leq V_{cc}$.

Equation (1.8) for the output current can be rewritten through the ratio between the quiescent current $I_q$ and the current amplitude $I$ as

$$\cos\theta = -\frac{I_q}{I}. \tag{1.30}$$

As a result, the basic definitions for nonlinear operation modes of a power amplifier through half the conduction angle $\theta$ can be introduced as

- When $\theta > 90°$, then $\cos\theta < 0$ and $I_q > 0$, corresponding to Class-AB operation.
- When $\theta = 90°$, then $\cos\theta = 0$ and $I_q = 0$, corresponding to Class-B operation.
- When $\theta < 90°$, then $\cos\theta > 0$ and $I_q < 0$, corresponding to Class-C operation.

The periodic pulsed output current $i(\omega t)$ can be represented as a Fourier-series expansion

$$i(\omega t) = I_0 + I_1\cos\omega t + I_2\cos 2\omega t + I_3\cos 3\omega t + \cdots \tag{1.31}$$

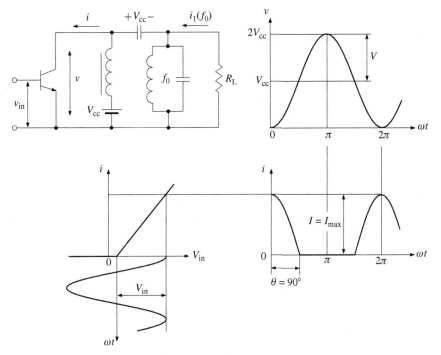

**FIGURE 1.5**

Voltage and current waveforms in a Class-B operation.

where the dc and fundamental-frequency components can be obtained by

$$I_0 = \frac{1}{2\pi} \int_{-\theta}^{\theta} I(\cos\omega t - \cos\theta)d\omega t = I\gamma_0 \tag{1.32}$$

$$I_1 = \frac{1}{\pi} \int_{-\theta}^{\theta} I(\cos\omega t - \cos\theta)\cos\omega t d\omega t = I\gamma_1, \tag{1.33}$$

respectively, where

$$\gamma_0 = \frac{1}{\pi}(\sin\theta - \theta\cos\theta) \tag{1.34}$$

$$\gamma_1 = \frac{1}{\pi}(\theta - \sin\theta\cos\theta). \tag{1.35}$$

From Eq. (1.32), it follows that the dc current component is a function of $\theta$ in the operation modes with $\theta < 180°$, in contrast to a Class-A operation mode where $\theta = 180°$ and the dc current is equal to the quiescent current during the entire period.

The collector efficiency of a power amplifier with resonant circuit, biased to operate in the nonlinear modes, can be obtained by

$$\eta = \frac{P_1}{P_0} = \frac{1}{2}\frac{I_1}{I_0}\xi = \frac{1}{2}\frac{\gamma_1}{\gamma_0}\xi \tag{1.36}$$

which is a function of $\theta$ only, where

$$\frac{\gamma_1}{\gamma_0} = \frac{\theta - \sin\theta\cos\theta}{\sin\theta - \theta\cos\theta}. \tag{1.37}$$

The Class-B power amplifiers had been defined as those which operate with a negative grid bias such that the anode current is practically zero with no excitation grid voltage, and in which the output power is proportional to the square of the excitation voltage [7]. If $\xi = 1$ and $\theta = 90°$, then from Eqs (1.36) and (1.37) it follows that the maximum collector efficiency in a Class-B operation mode is equal to

$$\eta = \frac{\pi}{4} \cong 78.5\%. \tag{1.38}$$

The fundamental-frequency power delivered to the load $P_L = P_1$ is defined as

$$P_1 = \frac{VI_1}{2} = \frac{VI\gamma_1(\theta)}{2} \tag{1.39}$$

showing its direct dependence on the conduction angle $2\theta$. This means that reduction in $\theta$ results in lower $\gamma_1$, and, to increase the fundamental-frequency power $P_1$, it is necessary to increase the current amplitude $I$. Since the current amplitude $I$ is determined by the input voltage amplitude $V_{in}$, the input power $P_{in}$ must be increased. The collector efficiency also increases with reduced value of $\theta$ and becomes maximum when $\theta = 0°$, where a ratio $\gamma_1/\gamma_0$ is maximal, as follows from Fig. 1.3($a$). For example, the collector efficiency $\eta$ increases from 78.5% to 92% when $\theta$ reduces from 90° to 60°. However, it requires increasing the input voltage amplitude $V_{in}$ by 2.5 times, resulting in a lower value of the power-added efficiency (*PAE*), which is defined as

$$PAE = \frac{P_1 - P_{in}}{P_0} = \frac{P_1}{P_0}\left(1 - \frac{1}{G_p}\right) \tag{1.40}$$

where $G_p = P_1/P_{in}$ is the operating power gain.

The Class-C power amplifiers had been defined as those that operate with a negative grid bias more than sufficient to reduce the anode current to zero with no excitation grid voltage, and in which the output power varies as the square of the anode voltage between limits [7]. The main distinction between Class B and Class C is in the duration of the output current pulses, which are shorter for Class C when the active device is biased beyond the cutoff point. It should be noted that, for the device transfer characteristic, which can be ideally represented by a square-law approximation, the odd-harmonic current coefficients $\gamma_n(\theta)$ are not equal to zero in this case, although there is no significant difference between the square-law and linear cases [8]. To achieve the maximum anode efficiency in Class C, the active device should be biased (negative) considerably past the cutoff (pinch-off) point to provide the sufficiently low conduction angles [9].

In order to obtain an acceptable trade-off between a high power gain and a high power-added efficiency in different situations, the conduction angle should be chosen within the range of $120° \leq 2\theta \leq 190°$. If it is necessary to provide high collector efficiency of the active device having a high gain capability, it is necessary to choose a Class-C operation mode with $\theta$ close to 60°. However, when the input power is limited and power gain is not sufficient, a Class-AB operation mode is recommended with small quiescent current when $\theta$ is slightly greater than 90°. In the latter case, the linearity of the power amplifier can be significantly improved. From Eq. (1.37) it follows that that the ratio of the fundamental-frequency component of the anode current to the dc current is a function of $\theta$ only, which means that, if the operating angle is maintained constant, the fundamental component of the anode current will replicate linearly to the variation of the dc current, thus providing the linear operation of the Class-C power amplifier when dc current is directly proportional to the grid voltage [10].

## 1.3 Load line and output impedance

The graphical method of laying down a load line on the family of the static curves representing anode current against anode voltage for various grid potentials was already well known in the 1920s [11]. If an active device is connected in a circuit in which the anode load is a pure resistance, the performance may be analyzed by drawing the load line where the lower end of the line represents the anode supply voltage and the slope of the line is established by the load resistance, i.e. the load resistance is equal to the value of the intercept on the voltage axis divided by the value of the intercept on the current axis.

In a Class A, the output voltage $v$ across the device anode (collector or drain) represents a sum of the dc supply voltage $V_{cc}$ and cosine voltage across the load resistance $R_L$, and can be defined by Eq. (1.22). In this case, the power dissipated in the load and the power dissipated in the device is equal when $V_{cc} = V$, and the load resistance $R_L = V/I$ is equal to the device output resistance $R_{out}$ [7]. In a pulsed operation mode (Class AB, B, or C), since the parallel $LC$ circuit is tuned to the fundamental frequency, ideally the voltage across the load resistor $R_L$ represents a cosine waveform. By using Eqs (1.7), (1.22), and (1.33), the relationship between the collector current $i$ and voltage $v$ during a time period of $-\theta \leq \omega t < \theta$ can be expressed by

$$i = \left(I_q + \frac{V_{cc}}{\gamma_1 R_L}\right) - \frac{v}{\gamma_1 R_L} \tag{1.41}$$

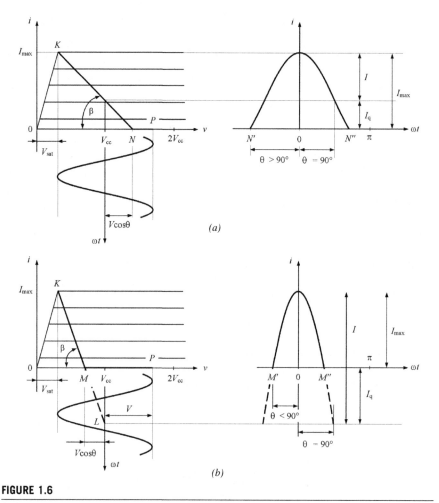

**FIGURE 1.6**

Collector current waveforms in Class-AB and Class-C operations.

where the fundamental current coefficient $\gamma_1$ as a function of $\theta$ is determined by Eq. (1.35), and the load resistance is defined by $R_L = V/I_1$, where $I_1$ is the fundamental current amplitude. Equation (1.41) determining the dependence of the collector current on the collector voltage for any values of conduction angle in the form of a straight-line function is called the *load line* of the active device. For a Class-A operation mode with $\theta = 180°$ when $\gamma_1 = 1$, the load-line defined by Eq. (1.41) is identical to the load-line defined by Eq. (1.24).

Figure 1.6 shows the idealized active device output $I-V$ curves and load lines for different conduction angles according to Eq. (1.41) with the corresponding collector and current waveforms. From Fig. 1.6, it follows that the maximum collector current amplitude $I_{max}$ corresponds to the minimum collector voltage $V_{sat}$ when $\omega t = 0$, and is the same for any conduction angle. The slope of the load line defined by its slope angle $\beta$ is different for different conduction angles and values of the load resistance, and can be obtained by

$$\tan \beta = \frac{I}{V(1 - \cos \theta)} = \frac{1}{\gamma_1 R_L} \tag{1.42}$$

from which it follows that a greater slope angle $\beta$ of the load line results in a smaller value of the load resistance $R_L$ for the same $\theta$.

The load resistance $R_L$ for the active device as a function of $\theta$, which is required to terminate the device output to deliver the maximum output power to the load can be written in a general form as

$$R_L(\theta) = \frac{V}{\gamma_1(\theta)I} \tag{1.43}$$

which is equal to the device equivalent output resistance $R_{out}$ at the fundamental frequency [7]. The term "equivalent" means that this is not a real physical device resistance as in a Class-A mode, but its equivalent output resistance, the value of which determines the optimum load, which should terminate the device output to deliver maximum fundamental-frequency output power. The equivalent output resistance is calculated as a ratio between the amplitudes of the collector cosine voltage and fundamental-frequency collector current component, which depends on the angle $\theta$.

In a Class-B mode when $\theta = 90°$ and $\gamma_1 = 0.5$, the load resistance $R_L^B$ is defined as $R_L^B = 2V/I_{max}$. Alternatively, taking into account that $V_{cc} = V$ and $P_{out} = I_1 V$ for the fundamental-frequency output power, the load resistance $R_L^B$ can be written in a simple idealized analytical form with zero saturation voltage, $V_{sat}$, as

$$R_L^B = \frac{V_{cc}^2}{2P_{out}}. \tag{1.44}$$

In general, the entire load line represents a broken line $PK$ including a horizontal part, as shown in Fig. 1.6. Figure 1.6(a) represents a load line $PNK$ corresponding to a Class-AB mode with $\theta > 90°$, $I_q > 0$, and $I < I_{max}$. Such a load line moves from point $K$ corresponding to the maximum output-current amplitude $I_{max}$ at $\omega t = 0$ and determining the device saturation voltage $V_{sat}$ through the point $N$ located at the horizontal axis $v$ where $i = 0$ and $\omega t = \theta$. For a Class-AB operation, the conduction angle for the output-current pulse between points $N'$ and $N''$ is greater than $180°$. Figure 1.6(b) represents a load line $PMK$ corresponding to a Class-C mode with $\theta < 90°$, $I_q < 0$, and $I > I_{max}$. For a Class-C operation, the load line intersects a horizontal axis $v$ in a point $M$, and the conduction angle for the output-current pulse between points $M'$ and $M''$ is smaller than $180°$. Hence, generally the load line represents a broken line with the first section having a slope angle $\beta$ and the other horizontal section with zero current $i$. In a Class-B mode,

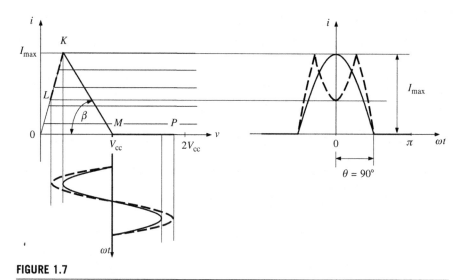

**FIGURE 1.7**

Collector current waveforms for the device operating in saturation, active, and pinch-off regions.

the collector current represents half-cosine pulses with the conduction angle of $2\theta = 180°$ and $I_q = 0$.

Now let us consider a Class-B operation with increased amplitude of the cosine collector voltage. In this case, as shown in Fig. 1.7, an active device is operated in the saturation, active, and pinch-off regions, and the load line represents a broken line *LKMP* with three linear sections (*LK*, *KM*, and *MP*). The new section *KL* corresponds to the saturation region, resulting in the half-cosine output-current waveform with a depression in the top part. With further increase of the output-voltage amplitude, the output-current pulse can be split into two symmetrical pulses containing a significant level of the higher-order harmonic components. The same result can be achieved by a increasing a value of the load resistance $R_L$ when the load line is characterized by a smaller slope angle $\beta$.

The collector current waveform becomes asymmetrical for the complex load, the impedance of which represents the load resistance and capacitive or inductive reactances. In this case, the Fourier-series expansion of the output current given by Eq. (1.31) includes a particular phase for each harmonic component. Then, the output voltage at the device collector is written as

$$v = V_{cc} - \sum_{n=1}^{\infty} I_n |Z_n| \cos(n\omega t + \phi_n) \tag{1.45}$$

where $I_n$ is the amplitude of the $n$th output-current harmonic component, $|Z_n|$ is the magnitude of the load-network impedance at the $n$th output-current harmonic

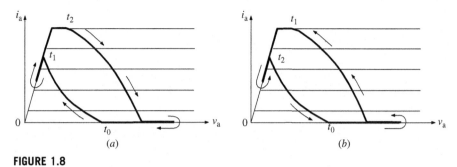

**FIGURE 1.8**

Load lines for (a) inductive and (b) capacitive load impedances.

component, and $\phi_n$ is the phase of the $n$th output current harmonic component. Assuming that $Z_n$ is zero for $n = 2, 3, \ldots$, which is possible for a resonant load network having negligible impedance at any harmonic component except the fundamental, Eq. (1.45) can be rewritten as

$$v = V_{cc} - I_1|Z_1|\cos(\omega t + \phi_1). \tag{1.46}$$

As a result, for the inductive load impedance, the depression in the collector current waveform reduces and moves to the left side of the waveform, whereas the capacitive load impedance causes the depression to deepen and shift to the right side of the collector current waveform [12]. This effect can simply be explained by the different phase conditions for fundamental and higher-order harmonic components composing the collector current waveform and is illustrated by the different load lines for (a) inductive and (b) capacitive load impedances shown in Fig. 1.8. Note that now the load line represents a two-dimensional curve with a complicated behavior.

## 1.4 Classes of operation based upon a finite number of harmonics

Figure 1.9(a) shows the block diagram of a generic power amplifier, where the active device (which is shown as a MOSFET device but can be a bipolar transistor or any other suitable device) is controlled by its drive and bias to operate as a multiharmonic current source or switch, $V_{dd}$ is the supply voltage, and $I_0$ is the dc current flowing through the RF choke [13]. The load-network bandpass filter is assumed linear and lossless and provides the drain load impedance $R_1 + jX_1$ at the fundamental frequency and pure reactances $X_k$ at each $k$th-harmonic component. For analysis simplicity, the load-network filter can incorporate the reactances of the RF choke and device drain-source capacitance which is considered voltage

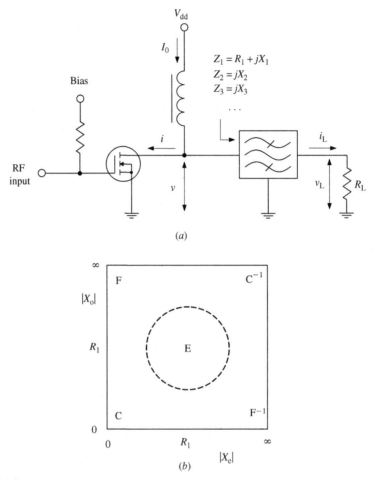

**FIGURE 1.9**

Basic power-amplifier structure and classes of amplification.

independent. Since such a basic power amplifier is assumed to generate power at only the fundamental frequency, harmonic components can be present generally in the voltage and current waveforms depending on class of operation. In a Class-AB, -B, or -C operation, harmonics are present only in the drain current. However, in a Class-F mode, a given harmonic component is present in either drain voltage or drain current, but not both, and all or most harmonics are present in both the drain voltage and current waveforms in a Class-E mode. The required harmonics with optimum or near-optimum amplitudes can be produced by driving the power amplifier to saturation. The analysis based on a Fourier-series expansion of the drain voltage and current waveforms shows that maximum achievable efficiency depends not upon the class of operation, but upon the number of harmonics employed [13,14]. For any set of harmonic reactances, the same maximum efficiency can be achieved by proper adjustment of the waveforms and the fundamental-frequency load reactance.

A mechanism for differentiating the various classes of power-amplifier operation implemented with small numbers of harmonic components is shown in Fig. 1.9($b$) [13]. It is based on the relative magnitudes of the even ($X_e$) and odd ($X_o$) harmonic impedances relative to the fundamental-frequency load resistance $R_1$. In this case, the classes of operation can be characterized in terms of a small number of harmonics as follows:

- Class-F: even-harmonic reactances are low and odd-harmonic reactances are high so that the drain voltage is shaped toward a square wave and drain current is shaped toward a half-sine wave.
- Inverse Class-F (Class-F$^{-1}$): even-harmonic reactances are high and odd-harmonic reactances are low so that the drain voltage is shaped toward a half-sine wave and drain current is shaped toward a square wave.
- Class C: all harmonic reactances are low so that the drain current is shaped toward a narrow pulse.
- Inverse Class C (Class C$^{-1}$): all harmonic reactances are high so that the drain voltage is shaped toward a narrow pulse.
- Class-E: all harmonic reactances are negative and comparable in magnitude to the fundamental-frequency load resistance.

The transition from 'low' to 'comparable' occurs in the range from $R_1/3$ to $R_1/2$, whereas the transition from 'comparable' to 'high' similarly occurs in the range from $2R_1$ to $3R_1$. In this case, the circular boundary is for illustration only, and the point at which an amplifier transitions from one class to another is somewhat judgmental and arbitrary, as there is not an abrupt change in the mode of operation. All power amplifiers degenerate to a Class-A operation when there is only a single (fundamental) frequency component. Class B is the special case of a pulsed operation with a conduction angle of 180°, which is represented by a half-sine current waveform based upon even harmonics. Class-D can be considered as a push−pull Class-F power amplifier, in which the two active devices provide each other with paths for the even harmonics.

The transition from Class-F to Class-E and then to Class-F$^{-1}$ moves diagonally in Fig. 1.9($b$) by progressively increasing $X_2$ from zero to $\infty$ while decreasing $X_3$ from $\infty$ to zero so that $X_3 = 1/X_2$. In a Class-F with $X_2 = 0$ and $X_3 = \infty$, the voltage is a third-harmonic maximum-power waveform, while the current is a second-harmonic maximum-power waveform. For $X_2 = X_3 = -1$, the voltage waveform leans leftward and the current waveform leans rightward, thus approximating the all-harmonic Class-E waveforms. Finally, when $X_2 = \infty$ and $X_3 = 0$, the power amplifier operates in an inverse Class-F (Class-F$^{-1}$). The transition from Class-F to Class C moves down to the left-hand side of Fig. 1.9($b$) by setting $X_2$ at zero and progressively decreasing $X_3$ from $\infty$ to zero, and the waveforms remain almost unchanged for $X_3 \leq -3$. The explicit analytical expression for maximum achievable efficiency of finite-harmonic Class C with conduction angle $2\theta \rightarrow 0$ can be written as

$$\eta = \cos\left(\frac{\pi}{n+2}\right) \tag{1.47}$$

where $n$ is a number of harmonics [15].

## 1.5 Active device models

Generally, for an accurate power-amplifier simulation and circuit design for different operating frequencies and output-power levels, it is necessary to represent an active device in the form of a nonlinear equivalent circuit, which can adequately describe the small- and large-signal electrical behavior of the power amplifier up to the device transition frequency $f_T$ and higher to its maximum frequency $f_{max}$ that allows a sufficient number of harmonic components to be taken into account. Accurate device modeling is extremely important to develop monolithic integrated circuits. Better approximations of the final design can only be achieved if the nonlinear device behavior is described accurately.

### 1.5.1 LDMOSFETs

Figure 1.10(a) shows the cross-section of a physical structure of an LDMOSFET device where a heavily doped $p^+$-sinker is inserted between top source and $p^+$-substrate (source grounding) for low resistivity to provide high-current flow between the drain and source terminals [16]. The lightly doped $p$-epilayer and $n$-drift layer are required to provide sufficient distance between regions to prevent latchup (forward-biased $p$-$n$ diodes) and for the drain-source breakdown protection. The parasitic gate-drain capacitance is directly related to the overlap of the gate oxide onto the heavily doped $n^+$-source region. To describe accurately the nonlinear properties of the large-size MOSFET device, it is necessary to consider its two-dimensional gate-distributed nature along both the channel length and channel width, resulting in lower values of the intrinsic series gate and shunt gate-source resistances. Figure 1.10(b) shows the nonlinear MOSFET equivalent circuit with the extrinsic parasitic elements, which can properly describe the nonlinear behavior of both VDMOSFET and LDMOSFET devices [17,18].

The nonlinear current source $i(v_{gs}, v_{ds}, \tau)$ as a function of the input gate-source and output drain-source voltages incorporating self-heating effect can be described sufficiently simply and accurately using hyperbolic functions [18,19]. Careful analytical description of the transition from quadratic to linear regions of the device transfer characteristic enables the more accurate prediction of the

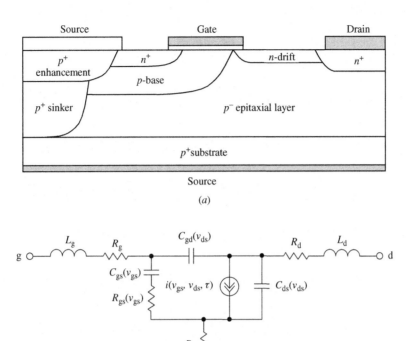

**FIGURE 1.10**

Nonlinear LDMOSFET model and its physical structure.

intermodulation distortion [20]. The overall channel carrier transit time $\tau$ includes also an effect of the transcapacitance required for charge conservation. The drain-source capacitance $C_{ds}$ and gate-drain capacitance $C_{gd}$ are considered as the junction capacitances that strongly depend on the drain-source voltage. The extrinsic parasitic elements are represented by the gate and drain bondwire inductances $L_g$ and $L_d$, source inductance $L_s$, source and drain bulk and ohmic resistances $R_s$ and $R_d$, and gate contact and ohmic resistance $R_g$. The effect of the gate-source channel resistance $R_{gs}$ becomes significant at higher frequencies close to transition frequency $f_T = g_m/(2\pi C_{gs})$, where $g_m$ is the device transconductance. To account for self-heating effects and substrate losses, a special four-port thermal circuit and a series combination of the resistance and capacitance between the external drain and source terminals can be included [21].

An empirical nonlinear model developed for silicon LDMOS transistors, which is single-piece and continuously differentiable, can be written as

$$I_{ds} = \beta V_{gst}^{V_{Gexp}}(1 + \lambda V_{ds})\tanh\left(\frac{\alpha V_{ds}}{V_{gst}}\right)[1 + K_1 \exp(V_{BReff1})] + I_{ss} \exp\left(\frac{V_{ds} - V_{BR}}{V_T}\right)$$

(1.48)

where

$$V_{gst} = V_{st} \ln\left[1 + \exp\left(\frac{V_{gst2}}{V_{st}}\right)\right]$$

$$V_{gst2} = V_{gst1} - \frac{1}{2}\left(V_{gst1} + \sqrt{(V_{gst1} - V_K)^2 + \Delta^2} - \sqrt{V_K^2 + \Delta^2}\right)$$

$$V_{gst1} = V_{gs} - V_{th0} - \gamma V_{ds}$$

$$V_{BReff1} = \frac{V_{ds} - V_{BReff}}{K_2} + M_3 \frac{V_{ds}}{V_{BReff}}$$

$$V_{BReff} = \frac{V_{BR}}{2}\left[1 + \tanh(M_1 - V_{gst}M_2)\right]$$

where $\lambda$ is the drain current slope parameter, $\beta$ is the transconductance parameter, $V_{th0}$ is the forward threshold voltage, $V_{st}$ is the subthreshold slope coefficient, $V_T$ is the temperature voltage, $I_{ss}$ is the forward diode leakage current, $V_{BR}$ is the breakdown voltage, $K_1$, $K_2$, $M_1$, $M_2$, and $M_3$ are the breakdown parameters, $V_K$, $V_{Gexp}$, $\Delta$, and $\gamma$ are the gate-source voltage parameters [19]. The gate-source capacitance $C_{gs}$ can be analytically described as a function of the gate-source voltage since it is practically independent on the drain-source voltage. It is equal to the oxide capacitance $C_{ox}$ in the accumulation region, slightly decreases in the weak-inversion region, significantly reduces in the moderate-inversion region and then becomes practically constant in the strong-inversion or saturation region, as shown in Fig. 1.11 [22]. The approximation function for the gate-source capacitance $C_{gs}$ as the dependence of $V_{gs}$ can be derived by using two components containing the hyperbolic functions as

$$C_{gs} = C_{gs1} + C_{gs2}\{1 + \tanh[C_{gs6}(V_{gs} + C_{gs3})]\} + C_{gs4}[1 - \tanh(C_{gs5}V_{gs})] \quad (1.49)$$

where $C_{gs1}$, $C_{gs2}$, $C_{gs3}$, $C_{gs4}$, $C_{gs5}$, and $C_{gs6}$ are the approximation parameters [19]. The gate-source resistance $R_{gs}$ is determined by the effect of the channel inertia in responding to rapid changes of the time varying gate-source voltage, and varies in such a manner that the charging time $\tau_g = R_{gs}C_{gs}$ remains approximately constant. Thus, the increase of $R_{gs}$ in the velocity saturation region, when the channel conductivity decreases, is partially compensated by the decrease of $C_{gs}$

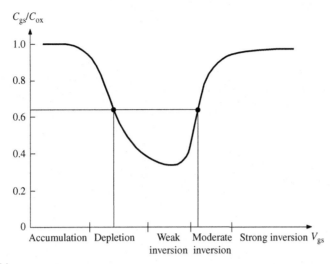

$C_{gs}/C_{ox}$

1.0

0.8

0.6

0.4

0.2

0

Accumulation  Depletion    Weak    Moderate   Strong inversion $V_{gs}$
inversion  inversion

**FIGURE 1.11**

Gate-source capacitance *versus* gate-source voltage.

due to nonuniform channel charge distribution [23]. The effect of $R_{gs}$ becomes significant at higher frequencies close to the transition frequency $f_T$ of the MOSFET and cannot be taken into consideration when designing RF circuits that operate below 2 GHz, as used for commercial wireless applications [24,25].

Figure 1.12(a) shows the equivalent representation of a loaded MOSFET input circuit derived from Fig. 1.10(b), where $\tau_g = C_{gs}R_{gs}$, $g_m(\theta)$ is the large-signal transconductance as a function of one-half the conduction angle $\theta$ and $R_L$ is the load resistance connected to the drain-source terminal. It is assumed that the series source resistance $R_s$ and inductance $L_s$ and transit time $\tau$ are sufficiently small for high-power MOSFETs in a frequency range up to $f \leq 0.3 f_T$, and the device drain-source capacitance $C_{ds}$ is inductively compensated. As a result, the equivalent input circuit shown in Fig. 1.12(a) can be significantly simplified to an equivalent input circuit shown in Fig. 1.12(b) with the input inductance $L_{in}$, resistance $R_{in}$, and capacitance $C_{in}$ are connected in series and defined as

$$L_{in} = L_g \tag{1.50}$$

$$R_{in} \cong R_g + R_{gs} \tag{1.51}$$

$$C_{in} \cong C_{gs} + C_{gd}(1 + g_m R_L). \tag{1.52}$$

Taking into account that usually $C_{gd} \ll C_{gs}$, the device equivalent output circuit can be represented by the series inductance $L_{out} = L_d$ and a parallel connection of the equivalent output resistance $R_{out}$ estimated by Eq. (1.44) at the fundamental frequency and output capacitance $C_{out} \cong C_{ds} + C_{gd}$.

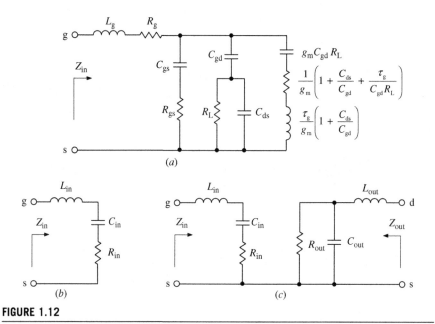

**FIGURE 1.12**

Equivalent circuits characterizing device input and output.

### 1.5.2 GaAs MESFETs and GaN HEMTs

Adequate representation for MESFETs and HEMTs in a frequency range up to at least 25 GHz can be provided using a nonlinear model shown in Fig. 1.13(a), which is very similar to a nonlinear MOSFET model [26,27]. The intrinsic model is described by the channel charging resistance $R_{gs}$, which represents the resistive path for the charging of the gate-source capacitance $C_{gs}$, the feedback gate-drain capacitance $C_{gd}$, the drain-source capacitance $C_{ds}$, the gate-source diode to model the forward conduction current $i_{gs}(v_{gs})$, and the gate-drain diode to account for the gate-drain avalanche current $i_{gd}(v_{gs}, v_{ds})$, which can occur at large-signal operation conditions. The gate-source capacitance $C_{gs}$ and gate-drain capacitance $C_{gd}$ represent the charge depletion region and can be treated as the voltage-dependent Schottky-barrier diode capacitances, being the nonlinear functions of the gate-source voltage $v_{gs}$ and drain-source voltage $v_{ds}$. For negative gate-source voltage and small drain-source voltage, these capacitances are practically equal. However, when the drain-source voltage is increased beyond the current saturation point, the gate-drain capacitance $C_{gd}$ is much more heavily back-biased than the gate-source capacitance $C_{gs}$. Therefore, the gate-source capacitance $C_{gs}$ is significantly more important and usually dominates the input impedance of the MESFET or HEMT device. The influence of the drain-source capacitance $C_{ds}$ on the device

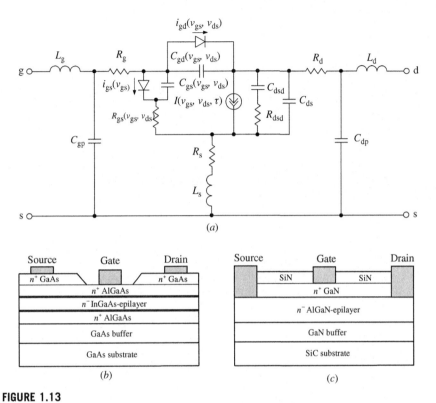

**FIGURE 1.13**

Nonlinear MESFET and HEMT model with HEMT physical structures.

behavior is insignificant and its value is bias independent. The capacitance $C_{dsd}$ and resistance $R_{dsd}$ model the dispersion of the MESFET or HEMT current–voltage characteristics due to trapping effect in the device channel, which leads to discrepancy between dc measurement and S-parameter measurements at higher frequencies [28,29]. A large-signal model for monolithic power-amplifier design should be accurate for all operating conditions. In addition, the model parameters should be easily extractable and the model must be as simple as possible. Various nonlinear MESFET and HEMT models with different complexity are available, and each one can be considered sufficiently accurate for a particular application. For example, although the Materka model does not fulfill charge conservation, it seems to be an acceptable compromise between accuracy and model simplicity for MESFETs, but not for HEMTs, where it is preferable to use the Angelov model [30,31]. For example, it can be used to predict the large-signal behavior of the pHEMT devices using high-power, high-efficiency 60-GHz MMICs [32]. By using three additional terms of a gate power-series function in the Angelov model, better accuracy can be achieved in large-signal modeling of AlGaN/GaN HEMT devices on SiC substrate [33]. This model can also be improved by incorporating two additional analytical expressions to model the device behavior in saturation regions which is important to design high-efficiency switchmode power amplifiers, for example, in inverse Class-F mode [34].

Figure 1.13(*b*) shows the cross-section of a physical structure of an InGaAs/AlGaAs HEMT device, where an undoped InGaAs *n*-epilayer is used as a channel and two heavily *n*-doped AlGaAs layers with a high energetic barrier for holes are necessary to maximize high electron mobility in the channel. In this case, spacing between AlGaAs layer and InGaAs channel is optimized to achieve high breakdown voltage. An example of the physical structure of a AlGaN/GaN HEMT device is shown in Fig. 1.13(*c*), where an undoped AlGaN *n*-epilayer is used as a channel, an *n*-type doped GaN layer can suppress dispersion in the device current−voltage characteristics, and a SiN passivation layer with optimized parameters contributes to a lower-trap device structure [35]. Thermal conductivity of a GaN HEMT device is improved by using a SiC substrate. Note that the GaN-based technology can provide wider bandwidth and higher efficiency of the power amplifier due to high charge density and the ability to operate at higher voltages for GaN HEMT devices which are characterized by lower output capacitance and on-resistance [36,37].

The basic electrical properties of the MESFET or HEMT device can be characterized by the admittance *Y*-parameters expressed through the device intrinsic small-signal equivalent circuit as

$$Y_{11} = \frac{j\omega C_{\text{gs}}}{1 + j\omega C_{\text{gs}} R_{\text{gs}}} + j\omega C_{\text{gd}} \tag{1.53}$$

$$Y_{12} = -j\omega C_{\text{gd}} \tag{1.54}$$

$$Y_{21} = \frac{g_{\text{m}} \exp(-j\omega\tau)}{1 + j\omega C_{\text{gs}} R_{\text{gs}}} + j\omega C_{\text{gd}} \tag{1.55}$$

$$Y_{22} = \frac{1}{R_{\text{ds}}} + j\omega(C_{\text{ds}} + C_{\text{gd}}) \tag{1.56}$$

where $g_{\text{m}}$ is the device transconductance and $R_{\text{ds}}$ is the differential drain resistance [18]. In this case, the dispersion effect, which is important at higher frequencies and modeled by $C_{\text{dsd}}$ and $R_{\text{dsd}}$, cannot be taken into account.

By separating Eqs (1.53) through (1.56) into their real and imaginary parts, the elements of the small-signal equivalent circuit can be analytically determined as

$$C_{\text{gd}} = -\frac{\text{Im}Y_{12}}{\omega} \tag{1.57}$$

$$C_{\text{gs}} = -\frac{\text{Im}Y_{11} - \omega C_{\text{gd}}}{\omega} \left[ 1 + \left( \frac{\text{Re}\,Y_{11}}{\text{Im}Y_{11} - \omega C_{\text{gd}}} \right)^2 \right] \tag{1.58}$$

$$R_{gs} = \frac{\text{Re } Y_{11}}{(\text{Im } Y_{11} - \omega C_{gd})^2 + (\text{Re } Y_{11})^2} \tag{1.59}$$

$$g_m = \sqrt{(\text{Re } Y_{21})^2 + (\text{Im } Y_{21} + \omega C_{gd})^2}\sqrt{1 + (\omega C_{gs} R_{gs})^2} \tag{1.60}$$

$$\tau = \frac{1}{\omega}\sin^{-1}\left(\frac{-\omega C_{gd} - \text{Im } Y_{21} - \omega C_{gs} R_{gs} \text{ Re } Y_{21}}{g_m}\right) \tag{1.61}$$

$$C_{ds} = \frac{\text{Im } Y_{22} - \omega C_{gd}}{\omega} \tag{1.62}$$

$$R_{ds} = \frac{1}{\text{Re } Y_{22}} \tag{1.63}$$

which are valid for a wide frequency range up to the transition frequency $f_T$ [38]. Assuming that all extrinsic parasitic elements are known, the only problem is to determine the admittance $Y$-parameters of the intrinsic two-port network from on-bias experimental data [39]. Consecutive steps shown in Fig. 1.14 can represent such a determination procedure [40]:

- Measurement of the $S$-parameters of the extrinsic device.
- Transformation of the $S$-parameters to the impedance $Z$-parameters with subtraction of the series inductances $L_g$ and $L_d$.
- Transformation of the impedance $Z$-parameters to the admittance $Y$-parameters with subtraction of the parallel capacitances $C_{gp}$ and $C_{dp}$.
- Transformation of the admittance $Y$-parameters to the impedance $Z$-parameters with subtraction of series resistances $R_g$, $R_s$, $R_d$, and inductance $L_s$.
- Transformation of the impedance $Z$-parameters to the admittance $Y$-parameters of the intrinsic device two-port network.

A simple and accurate nonlinear Angelov model is capable of modeling the drain current–voltage characteristics and its derivatives, as well as the gate-source and gate-drain capacitances, for different submicron gate-length HEMT devices and commercially available MESFETs. The drain current source is described by using the hyperbolic functions as

$$I_{ds} = I_{pk}(1 + \tanh\psi)(1 + \lambda V_{ds})\tanh\alpha V_{ds} \tag{1.64}$$

where $I_{pk}$ is the drain current at maximum transconductance with the contribution from the output conductance subtracted, $\lambda$ is the channel-length modulation parameter, and $\alpha = \alpha_0 + \alpha_1\tanh\psi$ is the saturation voltage parameter, where $\alpha_0$ is the saturation voltage parameter at pinch-off and $\alpha_1$ is the saturation voltage

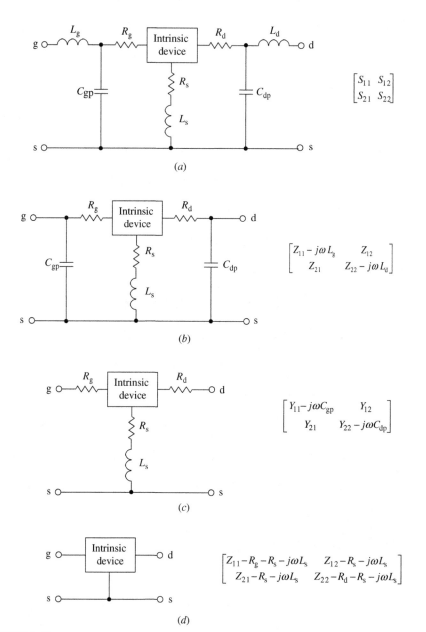

**FIGURE 1.14**

Method for extracting device intrinsic Z-parameters.

parameter at $V_{gs} > 0$. The parameter $\psi$ is a power-series function centered at $V_{pk}$ with bias voltage $V_{gs}$ as a variable,

$$\psi = P_1(V_{gs} - V_{pk}) + P_2(V_{gs} - V_{pk})^2 + P_3(V_{gs} - V_{pk})^3 + \cdots \qquad (1.65)$$

where $V_{pk}$ is the gate voltage for maximum transconductance $g_{mpk}$. The model parameters as a first approximation can be easily obtained from the experimental $I_{ds}(V_{gs}, V_{ds})$ dependences at a saturated channel condition when all higher terms in $\psi$ are assumed to be zero, and $\lambda$ is the slope of the $I_{ds} - V_{ds}$ characteristic.

The same hyperbolic functions can be used to model the intrinsic device capacitances $C_{gs}$ and $C_{ds}$. When an accuracy of $5-10\%$ is sufficient, the gate-source capacitance $C_{gs}$ and gate-drain capacitance $C_{gd}$ can be described by

$$C_{gs} = C_{gs0}[1 + \tanh(P_{1gsg}V_{gs})][1 + \tanh(P_{1gsd}V_{ds})] \qquad (1.66)$$

$$C_{gd} = C_{gd0}[1 + \tanh(P_{1gdg}V_{gs})][1 - \tanh(P_{1gdd}V_{ds} + P_{1cc}V_{gs}V_{ds})] \qquad (1.67)$$

where the product $P_{1cc}V_{gs}V_{ds}$ reflects the cross-coupling of $V_{gs}$ and $V_{ds}$ on $C_{gd}$, and the coefficients $P_{1gsg}$, $P_{1gsd}$, $P_{1gdg}$, and $P_{1gdd}$ are the fitting parameters.

### 1.5.3 Low- and high-voltage HBTs

Figure 1.15(a) shows the modified Gummel−Poon nonlinear model of the bipolar transistor with extrinsic parasitic elements [41,42]. Such a hybrid-$\pi$ equivalent circuit can model the nonlinear electrical behavior of bipolar transistors, in particular HBT devices, with sufficient accuracy up to about 20 GHz. The intrinsic model is described by the dynamic diode resistance $r_\pi$, the total base-emitter junction capacitance and base charging diffusion capacitance $C_\pi$, the base-collector diode required to account for the nonlinear effects at the saturation, the internal collector-base junction capacitance $C_{ci}$, the external distributed collector-base capacitance $C_{co}$, the collector-emitter capacitance $C_{ce}$, and the nonlinear current source $i(v_{be}, v_{ce})$. The lateral and base semiconductor resistances underneath the base contact, and the base semiconductor resistance underneath the emitter are combined into a base-spreading resistance $r_b$. The extrinsic parasitic elements are represented by the base bondwire inductance $L_b$, the emitter ohmic resistance $r_e$, the emitter inductance $L_e$, the collector ohmic resistance $r_c$, and the collector bondwire inductance $L_c$. To increase the usable operating frequency range of the device up to 50 GHz, it is necessary to include the collector current delay time $\tau$ in the collector current source as $g_m\exp(-j\omega\tau)$. The more complicated models, such as VBIC, HICUM, or MEXTRAM, include the effects of self-heating of a bipolar transistor, take into account the parasitic *p-n-p* transistor formed by the base, collector, and substrate regions, provide an improved description of depletion capacitances at large forward bias, take into account avalanche and tunneling currents and other nonlinear effects corresponding to distributed high-frequency effects [43].

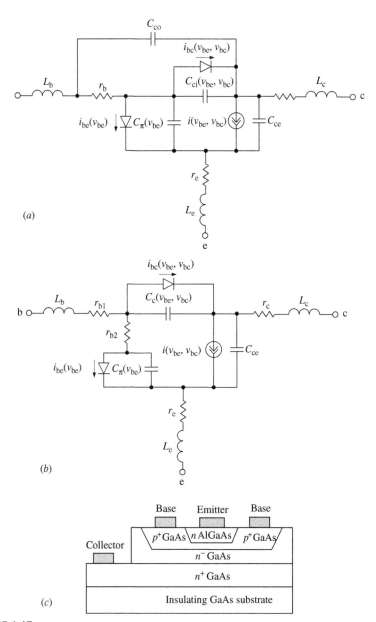

**FIGURE 1.15**

Nonlinear BJT and HBT models and HBT physical structure.

Figure 1.15(b) shows the modified version of a bipolar transistor equivalent circuit, where $C_c = C_{co} + C_{ci}$, $r_{b1} = r_b C_{ci}/C_c$, and $r_{b2} = r_b C_{co}/C_c$ [44]. Such an equivalent circuit becomes possible due to an equivalent $\pi$- to $T$-transformation of the elements $r_b$, $C_{co}$, and $C_{ci}$ and a condition $r_b \ll (C_{ci} + C_{co})/\omega C_{ci} C_{co}$, which is usually fulfilled over a frequency range close to the device maximum frequency $f_{max}$. Then, from a comparison of the transistor nonlinear models, for a bipolar transistor in Fig. 1.15(b), for a MOSFET device in Fig. 1.10(b), and for a MESFET device in Fig. 1.13(a), it is easy to detect the circuit similarity of all these equivalent circuits, which means that the basic circuit design procedure is very similar for any type of bipolar or field-effect transistors. The difference is in the device physics and values of the model parameters. However, techniques for the representation of the input and output impedances, stability analysis based on feedback effect, derivation of power gain and efficiency are very similar.

The cross-section of a physical structure of an AlGaAs/GaAs HBT device is shown in Fig. 1.15(c), with heavily $p$-doped base to reduce base resistance and lightly $n$-doped emitter to minimize emitter capacitance [45]. The lightly $n$-doped collector region allows the collector–base junction to sustain relatively high voltages without breaking down. The forward-bias emitter-injection efficiency is very high since the wider-bandgap AlGaAs emitter injects electrons into the GaAs $p$-base at a lower energy level, but the holes are prevented from flowing into the emitter by a high energy barrier, thus resulting in the ability to decrease base length, base-width modulation, and increase frequency response. By using a wide bandgap InGaP layer instead of an AlGaAs one, the device performance over temperature can be improved [46]. The high-linearity power performance in Class-AB condition at the backoff power level, the ruggedness under mismatch and overdrive condition, and the long lifetime of the InGaP/GaAs HBT technology makes it very attractive for the 28-V power amplifier applications [47]. The growth process used for a high-voltage HBT device is identical to the process used for the conventional low-voltage HBT device, which is widely used in handset power amplifiers, except for changes to the collector because of the higher voltage operating requirements. The epitaxial growth process starts with a highly doped $n$-type collector layer and lightly $n$-doped collector drift region, then followed by a heavily doped $p$-type base layer and an InGaP emitter layer, and finishes with an InGaAs cap layer [48]. As a result, the high-voltage HBT devices exhibit collector-base breakdown voltages higher than 70 V.

The bipolar transistor intrinsic $Y$-parameters can be written as

$$Y_{11} = \frac{1}{r_\pi} + j\omega(C_\pi + C_{ci}) \tag{1.68}$$

$$Y_{12} = -j\omega C_{ci} \tag{1.69}$$

$$Y_{21} = g_m \exp(-j\omega\tau) + j\omega C_{ci} \tag{1.70}$$

$$Y_{22} = \frac{1}{r_{ce}} + j\omega C_{ci} \tag{1.71}$$

where $r_{ce}$ is the output early resistance that models the effect of the base-width modulation on the transistor characteristics due to variations in the collector-base depletion region.

After separating Eqs (1.68) through (1.71) into their real and imaginary parts, the elements of the intrinsic small-signal equivalent circuit can be determined analytically as [49]

$$C_\pi = \frac{\mathrm{Im}(Y_{11} + Y_{12})}{\omega} \tag{1.72}$$

$$r_\pi = \frac{1}{\mathrm{Re}\, Y_{11}} \tag{1.73}$$

$$C_{ci} = -\frac{\mathrm{Im}\, Y_{12}}{\omega} \tag{1.74}$$

$$g_m = \sqrt{(\mathrm{Re}\, Y_{21})^2 + (\mathrm{Im}\, Y_{21} + \mathrm{Im}\, Y_{12})^2} \tag{1.75}$$

$$\tau_\pi = \frac{1}{\omega}\, \cos^{-1} \frac{\mathrm{Re}\, Y_{21} + \mathrm{Re}\, Y_{12}}{\sqrt{(\mathrm{Re}\, Y_{21})^2 + (\mathrm{Im}\, Y_{21} + \mathrm{Im}\, Y_{12})^2}} \tag{1.76}$$

$$r_{ce} = \frac{1}{\mathrm{Re}\, Y_{22}}. \tag{1.77}$$

A simple nonlinear HBT model for computer-aided simulations can be based on the representation of the collector current source through the power series and diffusion capacitances through the hyperbolic functions [50]. To equivalently represent the input impedance of a bipolar transistor, it needs to take into account that $C'_{ce}$ is usually much smaller than $C_c$. As a result, the equivalent output capacitance can be defined as $C_{out} \cong C_c$. The input equivalent $R_{in}$ can approximately be represented by the base resistance $r_b$, while the input equivalent capacitance can be defined as $C_{in} \cong C_\pi + C_c$. The feedback effect of the collector capacitance $C_c$ through $C_{c0}$ and $C_{ci}$ is sufficiently high when load variations are directly transferred to the device input with a significant extent.

## 1.6 High-frequency conduction angle

An idealized analysis of the physical processes in the nonlinear power amplifier based on the cosinusoidal input signal can be considered sufficiently accurate only at low frequencies when all phase delays due to the effect of the elements of the device equivalent circuit is neglected and it is represented as an ideal voltage or current-control current source. However, as seen from Figs. 1.10, 1.13, and 1.15, the electrical behavior of the transistor over its entire operating frequency range is described by the sufficiently complicated equivalent circuit, including linear and nonlinear internal and external parasitic elements. The bipolar transistor linear operation at low and medium frequencies up to approximately $(0.1 \div 0.2)f_T$, where $f_T$ is the transition frequency, can be adequately characterized by a Giacoletto equivalent circuit shown in Fig. 1.16($a$) [51]. In a linear small-signal mode, all elements of the device equivalent circuit are considered constant, including the base-emitter

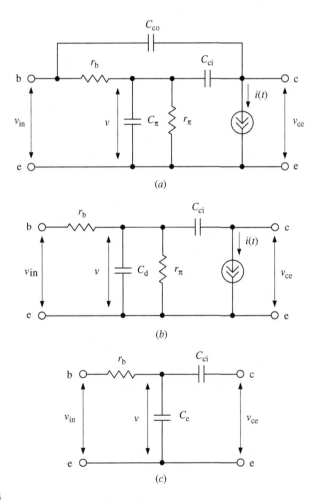

**FIGURE 1.16**

Giacoletto model for a bipolar transistor.

diffusion capacitance $C_d$ and differential resistance $r_\pi$. To analyze device behavior in a large-signal mode when the transistor is operated in the pinch-off and active regions, it is necessary to compose two equivalent circuits: the first should correspond to a linear-active region shown in Fig. 1.16(b) and the other corresponding to a pinch-off region as shown in Fig. 1.16(c). In this case, it is necessary to take into account that the capacitance $C_d$ and resistance $r_\pi$, whose values depend significantly on the driving signal amplitude, are set by their averaged values in an active mode. The external feedback capacitance $C_{co}$ can be included to external circuitry.

By applying a piecewise-linear approximation of the transistor transfer current–voltage characteristic, the current $i$ as a function of the driving junction voltage $v$ can be written as

$$i(v) = \begin{cases} g_m(v - V_p) & v \geq V_p \\ 0 & v < V_p \end{cases} \tag{1.78}$$

while the input driving signal applied to the device input port is defined as

$$v_{in}(\omega t) = V_p + V_{in}(\cos \omega t - \cos \theta) \tag{1.79}$$

with the base dc-bias voltage $V_{bias} = V_p$, where $V_p$ is the pinch-off voltage.

Let us assume that the current flowing through the internal feedback capacitance $C_{ci}$ is sufficiently small when its effect can be neglected. Then, to compose the differential equation describing the device behavior in an active mode, we can write

$$\frac{V_{in} - v}{r_b} = \frac{v}{r_\pi} + \omega C_d \frac{dv}{d\omega t}. \tag{1.80}$$

As a result, the transistor behavior for $v \geq V_p$ can be described by the first-order linear differential equation written as

$$\omega \tau_1 \frac{dv}{d\omega t} + v = kV_p - kV_{in} \ (\cos \omega t - \cos \theta) \tag{1.81}$$

where $\tau_1 = r_b k C_d$ and $k = r_\pi/(r_b + r_\pi)$.

Similarly, for $v < V_p$,

$$\omega \tau_2 \frac{dv}{d\omega t} + v = V_p - V_{in} \ (\cos \omega t - \cos \theta) \tag{1.82}$$

where $\tau_2 = r_b C_e$ and $C_e$ is the base–emitter junction capacitance.

Depending on the bipolar transistor type, the ratio between the differential capacitance $C_d$ and junction capacitance $C_e$ can be different. In most cases when $C_d$ is greater than $C_e$ by an order, the effect of the junction capacitance $C_e$ can be neglected for analysis simplicity. Then, assuming that $\tau_2 = 0$, Eq. (1.82) can be rewritten as

$$v = V_p - V_{in}(\cos \omega t - \cos \theta) \tag{1.83}$$

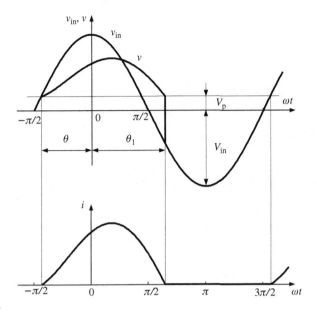

**FIGURE 1.17**

Device voltage and current waveforms when $C_e = 0$.

which means that the input driving signal is applied to the device junction without any changes in its voltage shape during a pinch-off mode.

Thus, the solution of Eq. (1.81) in the time domain for an initial condition $v(-\theta) = V_p$ can be obtained as

$$v = V_p + V_{in}\cos\phi_1 \left\{ \cos(\omega t + \phi_1) - \frac{\cos\theta}{\cos\phi_1} - \left[\cos(\theta - \phi_1) - \frac{\cos\theta}{\cos\phi_1}\right] \exp\left(-\frac{\omega t + \theta}{\omega\tau_1}\right) \right\}$$

(1.84)

where $\cos\phi_1 = 1/\sqrt{1 + (\omega\tau_1)^2}$ and $\phi_1 = -\tan^{-1}\omega\tau_1$ [44].

Figure 1.17 shows the time domain dependences of the periodical base input voltage $v_{in}$, base junction voltage $v$, and collector current $i$ for the base dc-bias voltage $V_{bias} = 0$. In this case, if the input driving voltage $v_{in}$ is cosinusoidal, the junction voltage $v$ contains an exponential component demonstrating a transient response that occurs when the transistor is turned on. At the time moment $\theta_1$ when the voltage $v$ becomes equal to $V_p$, the transistor is turned off and the voltage $v$ provides an instant step change to a certain negative value, since it was

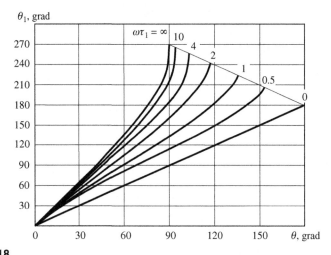

**FIGURE 1.18**

High-frequency conduction angle as function of $\omega\tau_1$.

assumed that $C_e = 0$ and any transient response during the pinch-off mode is not possible. Due to the effect of the diffusion capacitance $C_d$ during an active device mode, the transistor is turned off when the input driving voltage $v_{in}$ is negative at that time. In this case, the shape of the junction voltage $v$ is no longer cosinusoidal. Now it is characterized by a much smaller amplitude and is stretched to the right-hand side, thus making the conduction angle longer compared to a low-frequency case. According to Eq. (1.78), the time-domain behavior of the collector current $i$ represents the waveform similar to that of the junction voltage waveform during an active mode and equal to zero during a pinch-off mode.

The time moment $\theta_1$ can be defined from Eq. (1.84) by setting $\omega t = \theta_1$ and $v = V_p$ that results in a transcendental equation

$$\cos(\theta_1 + \phi_1) - \frac{\cos\theta}{\cos\phi_1} - \tan\phi_1 \sin(\theta - \phi_1)\exp\left(\frac{\theta_1 + \theta}{\tan\phi_1}\right) = 0 \tag{1.85}$$

from which the dependences $\theta_1(\theta)$ for different values of the input time constant $\omega\tau_1$ can be numerically calculated [44,52]. As seen from Fig. 1.18, $\theta_1 = \theta$ when $\omega\tau_1 = 0$, corresponding to a low-frequency case. However, at higher operating frequencies, the collector current pulses corresponding to the conduction state become longer with increased values of $\theta_1 > \theta$. Beginning from a boundary value of $\theta = 180° + \phi_1$, the transistor is operated in an active region only because the pinch-off operation region does not occur anymore. As a result, the sum $(\theta_1 + \theta)$, which is called the *high-frequency conduction angle*, is always greater than the low-frequency conduction angle $2\theta$ for any certain value of $\omega\tau_1$.

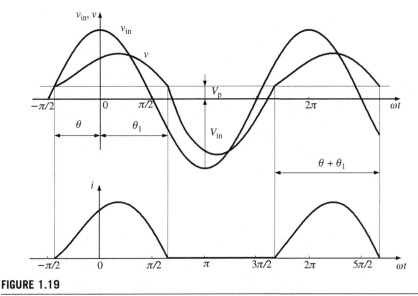

**FIGURE 1.19**

Device voltage and current waveforms when $C_e \neq 0$.

In a common case when it is impossible to neglect the effect of the junction capacitance $C_e$, Eq. (1.83) is rewritten by using Eq. (1.82) as

$$v = V_p + V_{in}\cos\phi_2 \left\{ \cos(\omega t + \phi_2) - \frac{\cos\theta}{\cos\phi_2} - \left[\cos(\theta_1 + \phi_2) - \frac{\cos\theta}{\cos\phi_2}\right] \exp\left(-\frac{\omega t - \theta_1}{\omega\tau_2}\right) \right\}$$

(1.86)

where $\cos\phi_2 = 1/\sqrt{1 + (\omega\tau_2)^2}$ and $\phi_2 = -\tan^{-1}\omega\tau_2$.

Equation (1.86) describes the transient response when $v < V_p(V_p > V_{bias} = 0)$, which occurs when the transistor is turned off. This transient response arises when the transistor turns off for the first time, as shown in Fig. 1.19. Then, when the junction voltage $v$ becomes equal to a pinch-off voltage $V_p$ again, the transistor turns on. The more the active-mode time constant $\tau_1$ exceeds the pinch-off-mode time constant $\tau_2$, the shorter is the transient response, resulting in a steady-state periodical pulsed current $i$ with fixed amplitude and conduction angle. The time constant $\tau_2$ directly affects the duration of the transient response. When $\tau_2 = 0$, the instant damping of the transient response will occur and voltage $v$ steps down to the value of the input voltage $v_{in}$. However, when $\tau_2 \neq 0$, the voltage dependence becomes smooth and more symmetrical.

For a boundary case of equal capacitances in active and pinch-off modes when $\tau_1 = \tau_2$, both the junction voltage and collector current pulses become fully symmetrical representing the truncated cosine waveform with a high-frequency conduction angle $(\theta_1 + \theta)$ different from a low-frequency conduction angle $2\theta$. The high-frequency conduction angle can be greater, equal, or smaller than its low-frequency counterpart depending on the device base dc-bias conditions. For a Class-C mode with $V_{bias} < V_p$, the junction voltage pulse when $v > V_p$ is shorter since $\theta_1 + \theta < 2\theta$, as shown in Fig. 1.20(a) for $V_p = 0$. For a Class-B mode when $V_{bias} = V_p$, both high-frequency and low-frequency conduction angles are equal and $\theta_1 = \theta$, as shown in Fig. 1.20(b) for $V_p = 0$. Finally, in a Class-AB mode when $V_{bias} > V_p$, the junction voltage pulse becomes longer when $v > V_p$ since $\theta_1 + \theta > 2\theta$, as shown in Fig. 1.20(c) for $V_p = 0$.

Analytically, the high-frequency angle can be easily calculated when $C_\pi = C_d = C_e$ and $\phi = \phi_1 = \phi_2$ from

$$\cos\theta_1 = -\frac{V_{bias} - V_p}{V_{in}}\frac{1}{\cos\phi} \tag{1.87}$$

where $\cos\phi = 1/\sqrt{1 + (\omega\tau_0)^2}$ and $\tau_0 = r_b C_\pi$.

To take into account the effect of the feedback collector capacitance $C_{ci}$, the equivalent input driving voltage $v'_{in}(\omega t)$ can be represented through the input base-emitter voltage $v_{in}$ and collector-emitter voltage $v_{ce}$ as

$$v'_{in}(\omega t) = v_{in}(\omega t) + j\omega\tau_c v_{ce}(\omega t) \tag{1.88}$$

where $\tau_c = r_b C_{ci}$.

## 1.7 Nonlinear effect of collector capacitance

Generally, the dependence of the collector capacitance on the output voltage represents a nonlinear function. To evaluate the influence of the nonlinear collector capacitance on electrical behavior of the power amplifier, let us consider the load network including a series resonant $L_0C_0$ circuit tuned to the fundamental frequency that provides open-circuit conditions for the second- and higher-order harmonic components of the output current and a low-pass $L$-type matching circuit with the series inductor $L$ and shunt capacitor $C$, as shown in Fig. 1.21(a). The matching circuit is needed to match the equivalent output resistance $R$, corresponding to the required output power at the fundamental frequency, with the standard load resistance $R_L$. Figure 1.21(b) shows the simplified output equivalent circuit of the bipolar power amplifier.

The total output current flowing through the device collector can be written as

$$i = I_0 + \sum_{n=1}^{\infty} I_n \cos(n\omega t + \phi_n) \tag{1.89}$$

where $I_n$ and $\phi_n$ are the amplitude and phase of the $n$th-harmonic component, respectively.

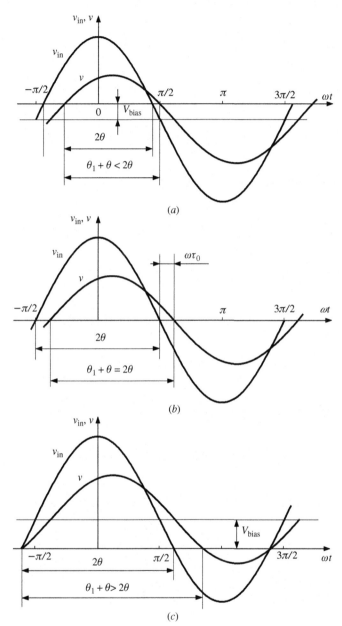

**FIGURE 1.20**

Device voltage and current waveforms when $C_d = C_e$.

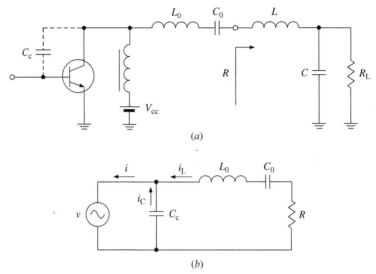

**FIGURE 1.21**

Circuit schematics of bipolar tuned power amplifier.

An assumption of a high-quality factor of the series resonant circuit allows the only fundamental-frequency current component to flow into the load. The current flowing through the nonlinear collector capacitance consists of the fundamental-frequency and higher-order harmonic components, which is written as

$$i_C = I_C \cos(\omega t + \phi_1) + \sum_{n=2}^{\infty} I_n \cos(n\omega t + \phi_n) \tag{1.90}$$

where $I_C$ is the fundamental-frequency capacitor current amplitude.

The nonlinear behavior of the collector junction capacitance is described by

$$C_c = C_0 \left( \frac{\varphi + V_{cc}}{\varphi + v} \right)^{\gamma} \tag{1.91}$$

where $C_0$ is the collector capacitance at $v = V_{cc}$, $V_{cc}$ is the supply voltage, $\varphi$ is the contact potential, and $\gamma$ is the junction sensitivity equal to 0.5 for abrupt junction.

As a result, the expression for charge flowing through collector capacitance can be obtained by

$$q = \int_0^v C(v)dv = \int_0^v \frac{C_0(\varphi + V_{cc})^{\gamma}}{(\varphi + v)^{\gamma}} dv. \tag{1.92}$$

When $v = V_{cc}$, then

$$q_0 = \frac{C_0(\varphi + V_{cc})}{1 - \gamma}\left[1 - \left(\frac{\varphi}{\varphi + V_{cc}}\right)^{1-\gamma}\right]. \tag{1.93}$$

Although the dc charge component $q_0$ is a function of the voltage amplitude, its variations at maximum voltage amplitude normally do not exceed 20% for $\gamma = 0.5$. Then, assuming $q_0$ is determined by Eq. (1.93) as a constant component, the total charge $q$ of the nonlinear capacitance can be represented by the dc component $q_0$ and ac component $\Delta q$ written as

$$q = q_0 + \Delta q = q_0\left(1 + \frac{\Delta q}{q_0}\right) = q_0\frac{(\varphi + v)^{1-\gamma} - \varphi^{1-\gamma}}{(\varphi + V_{cc})^{1-\gamma} - \varphi^{1-\gamma}}. \tag{1.94}$$

Since $V_{cc} \gg \varphi$ in the normal case, from Eq. (1.94) it follows that

$$\frac{v}{V_{cc}} = \left(1 + \frac{\Delta q}{q_0}\right)^{\frac{1}{1-\gamma}} \tag{1.95}$$

where $q_0 \cong C_0 V_{cc}/(1 - \gamma)$.

On the other hand, the charge component $\Delta q$ can be written using Eq. (1.90) as

$$\Delta q = \int i_C(t)dt = \frac{I_C}{\omega}\sin(\omega t + \phi_1) + \sum_{n=2}^{\infty}\frac{I_n}{n\omega}\sin(n\omega t + \phi_n). \tag{1.96}$$

As a result, substituting Eq. (1.96) into Eq. (1.95) yields

$$\frac{v}{V_{cc}} = \left[1 + \frac{I_C(1 - \gamma)}{\omega C_0 V_{cc}}\sin(\omega t + \phi_1) + \sum_{n=2}^{\infty}\frac{I_n(1 - \gamma)}{n\omega C_0 V_{cc}}\sin(n\omega t + \phi_n)\right]^{\frac{1}{1-\gamma}}. \tag{1.97}$$

After applying a Taylor-series expansion to Eq. (1.97), it is sufficient to be limited to its first three terms to reveal the parametric effect. Then, equating the fundamental-frequency collector voltage components gives

$$\frac{v_1}{V_{cc}} = \frac{I_C}{\omega C_0 V_{cc}}\sin(\omega t + \phi_1) + \frac{I_C I_2\gamma}{(2\omega C_0 V_{cc})^2}\cos(\omega t + \phi_2 - \phi_1)$$
$$+ \frac{I_2 I_3\gamma}{12(\omega C_0 V_{cc})^2}\cos(\omega t + \phi_3 - \phi_2). \tag{1.98}$$

Consequently, by taking into account that $v_1 = V_1\sin(\omega t + \phi_1)$, the fundamental voltage amplitude $V_1$ can be obtained from Eq. (1.98) as

$$\frac{V_1}{V_{cc}} = \frac{I_C}{\omega C_0 V_{cc}}\left[1 + \frac{I_2\gamma}{4\omega C_0 V_{cc}}\cos(90° + \phi_2 - 2\phi_1)\right.$$
$$\left. + \frac{I_2 I_3\gamma}{12\omega C_0 V_{cc} I_C}\cos(90° + \phi_3 - \phi_2 - \phi_1)\right]. \tag{1.99}$$

Since a large-signal value of the abrupt-junction collector capacitance usually does not exceed 20%, the fundamental-frequency capacitor current amplitude $I_C$ as a first-order approximation can be written as

$$I_C \cong \omega C_0 V_1. \tag{1.100}$$

As a result, from Eq. (1.99) it follows that, because of the parametric transformation due to the collector capacitance nonlinearity, the fundamental-frequency collector voltage amplitude increases by $\sigma_p$ times according to

$$\sigma_p = 1 + \frac{I_2 \gamma}{4\omega C_0 V_{cc}} \cos(90° + \phi_2 - 2\phi_1) + \frac{I_2 I_3 \gamma}{12(\omega C_0)^2 V_1 V_{cc}} \cos(90° + \phi_3 - \phi_2 - \phi_1) \tag{1.101}$$

where $\sigma_p = \xi_p/\xi$ and $\xi_p$ is the collector voltage peak factor with parametric effect [12].

From Eq. (1.101), it follows that to maximize the collector voltage peak factor and consequently the collector efficiency for a given value of the supply voltage $V_{cc}$, it is necessary to provide the following phase conditions:

$$\phi_2 = 2\phi_1 - 90° \tag{1.102}$$

$$\phi_3 = 3\phi_1 - 180°. \tag{1.103}$$

Then, for $\gamma = 0.5$,

$$\sigma_p = 1 + \frac{I_2}{8\omega C_0 V_{cc}} + \frac{I_2 I_3}{24(\omega C_0)^2 V_1 V_{cc}}. \tag{1.104}$$

Equation (1.104) shows the theoretical possibility to increase the collector voltage peak factor by 1.1–1.2 times, thus achieving collector efficiency of 85–90%. Physically, the improved efficiency can be explained by the transformation of powers corresponding to the second and higher-order harmonic components into the fundamental-frequency output power because of the nonlinearity of the collector capacitance. However, this becomes effective only in the case of the load network with a series resonant circuit, since it ideally provides infinite impedance at the second- and higher-order harmonics unlike the load network with a parallel resonant circuit having ideally zero impedance at these harmonics.

---

## 1.8 Push–pull power amplifiers

Generally, if it is necessary to increase an overall output power of the power amplifier, several active devices can be used in parallel or push–pull configurations. In a parallel configuration, the active devices are not isolated from each other, which requires a very good circuit symmetry, and output impedance becomes too small in the case of high output power. The latter drawback can be eliminated in a push–pull configuration, which provides increased values of the input and output impedances. For the same output power level, the input

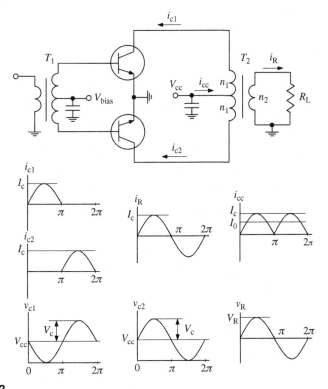

**FIGURE 1.22**

Basic concept of push–pull operation.

impedance $Z_{in}$ and output impedance $Z_{out}$ under a push–pull operation mode are approximately four times as high as for a parallel connection of the active devices. At the same time, the loaded quality factors of the input and output matching circuits remain unchanged because both the real and reactive parts of these impedances are increased by the factor of four. Very good circuit symmetry can be provided using balanced active devices with common emitters in a single package. The basic concept of a push–pull operation can be analyzed by using the corresponding circuit schematic shown in Fig. 1.22 [53].

It is most convenient to consider an ideal Class-B operation, which means that each transistor conducts exactly half a cycle (180°) with zero quiescent current. Let us also assume that the number of turns of both primary and secondary windings of the output transformer $T_2$ is equal ($n_1 = n_2$) and the collector current of each transistor can be presented in the following half-sinusoidal form: for the first transistor

$$i_{c1} = \begin{cases} + I_c \sin \omega t & 0 \le \omega t < \pi \\ 0 & \pi \le \omega t < 2\pi \end{cases} \qquad (1.105)$$

for the second transistor

$$i_{c2} = \begin{cases} 0 & 0 \le \omega t < \pi \\ -I_c \sin \omega t & \pi \le \omega t < 2\pi \end{cases} \tag{1.106}$$

where $I_c$ is the output current amplitude.

Being transformed through the output transformer $T_2$ with the appropriate phase conditions, the total current flowing through the load $R_L$ is obtained as

$$i_R(\omega t) = i_{c1}(\omega t) - i_{c2}(\omega t) = I_c \sin \omega t. \tag{1.107}$$

The current flowing into the center tap of the primary windings of the output transformer $T_2$ is the sum of the collector currents, resulting in

$$i_{cc}(\omega t) = i_{c1}(\omega t) + i_{c2}(\omega t) = I_c |\sin \omega t|. \tag{1.108}$$

Ideally, even-order harmonics being in phase are canceled out and should not appear at the load. In practice, a level of the second-harmonic component of $30-40$ dB below the fundamental is allowable. However, it is necessary to connect a bypass capacitor to the center tap of the primary winding to exclude power losses due to even-order harmonics. The current $i_R(\omega t)$ produces the load voltage $v_R(\omega t)$ onto the load $R_L$ as

$$v_R(\omega t) = I_c R_L \sin(\omega t) = V_R \sin(\omega t) \tag{1.109}$$

where $V_R$ is the load voltage amplitude.

The total dc collector current is defined as the average value of $i_{cc}(\omega t)$, which yields

$$I_0 = \frac{1}{2\pi} \int_0^{2\pi} i_{cc}(\omega t) d\omega t = \frac{2}{\pi} I_c. \tag{1.110}$$

For the ideal case of zero saturation voltage of both transistors when $V_c = V_{cc}$ and taking into account that $V_R = V_c$ for equal turns of windings when $n_1 = n_2$, the total dc power $P_0$ and fundamental-frequency output power $P_{out}$ are obtained by

$$P_0 = \frac{2}{\pi} I_c V_{cc} \tag{1.111}$$

$$P_{out} = \frac{I_c V_{cc}}{2}. \tag{1.112}$$

Consequently, the maximum theoretical collector efficiency that can be achieved in a push−pull Class-B operation is equal to

$$\eta = \frac{P_{out}}{P_0} = \frac{\pi}{4} \cong 78.5\%. \tag{1.113}$$

In a balanced circuit, identical sides carry 180° out-of-phase signals of equal amplitude. If perfect balance is maintained on both sides of the circuit, the

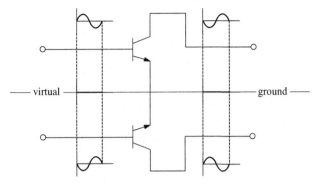

**FIGURE 1.23**

Basic concept of a balanced transistor.

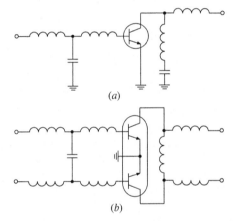

(a)

(b)

**FIGURE 1.24**

Matching technique for single-ended and balanced transistors.

difference between signal amplitudes becomes equal to zero in each midpoint of the circuit, as shown in Fig. 1.23. This effect is called the *virtual grounding*, and this midpoint line is referred to as the *virtual ground*. The virtual ground, being actually inside the device package, reduces a common mode inductance and results in better stability and usually higher power gain.

When using a balanced transistor, new possibilities for both internal and external impedance matching procedure emerge. For instance, for a push–pull operation mode of two single-ended transistors, it is necessary to provide reliable grounding for input and output matching circuits for each device, as shown in Fig. 1.24(a). Using the balanced transistors simplifies significantly the matching circuit topologies with the series inductors and parallel capacitors connected

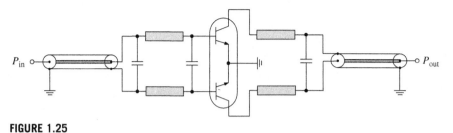

**FIGURE 1.25**

Push−pull power amplifier with balanced-to-unbalanced transformers.

between amplifying paths, as shown in Fig. 1.24(b), and dc-blocking capacitors are not needed.

For a push−pull operation of the power amplifier with a balanced transistor, it is also necessary to provide the unbalanced-to-balanced transformation referenced to the ground both at the input and at the output of the power amplifier. The most suitable approach to solve this problem in the best possible manner at high frequencies and microwaves is to use the transmission-line transformers, as shown in Fig. 1.25. If the characteristic impedance $Z_0$ of the coaxial transmission line is equal to the input impedance at the unbalanced end of the transformer, the total impedance from both devices seen at the balanced end of the transformer will be equal to the input impedance. Hence, such a transmission-line transformer can be used as a 1:1 balanced-to-unbalanced transformer (balun). For the standard input impedance of 50 Ω, if $Z_0 = 50$ Ω, the impedance seen at each balanced part is equal to 25 Ω, which then is necessary to match with the appropriate input impedance of each part of a balanced transistor. The input and output matching circuits can easily be realized by using the series microstrip lines with parallel capacitors.

The miniaturized compact input unbalanced-to-balanced transformer shown in Fig. 1.26 covers the frequency bandwidth up to an octave with well-defined rejection-mode impedances [18,54]. To avoid the parasitic capacitance between the outer conductor and the ground, the coaxial semirigid transformer $T_1$ is mounted atop microstrip shorted stub $l_1$ and soldered continuously along its length. The electrical length of this stub is usually chosen from the condition of $\theta \leq \pi/2$ on the high bandwidth frequency depending on the matching requirements. To maintain circuit symmetry on the balanced side of the transformer network, another semirigid coaxial section $T_2$ with an unconnected center conductor is soldered continuously along the microstrip shorted stub $l_2$. The lengths of $T_2$ and $l_2$ are equal to the lengths of $T_1$ and $l_1$, respectively. Because the input short-circuited microstrip stubs provide inductive impedances, the two series capacitors $C_1$ and $C_2$ of the same value are used for matching purposes, thereby forming the first high-pass matching section and providing dc blocking at the same time. The practical circuit realization of the output matching circuit and balanced-to-unbalanced transformer can be the same as for the input matching circuit.

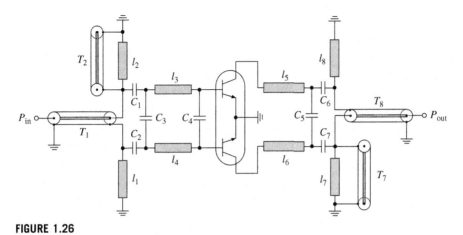

**FIGURE 1.26**

Push—pull power amplifier with compact balanced-to-unbalanced transformers.

## 1.9 **Power gain and impedance matching**

Power amplifier design aims for maximum power gain and efficiency for a given value of output power with a predictable degree of stability. In order to extract the maximum power from a generator, it is a well-known fact that the external load should have a vector value which is a conjugate of the internal impedance of the source [55]. The power delivered from a generator to a load, when matched on this basis, will be called the available power of the generator [56]. In this case, the power gain of the four-terminal network is defined as the ratio of the power delivered to the load impedance connected at the output terminals to the power available from the generator connected to the input terminals, usually measured in decibels, and this ratio is called the *power gain* irrespective of whether it is greater or less than one [57,58].

Figure 1.27 shows the basic block schematic of a single-stage power-amplifier circuit, which includes an active device, an input matching circuit to match with the source impedance, and an output matching circuit to match with the load impedance. Generally, the two-port active device is characterized by a system of the immittance $W$-parameters, i.e. any system of impedance $Z$-parameters, hybrid $H$-parameters, or admittance $Y$-parameters [59,60]. The input and output matching circuits transform the source and load immittances $W_S$ and $W_L$ into specified values between points 1—2 and 3—4, respectively, by means of which the optimal design operation mode of the power amplifier is realized.

The operating power gain $G_P$, which represents the ratio of power dissipated in the active load $\mathrm{Re}W_L$ to the power delivered to the input port of the active device, can be expressed in terms of the immittance $W$-parameters as

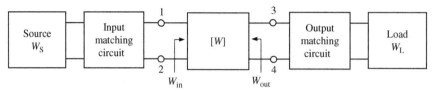

**FIGURE 1.27**

Block schematic of single-stage power amplifier.

$$G_P = \frac{|W_{21}|^2 \text{Re} W_L}{|W_{22} + W_L|^2 \text{Re} W_{in}} \tag{1.114}$$

where

$$W_{in} = W_{11} - \frac{W_{12} W_{21}}{W_{22} + W_L} \tag{1.115}$$

is the input immittance and $W_{ij}(i, j = 1, 2)$ are the immittance two-port parameters of the active device equivalent circuit.

The transducer power gain $G_T$, which represents the ratio of power dissipated in the active load $\text{Re} W_L$ to the power available from the source, can be expressed in terms of the immittance $W$-parameters as

$$G_T = \frac{4|W_{21}|^2 \text{Re} W_S \ \text{Re} W_L}{\left|(W_{11} + W_S)(W_{22} + W_L) - W_{12} W_{21}\right|^2}. \tag{1.116}$$

The operating power gain $G_P$ does not depend on the source parameters and characterizes only the effectiveness of the power delivery from the input port of the active device to the load. This power gain helps to evaluate the gain property of a multistage amplifier when the overall operating power gain $G_{P(total)}$ is equal to the product of each stage $G_P$. The transducer power gain $G_T$ includes an assumption of conjugate matching of the load and the source.

The bipolar transistor simplified small-signal $\pi$-hybrid equivalent circuit shown in Fig. 1.28 provides an example for a conjugate-matched bipolar power amplifier. The impedance $Z$-parameters of the equivalent circuit of the bipolar transistor in a common-emitter configuration can be written as

$$Z_{11} = r_b + \frac{1}{g_m + j\omega C_\pi} \qquad Z_{12} = \frac{1}{g_m + j\omega C_\pi}$$

$$Z_{21} = -\frac{1}{j\omega C_c} \frac{g_m - j\omega C_c}{g_m + j\omega C_\pi} \qquad Z_{22} = \left(1 + \frac{C_\pi}{C_c}\right) \frac{1}{g_m + j\omega C_\pi} \tag{1.117}$$

where $g_m$ is the transconductance, $r_b$ is the series base resistance, $C_\pi$ is the base-emitter capacitance including both diffusion and junction components, and $C_c$ is the feedback collector capacitance.

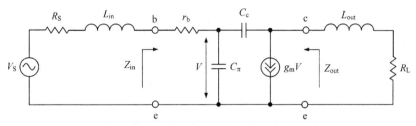

**FIGURE 1.28**

Simplified equivalent circuit of matched bipolar power amplifier.

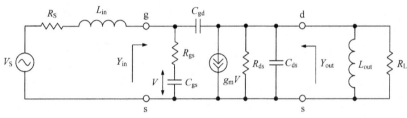

**FIGURE 1.29**

Simplified equivalent circuit of matched FET power amplifier.

By setting the device feedback impedance $Z_{12}$ to zero and complex conjugate-matching conditions at the input of $R_S = \mathrm{Re}Z_{in}$ and $L_{in} = -\mathrm{Im}Z_{in}/\omega$ and at the output of $R_L = \mathrm{Re}Z_{out}$ and $L_{out} = -\mathrm{Im}Z_{out}/\omega$, the small-signal transducer power gain $G_T$ can be calculated from

$$G_T = \left(\frac{f_T}{f}\right)^2 \frac{1}{8\pi\, f_T r_b C_c} \tag{1.118}$$

where $f_T = g_m/2\pi C_\pi$ is the device transition frequency.

Figure 1.29 shows the simplified circuit schematic for a conjugate-matched FET (field-effect transistor) power amplifier. The admittance $Y$-parameters of the small-signal equivalent circuit of any FET device in a common-source configuration can be written as

$$Y_{11} = \frac{j\omega C_{gs}}{1 + j\omega C_{gs}R_{gs}} + j\omega C_{gd} \qquad Y_{12} = -j\omega C_{gd}$$

$$\tag{1.119}$$

$$Y_{21} = \frac{g_m}{1 + j\omega C_{gs}R_{gs}} - j\omega C_{gd} \qquad Y_{22} = \frac{1}{R_{ds}} + j\omega(C_{ds} + C_{gd})$$

where $g_m$ is the transconductance, $R_{gs}$ is the gate-source resistance, $C_{gs}$ is the gate-source capacitance, $C_{gd}$ is the feedback gate-drain capacitance, $C_{ds}$ is the drain-source capacitance, and $R_{ds}$ is the differential drain-source resistance.

Since the value of the gate-drain capacitance $C_{gd}$ is normally relatively small, the effect of the feedback admittance $Y_{12}$ can be neglected in a simplified case. Then, it is necessary to set $R_S = R_{gs}$ and $L_{in} = 1/\omega^2 C_{gs}$ for input matching, while $R_L = R_{ds}$ and $L_{out} = 1/\omega^2 C_{ds}$ for output matching. Hence, the small-signal transducer power gain $G_T$ can approximately be calculated from

$$G_T(C_{gd} = 0) = MAG = \left(\frac{f_T}{f}\right)^2 \frac{R_{ds}}{4R_{gs}} \qquad (1.120)$$

where $f_T = g_m/2\pi C_{gs}$ is the device transition frequency and $MAG$ is the maximum available gain representing a theoretical limit on the power gain that can be achieved under complex conjugate-matching conditions.

From Eqs (1.118) and (1.120), it follows that the small-signal power gain of a conjugate-matched power amplifier for any type of the active device drops off as $1/f^2$ or 6 dB per octave. Therefore, $G_T(f)$ can readily be predicted at a certain frequency $f$, if a power gain is known at the transition frequency $f_T$, by

$$G_T(f) = G_T(f_T) \left(\frac{f_T}{f}\right)^2. \qquad (1.121)$$

It should be noted that previous analysis is based upon the linear small-signal consideration when generally nonlinear device current source as a function of both input and output voltages can be characterized by the linear transconductance $g_m$ as a function of the input voltage and the output differential resistance $R_{ds}$ as a function of the output voltage. This is a result of a Taylor-series expansion of the output current as a function of the input and output voltages with maintaining only the dc and linear components. Such an approach helps one to understand and derive the maximum achievable power-amplifier parameters in a linear approximation. In this case, an active device is operated in a Class-A mode when one-half of the dc power is dissipated in the device, while the other half is transformed to the fundamental-frequency output power flowing into the load, resulting in a maximum ideal collector efficiency of 50%. The device output resistance $R_{out}$ remains constant and can be calculated as a ratio of the dc supply voltage to the dc current flowing through the active device. In a common case, for a complex conjugate-matching procedure, the device output immittance under large-signal consideration should be calculated using a Fourier-series analysis of the output current and voltage fundamental components. This means that, unlike a linear Class-A mode, an active device is operated in a device linear region only part of the entire period, and its output resistance is defined as a ratio of the fundamental-frequency output voltage to the fundamental-frequency output current. This is not a physical resistance resulting in a power loss inside the device, but an equivalent resistance required for a conjugate matching procedure. In this case, the complex conjugate matching is valid and necessary, firstly, to compensate for the reactive part of the device output impedance and, secondly, to provide a proper load resistance resulting in a maximum power gain for a given supply voltage and required output power delivered to the load. Note that this is not a maximum available small-signal power gain which can be achieved in a linear operation mode, but a maximum achievable large-signal power gain that can be achieved for a particular operation mode with a certain conduction angle. Of course, the maximum large-signal power gain is smaller than the small-signal power gain for the same input power, since the output power in a nonlinear operation mode also includes the powers at the harmonic components of the fundamental frequency.

Therefore, it makes more practical sense not to separately introduce the concepts of the gain match with respect to the linear power amplifiers and the power match in nonlinear power-amplifier circuits since the maximum large-signal power gain, being a function of the angle $\theta$, corresponds to the maximum fundamental frequency output power delivered to the load due to large-signal conjugate output matching. It is very important to provide a conjugate match at both input and output device ports to achieve maximum power gain in a large-signal mode. In a Class-A mode, the maximum small-signal power gain ideally remains constant regardless of the output power level.

The transistor characterization in a large-signal mode can be done based on equivalent quasi-harmonic nonlinear approximation under the condition of sinusoidal port voltages [61]. In this case, the large-signal impedances are generally determined in the following manner. The designer tunes the load network (often by trial and error) to maximize the output power to the required level using a particular transistor at a specified frequency and supply voltage. Then, the transistor is removed from the circuit and the impedance seen by the collector is measured at the carrier frequency. The complex-conjugate of the measured impedance then represents the equivalent large-signal output impedance of the transistor at that frequency, supply voltage, and output power. Similar design process is used to measure the input impedance of the transistor and to maximize power-added efficiency of the power amplifier.

To deliver maximum power to the load, it is necessary to use some impedance matching network which can modify the load as viewed from the generator [55]. The design of reactance networks to connect a resistive load efficiently to a source of power can be carried out most conveniently by the theory of image impedances. In this case, if the image impedances of such a network are pure resistances and it is connected between a generator and a load whose impedances are equal to these image impedances, the impedances will match at both junctions. Under these conditions with an assumption of pure reactances of the arms of the connecting network, no power will dissipate during transmission and so this maximum power will be delivered to the load. If the terminating impedances are not pure resistances, they can be made so at any single frequency by additional reactance in series with them. Such reactance networks can provide not only for high efficiency but can also attenuate undesired harmonics. A variety of configurations can be designed to accomplish the desired result [18]. Generally, $L$-type, $T$-type, and $\pi$-type configurations of reactances are used for matching networks depending on convenience of the circuit implementation and particular application technology. They can be implemented in both low-pass and high-pass topologies.

## 1.10 **Load—pull characterization**

In designing power amplifiers, it is important to know the transistor input impedance and load characteristics at a high input-driving power level and to optimize
the output matching circuit on the basis of these large-signal characteristics.
A computer-controlled technique for large-signal characterization of microwave
power transistors used to map contours of constant power and efficiency on a
Smith chart for dynamic matching of both input and output circuits was originally
developed for the interstage matching between a varactor multiplier and a transistor power stage and then successfully applied to the broadband optimization of
Class-A and Class-C transistor power amplifiers [62,63]. The power-load contours
consist of a series of curves on a Smith chart representing constant-output power,
approaching the point of maximum output power, as shown in Fig. 1.30($a$). When
drawn at several frequencies over the band of interest, they represent loci of
required output impedance for various output power levels. If constant-efficiency
contours are overlaid on the constant-power contours, the efficiency is also known
at each of the load impedances. Generally, the points for maximum output power
and maximum efficiency can be located at different positions, each corresponding
to its optimum impedance, as shown in Fig. 1.30($b$), since the maximum power
is achieved at the fundamental frequency, whereas the maximum efficiency significantly depends on the effect of the second- and higher-order harmonic
components.

A power transistor operated in a Class-C mode has an output equivalent circuit
which is represented by a nonlinear multiharmonic current source and a nonlinear
reactance. To obtain maximum output power, the load impedance must produce
the maximum current and voltage swing. The relationship between the required
load impedance and nonlinear output impedance determines the shape of the

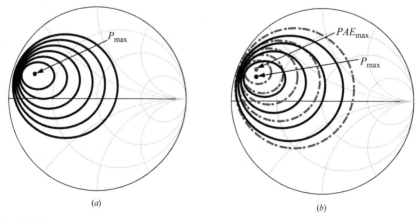

($a$)  ($b$)

**FIGURE 1.30**

Constant-power and constant-efficiency load—pull contours.

power-load contour which can vary with input drive level from elliptical to circle contours [64]. With the conventional passive load—pull technique, external tuners are adjusted, demounted, and measured. In this case, the dc bias-dependent transistor small-signal impedances can be used as references [65]. To get a variable load, in addition to the traditional passive-network technique with variable elements, an active load—pull characterization can be used where the reflection coefficient is obtained using an auxiliary signal derived from the same test generator to inject into the output port [66,67]. One of the advantages of this technique lies in the inherent simplicity of the calibration procedure necessary to correct the transmission-line losses in the measurement system. However, its accuracy critically depends on the effective directivities of the directional couplers in the system [68]. The accuracy of the load—pull measurements can be improved by using ultralow-loss broadband tuners based on nonuniform, nonsymmetrical rectangular coaxial-to-microstrip directional couplers [69]. Independent tuning of the fundamental frequency and its second harmonic component is possible by using a scheme with frequency-selective tuners [70].

The load impedance $Z_L$, incident and reflected traveling waves $a_2$ and $b_2$, and reflection coefficient $\Gamma_L$ at the output port of the transistor are related as

$$\Gamma_L = \frac{a_2}{b_2} = \frac{Z_L - Z_0}{Z_L + Z_0} \tag{1.122}$$

where $Z_0$ is the characteristic impedance of the system, in which the device under test (DUT) is used. The function of the tuners is to vary the magnitude and phase of the reflected signal so as to synthesize an appropriate $\Gamma_L$. This functionality can be obtained by altering the tuner setting by way of moving the slug or stub up and down and back and forth in a passive tuner, as shown in Fig. 1.31, or by actively injecting a magnitude- and phase-controlled signal in an active load—pull system [71]. In this case, as the stub (probe or slug) is inserted into the tuner transmission line, it introduces mismatch by adding parallel susceptance, and the parallel susceptance increases as the stub approaches the line and aids in the synthesis of the desired reflection coefficient. As a result, the magnitude of the impedance mismatch is determined by the stub position (depth) and the phase of the impedance mismatch is determined by the carriage position (length).

Generally, tuning with a load—pull system (mechanical or fully automatic) is a very complicated procedure, especially if it is necessary to take into account several harmonics of the fundamental frequency. Besides, it can only give the optimum input and output impedances, which are usually incorporated in the datasheet for power transistors for specified output power, supply voltage, and frequency range. In a monolithic implementation, however, this is difficult to physically realize and this can be done based on the load—pull setup incorporated into the simulation tool. But, in all cases, this measurement should be followed by the subsequent load—network design with real elements. Therefore, in most cases where the device nonlinear model is available, the device equivalent output circuit can be represented by the output admittance where the equivalent output

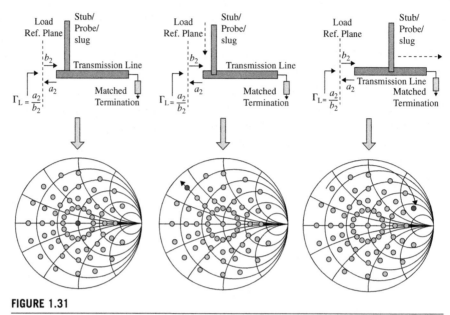

**FIGURE 1.31**

Passive tuner and reflection coefficient position.

resistance can be estimated from Eq. (1.44) for fixed conduction angle $2\theta$ and supply voltage $V_{cc}$, which is equal in an ideal case to the cosine voltage amplitude (the corresponding device saturation voltage can be subtracted). For example, in a Class-B mode with $\theta = 90°$, it is written as $R_L = R_{out} = V/I_1 = V_{cc}^2/2P_{out}$, where $P_{out}$ is the maximum fundamental-frequency power delivered to the load, and the output shunt reactance is represented by the sum of the output and feedback capacitances.

## 1.11 Amplifier stability

In early radio-frequency vacuum-tube transmitters, it was observed that the tubes and associated circuits may have damped or undamped oscillations depending upon the circuit losses, the feedback coupling, the grid and anode potentials, and the reactance or tuning of the parasitic circuits [72,73]. Various parasitic oscillator circuits such as the tuned-grid—tuned-anode circuit with capacitive feedback, Hartley, Colpitts, or Meissner oscillators can be realized at high frequencies, which potentially can be eliminated by adding a small resistor close to the grid or anode connections of the tubes for damping the circuits. Inductively coupled rather than capacitively coupled input and output circuits should be used wherever possible.

According to the immittance approach to the stability analysis of the active non-reciprocal two-port network, it is necessary and sufficient for its unconditional stability if the following system of equations can be satisfied for the given active device:

$$\mathrm{Re}[W_S(\omega) + W_{in}(\omega)] > 0 \qquad (1.123)$$

$$\mathrm{Im}[W_S(\omega) + W_{in}(\omega)] = 0 \qquad (1.124)$$

or

$$\mathrm{Re}[W_L(\omega) + W_{out}(\omega)] > 0 \qquad (1.125)$$

$$\mathrm{Im}[W_L(\omega) + W_{out}(\omega)] = 0 \qquad (1.126)$$

where $\mathrm{Re}W_S$ and $\mathrm{Re}W_L$ are considered to be greater than zero [74,75]. The active two-port network can be treated as unstable or potentially unstable in the case of the opposite signs in Eqs (1.123) and (1.125).

Analysis of Eq. (1.123) or Eq. (1.125) on extremum results in a special relationship between the device immittance parameters called the device stability factor

$$K = \frac{2\mathrm{Re}W_{11}\,\mathrm{Re}W_{22} - \mathrm{Re}(W_{12}W_{21})}{|W_{12}W_{21}|} \qquad (1.127)$$

which shows a stability margin indicating how far from zero value are the real parts in Eqs (1.123) and (1.125) if they are positive [75]. An active device is unconditionally stable if $K \geq 1$ and potentially unstable if $K < 1$.

When the active device is potentially unstable, an improvement of the power amplifier stability can be provided with the appropriate choice of the source and load immittances $W_S$ and $W_L$. In this case, the circuit stability factor $K_T$ is defined in the same way as the device stability factor $K$, taking into account $\mathrm{Re}W_S$ and $\mathrm{Re}W_L$ along with the device $W$-parameters, and written as

$$K_T = \frac{2\,\mathrm{Re}(W_{11} + W_S)\mathrm{Re}(W_{22} + W_L) - \mathrm{Re}(W_{12}W_{21})}{|W_{12}W_{21}|}. \qquad (1.128)$$

If the circuit stability factor $K_T \geq 1$, the power amplifier is unconditionally stable. However, the power amplifier becomes potentially unstable if $K_T < 1$. The value of $K_T = 1$ corresponds to the border of the circuit's unconditional stability. The values of the circuit stability factor $K_T$ and device stability factor $K$ become equal if $\mathrm{Re}W_S = \mathrm{Re}W_L = 0$.

For the active device stability factor $K > 1$, the operating power gain $G_P$ has to be maximized. By analyzing Eq. (1.114) in extremum, it is possible to find optimum values $\mathrm{Re}W_L^o$ and $\mathrm{Im}W_L^o$ when the operating power gain $G_P$ is maximal [76,77]. As a result,

$$G_{Pmax} = \left|\frac{W_{21}}{W_{12}}\right| / \left(K + \sqrt{K^2 - 1}\right). \qquad (1.129)$$

The power amplifier with an unconditionally stable active device provides a maximum power gain operation only if the input and output of the active device are conjugate-matched with the source and load impedances, respectively. For the lossless input matching circuit when the power available at the source is equal to the power delivered to the input port of the active device, i.e. $P_S = P_{in}$, the maximum operating power gain is equal to the maximum transducer power gain, i.e. $G_{Pmax} = G_{Tmax}$.

Domains of the device's potential instability include the operating frequency ranges where the active device stability factor is equal to $K < 1$. Within the bandwidth of such a frequency domain, parasitic oscillations can occur, defined by internal positive feedback and operating conditions of the active device. The instabilities may not be self-sustaining, induced by the RF drive power but remaining on its removal. One of the most serious cases of power amplifier instability can occur when there is a variation of the load impedance. Under these conditions, the transistor may be destroyed almost instantaneously. However, even if it is not destroyed, the instability can result in an increased level of spurious emissions in the output spectrum of the power amplifier. Generally, the following classification for linear instabilities can be made [78]:

- Low-frequency oscillations produced by thermal feedback effects.
- Oscillations due to internal feedback.
- Negative resistance or conductance-induced instabilities due to transit-time effects, avalanche multiplication, etc.
- Oscillations due to external feedback as a result of insufficient decoupling of the dc supply, etc.

Therefore, it is very important to determine the effect of the device feedback parameters on the origin of the parasitic self-oscillations and to establish possible circuit configurations of the parasitic oscillators. Based on the simplified bipolar equivalent circuit shown in Fig. 1.28, the device stability factor can be expressed through the parameters of the transistor equivalent circuit as

$$K = 2r_b g_m \frac{1 + \dfrac{g_m}{\omega_T C_c}}{\sqrt{1 + \left(\dfrac{g_m}{\omega C_c}\right)^2}} \qquad (1.130)$$

where $\omega_T = 2\pi f_T$ [18,44].

At very low frequencies, the bipolar transistors are potentially stable and the fact that $K \to 0$ when $f \to 0$ can be explained by simplifying the bipolar equivalent circuit. In practice, at low frequencies, it is necessary to take into account the dynamic base-emitter resistance $r_\pi$ and early collector-emitter resistance $r_{ce}$, the presence of which substantially increase the value of the device stability factor. This gives only one unstable frequency domain with $K < 1$ and low-boundary frequency $f_{p1}$. However, an additional region of possible low-frequency oscillations

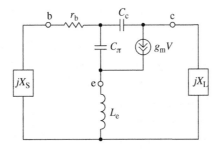

**FIGURE 1.32**

Simplified bipolar $\pi$-hybrid equivalent circuit with emitter lead inductance.

can occur due to thermal feedback where the collector junction temperature becomes frequently dependent, and the common-base configuration is especially affected by this [79].

Equating the device stability factor $K$ with unity allows us to determine the high-boundary frequency of a frequency domain of the bipolar transistor potential instability as

$$f_{p2} = \frac{g_m}{2\pi C_c} \bigg/ \sqrt{(2r_b g_m)^2 \left(1 + \frac{g_m}{\omega_T C_c}\right)^2 - 1}. \tag{1.131}$$

When $r_b g_m > 1$ and $g_m \gg \omega_T C_c$, Eq. (1.131) is simplified to

$$f_{p2} \approx \frac{1}{4\pi r_b C_\pi}. \tag{1.132}$$

At higher frequencies, a presence of the parasitic reactive intrinsic transistor parameters and package parasitics can be of great importance in view of power amplifier stability. The parasitic series emitter lead inductance $L_e$ shown in Fig. 1.32 has a major effect on the device stability factor. The presence of $L_e$ leads to the appearance of the second frequency domain of potential instability at higher frequencies. The circuit analysis shows that the second frequency domain of potential instability can be realized only under the particular ratios between the normalized parameters $\omega_T L_e/r_b$ and $\omega_T r_b C_c$ [18,44]. For example, the second domain does not occur for any values of $L_e$ when $\omega_T r_b C_c \geq 0.25$.

An appearance of the second frequency domain of the device potential instability is the result of the corresponding changes in the device feedback phase conditions and takes place only under a simultaneous effect of the collector capacitance $C_c$ and emitter lead inductance $L_e$. If the effect of one of these factors is lacking, the active device is characterized by only the first domain of its potential instability.

Figure 1.33 shows the potentially realizable equivalent circuits of the parasitic oscillators. If the value of a series-emitter inductance $L_e$ is negligible, the parasitic

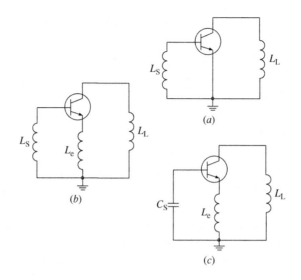

**FIGURE 1.33**

Equivalent circuits of parasitic bipolar oscillators.

oscillations can occur only when the values of the source and load reactances are positive, i.e. $X_S > 0$ and $X_L > 0$. In this case, the parasitic oscillator shown in Fig. 1.33(a) represents the inductive three-point circuit, where the inductive elements $L_S$ and $L_L$ in combination with the collector capacitance $C_c$ form a Hartley oscillator. From a practical point of view, the more the value of the collector dc-feed inductance exceeds the value of the base-bias inductance, the more likely low-frequency parasitic oscillators can be created. It was observed that a very low inductance, even a short between the emitter and the base, can produce very strong and dangerous oscillations which may easily destroy a transistor [78]. Therefore, it is recommended to increase a value of the base choke inductance and to decrease a value of the collector choke inductance.

The presence of $L_e$ leads to narrowing of the first frequency domain of the potential instability, which is limited to the high-boundary frequency $f_{p2}$, and can contribute to the appearance of the second frequency domain of the potential instability at higher frequencies. The parasitic oscillator that corresponds to the first frequency domain of the device potential instability can be realized only if the source and load reactances are inductive, i.e. $X_S > 0$ and $X_L > 0$, with the equivalent circuit of such a parasitic oscillator shown in Fig. 1.33(b). The parasitic oscillator corresponding to the second frequency domain of the device potential instability can be realized only if the source reactance is capacitive and the load reactance is inductive, i.e. $X_S < 0$ and $X_L > 0$, with the equivalent circuit shown in Fig. 1.33(c). The series emitter inductance $L_e$ is an element of fundamental importance for the parasitic oscillator that corresponds to the second frequency domain of the device potential instability. It changes the circuit phase conditions so that it becomes possible to establish the oscillation phase-balance condition at high frequencies. However, if it is possible to eliminate the parasitic oscillations at high frequencies by other means, increasing $L_e$ will result in the narrowing of a low-frequency domain of potential instability, thus making the power amplifier potentially more stable, although at the expense of reduced power gain.

Similar analysis of the MOSFET power amplifier also shows two frequency domains of MOSFET potential instability due to the internal feedback gate-drain capacitance $C_{gd}$ and series source inductance $L_s$ [18]. Because of the very high gate-leakage resistance, the value of the low-boundary frequency $f_{p1}$ is sufficiently small. For usually available conditions for power MOSFET devices when $g_m R_{ds} = 10 \div 30$ and $C_{gd}/C_{gs} = 0.1 \div 0.2$, the high boundary frequency $f_{p2}$ can approximately be calculated from

$$f_{p2} \approx \frac{1}{4\pi R_{gs} C_{gs}}. \tag{1.133}$$

It should be noted that power MOSFET devices have a substantially higher value of $g_m R_{ds}$ at small values of the drain current than at its high values. Consequently, for small drain current, the MOSFET device is characterized by a wider domain of potential instability. This domain is significantly wider than the same first domain of the potential instability of the bipolar transistor. The series source inductance $L_s$ contributes to the appearance of the second frequency domain of the device potential instability. The potentially realizable equivalent circuits of the MOSFET parasitic oscillators are the same as for the bipolar transistor shown in Fig. 1.33 [18].

Thus, to prevent the parasitic oscillations and to provide a stable operation mode of any power amplifier, it is necessary to take into consideration the following common requirements:

- Use an active device with stability factor $K > 1$.
- If it is impossible to choose an active device with $K > 1$, it is necessary to provide the circuit stability factor $K_T > 1$ by the appropriate choice of the real parts of the source and load immittances.
- Disrupt the equivalent circuits of the possible parasitic oscillators.
- Choose proper reactive parameters of the matching circuit elements adjacent to the input and output ports of the active device, which are necessary to avoid the self-oscillation conditions.

Generally, the parasitic oscillations can arise on any frequency within the potential instability domains for particular values of the source and load immittances $W_S$ and $W_L$. The frequency dependences of $W_S$ and $W_L$ are very complicated and very often cannot be predicted exactly, especially in multistage power amplifiers. Therefore, it is very difficult to propose a unified approach to provide a stable operation mode of the power amplifiers with different circuit configurations and operation frequencies. In practice, the parasitic oscillations can arise close to the operating frequencies due to the internal positive feedback inside the transistor and at the frequencies sufficiently far from the operating frequencies due to the external positive feedback created by the surface mounted elements. As a result, the stability analysis of the power amplifier must include the methods to prevent the parasitic oscillations in different frequency ranges.

It should be noted that expressions in Eqs (1.123) to (1.129) are given through the device immittance $Z$- or $Y$-parameters that allow the power gain and stability to be calculated using the parameters of the device equivalent circuit and to physically understand the corresponding effect of each circuit parameter, but not through the scattering $S$-parameters, which are very convenient during the measurement procedure required for device modeling. Moreover, using modern simulation tools, there is no need to even draw stability circles on a Smith chart or analyze stability factors across the wide frequency range since $K$-factor is just a derivation from the basic stability conditions and usually is a function of linear parameters, which can only reveal linear instabilities. Besides, it is difficult to predict unconditional stability for a multistage power amplifier because parasitic oscillations can be caused by the interstage circuits. In this case, the easiest and most effective way to provide stable operation of the multistage power amplifier (or single-stage power amplifier) is to simulate the real part of the device input impedance $Z_{in} = V_{in}/I_{in}$ at the input terminal of each transistor as a ratio between the input voltage and current by placing a voltage node and a current meter, as shown in Fig. 1.34($a$). If Re$Z_{in} < 0$, then either a small series resistor must be added to the device base terminal as a part of the input matching circuit or a load-network configuration can be properly chosen to provide the resulting positive value of Re$Z_{in}$. In this case, not only linear instabilities with small-signal soft startup oscillation conditions but also nonlinear instabilities with large-signal hard startup oscillation conditions or parametric oscillations can be identified around the operating region. Figure 1.34($b$) shows the parallel $RC$ stabilizing circuit with a bypass capacitor $C_{bypass}$ connected in series to the input port of a GaN HEMT device [80]. In this case, using a stabilizing resistor $R_{gate}$ and a low-value gate-bias resistor $R_{bias}$ improves the stability factor considerably at low frequencies without affecting the device performance at higher frequencies.

Figure 1.35 shows the example of a stabilized bipolar VHF power amplifier configured to operate in a zero-bias Class-C mode. Conductive input and output loading due to resistances $R_1$ and $R_2$ eliminate a low-frequency instability domain. The series inductors $L_3$ and $L_4$ contribute to higher power gain if the resistance values are too small, and can compensate for the capacitive input and output device impedances. To provide a negative-bias Class-C mode, the shunt inductor $L_2$ can be removed. The equivalent circuit of the potential parasitic oscillator at higher frequencies is realized by means of the parasitic reactive parameters of the transistor and external circuitry. The only possible equivalent circuit of such a parasitic oscillator at these frequencies is shown in Fig. 1.33($c$). It can only be realized if the series-emitter lead inductance is present. Consequently, the electrical length of the emitter lead should be reduced as much as possible, or, alternatively,

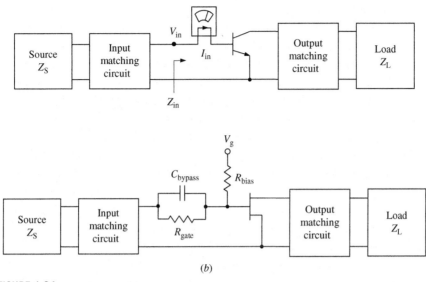

**FIGURE 1.34**

Single-stage power amplifiers with measured device input impedance.

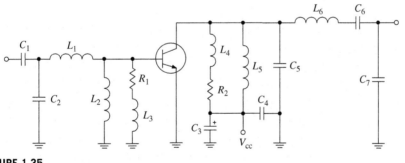

**FIGURE 1.35**

Stabilized bipolar Class-C VHF power amplifier.

the appropriate reactive immittances at the input and output transistor ports are provided. For example, it is possible to avoid the parasitic oscillations at these frequencies if the inductive immittance is provided at the input of the transistor and capacitive reactance is provided at the output of the transistor. This is realized by an input series inductance $L_1$ and an output shunt capacitance $C_5$.

The collector efficiency of the power amplifier can be increased by removing the shunt capacitor and series $RL$-circuit in the load network. The remaining series $LC$ circuit provides high impedances at the second- and higher-order

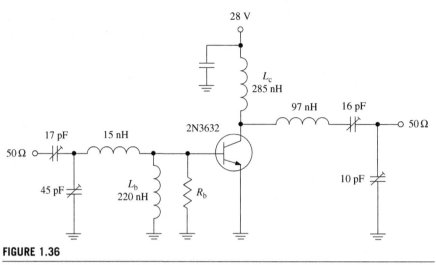

**FIGURE 1.36**

High-efficiency VHF bipolar Class-C power amplifier.

harmonic components of the output current, which flows now through the device collector capacitance unlike being grounded by the shunt capacitance. As a result, the bipolar Class-C power amplifier, whose circuit schematic is shown in Fig. 1.36, achieved a collector efficiency of 73% and a power gain of 9 dB with an output power of 13.8 W at an operating frequency of 160 MHz [81]. However, special care must be taken to eliminate parasitic spurious oscillations. In this case, the most important element in preventing the potential instability is the base bias resistor $R_b$. For example, for a relatively large base choke inductance $L_b$ and $R_b = 1$ k$\Omega$, spurious oscillations exist at any tuning. Tuning becomes possible with no parasitic oscillations for the output voltage standing wave ratio (*VSWR*) less than 1.3 or supply voltage more than 22 V when $R_b$ is reduced to 470 $\Omega$. However, a very small reduction in input drive power causes spurious oscillations. Further reduction of $R_b$ to 47 $\Omega$ provides a stable operation for output *VSWR* $\leq 7$ and supply voltages down to 7 V. Finally, no spurious oscillations occur at any load, supply voltage, and drive power level for $R_b = 26$ $\Omega$.

## 1.12 Parametric oscillations

Since the transistor used as an active device in power amplifiers is characterized by substantially nonlinear behavior, this can result in nonlinear instabilities, which provide generally the parametric generation of both harmonic and subharmonic components. The presence of subharmonics can be explained by the parametric varactor junction action of the collector-base voltage-dependent capacitance when the large-signal driving acts like pumping a varactor diode, as in a parametric amplifier [82,83]. Such an amplifier exhibits negative resistance under certain conditions when a circuit starts oscillating at subharmonics or rational fractions of the operating frequency [84]. Generally, the parametric oscillations are the result of the external force impact on the element of the oscillation system by varying its parameter. Understanding of the physical origin of this parametric effect is very important in order to disrupt any potentially realizable parametric oscillator circuits. It is an especially serious concern for high-efficiency power amplifiers with very high voltage peak factors and voltage swing across the device nonlinear output capacitance, since the transistor is operated in pinch-off, active, and saturation regions.

Figure 1.37 shows the simplified large-signal equivalent circuit of the (*a*) MOSFET or (*b*) bipolar device with a nonlinear current source $i(t)$, respectively. The most nonlinear capacitances are the bipolar collector capacitance $C_c$ and the MOSFET gate-drain capacitance $C_{gd}$ and drain-source capacitance $C_{ds}$, which can be modeled as junction capacitances with different sensitivities. However, since the drain-source capacitance $C_{ds}$ is normally greater by about the order than the gate-drain capacitance $C_{gd}$, the parametric effect due to $C_{ds}$ causes a major effect on potential parametric oscillations in a MOSFET power amplifier. The value of the collector capacitance $C_c$ is by an order smaller than the value of the base-emitter capacitance $C_\pi$ in active mode. In this case, the circuit of a potential parametric oscillator represents a system with one degree of freedom, as shown in Fig. 1.37(*c*), where $V_{cc}$ is the supply voltage applied to the varying output capacitance $C_{out}$ ($C_{out} \approx C_c$ for a bipolar device and $C_{out} \approx C_{ds}$ for a MOSFET device), while capacitor $C_0$ and inductor $L_0$ represent the series high-$Q$ resonant circuit tuned to the fundamental frequency and having high impedances at the second- and higher-order harmonics.

The theoretical analysis can be simplified by representing the output power-amplifier circuit in the form of a basic series $RLC$ circuit shown in Fig. 1.38(*a*) [85]. Let us assume that the nonlinear capacitance $C$ varies in time relatively to its average value $C_0$ due to external large-signal voltage drive representing a pulsed function shown in Fig. 1.38(*b*), while the charge $q(t)$ generally represents a sinusoidal function of time shown in Fig. 1.38(*c*). When the capacitance $C$ decreases by $2\Delta C$, the voltage amplitude $V_0 = q_0 C_0$ across the capacitor, and the energy $W_0 = q_0^2/2C_0$ stored by the capacitor just prior to its stepped change, increase. The charge $q(t)$ during these rapid capacitance variations in time does not change its behavior as it is a slowly time-varying parameter. In this case, if $\Delta C \ll C_0$, the increment of the energy obtained by the capacitor at the single step moment is defined as $\Delta W = W_0 2\Delta C/C_0 = 4mq_0^2/2C_0$, where $m = \Delta C/C_0$ is the parameter modulation factor.

The maximum energy contribution into the oscillation system will be at the times of maximum charge amplitude $q_0$ and no energy contribution will be at zero crossing. This means that the capacitance as a parameter changes two times faster than the oscillation frequency. The entire energy increment into the system

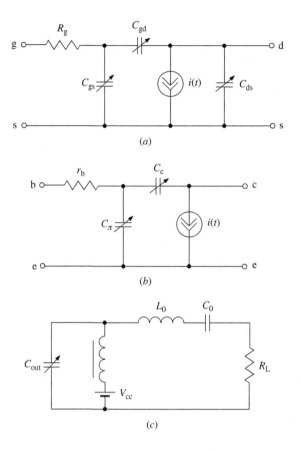

**FIGURE 1.37**

Simplified nonlinear transistor models and output amplifier circuit.

for a period will be $2\Delta W$. At the same time, the energy lost during the period $T = 2\pi/\omega$ for the sinusoidal charge variation with amplitude $q_0$ is defined as $0.5R\omega^2 q_0^2 T = \pi\omega q_0^2 R$. If the losses in the oscillation system are smaller than the energy input into the system when $4mq_0^2/2C_0 > \pi\omega q_0^2 R$ or $m > 0.5\pi\omega RC_0$, the build-up of the self-oscillations can occur. Such a process of the excitation of self-oscillations due to periodic changes of the energy-storing parameter of an oscillation system is called the *parametric excitation of self-oscillations* or *parametric resonance*. If the capacitance $C$ varies with the same periodicity but having a different behavior, quantitatively the result will be the same.

Now assume that the nonlinear capacitance $C$ in the oscillation system is time varying according to

$$C(t) = \frac{C_0}{1 + m\cos pt} \tag{1.134}$$

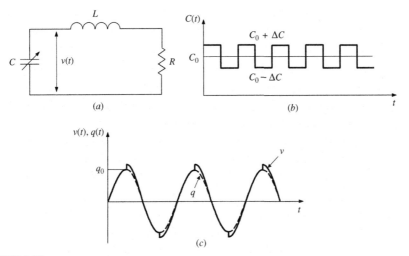

**FIGURE 1.38**

Nonlinear resonant circuit with parametric pumping.

where $p = 2\omega/n$ is the frequency of the parameter variation, $\omega$ is the frequency of the self-oscillations, and $n = 1, 2, 3, \ldots$.

The voltage across the nonlinear capacitance $C$ with abrupt junction sensitivity $\gamma = 0.5$ can be written as

$$v(t) = \frac{1}{C_0}(q - \beta q^2) \tag{1.135}$$

where $C_0$ is the small-signal capacitance corresponding to the dc-bias condition and $\beta$ is the coefficient responsible for the capacitance nonlinear behaviour. The mathematical description of parametric oscillations in a single-resonant oscillation system with a time-varying junction capacitance $C(t)$ can be done based on the second-order differential equation characterizing this circuit, which can generally be written in the form

$$\frac{d^2q}{dt^2} + 2\delta\frac{dq}{dt} + \omega_0^2(1 + m\cos pt)(q - \beta q^2) = 0 \tag{1.136}$$

where $\omega_0 = 1/\sqrt{LC_0}$ is the small-signal resonant frequency, $2\delta = \omega_0/Q$ is the dissipation factor, and $Q = \omega_0 L/R$ is the quality factor of the resonant circuit [85]. In a linear case when $\beta = 0$, Eq. (1.136) simplifies to a well-known Mathieu equation, the stable and unstable solutions and important properties of which are thoroughly developed and analyzed.

The basic results in a graphical form presenting the domains of the potential parametric instability as a function of the parameter $n = 2\omega_0/p$ are plotted in

Fig. 1.39. The shaded areas corresponding to a growing self-oscillating process with a frequency $\omega = np/2 \approx \omega_0$ rise, and their ends are located on the line having an angle $\varphi$ with a horizontal axis. This means that the greater the dissipation factor $2\delta$ of the oscillation system, the greater the modulation factor $m$ is necessary to realize the parametric oscillations. For a fixed modulation factor $m_0$, the width of the instability domain for different $n$ is different, being smaller for higher $n$. As this number grows due to more seldom energy input into the system ($p = 2\omega/n$), it is necessary to increase the modulation factor. The effect of nonzero $\beta$ in the nonlinear voltage-charge dependence given in Eq. (1.135) leads to a deviation of the oscillation frequency in a steady-state mode from its start-up value since the averaged junction capacitance differs from its small-signal value for a large-signal mode. The difference becomes greater with a growth of the oscillation amplitude reaching the border of the instability domain. This decreases energy input into the oscillation, thus limiting the increase in the amplitude.

Consequently, the most probable parametric oscillations in the nonlinear power amplifier can occur at a subharmonic frequency $\omega_{1/2} = \omega_p/2$, where $\omega_p$ is the operating frequency varying the device capacitance. In this case, the subharmonic frequency $\omega_{1/2}$ corresponds to the resonant frequency $\omega_0$ of the circuit shown in Fig. 1.38(a), being equal to half the operating frequency $\omega_p$, i.e. $n = 1$ and $p = \omega_p = 2\omega_0$. Hence, to eliminate such a parasitic sub-harmonic parametric oscillation, it is necessary to provide the circuit design solution when the device output can see very high impedance at a subharmonic frequency $\omega_{1/2}$. Alternatively, an additional lossy element in the subharmonic circuit with its proper isolation from the fundamental circuit can be incorporated. In other words, it is necessary to break out any possible resonant conditions at the subharmonic frequency component, which can cause the parametric oscillations.

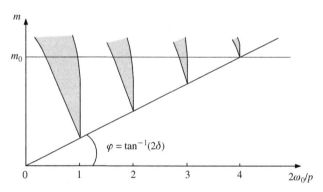

**FIGURE 1.39**

Domains of parametric instability for damping systems.

## 1.13 **Bias circuits**

The simplest way to provide a Class-AB dc-bias condition for a power MOSFET device is to use the potentiometer-type voltage divider for the gate bias, with a choke inductor in the drain circuit, as shown in Fig. 1.40(a). In this case, any variation of the ambient temperature or supply voltage leads to variations of quiescent current and, as a result, to appropriate variations of the output power, linearity, drain efficiency, and power gain of the power amplifier. The threshold voltage of the MOSFET transistor $V_{th}$ varies *versus* temperature $T$ linearly with approximate velocity of $\Delta V_{th}/\Delta T \cong 2 \, mV/°C$. The simple addition of a diode in series to the variable resistor allows the quiescent current variation to be reduced substantially over temperature. A bias circuit corresponding to this stabilizing condition is shown in Fig. 1.40(b). For a high value of $V_{th}$, several diodes can be connected in series. Such a simple bias circuit configuration for power MOSFET transistors is possible when an extremely small value of the dc gate current, which is restricted to a value of the gate leakage current, is present.

In contrast to MOSFET devices, where it can be possible to choose the optimum operating point with practically zero temperature coefficient or to be limited to connection of an additional diode only, the bipolar transistors require the more complicated approach of dc biasing depending on the class of operation. In a Class-AB bias mode, the bias circuit has to deliver a constant voltage, slightly adjustable approximately within limits of 0.7−0.8 V, with a wide range

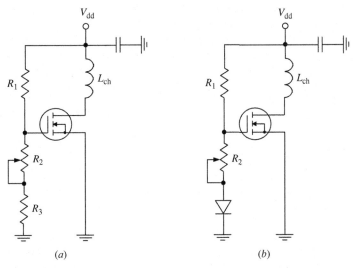

(a)                    (b)

**FIGURE 1.40**

MOSFETs with simple bias circuits.

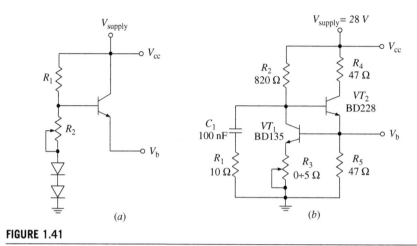

**FIGURE 1.41**

Typical bipolar Class-AB bias circuits.

of the current values to stabilize the base current of the bipolar transistor. Besides, it is necessary to provide an operation mode of the power amplifier with temperature compensation (collector current stabilization over temperature) and minimum possible reference current (dc current from the reference dc voltage supply). One of the simplest versions of such a bias circuit with diode temperature compensation is shown in Fig. 1.41(a). In this bias circuit, each diode can be replaced by the *n-p-n* diode-connected transistor, the collector and the base of which are directly connected between each other. A better temperature-compensating result can be achieved using the same transistors as for the RF power transistor for each bias circuit transistor, but with reduced area size. Such an approach is usually used in monolithic integrated circuit design when transistor cells with different area sizes are used for both an RF power device and bias circuit transistors.

Figure 1.41(b) shows the more complicated bias circuit that is commonly used for biasing the high-power bipolar transistors to provide their temperature-stable and reliable operation mode [86]. The temperature stabilization is provided by the parallel connection of the base-emitter diode junction of the transistor $VT_1$, whereas high value of the bias drive current for the RF power transistor is delivered by the transistor $VT_2$. If the dc collector current of an RF power transistor is 5 A and a value of its forward current gain $\beta_F$ is approximately equal to 10, then the maximum base current of the RF power transistor can be 0.5 A. In this bias circuit, the resistor $R_5$ is used to reduce the base current variations. At $V_b = 0.7$ V, for a current of 15 mA, the value of $R_5$ should be equal to 0.7 V/15 mA = 47 $\Omega$. Suppose that a value of the collector current of $VT_1$ is 30 mA. Then, if the base-emitter junction voltage of $VT_2$ is equal to 0.8 V with a voltage across the resistor

$R_2$ of 28 V − 1.5 V = 26.5 V, its value is 26.5 V/30 mA $\cong$ 820 $\Omega$. The variable resistor $R_3$ serves to adjust the output voltage in limits of 0.1 V. To limit the maximum collector current of $VT_2$ by a value of 0.5 A, it is advisable to use the resistor $R_4$. Since $VT_2$ has a value of the saturation voltage of 0.8 V, it follows that the maximum value of $R_4$ is 26.5 V/0.5 A = 53 $\Omega$. It is sufficient to use its value of 47 $\Omega$ with a power dissipation of $(0.5 \text{ A})^2 \times 47 \, \Omega = 11.75$ W. Such a bias circuit can develop the parasitic oscillations near 1 MHz with highly capacitive loads. Therefore, to prevent these oscillations, it is necessary to connect the $RC$ circuit between the collector of $VT_1$ and ground.

In most wireless communication systems, it is preferable that the power amplifier operates with high efficiency, maintaining an acceptable linearity over the desired supply voltage and output power ranges. However, there is a tradeoff between efficiency and linearity, with improvement in one coming at the expense of another. This means that it is necessary to provide an optimum fixed or adaptive bias point over wide temperature ranges and process variations. As a current-controlled device, the bipolar transistor in RF operation requires a dc base driving current, the value of which depends on the output power and device parameters. Because technologically the bipolar device represents a parallel connection of the basic cells, it is important to use the ballast series resistors to avoid current imbalance and possible device collapse at higher current-density levels. Another important aspect is to keep the dc base-emitter bias point constant (or properly variable) over any RF input power variations to prevent the linearity worsening at the maximum output power for power amplifiers with a variable envelope signal (such as GSM/EDGE, WCDMA/LTE, or CDMA2000).

The typical temperature-compensation current-mirror bias circuit with one reference transistor $Q_1$ and one driving transistor $Q_2$ is shown in Fig. 1.42(a) [18]. This circuit keeps the quiescent current for the RF device $Q_0$ more or less constant over temperature variations, and the current flowing through the resistor $R_2$ is sufficiently small. It is very important to provide the proper ratio between the ballast resistors $R_1$ and $R_0$, equal to the ratio of the device areas $Q_0/Q_1$. This can minimize the overall performance variation with temperature as well as stabilize the dc bias point. The latter case is very important for the variable-envelope signals, as the dc bias voltage $V_{be0}$ establishes the conduction angle and operation class for the RF device. If the dc base-emitter bias voltage reduces with the increase of RF input power, the Class-AB bias mode required for linear operation changes to a Class-C bias point corresponding to nonlinear operation with zero quiescent current.

Figure 1.42(b) shows the dependence of the dc base-emitter bias voltage $V_{be0}$ *versus* input power $P_{in}$ for the second stage of a WCDMA InGaP/GaAs HBT power amplifier for three cases, including two of them with ballast resistor $R_1 = 0$ (curve 1) and optimum ballast resistor $R_1$ (curve 2). From Fig. 1.42(b) it follows that including the ballast resistor with an optimum value results in a more constant base-emitter dc bias voltage over a wider range of input powers, thus

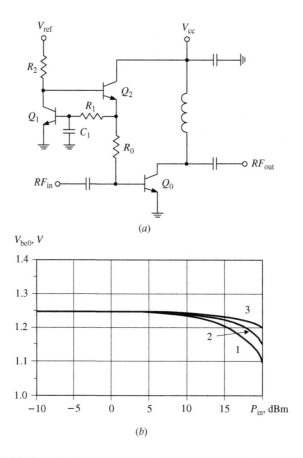

**FIGURE 1.42**

Bipolar power amplifier stage with current-mirror bias circuit and its performance.

improving the linearity performance of the power amplifier at high power levels. In addition, it is advisable to use a shunt capacitance $C_1$ connected to the base terminal of the device $Q_1$ to form a low-pass $RC$ filter, which provides better isolation of the bias circuit from the RF signal, with a much more constant base-emitter dc bias voltage (curve 3).

Figure 1.43 shows the emitter-follower bias circuit that provides temperature compensation and minimizes reference current requirements [87]. The emitter-follower bias circuit requires only several tens of microamperes of reference current, whereas the current-mirror bias circuit requires a few milliamperes of reference current. Both the current-mirror and emitter-follower bias circuits have the same current–voltage behavior but, for the same circuit parameters ($R_0$, $R_1$, and $R_2$) and device areas for $Q_0$, $Q_1$, and $Q_2$, the emitter-follower bias circuit is the less

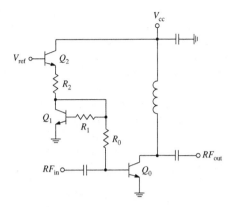

**FIGURE 1.43**

Bipolar power amplifier stage with emitter-follower bias circuit.

sensitive to the reference voltage variations. Variations of the collector supply voltage $V_{cc}$ in limits of 3.0–5.0 V have no effect on the quiescent current set by the reference voltage $V_{ref}$.

GaN HEMTs are depletion-mode devices which require a negative voltage to be applied to their gate terminal. Therefore, it is very important to keep the bias sequence when the negative gate bias voltage corresponding to the required quiescent current is applied earlier than the nominal positive drain supply voltage before turning on the RF power. In addition, attention needs to be paid to how to deal with a positive gate current, which will arise when the device is driven into saturation. Figure 1.44 shows the gate bias circuit which provides simultaneously the bias sequencing and temperature compensation to maintain a constant quiescent current over temperature, where a gate resistor $R_{gate}$ along the gate bias line is required to suppress parasitic oscillations [88]. The bias sequence and negative gate voltage is generated by the MAX881R bias supply IC which contains an integrated charge pump to supply the necessary negative voltage rail to the operational amplifier MIC7300 and generates a power-OK signal ($\overline{POK}$) used to turn on the drain switch after the negative supply is stable. The operational amplifier must be capable of supplying the maximum negative and positive gate current for the GaN HEMT being biased, and the charge pump must be capable of supplying the maximum negative current needed for that circuit. The quiescent current which is maintained constant over temperature is set by the 20-k$\Omega$ potentiometer. The operational amplifier is configured as an inverting amplifier with the positive terminal grounded, while the negative terminal is fed from +5 V voltage reference and a feedback circuit where the 100-k$\Omega$ thermistor provides a temperature-dependent resistance and is mounted near the active GaN HEMT device.

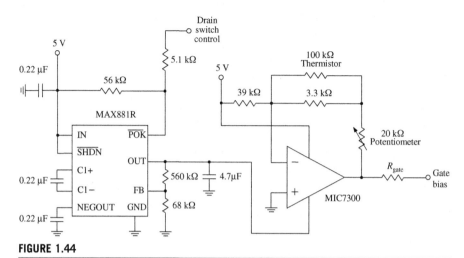

**FIGURE 1.44**

Gate bias circuit for a GaN HEMT with temperature compensation.

## 1.14 Distortion fundamentals
### 1.14.1 Linearity

The relationship between the output signal $y(t)$ and the input signal $x(t)$ of a system can be written as

$$y(t) = T[x(t)] \tag{1.137}$$

where $T[x(t)]$ is the transfer function that characterizes the system and $t$ is the time.

A system is considered linear if it satisfies the following two conditions:

$$T\left[\sum_i x_i(t)\right] = \sum_i T[x_i(t)] \tag{1.138}$$

$$T[kx(t)] = kT[x(t)]. \tag{1.139}$$

The first condition requires that the response of the system to the sum of many input signals must be the same as the sum of the responses of the system to the individual signals. This concept is also known as superposition. For example, if a multi-tone signal is applied to a linear amplifier, the effect of the amplifier on each tone must be the same as if each tone was applied individually, i.e. there would be no interaction between the tones, which will result in no additional frequencies appearing at the output with respect to the input.

The second condition requires that any scaling of the input signal by a constant factor $k$ must result in the scaling of the output signal by the same scale factor. If the amplitude of a single tone applied to the input of a linear amplifier is doubled, the amplitude of the output tone must also double. If the gain of the system is defined as the ratio between the output and the input signal amplitude, it can be concluded that the gain of a linear system must be independent of the input signal amplitude.

Both Eqs (1.138) and Eq. (1.139) can be combined into a single expression for $n$ input signals as

$$T[k_1 x_1(t) + k_2 x_2(t) \cdots + k_n x_n(t)] = k_1 T[x_1(t)] + k_2 T[x_2(t)] + \cdots + k_n T[x_n(t)].$$
(1.140)

## 1.14.2 Time variance

The input/output relationship of the system described by Eq. (1.137) is given by the operator $T$. If the system is observed at any given time and $T$ remains unchanged, the system is said to be *time invariant*. As a consequence, if the input signal is delayed by a constant time $\tau$, then the only consequence will be that the output will also be delayed by the same amount of time. Analytically,

$$y(t - \tau) = T[x(t - \tau)].$$
(1.141)

## 1.14.3 Memory

Memory is the dependence of a system transfer function on the input signal past values. The transfer function $T$ of a system can be approximated by a polynomial of degree $n$ and coefficients $a_i$, which is written as

$$y(t) = \sum_{i=0}^{n} a_i[x^i(t)].$$
(1.142)

If the system is linear, the polynomial must be of the first degree. This polynomial approximation can only describe the system at an instant, without the possibility of specifying any initial conditions. If the system under consideration contains a capacitor, the system response at a given instant will depend on the charge stored in the capacitor, which will itself depend on the past value of the input signal. For that reason, systems with memory cannot be described with simple algebra but with integro-differential equations of time.

## 1.14.4 Distortion of electrical signals

Distortion of an electrical signal is any change in its waveform, with the exception of scaling and a constant time delay. Scaling is a fixed amplitude change regardless of time and frequency, and it can be produced by a linear attenuator or amplifier. A constant delay assures a linear change in phase as a function of frequency, like when a signal travels through an ideal transmission line.

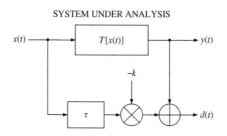

**FIGURE 1.45**

Block diagram used for the definition of distortion in a system.

A system with a transfer function $T[x(t)]$ is considered to have distortion if, for any given input signal $x(t)$, its output signal $y(t)$ differs from the input by other than a multiplying constant $k$ and a positive finite time delay $\tau$.

The system shown in Fig. 1.45 will be distortionless if, for all time $t$, the distortion $d(t)$ equals zero,

$$d(t) = T[x(t)] - kx(t - \tau) = 0. \tag{1.143}$$

### 1.14.5 Types of distortion

If the distortion signal $d(t)$ from Eq. (1.143) does not equal 0, the system generates distortion, which can be appropriately classified for linear and nonlinear systems.

For linear systems, the transfer function of the system of Fig. 1.45 can be written as a function of the angular frequency $\omega$ as

$$T(\omega) = |T(\omega)| e^{j\theta(\omega)}. \tag{1.144}$$

*Amplitude distortion* occurs when

$$|T(\omega)| \neq \text{constant for all } \omega. \tag{1.145}$$

*Phase or delay distortion* occurs when

$$\theta(\omega) \neq \omega\tau \text{ for all } \omega \tag{1.146}$$

which means that the system does not have a constant delay as a function of frequency. Amplitude and phase distortions are often called *linear distortions*, because they are linear processes and can be corrected by using a linear filter.

*Nonlinear distortion* occurs when the system includes nonlinear elements. If the system of Fig. 1.45 is memoryless, its transfer function $y(t) = T[x(t)]$ can be approximated by the power series given by Eq. (1.142). All powers of $x(t)$ which are greater than 1 in Eq. (1.142) give rise to nonlinear distortion.

### 1.14.6 Nonlinear distortion analysis for sinusoidal signals — measures of nonlinear distortion

Given a single-tone sinusoidal signal applied to a system, its nonlinear distortion can be measured as the *total harmonic distortion* (THD) defined as the ratio of (*a*) the sum of the powers of all harmonic frequencies above the fundamental frequency of the input signal to (*b*) the power of the fundamental frequency, all measured at the output of the system. If $P_1$ is the fundamental signal power and $P_2$ to $P_n$ are the powers of the nth harmonics of the input signal, the THD can be written as

$$\text{THD} = \frac{\sum_{i=2}^{n} P_i}{P_1}. \tag{1.147}$$

Another type of nonlinear distortion occurs when two or more sinusoidal signals of different frequencies are applied to the input of a nonlinear system. In this case, the output will contain additional signals not existent at the input, whose frequencies are not only harmonically related to the fundamental tones. If a power series such as Eq. (1.142) is used to approximate the transfer function of a memoryless system, the output of the system $y(t)$ can be found by replacing $x(t)$ with a two-tone test signal of amplitude $A = 1$,

$$x(t) = A[\cos \alpha + \cos \beta] \tag{1.148}$$

where $\alpha = \omega_1 t$, $\beta = \omega_2 t$, $\omega_1$ and $\omega_2$ are the frequencies of each tone, and $t$ is the time. The output $y(t)$ will contain the two input tones plus additional signals of different frequencies (some very close to the input tones). Those frequencies are given by $2\omega_2 - \omega_1$, $2\omega_1 - \omega_2$, $3\omega_2 - 2\omega_1$, $3\omega_1 - 2\omega_2$, etc., and they are called *intermodulation distortion* (IMD) products. Based on the sum of the coefficients of each term, the coefficient $2\omega_2 - \omega_1$ is defined as a third-order intermodulation coefficient $IM_3$, the coefficient $3\omega_2 - 2\omega_1$ is defined as a fifth-order intermodulation coefficient $IM_5$, etc.

IMD products are a major problem in communication systems because their frequencies fall very close to the fundamental signal and it is not feasible to implement linear filters that are capable of suppressing them. It is customary to measure intermodulation distortion as a dB ratio between the power of one of the IMD products and the power of the fundamental tones.

As an example, the distortion $d(t)$ in Eq. (1.143) of the system shown in Fig. 1.45 with $\tau = 0$, for a third-degree nonlinearity represented by the power series of Eq. (1.142) and the two-tone input signal of Eq. (1.148), will be written as

$$d(t) = \left[ a_1(\cos \alpha + \cos \beta) + a_2(\cos \alpha + \cos \beta)^2 + a_3(\cos \alpha + \cos \beta)^3 \right] - k\cos \alpha - k\cos \beta \tag{1.149}$$

$$d(t) = \{ a_1 \cos \alpha + a_1 \cos \beta + a_2 \left[ \cos^2\alpha + 2\cos \alpha \cos \beta + \cos^2\beta \right]$$
$$+ a_3 \left[ \cos^3\alpha + 3\cos^2\alpha \cos \beta + 3\cos \alpha \cos^2 \beta + \cos^3\beta \right] \} - k\cos \alpha - k\cos \beta \tag{1.150}$$

$$d(t) = \left\{ a_1 \cos \alpha + a_1 \cos \beta \right.$$

$$+ a_2 \left[ \left( \frac{1}{2} + \frac{1}{2} \cos 2\alpha \right) + \cos(\beta - \alpha) + \cos(\beta + \alpha) + \left( \frac{1}{2} + \frac{1}{2} \cos 2\beta \right) \right]$$

$$+ a_3 \left[ \frac{3}{4} \cos \alpha + \frac{1}{4} \cos 3\alpha + 3 \left( \frac{1}{2} + \frac{1}{2} \cos 2\alpha \right) \cos \beta + 3\cos\alpha \left( \frac{1}{2} + \frac{1}{2} \cos 2\beta \right) \right.$$

$$\left. \left. + \frac{3}{4} \cos \beta + \frac{1}{4} \cos 3\beta \right] \right\} - k \cos \alpha - k \cos \beta \tag{1.151}$$

$$d(t) = \left\{ a_1 \cos\alpha + a_1 \cos\beta + a_2 \left[ 1 + \frac{1}{2}\cos 2\alpha + \frac{1}{2}\cos 2\beta + \cos(\beta - \alpha) + \cos(\beta + \alpha) \right] \right.$$

$$+ a_3 \left[ \frac{3}{4}\cos\alpha + \frac{1}{4}\cos 3\alpha + \frac{3}{2}\cos\beta + \frac{3}{4}\cos(2\alpha - \beta) + \frac{3}{4}\cos(2\alpha + \beta) \right.$$

$$\left. \left. + \frac{3}{2}\cos\alpha + \frac{3}{4}\cos(2\beta - \alpha) + \frac{3}{4}\cos(2\beta + \alpha) + \frac{3}{4}\cos\beta + \frac{1}{4}\cos 3\beta \right] \right\}$$

$$- k\cos\alpha - k\cos\beta \tag{1.152}$$

$$d(t) = \left( a_1 + \frac{9a_3}{4} - k \right) \cos\alpha + \left( a_1 + \frac{9a_3}{4} - k \right) \cos\beta + a_2 \frac{a_2}{2}\cos 2\alpha + \frac{a_2}{2}\cos 2\beta$$

$$+ a_2\cos(\beta - \alpha) + a_2\cos(\beta + \alpha) + \frac{a_3}{4}\cos 3\alpha + \frac{3a_3}{4}\cos(2\alpha - \beta)$$

$$+ \frac{3a_3}{4}\cos(2\alpha + \beta) + \frac{3a_3}{4}\cos(2\beta - \alpha) + \frac{3a_3}{4}\cos(2\beta + \alpha) + \frac{a_3}{4}\cos 3\beta. \tag{1.153}$$

If the input tones are removed by making $k$ equal $a_1 + 9a_3/4$, the resultant distortion $d(t)$ is written as

$$d(t) = a_2 + \frac{a_2}{2}\cos 2\alpha + \frac{a_2}{2}\cos 2\beta + a_2\cos(\alpha - \beta) + a_2\cos(\alpha + \beta) + \frac{a_3}{4}\cos 3\alpha$$

$$+ \frac{3a_3}{4}\cos(2\alpha - \beta) + \frac{3a_3}{4}\cos(2\alpha + \beta) + \frac{3a_3}{4}\cos(2\beta - \alpha)$$

$$+ \frac{3a_3}{4}\cos(2\beta + \alpha) + \frac{a_3}{4}\cos 3\beta. \tag{1.154}$$

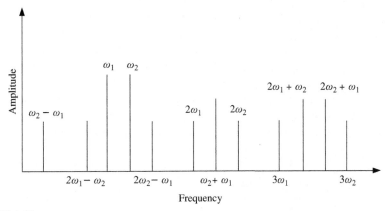

**FIGURE 1.46**

Frequency spectrum of third-order nonlinearity with two-tone sinusoidal excitation.

Replacing $\alpha$ with $\omega_1 t$ and $\beta$ with $\omega_2 t$ in Eq. (1.148), the intermodulation distortion products will be written as

Second-order components:

$$a_2 \, \cos(\omega_2 - \omega_1)t \ \text{(envelope)} \tag{1.155}$$

$$a_2 \, \cos(\omega_2 + \omega_1)t \ \text{(sum)} \tag{1.156}$$

Third-order components:

$$\frac{3a_3}{4} \cos(2\omega_2 - \omega_1)t \tag{1.157}$$

$$\frac{3a_3}{4} \cos(2\omega_1 - \omega_2)t \tag{1.158}$$

$$\frac{3a_3}{4} \cos(2\omega_1 + \omega_2)t \tag{1.159}$$

$$\frac{3a_3}{4} \cos(2\omega_2 + \omega_1)t. \tag{1.160}$$

Figure 1.46 represents the frequency spectrum of the third-order nonlinearity with the two-tone excitation generated by Eq. (1.154).

# REFERENCES

1. Berg AI. *Theory and Design of Vacuum-Tube Generators* (in Russian). Moskva: GEI; 1932.
2. Osborn PH. A study of class B and C amplifier tank circuits. *Proc. IRE*. May 1932;20:813−834.
3. Morecroft JH, Friis HT. The vacuum tubes as a generator of alternating-current power. *Trans. AIEE*. October 1919;38:1415−1444.
4. Prince DC. Vacuum tubes as power oscillators, Part I. *Proc. IRE*. June 1923;11:275−313.
5. Oswald AA. Power amplifiers in trans-atlantic radio telephony. *Proc. IRE*. June 1925;13:313−324.
6. Barton LE. High audio power from relatively small tubes. *Proc. IRE*. July 1931;19:1131−1149.
7. Fay CE. The operation of vacuum tubes as class B and class C amplifiers. *Proc. IRE*. March 1932;20:548−568.
8. Terman FE, Ferns JH. The calculation of class C amplifier and harmonic generator performance of screen-grid and similar tubes. *Proc. IRE*. March 1934;22:359−373.
9. Hallman LB. A fourier analysis of radio-frequency power amplifier wave forms. *Proc. IRE*. October 1932;20:1640−1659.
10. Everitt WL. Optimum operating conditions for class C amplifiers. *Proc. IRE*. February 1934;22:152−176.
11. Kilgour CE. Graphical analysis of output tube performance. *Proc. IRE*. January 1931;19:42−50.
12. Kaganov VI. *Transistor Radio Transmitters* (in Russian). Moskva: Energiya; 1976.
13. Raab FH. Class-E, Class-C, and Class-F power amplifiers based upon a finite number of harmonics. *IEEE Trans. Microwave Theory Tech*. August 2001;MTT-49:1462−1468.
14. Raab FH. Maximum efficiency and output of Class-F power amplifiers. *IEEE Trans. Microwave Theory Tech*. June 2001;MTT-49:1162−1166.
15. Juhas A, Novak LA. Comments on "Class-E, Class-C, and Class-F power amplifiers based upon a finite number of harmonics. *IEEE Trans. Microwave Theory Tech*. June 2009;MTT-57:1623−1625.
16. Chevaux N, De Souza MM. Comparative analysis of VDMOS/LDMOS power transistors for RF amplifiers. *IEEE Trans. Microwave Theory Tech*. November 2009; MTT-57:2643−2651.
17. Holle GA, Reader HC. Nonlinear MOSFET model for the design of RF power amplifiers. *IEE Proc. Circuits Devices Syst*. October 1992;139:574−580.
18. Grebennikov A. *RF and Microwave Power Amplifier Design*. New York: McGraw-Hill; 2004.
19. Curtice WR, Pla JA, Bridges D, Liang T, Shumate EE. A new dynamic electro-thermal nonlinear model for silicon RF LDMOS FETs. *1999 IEEE MTT-S Int. Microwave Symp. Dig*. 1999;2:419−422.
20. Fager C, Pedro JC, Carvalho NB, Zirath H. Prediction of IMD in LDMOS transistor amplifiers using a new large-signal model. *IEEE Trans. Microwave Theory Tech*. December 2002;MTT-50:2834−2842.
21. Curtice WR, Dunleavy L, Claussen W, Pengelly R. New LDMOS model delivers powerful transistor library: the CMC model. *High Frequency Electronics*. October 2004;3:18−25.

22. Tsividis YP. *Operation and Modeling of the MOS Transistor.* New York: McGraw-Hill; 1987.

23. Sung R, Bendix P, Das MB. Extraction of high-frequency equivalent circuit parameters of submicron gate-length MOSFET's. *IEEE Trans. Electron Devices.* August 1998;ED-45:1769−1775.

24. Ho MC, Green K, Culbertson R, Yang JY, Ladwig D, Ehnis P. A physical large signal Si MOSFET model for RF circuit design. *1997 IEEE MTT-S Int. Microwave Symp. Dig.* June 1997;2:391−394.

25. Cheu BJ, Ko PK. Measurement and modeling of short-channel MOS transistor gate capacitances. *IEEE J. Solid-State Circuits.* June 1987;SC-22:464−472.

26. Dortu J-M, Muller J-E, Pirola M, Ghione G. Accurate large-signal GaAs MESFET and HEMT modeling for power MMIC amplifier design. *Int. J. Microwave and Millimeter-Wave Computer-Aided Eng.* September 1995;5:195−208.

27. Liu L-S, Ma J-G, Ng G-I. Electrothermal large-signal model for III-V FETs including frequency dispersion and charge conservation. *IEEE Trans. Microwave Theory Tech.* June 1990;MTT-38:822−824.

28. Wei C-J, Tkachenko Y, Bartle D. An accurate large-signal model of GaAs MESFET which accounts for charge conservation, dispersion, and self-heating. *IEEE Trans. Microwave Theory Tech.* November 1998;MTT-46:1638−1644.

29. Jarndal A, Kompa G. Large-signal model for AlGaN/GaN HEMTs accurately predicts trapping- and self-heating-induced dispersion and intermodulation distortion. *IEEE Trans. Electron Devices.* November 2007;ED-54:2830−2836.

30. Kacprzak T, Materka A. Compact DC model of GaAs FET's for large-signal computer calculation. *IEEE J. Solid-State Circuits.* April 1983;SC-18:211−213.

31. Angelov I, Zirath H, Rorsman N. A new empirical nonlinear model for HEMT and MESFET devices. *IEEE Trans. Microwave Theory Tech.* December 1992;MTT-40:2258−2266.

32. Tang OSA, Duh KHG, Liu SMJ, et al. Design of high-power, high-efficiency 60-GHz MMIC's using an improved nonlinear PHEMT model. *IEEE J. Solid-State Circuits.* September 1997;SC-32:1326−1333.

33. Angelov I, Desmaris V, Dynefors K, Nilsson PA, Rorsman N, Zirath H. On the large-signal modelling of AlGaN/GaN HEMTs and SiC MESFETs. *13th Europ. GAAS Symp. Dig.* 2005:309−312.

34. Garcia-Osorio A, Loo-Yau JR, Reynoso-Hernandez JA, Ortega S, del Valle-Padilla JL. An empirical $I-V$ nonlinear model suitable for GaN FET Class F PA design. *Microwave and Optical Technology Lett.* June 2011;53:1256−1259.

35. Joshin K, Kikkawa T. High-power and high-efficiency GaN HEMT amplifiers. *2008 IEEE Radio and Wireless Symp. Dig.* January 2008:65−68.

36. Ishida T. GaN HEMT technologies for space and radio applications. *Microwave J.* August 2011;54:56−66.

37. Mishra UK, Parikh P, Wu Y-F. AlGan/GaN HEMTs−an overview of device operation and applications. *Proc. IEEE.* June 2002;90:1022−1031.

38. Berroth M, Bosch R. Broad-band determination of the FET small-signal equivalent circuit. *IEEE Trans. Microwave Theory Tech.* July 1990;MTT-38:891−895.

39. Fan Q, Leach JH, Morkoc H. Small signal equivalent circuit modeling for AlGaN/GaN HFET: hybrid extraction method for determining circuit elements of AlGaN/GaN HFET. *Proc. IEEE.* July 2010;98:1140−1150.

40. Dambrine G, Cappy A, Heliodore F, Playez E. A new method for determining the FET small-signal equivalent circuit. *IEEE Trans. Microwave Theory Tech.* July 1988; MTT-36:1151−1159.
41. Rohringer NM, Kreuzgruber P. Parameter extraction for large-signal modeling of bipolar junction transistors. *Int. J. Microwave and Millimeter-Wave Computer-Aided Eng.* September 1995;5:161−272.
42. Fraysee JP, Floriot D, Auxemery P, Campovecchio M, Quere R, Obregon J. A non-quasi-static model of GaInP/AlGaAs HBT for power applications. *1997 IEEE MTT-S Int. Microwave Symp. Dig.* June 1997;2:377−382.
43. Reisch M. *High-Frequency Bipolar Transistors*. Berlin: Springer; 2003.
44. Bogachev VM, Nikiforov VV. *Transistor Power Amplifiers* (in Russian). Moskva: Energiya; 1978.
45. Asbeck PM, Chang MF, Wang K-C, et al. Heterojunction bipolar transistors for microwave and millimeter-wave integrated circuits. *IEEE Trans. Microwave Theory Tech.* December 1987;MTT-35:1462−1468.
46. Lin Y-S, Jiang J-J. Temperature dependence of current gain, ideality factor, and offset voltage of AlGaAs/GaAs and InGaP/GaAs HBTs. *IEEE Trans. Electron Devices.* December 2009;ED-56:2945−2951.
47. Wang NL, Ma W, Xu S, et al. 28-V high-linearity and rugged InGaP/GaAs HBT. *2006 IEEE MTT-S Int. Microwave Symp. Dig.* June 2006:881−884.
48. Steinberger C, Landon T, Suckling C, et al. 250 W HVHBT doherty with 57% WCDMA efficiency linearized to −55 dBc for 2c11 6.5 dB PAR. *IEEE J. Solid-State Circuits.* October 2008;SC-43:2218−2228.
49. Costa D, Liu WU, Harris JS. Direct extraction of the AlGaAs/GaAs heterojunction bipolar transistor small-signal equivalent circuit. *IEEE Trans. Electron Devices.* September 1991;ED-38:2018−2024.
50. Angelov I, Choumei K, Inoue A. An empirical HBT large-signal model for CAD. *Int. J. RF and Microwave Computer-Aided Eng.* November 2003;13:518−533.
51. Giacoletto LJ. Study of n-p-n alloy junction transistors from DC through medium frequencies. *RCA Rev.* December 1954;15:506−562.
52. Rudiakova AN. BJT Class-F power amplifier near transition frequency. *IEEE Trans. Microwave Theory Tech.* September 2005;MTT-53:3045−3050.
53. Krauss HL, Bostian CW, Raab FH. *Solid State Radio Engineering*. New York: John Wiley & Sons; 1980.
54. Lee BM. Apply wideband techniques to balanced amplifiers. *Microwaves.* April 1980;19:83−88.
55. Everitt WL. Output networks for radio-frequency power amplifiers. *Proc. IRE.* May 1931;19:725−737.
56. Friis HT. Noise figure of radio receivers. *Proc. IRE.* July 1944;32:419−422.
57. Roberts S. Conjugate-image impedances. *Proc. IRE.* April 1946;34:198−204.
58. Haefner SJ. Amplifier-gain formulas and measurements. *Proc. IRE.* July 1946;34:500−505.
59. Pritchard RL. High-frequency power gain of junction transistors. *Proc. IRE.* September 1955;43:1075−1085.
60. Stern AR. Stability and power gain of tuned power amplifiers. *Proc. IRE.* March 1957;45:335−343.

61. Houselander LS, Chow HY, Spense R. Transistor characterization by effective large-signal two-port parameters. *IEEE J. Solid-State Circuits*. April 1970;SC-5:77−79.

62. Belohoubek EF, Rosen A, Stevenson DM, Presser A. Hybrid integrated 10-Watt CW broad-band power source at *S* band. *IEEE J. Solid-State Circuits*. December 1969;SC-4:360−366.

63. Presser A, Belohoubek EF. 1−2 GHz high power linear transistor amplifier. *RCA Rev.* December 1972;33:737−751.

64. Cusack JM, Perlow SM, Perlman BS. Automatic load contour mapping for microwave power transistors. *IEEE Trans. Microwave Theory Tech.* December 1974;MTT-12:1146−1152.

65. Abe H, Aono Y. 11-GHz GaAs power MESFET load−pull measurements utilizing a new method of determining tuner *Y* parameters. *IEEE Trans. Microwave Theory Tech.* May 1979;MTT-27:394−399.

66. Takayama Y. A new load−pull characterization method for microwave power transistors. *1976 IEEE MTT-S Int. Microwave Symp. Dig.* June 1976:218−220.

67. Bava GP, Pisani U, Pozzolo V. Active load technique for load−pull characterization at microwave frequencies. *Electronics Lett.* February 1982;18:178−180.

68. Rauscher C, Willing HA. Simulation of nonlinear microwave FET performance using a quasi-static model. *IEEE Trans. Microwave Theory Tech.* October 1979;MTT-27:834−840.

69. Teppati V, Ferrero A, Pisani U. Recent advances in real-time load−pull systems. *IEEE Trans. Instrumentation and Measurements*. November 2008;IM-57:2640−2646.

70. Stancliff RB, Poulin DD. Harmonic load−pull. *1979 IEEE MTT-S Int. Microwave Symp. Dig.* April/May 1979:185−187.

71. Hashmi MS, Ghannouchi FM, Tasker PJ, Rawat K. Highly reflective load−pull. *IEEE Microwave Mag.* June 2011;12:96−107.

72. Thompson BJ. Oscillations in tuned radio-frequency amplifiers. *Proc. IRE*. March 1931;19:421−437.

73. Fyler GW. Parasites and instability in radio transmitters. *Proc. IRE*. September 1935;23:985−1012.

74. Llewellyn FB. Some fundamental properties of transmission systems. *Proc. IRE*. March 1952;40:271−283.

75. Page DF, Boothroyd AR. Instability in two-port active networks. *IRE Trans. Circuit Theory*. June 1958;CT-5:133−139.

76. Rollett JM. Stability and power gain invariants of linear two-ports. *IRE Trans. Circuit Theory Appl*. January 1962;CT-9:29−32.

77. Linvill JG, Schimpf LG. The design of tetrode transistor amplifiers. *Bell Syst. Tech. J.* April 1956;35:813−840.

78. Muller O, Figel WG. Stability problems in transistor power amplifiers. *Proc. IEEE*. August 1967;55:1458−1466.

79. Muller O. Internal thermal feedback in fourpoles, especially in transistors. *Proc. IEEE*. August 1964;52:924−930.

80. Application Note AN-010, *GaN for LDMOS Users*, Nitronex Corp., 2008.

81. Vidkjaer J. Instabilities in RF-power amplifiers caused by a self-oscillation in the transistor bias network. *IEEE J. Solid-State Circuits*. October 1976;SC-11:703−712.

82. Lohrmann DR. Parametric oscillations in VHF transistor power amplifiers. *Proc. IEEE*. March 1966;54:409−410.

83. Lohrmann DR. Amplifier has 85% efficiency while providing up to 10 watts power over a wide frequency band. *Electronic Design*. March 1966;14:38−43.

84. Penfield P, Rafuse RP. *Varactor Applications*. Cambridge: The M.I.T. Press; 1962.

85. Migulin VV, Medvedev VI, Mustel ER, Parygin VN. *Basic Theory of Oscillations*. Moscow: Mir Publishers; 1983.

86. *RF Transmitting Transistor and Power Amplifier Fundamentals*, Philips Semiconductors, 1998.

87. Sato T, Grigorean C. Design advantages of CDMA power amplifiers built with MOSFET technology. *Microwave J*. October 2002;45:64−78.

88. Application Note AN-009, Bias Sequencing and Temperature Compensation for GaN HEMTs, Nitronex Corp., 2008.

# Class-D Power Amplifiers

In this chapter, different configurations of Class-D power amplifiers are presented, the increased efficiency of which is a result of employing the active device as switches. First, the switchmode power amplifiers with resistive load of different configurations, which can be considered driver power amplifiers, are discussed. Then, the current-switching and voltage-switching configurations based on complementary and transformer-coupled topologies are analyzed. The effect of the saturation resistance, rectangular and sinusoidal driving signals, finite transition time, and parasitic shunt capacitance and series inductance is demonstrated. The practical design examples of voltage-switching and current-switching Class-D power amplifiers intended to operate at high frequencies and microwaves are described. Finally, a high-frequency Class-S amplification architecture employing a switchmode Class-D power amplifier to efficiently amplify a digitally modulated transmitting signal is shown and discussed.

## 2.1 Switchmode power amplifiers with resistive load

The efficiency of a power amplifier can be maximized if the active device is operated as a switch. When the transistor is turned on, the voltage is nearly zero and high current is flowing through the device; that is, the transistor acts as a low resistance (closed switch) during this part of a period. When the transistor is turned off, the current is zero and there is high voltage across the device, i.e. the transistor acts as an open switch during the other part of a period.

Let us consider the three types of a switchmode power amplifier with resistive load: a single-ended amplifier, a complementary voltage-switching push−pull amplifier, and a transformer-coupled current-switching push−pull amplifier [1]. Figure 2.1(*a*) shows the simplified circuit schematic of a switchmode single-ended

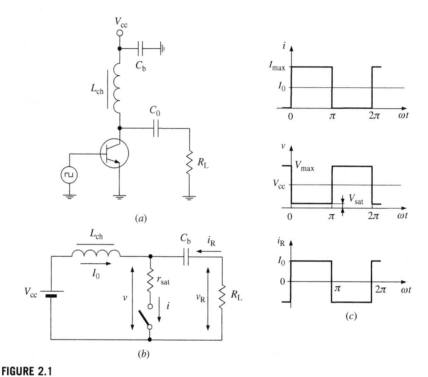

**FIGURE 2.1**

Switchmode single-ended pulsed power amplifier.

bipolar power amplifier, where $L_{ch}$ is the RF choke required to isolate a dc power supply from an RF circuit, $C_b$ is the bypass capacitor, and $C_0$ is the blocking capacitor with the supply voltage $V_{cc}$ applied to its plates.

The theoretical analysis of the operation conditions of a single-ended switchmode power amplifier can be carried out based on an equivalent circuit shown in Fig. 2.1(b), where the active device is considered a switch with a saturation resistance $r_{sat}$ driven in such a way as to provide the device with instantaneous switching between its on-state and off-state operation conditions under an assumption of a 50% duty ratio.

When the switch is turned on for $0 \leq \omega t \leq \pi$,

$$i(\omega t) = I_{max} = \frac{V_{cc} + R_L I_0}{R_L + r_{sat}} \tag{2.1}$$

$$v(\omega t) = V_{sat} = r_{sat} I_{max} \tag{2.2}$$

$$i_R(\omega t) = I_0 \tag{2.3}$$

$$v_R(\omega t) = R_L I_0 \tag{2.4}$$

where $I_0$ is the dc current, $I_{max}$ is the peak collector current, $V_{sat}$ is the saturation voltage, and $R_L$ is the load resistance.

When the switch is turned off for $\pi \leq \omega t \leq 2\pi$,

$$i(\omega t) = 0 \tag{2.5}$$

$$v(\omega t) = V_{max} = V_{cc} + R_L I_0 \tag{2.6}$$

$$i_R(\omega t) = -I_0 \tag{2.7}$$

$$v_R(\omega t) = -R_L I_0 \tag{2.8}$$

where $V_{max}$ is the peak collector voltage.

The rectangular collector voltage, collector current, and load current waveforms are shown in Fig. 2.1(c) demonstrating that zero collector current corresponds to maximum collector voltage, and minimum collector voltage corresponds to maximum collector current. However, since the current flowing to the load is not sinusoidal, its harmonic components are also presented in the output spectrum.

The dc current $I_0$ can be calculated by applying a Fourier transform to Eq. (2.1) as

$$I_0 = \frac{1}{2\pi} \int_0^\pi i(\omega t) d\omega t = \frac{V_{cc}}{R_L + r_{sat}} \left( 1 + \frac{r_{sat}}{R_L + r_{sat}} \right)^{-1}. \tag{2.9}$$

Similarly, the fundamental-frequency current amplitude $I$ can be calculated as

$$I = \frac{1}{\pi} \int_0^\pi i(\omega t) \sin \omega t \, d\omega t = \frac{4}{\pi} I_0. \tag{2.10}$$

Taking into account that $I_{max} = 2 I_0$ for a duty ratio of 50% or a conduction angle of 180°, the fundamental-frequency output power $P$ can be obtained by

$$P = \frac{1}{2} I^2 R_L = \frac{8}{\pi^2} \frac{R_L}{(R_L + r_{sat})^2} \frac{V_{cc}^2}{\left( 1 + \frac{r_{sat}}{R_L + r_{sat}} \right)^2}. \tag{2.11}$$

Figure 2.2(a) shows the simplified circuit schematic of a quasi-complementary voltage-switching push–pull bipolar power amplifier, where $C_b$ is the bypass capacitor, $C_0$ is the blocking capacitor, and $R_L$ is the load resistance. The input transformer causes both active devices to be driven with currents that are 180° out of phase by reversing one secondary winding on the transformer. However, there is no need for phase reversing if the transistors are true-complementary with different base or channel majority-carrier types that simplify the circuit schematic. Note than any type of the vacuum tubes, bipolar, and MOSFET transistors can be used in this circuit if suitable drive is applied. Due to the grounding effect of a bypass capacitor $C_b$, the RF connection of the transistor outputs is parallel, thus resulting in an equivalent load resistance equal to $2R_L$ for each device.

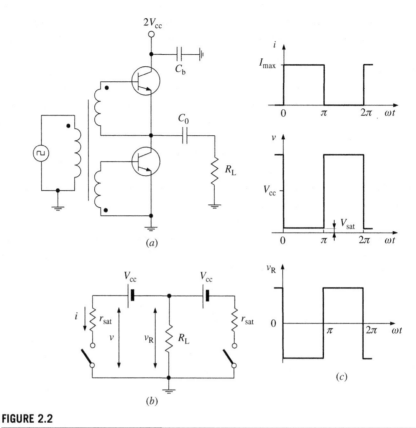

**FIGURE 2.2**

Voltage-switching push—pull power amplifier.

The theoretical analysis of the operation conditions of a quasi-complementary voltage-switching push—pull power amplifier is based on its equivalent circuit shown in Fig. 2.2(b), where each active device is considered a switch with a saturation resistance $r_{sat}$ driven in such a way as to provide an alternating switching between the on-state and off-state operation conditions of the transistor under an assumption of a 50% duty ratio. It should be noted that, for a voltage-switching push—pull power amplifier, only operation conditions with duty ratios equal or less than 50% are acceptable, because when both devices are turned on for duty ratios greater than 50%, both dc power supplies are connected to each other through the small saturation resistances of the identical transistors equal to $2r_{sat}$, thus resulting in significant efficiency reduction due to an increased total current flowing through both transistors.

When the left-hand switch is turned on for $0 \leq \omega t \leq \pi$,

$$i(\omega t) = I_{max} = \frac{V_{cc}}{R_L + r_{sat}} \qquad (2.12)$$

$$v(\omega t) = V_{sat} = r_{sat} I_{max} \qquad (2.13)$$

$$v_R(\omega t) = -R_L I_{max}. \qquad (2.14)$$

When the left-hand switch is turned off for $\pi \leq \omega t \leq 2\pi$,

$$i(\omega t) = 0 \qquad (2.15)$$

$$v(\omega t) = V_{max} = V_{cc} \left( 2 - \frac{r_{sat}}{R_L + r_{sat}} \right) \qquad (2.16)$$

$$v_R(\omega t) = V_R = R_L I_{max}. \qquad (2.17)$$

The dc current $I_0$ can be calculated by applying a Fourier transform to Eq. (2.12) as

$$I_0 = \frac{1}{2} \frac{V_{cc}}{R_L + r_{sat}}. \qquad (2.18)$$

The fundamental-frequency output power $P$ can be obtained using Eqs (2.12) and (2.17) by

$$P = \frac{8}{\pi^2} \frac{V_R^2}{R_L} = \frac{8}{\pi^2} \frac{R_L}{(R_L + r_{sat})^2} V_{cc}^2. \qquad (2.19)$$

The rectangular collector current, collector voltage, and load voltage waveforms are shown in Fig. 2.2(c) demonstrating that zero collector current corresponds to maximum collector voltage and minimum collector voltage corresponds to maximum collector current. However, since the voltage on the load is not sinusoidal, its harmonic components are also presented in the output spectrum. The maximum collector voltage peak factor for an idealized case of zero saturation resistance is equal to $v/V_{cc} = 2$.

Figure 2.3(a) shows the simplified circuit schematic of a transformer-coupled current-switching push−pull bipolar power amplifier, where $C_b$ is the bypass capacitor, $C_0$ is the blocking capacitor, $L_{ch}$ is the RF choke, and $R_L$ is the load resistance. As before, the input transformer causes both active devices to be driven with currents that are 180° out of phase. However, the transistors of the same type have the series output RF connection with an equivalent load resistance equal to $R_L/2$ for each device. The RF chokes provide bipolar $\pm I_0$ pulses of the current flowing to the load through the output balanced-to-unbalanced transformer.

The theoretical analysis of the operation conditions of a transformer-coupled current-switching push−pull power amplifier is based on its equivalent circuit shown in Fig. 2.3(b), where each active device is considered a switch with a saturation resistance $r_{sat}$ driven in such a way to provide an alternating switching

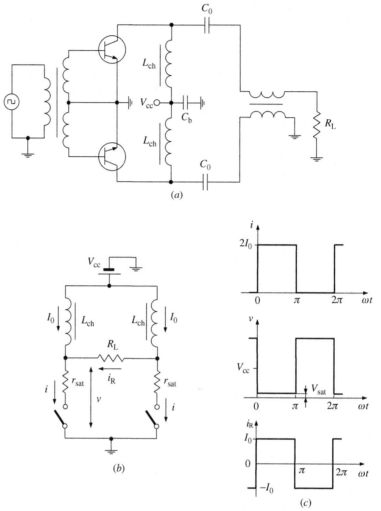

**FIGURE 2.3**

Current-switching push–pull power amplifier.

between the on-state and off-state operation conditions of the transistor under an assumption of a 50% duty ratio. It should be noted that, for a current-switching push–pull power amplifier, only operation conditions with duty ratios equal or greater than 50% are acceptable, because when both devices are turned off for duty ratios less than 50%, the choke currents will result in a significant voltage increase at the device collectors that may cause their breakdown.

When the left-hand switch is turned on for $0 \leq \omega t \leq \pi$,

$$i(\omega t) = I_{max} = 2I_0 \tag{2.20}$$

$$v(\omega t) = V_{sat} = r_{sat}I_{max} \tag{2.21}$$

$$i_R(\omega t) = I_R = I_0. \tag{2.22}$$

When the left-hand switch is turned off for $\pi \leq \omega t \leq 2\pi$,

$$i(\omega t) = 0 \tag{2.23}$$

$$v(\omega t) = V_{max} = (R_L + 2r_{sat})I_0 \tag{2.24}$$

$$i_R(\omega t) = -I_0. \tag{2.25}$$

The dc supply voltage $V_{cc}$ can be calculated by applying a Fourier transform to Eqs (2.21) and (2.24) as

$$V_{cc} = \frac{1}{2\pi} \int_0^{2\pi} v(\omega t)\, d\omega t = \frac{V_{max} + V_{sat}}{2}. \tag{2.26}$$

The dc current $I_0$ can then be obtained by

$$I_0 = \left( \frac{1}{2} + 2\frac{r_{sat}}{R_L} \right)^{-1} \frac{V_{cc}}{R_L}. \tag{2.27}$$

Finally, the fundamental-frequency output power $P$ can be written using Eqs (2.22) and (2.27) as

$$P = \frac{8}{\pi^2} I_R^2 R_L = \frac{8}{\pi^2} \left( \frac{1}{2} + 2\frac{r_{sat}}{R_L} \right)^{-2} \frac{V_{cc}^2}{R_L}. \tag{2.28}$$

To provide sufficient isolation between the two transistors, it is advisable to use the balanced circuit schematic of a switchmode push−pull power amplifier with an additional broadband transformer $T_1$ and a ballast resistor $R_{bal}$ connected to its midpoint through the blocking capacitor $C_0$, as shown in Fig. 2.4. In this case, when $R_{bal} \rightarrow 0$, the balanced push−pull schematic transforms to a transformer-coupled voltage-switching push−pull circuit, while when $R_{bal} \rightarrow \infty$, this schematic represents a transformer-coupled current-switching push−pull power amplifier. If both transistors are identical, they operate independently from each other, being fully isolated, and such a switchmode balanced power amplifier can be considered as the two switchmode single-ended power amplifiers operating independently with opposite phases on the same load through the balanced power combiner including a balanced-to-unbalanced transformer $T_2$. Since the output currents from both transistors are combined at the load resistor $R_L$ and subtracted at the ballast resistor $R_{bal}$, the load output power spectrum contains the powers

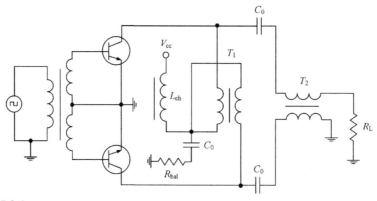

**FIGURE 2.4**

Transformer-coupled current-switching push–pull power amplifier.

corresponding to the fundamental-frequency and odd harmonic components, while the even harmonic components are combined at the ballast resistor.

Generally, the choice of a conduction angle of 180° is optimal since, in this case, it provides the collector efficiency close to maximum, the maximum voltage peak factor of less than 2, the maximum fundamental-frequency power delivered to the load, and zero power at even harmonics for all types of the switchmode power amplifiers with resistive load. The maximum collector efficiency when the saturation $r_{sat}$ is equal to zero can be obtained using Eqs (2.9), (2.11), (2.18), (2.19), (2.27), and (2.28) by

$$\eta = \frac{P}{P_0} = \frac{8}{\pi^2} \cong 81\%. \tag{2.29}$$

Consequently, with a rectangular voltage waveform on the load, the relative harmonic power is about 19% of the total output power. As a result, these switchmode single-ended and push–pull power amplifiers can be directly used as the driver power amplifiers when the harmonic level and power loss at the harmonics are not so significant. To compensate for the transformer magnetizing (leakage) and device stray input inductances, the external shunt capacitors can be used in addition to the device input capacitances, thus extending an operating high-bandwidth frequency [2]. If it is intended to be used as a final power amplifier stage, the special filter sections must be included between the device output and the load to suppress the harmonic components down to the required level.

For example, a filter-diplexer can be used to separate the signals with fundamental frequency and other harmonic components, as shown in Fig. 2.5(a). Figure 2.5(b) shows the simplest representation of a filter-diplexer is a parallel connection of a single low-pass filter section and a single high-pass filter section. To further improve isolation between these two branches, the number of the

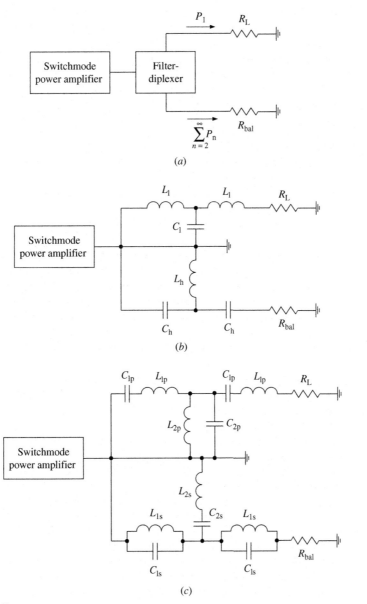

**FIGURE 2.5**

Switchmode power amplifiers with filter-diplexer.

low-pass and high-pass filter sections can be increased. The better passband and stopband performance with higher cutoff rate can be obtained by using the elliptic function or Chebyshev polynomial filters. Figure 2.5(c) shows an example of a filter-diplexer using a single passband filter section for a fundamental-frequency signal having very high impedance at the second- and higher-order harmonic components, and a single stopband filter section for a fundamental-frequency signal with minimum attenuation at the second- and higher-order harmonic components. In addition, it is possibile to improve efficiency of the power amplifier by using a power-recycling technique when the harmonic power can be transformed to the dc power by means of the RF-to-dc diode converter and returned back to the dc power supply.

## 2.2 Complementary voltage-switching configuration

The switchmode RF power amplifiers with output filters tuned to the fundamental frequency, which were called Class-D power amplifiers, ideally transform the total dc power into a fundamental-frequency power delivered to the load without power losses at the harmonics, unlike the switchmode power amplifiers with resistive load [3,4]. Conceptually, a Class-D power amplifier employs a pair of active devices operating in a push–pull mode and a tuned output circuit. The transistors are driven to act as a two-pole switch that defines either a rectangular voltage or rectangular current collector (or drain) waveforms. The output circuit is tuned to the switching frequency and removes ideally its higher-order harmonic components resulting in a purely sinusoidal signal delivered to the load. Consequently, the theoretical efficiency of an idealized Class-D power amplifier achieves 100%. Let us consider the basic principles, circuit schematics, and voltage/current waveforms corresponding to the different types of a Class-D power amplifier with output filters [1,5−7].

Figure 2.6(a) shows the simplified circuit schematic of a quasi-complementary voltage-switching Class-D bipolar power amplifier consisting of the same type of the active devices, fundamentally tuned series $L_0C_0$ filter, and load resistance $R_L$. The large-value bypass capacitor $C_b$ is necessary to isolate the dc power supply by bypassing the RF current to ground. The input transformer causes both active devices to be driven with currents that are 180° out of phase by reversing one secondary winding on the transformer. However, there is no need for phase reversing if the transistors are true-complementary with different base or channel majority-carrier type that simplifies the circuit schematic. Note than any type of the vacuum tubes, bipolar, and MOSFET transistors can be used in this circuit if suitable drive is applied. Due to the grounding effect of a bypass capacitor $C_b$, the RF connection of the transistor outputs is parallel, thus resulting in an equivalent load resistance equal to $2R_L$ for each device.

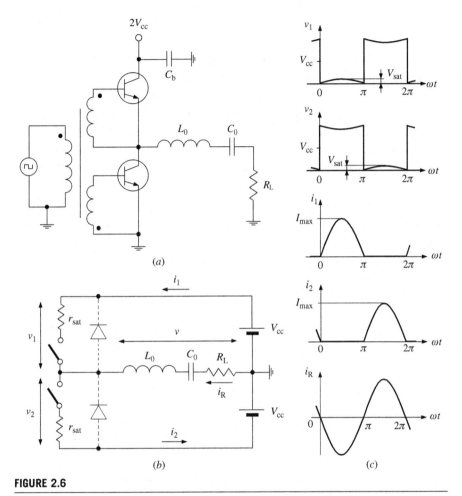

**FIGURE 2.6**

Quasi-complementary voltage-switching configuration with series filter.

To determine the collector voltage and current waveforms and to calculate the output power and collector efficiency, the following assumptions are taken into account:

- Power loss due to flow of leakage current during transistor pinch-off is negligible.
- Power loss due to non-ideal tuning is negligible.
- Power loss during switching transitions is negligible.

In this case, each active device is considered a switch with a saturation resistance $r_{sat}$ shown in Fig. 2.6(b) that is driven in such a way as to provide an alternating switching between the on-state and off-state operation conditions of the transistor under an assumption of a 50% duty ratio. The alternating half-period switching of the two transistors between their pinch-off mode and voltage-saturation mode results in rectangular collector-voltage pulses with a maximum amplitude of $2V_{cc}$, as shown in Fig. 2.6(c). The collector-voltage pulses, which contain only the odd-harmonic components, are applied to the series $L_0C_0$ filter with high loaded quality factor $Q_L = \omega L_0/R_L \gg 1$ tuned to the fundamental frequency $f_0 = \omega_0/2\pi = 1/2\pi\sqrt{L_0C_0}$, resulting in the fundamental-frequency sinusoidal current $i_R = -I_R \sin \omega t$ which flows to the load $R_L$. The half waves of this current flow through the transistors representing the half-sinusoidal collector-current pulses that contain the fundamental-frequency, second-, and higher-order even harmonic components only. The shape of the saturation voltage with a maximum amplitude $V_{sat}$ is fully determined by the collector current waveform when $i(\omega t) > 0$ according to $v_{sat}(\omega t) = r_{sat}\, i(\omega t)$, where $i(\omega t) = i_1(\omega t) = i_2(\omega t)$ for a symmetrical circuit with identical transistors. The collector voltage peak factor is equal to $v/V_{cc} = 2$.

It should be noted that, for an operation with conduction angles less than 180° when both transistors are turned on for duty ratios less than 50%, there is a time period when both transistors are turned off simultaneously. Therefore, in order that the load current $i_R(\omega t)$ can flow continuously, it is necessary to include a diode in parallel to each device, as shown in Fig. 2.6(b). The operation conditions with conduction angles greater than 180° are unacceptable, since both dc power supplies are connected to each other through the small saturation resistances of the identical transistors equal to $2r_{sat}$, thus resulting in the significant efficiency reduction due to an increased total current flowing through both transistors.

Now let us determine the voltage $v(\omega t) = v_1(\omega t) - V_{cc}$ at the input of a series $L_0C_0$ circuit and collector current $i_1(\omega t)$ for the first transistor operating as a switch with a saturation resistance $r_{sat}$. When the switch is turned on for $0 \leq \omega t \leq \pi$,

$$v(\omega t) = -V_{cc} + r_{sat}\, I_R \sin \omega t \tag{2.30}$$

$$i_1(\omega t) = I_R \sin \omega t \tag{2.31}$$

where $I_R$ is the load current amplitude.

When the switch is turned off for $\pi \leq \omega t \leq 2\pi$,

$$v(\omega t) = V_{cc} + r_{sat}I_R \sin \omega t \tag{2.32}$$

$$i_1(\omega t) = 0. \tag{2.33}$$

The fundamental-frequency voltage amplitude $V$ of the voltage $v(\omega t)$ can be calculated by applying a Fourier transform to Eqs (2.30) and (2.32) as

$$V = -\frac{1}{\pi} \int_0^{2\pi} v(\omega t)\sin \omega t \, d\,\omega t = \frac{4}{\pi} V_{cc} - r_{sat}I_R. \tag{2.34}$$

Similarly, the dc current $I_0$ can be obtained from Eq. (2.31) by

$$I_0 = \frac{1}{2\pi} \int_0^{\pi} i_1(\omega t) d\,\omega t = \frac{I_R}{\pi}. \tag{2.35}$$

Taking into account that $I_R = V/R_L$ and the fact that the sinusoidal output current flows through either one or another transistor depending on which device is turned on and has a half-sinusoidal waveform, using Eq. (2.34) will result in

$$I_R = I_{max} = \frac{4}{\pi} \frac{V_{cc}}{R_L} \frac{1}{1 + \frac{r_{sat}}{R_L}}. \tag{2.36}$$

The dc power $P_0$ and fundamental-frequency output power $P$ and can be obtained using Eqs (2.35) and (2.36) by

$$P_0 = 2V_{cc}I_0 = \frac{8}{\pi^2} \frac{V_{cc}^2}{R_L} \frac{1}{1 + \frac{r_{sat}}{R_L}} \tag{2.37}$$

$$P = \frac{1}{2} I_R^2 R_L = \frac{8}{\pi^2} \frac{V_{cc}^2}{R_L} \frac{1}{\left(1 + \frac{r_{sat}}{R_L}\right)^2}. \tag{2.38}$$

As a result, the collector efficiency $\eta$ of a quasi-complementary voltage-switching push−pull power amplifier with a series filter can be written as

$$\eta = \frac{P}{P_0} = \frac{1}{1 + \frac{r_{sat}}{R_L}}. \tag{2.39}$$

From Eq. (2.39), it follows that the collector efficiency is equal to 100% for an idealized case of the lossless active devices with zero saturation resistance. It should be mentioned that the dc current drawn from the power supply represents the form of a half-sinusoidal pulse train. Therefore, it is very important to provide the proper RF bypassing circuit representing either a single large-value capacitor or an additional filter with a low cutoff frequency in the power supply line including the series RF choke and shunt capacitors. In addition, a loaded quality factor of $>5$ must be chosen for a simple series-tuned $L_0C_0$ circuit to provide a good compromise between the prevention of harmonic current and coil losses. An additional harmonic suppression can be obtained by inserting standard filters between the series-tuned circuit and the load. Note that the collector efficiency degrades with increasing operating frequency where the switching transitions become an appreciable fraction of the signal period. In practice, it was found that high efficiency of a Class-D power amplifier can be maintained to frequencies of the order of $0.1f_T$ for low-power transistors and to $0.01f_T$ for high-power transistors rated at greater than 10 W, where $f_T$ is the device transition frequency [8].

The circuit schematic of a voltage-switching Class-D power amplifier shown in Fig. 2.6(*a*) is called a quasi-complementary circuit because it is based on the

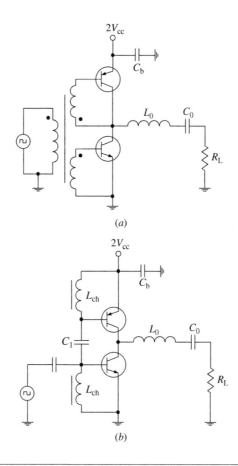

**FIGURE 2.7**

True-complementary voltage-switching Class-D power amplifier circuits.

same type of the two identical transistors, *n-p-n* bipolar or *n*-channel MOSFET devices, each in a common-emitter or a common-source configuration [9]. A true-complementary power-amplifier circuit configuration requires two transistors of different types, *n-p-n* and *p-n-p* bipolar or *n*- and *p*-channel MOSFET devices. Figure 2.7 shows the two examples of such a true-complementary voltage-switching Class-D push—pull power amplifier, a configuration with input transformer shown in Fig. 2.7(*a*) and a transformerless configuration shown in Fig. 2.7(*b*). In the latter case, there is no need for a driving transformer because both transistors can be mounted on a single heatsink without electrical insulation [10]. The coupling capacitor $C_1$ must have a very low reactance to make a voltage drop at carrier frequency across it negligible compared with the input signal amplitude. Its value can be chosen as

$$C_1 \gg \frac{1}{2\pi f_0 r_b} \tag{2.40}$$

where $f_0$ is the fundamental frequency of the input voltage and $r_b$ is the series base resistance of the bipolar transistor.

However, in practice such circuits can be used at sufficiently low frequencies because of the much higher transition losses resulting from performance mismatch between the complementary transistors. Consequently, further development in technology to provide similar electrical characteristics of truly complementary transistors would offer a new possibility for these Class-D power amplifiers, especially for a transformerless configuration, which can easily be integrated in a monolithic structure.

## 2.3 Transformer-coupled voltage-switching configuration

The broadband center-tapped and balanced-to-unbalanced (balun) transformers can also be used in voltage-switching Class-D power amplifiers in much the same manner as they are used in the switchmode power amplifiers with resistive load or conventional Class-B push–pull power amplifiers. In this case, such a configuration is called the transformer-coupled voltage-switching Class-D power amplifier. Figure 2.8(a) shows the simplified circuit schematic of a transformer-coupled voltage-switching Class-D bipolar power amplifier including an output series-tuned $L_0C_0$ circuit and a load resistance $R_L$. The output transformer $T_2$ is considered as ideal, having $m$ turns in each half of the primary winding and $n$ turns in the secondary winding. As in the complementary voltage-switching configuration, the input transformer $T_1$ causes both active devices $Q_1$ and $Q_2$ to be driven with currents that are 180° out-of-phase to switch on and off alternately. During the first half-cycle when the transistor $Q_1$ is turned on, its collector voltage $v_1(\omega t)$ is equal to zero, assuming zero saturation resistance. As a result, the dc supply voltage $V_{cc}$ is placed across one-half of the primary winding of the transformer $T_2$, being then transformed to the voltage $(-n/m)V_{cc}$ on its secondary winding. When the transistor $Q_2$ is turned on, the dc supply voltage $V_{cc}$ is applied to the other half of the primary winding causing voltage $(n/m)V_{cc}$ to appear on the secondary winding, as shown in Fig. 2.8(b).

Consequently, when the switch $Q_1$ is turned on for $0 \leq \omega t \leq \pi$,

$$v_1(\omega t) = 0 \tag{2.41}$$

$$v_2(\omega t) = 2V_{cc} \tag{2.42}$$

$$v(\omega t) = -\frac{n}{m} V_{cc}. \tag{2.43}$$

When the switch $Q_1$ is turned off for $\pi \leq \omega t \leq 2\pi$,

$$v_1(\omega t) = 2V_{cc} \tag{2.44}$$

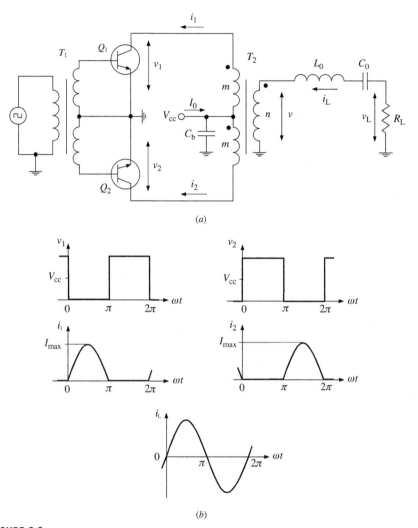

**FIGURE 2.8**

Transformer-coupled voltage-switching push—pull configuration with series filter.

$$v_2(\omega t) = 0 \tag{2.45}$$

$$v(\omega t) = \frac{n}{m} V_{cc}. \tag{2.46}$$

Thus, the resulting secondary voltage $v(\omega t)$ represents a square wave with levels of $\pm (n/m)V_{cc}$, while the collector voltages are square waves with levels of

0 and $+2V_{cc}$. The fundamental-frequency voltage amplitude $V$ of the voltage $v(\omega t)$ can be calculated as

$$V_L = V = -\frac{1}{\pi}\int_0^{2\pi} v(\omega t)\sin \omega t \ d \ \omega t = \frac{4}{\pi}\frac{n}{m}V_{cc} \tag{2.47}$$

where $V_L$ is the fundamental-frequency voltage amplitude on the load $R_L$.

Then, the fundamental-frequency output power $P$ can be obtained by

$$P = \frac{1}{2}\frac{V_L^2}{R_L} = \frac{8}{\pi^2}\left(\frac{n}{m}\right)^2\frac{V_{cc}^2}{R_L} = \frac{8}{\pi^2}\frac{V_{cc}^2}{R}. \tag{2.48}$$

where

$$R = \left(\frac{m}{n}\right)^2 R_L \tag{2.49}$$

is the equivalent fundamental-frequency resistance across one-half of the primary winding of the output transformer $T_2$ seen by each device output, with the other one-half of the primary winding open.

The amplitude $I_L$ of a sinusoidal current $i_L(\omega t)$ flowing to the load is given by

$$I_L = \frac{V_L}{R_L} = \frac{4}{\pi}\frac{V_{cc}}{R}. \tag{2.50}$$

Consequently, the collector currents of each transistor represent the half-sinu-soidal waveforms with an opposite phase of 180° between each other whose peak amplitudes are $(4/\pi)(V_{cc}/R)$. The dc supply current $I_0$ represents a sum of two collector currents drawn into the center tap, hence

$$I_0 = \frac{1}{2\pi}\int_0^{2\pi} [i_1(\omega t) + i_2(\omega t)]d\omega t = \frac{8}{\pi^2}\left(\frac{n}{m}\right)^2\frac{V_{cc}}{R_L} \tag{2.51}$$

resulting in the collector efficiency of 100%, because the fundamental-frequency power and dc power are equal. Note that high efficiency can only be achieved at lower frequencies because the presence of parasitic device output capacitance during the transition from the off-state to on-state in addition to the requirement of a wideband balun rapidly reduces efficiency with frequency. However, this performance degradation is less severe in current-mode Class-D power amplifiers.

## 2.4 **Transformer-coupled current-switching configuration**

Figure 2.9(a) shows the circuit schematic of a current-switching Class-D power amplifier where the output balanced-to-unbalanced transformer is used to connect to a standard load by providing a push–pull operation. In this case, such a

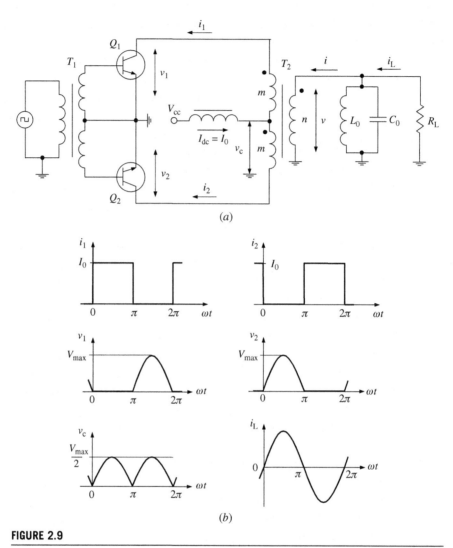

**FIGURE 2.9**

Transformer-coupled current-switching configuration with a parallel filter.

configuration is called the transformer-coupled current-switching Class-D power amplifier, which is the dual of the transformer-coupled voltage-switching Class-D power amplifier because the collector voltage and current waveforms are interchanged. As in the transformer-coupled voltage-switching configuration, the input transformer $T_1$ is necessary to drive both active devices with currents having opposite phases for the on-to-off alternate device switching. The output transformer $T_2$ is considered ideal, having $m$ turns in each half of the primary winding and $n$ turns in the secondary winding. However, the dc current supply is connected to the center tap of the transformer primary winding through the RF choke. The wideband balun at the output provides an open-circuit condition at even harmonics.

The load network of the transformer-coupled current-switching configuration requires a parallel fundamentally tuned resonant $L_0 C_0$ circuit connected in parallel to the load $R_L$ which provides low impedance to the two active devices at odd harmonics, instead of a series fundamentally tuned resonant $L_0 C_0$ circuit required for the transformer-coupled voltage-switching configuration. Whichever device is turned on during the first half-cycle, it takes the entire dc current $I_0$ resulting in a rectangular collector current waveform with levels of 0 and $I_0$, as shown in Fig. 2.9(b). Its collector voltage is equal to zero when it is turned on, assuming zero saturation resistance. Transformation of the rectangular collector currents from half of the primary winding to the secondary winding produces a rectangular current $i(\omega t)$ with levels of $\pm (m/n)I_0$.

Consequently, when the switch $Q_1$ is turned on for $0 \leq \omega t \leq \pi$,

$$i_1(\omega t) = I_0 \tag{2.52}$$

$$i_2(\omega t) = 0 \tag{2.53}$$

$$i(\omega t) = \frac{m}{n} I_0. \tag{2.54}$$

When the switch $Q_1$ is turned off for $\pi \leq \omega t \leq 2\pi$,

$$i_1(\omega t) = 0 \tag{2.55}$$

$$i_2(\omega t) = I_0 \tag{2.56}$$

$$i(\omega t) = -\frac{m}{n} I_0. \tag{2.57}$$

Since the output parallel-tuned resonant circuit suppresses the harmonic components of the secondary current $i(\omega t)$ allowing only the fundamental-frequency component $i_L(\omega t)$ to flow to the load $R_L$, its amplitude $I_L$ can be calculated as

$$I_L = \frac{1}{\pi} \int_0^{2\pi} i(\omega t)\sin \omega t \, d\omega t = \frac{4}{\pi}\frac{m}{n} I_0. \tag{2.58}$$

The fundamental-frequency sinusoidal current $i_L(\omega t)$ produces the sinusoidal voltage $v(\omega t) = v_L(\omega t) = i_L(\omega t)R_L$ on the load resistance $R_L$ with amplitude

$$V_L = V = \frac{4}{\pi}\frac{m}{n} I_0 R_L \tag{2.59}$$

where $V$ is the sinusoidal secondary voltage amplitude.

The sinusoidal voltage $v(\omega t)$ across the secondary winding is transformed to the primary winding, where it produces the sinusoidal voltage across each half of the primary winding with amplitude of $(m/n)V$. Because, during half a period, one of the two transistors is turned on and the corresponding end of the primary winding is grounded, the peak amplitude of the voltage at the collector of the other transistor can be obtained by

$$V_{\max} = 2\left(\frac{m}{n}\right)V_{\mathrm{L}} = \frac{8}{\pi}I_0 R \tag{2.60}$$

where

$$R = \left(\frac{m}{n}\right)^2 R_{\mathrm{L}} \tag{2.61}$$

is the equivalent fundamental-frequency resistance across one-half of the primary winding of the output transformer $T_2$ seen by each device output, with the other one-half of the primary winding open. Thus, the sinusoidal voltage $v(\omega t)$ being transformed to the primary winding produces two half-sinusoidal collector voltages with opposite phases.

Since the center-tap voltage $v_{\mathrm{c}}(\omega t)$ represents half of the sum of the two collector voltages, its average value can be calculated using Eq. (2.60) as

$$V_{\mathrm{cc}} = \frac{1}{2\pi}\int_0^{2\pi}\left[\frac{v_1(\omega t)}{2} + \frac{v_2(\omega t)}{2}\right]d\omega t = \frac{8}{\pi^2}\left(\frac{m}{n}\right)^2 I_0 R_{\mathrm{L}} \tag{2.62}$$

where $v_1(\omega t) = -V_{\max}\sin\omega t$ for $\pi \le \omega t \le 2\pi$ and $v_2(\omega t) = V_{\max}\sin\omega t$ for $0 \le \omega t \le \pi$.

As a result,

$$I_0 = \frac{\pi^2}{8}\left(\frac{n}{m}\right)^2\frac{V_{\mathrm{cc}}}{R_{\mathrm{L}}} = \frac{\pi^2}{8}\frac{V_{\mathrm{cc}}}{R}. \tag{2.63}$$

Then, the peak collector voltage $V_{\max}$ given by Eq. (2.60) can be rewritten as a function of the dc supply voltage $V_{\mathrm{cc}}$ only as

$$V_{\max} = \pi V_{\mathrm{cc}}. \tag{2.64}$$

Since there is no voltage drop across RF choke, the peak value of the voltage $v_{\mathrm{c}}(\omega t)$ having a full-wave rectified shape is equal to $(\pi/2)V_{\mathrm{cc}}$. By using Eqs (2.59) and (2.63), the fundamental-frequency output power $P$ and dc supply power $P_0$ can be obtained by

$$P = \frac{1}{2}\frac{V_{\mathrm{L}}^2}{R_{\mathrm{L}}} = \frac{\pi^2}{8}\left(\frac{n}{m}\right)^2\frac{V_{\mathrm{cc}}^2}{R_{\mathrm{L}}} \tag{2.65}$$

$$P_0 = V_{\mathrm{cc}}I_0 = \frac{\pi^2}{8}\left(\frac{n}{m}\right)^2\frac{V_{\mathrm{cc}}^2}{R_{\mathrm{L}}} \tag{2.66}$$

resulting in the collector efficiency of 100%, because the fundamental-frequency power and dc power are equal.

The transformer-coupled current-switching Class-D power amplifier shown in Fig. 2.9(a) requires a wideband balun, which has to have 180° phase shift between its two ports at least up to the third harmonic. Therefore, this topology is mainly used at low frequencies where wideband transformers can easily be designed. To avoid this problem of designing a wideband balun at higher frequencies, a symmetrical current-switching structure can be used where the output parallel-tuned tank is moved from the unbalanced load side to the balanced side that reduces bandwidth requirement for the balun to one octave instead of multiple octaves.

## 2.5 Symmetrical current-switching configuration

Figure 2.10(a) shows the simplified circuit schematic of a symmetrical current-switching Class-D bipolar power amplifier consisting of the same type of the active devices, fundamentally tuned parallel $L_0C_0$ filter, and balanced load resistance $R_L$. The RF choke $L_{ch}$ connected to the center point of the inductor $L_0$ is necessary to isolate the dc power supply and make the circuit symmetrical. Each active device is considered a switch with the saturation resistance $r_{sat}$ shown in Fig. 2.10(b) that is driven in such a way as to provide an alternating switching between the on-state and off-state operation conditions of the transistor under an assumption of a 50% duty ratio. Since the only dc current $I_0$ is flowing through the RF choke $L_{ch}$, the alternating half-period device switching between their pinch-off mode and saturation mode results in rectangular collector current pulses with a maximum amplitude of $2I_0$, as shown in Fig. 2.10(c).

The collector current pulses, which contain the odd harmonic components only, are applied to the parallel $L_0C_0$ filter with high loaded quality factor $Q_L = \omega L_0/R_L \gg 1$ tuned to the fundamental frequency $f_0 = \omega_0/2\pi = 1/2\pi\sqrt{L_0C_0}$ resulting in the fundamental-frequency sinusoidal voltage $v_R = V_R\sin\omega t$ across the load $R_L$. During half a period when one transistor is turned on, the half wave of this voltage is applied to the other transistor representing the half-sinusoidal collector voltage pulses, which contain the fundamental-frequency, second-, and higher-order even harmonic components only. The flat shape of the saturation voltage with maximum amplitude $V_{sat}$ is fully determined by the collector current waveform when $i_1(\omega t)$, $i_2(\omega t) > 0$ according to $v_{sat}(\omega t) = r_{sat}\, i_1(\omega t) = r_{sat}\, i_2(\omega t)$ for a symmetrical circuit with identical transistors. Unlike a complementary voltage-switching configuration, the transistor outputs are connected in series, thus resulting in an equivalent load resistance equal to $R_L/2$ for each device.

It should be noted that the symmetrical current-switching Class-D power amplifier, the collector voltage and current waveforms of which are shown in Fig. 2.10(c), is the dual of the complementary voltage-switching Class-D power amplifier, the collector voltage and current waveforms of which are shown in Fig. 2.6(c), because the voltage and current waveforms are interchanged.

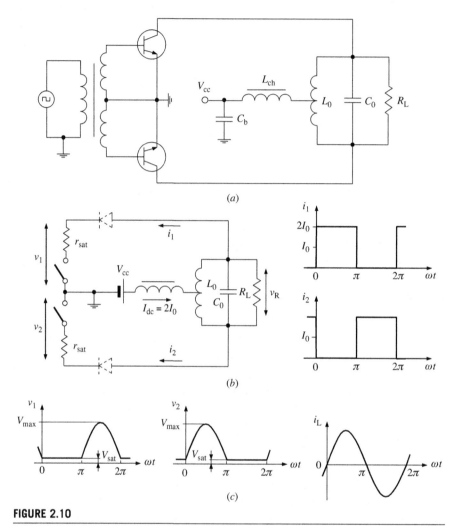

**FIGURE 2.10**

Symmetrical current-switching configuration with a parallel filter.

However, if, in the case of a voltage-switching configuration, the collector peak voltage is defined by the dc supply voltage $V_{cc}$ only, then, in the latter case of a current-switching configuration, the transistors represent current switches with the current amplitude defined by the dc voltage supply $V_{cc}$, saturation resistance $r_{sat}$, and load resistance $R_L$. For an operation mode with conduction angles greater than $180°$, there is a time period when both active devices are turned on simultaneously, and the parallel-tuned circuit is shunted by a small resistance $2r_{sat}$. Therefore, to eliminate this shunting effect accompanied by power losses in both transistors, it is necessary to include a diode in series to each device collector, as shown in Fig. 2.10(*b*). However, the operation conditions with conduction angles less than 180° are unacceptable, since there are time intervals when both transistors are turned off simultaneously causing the significant increase in the collector voltage amplitude due to the growth of the current flowing through RF choke.

Now let us determine the current $i(\omega t) = i_1(\omega t) - I_0$ in a parallel $L_0C_0$ circuit and collector voltage $v_1(\omega t)$ for the first transistor operating as a current switch with a saturation resistance $r_{sat}$. When the first switch is turned on for $0 \leq \omega t \leq \pi$, then $i_1(\omega t) = 2I_0$ resulting in

$$i(\omega t) = I_0 \tag{2.67}$$

$$v_1(\omega t) = V_{sat} = 2r_{sat}I_0 \tag{2.68}$$

where $I_0$ is the dc supply current of each device.

When the first switch is turned off for $\pi \leq \omega t \leq 2\pi$, then $i_1(\omega t) = 0$ resulting in

$$i(\omega t) = -I_0 \tag{2.69}$$

$$v_1(\omega t) = -(V_{max} - 2r_{sat}I_0)\sin \omega t + 2r_{sat}I_0 \tag{2.70}$$

where $V_{max}$ is the collector peak voltage.

The fundamental-frequency current amplitude $I$ of the current $i(\omega t)$ can be calculated by applying a Fourier transform to Eqs (2.67) and (2.69) as

$$I = \frac{1}{\pi} \int_0^{2\pi} i(\omega t)\sin \omega t \, d \, \omega t = \frac{4}{\pi} I_0. \tag{2.71}$$

Similarly, the dc supply voltage $V_{cc}$ can be written from Eqs (2.68) and (2.70) as

$$V_{cc} = \frac{1}{2\pi} \int_0^{\pi} v_1(\omega t)d \, \omega t = \frac{1}{\pi}(V_{max} - 2r_{sat}I_0). \tag{2.72}$$

Taking into account that $V_1 = V_{max} - 2r_{sat} I_0 = IR_L$, the dc current $I_0$ and fundamental-frequency collector voltage $V_1$ can be obtained respectively by

$$I_0 = \left(\frac{\pi}{2}\right)^2 \left(1 + \frac{\pi^2}{2} \frac{r_{sat}}{R_L}\right)^{-1} \frac{V_{cc}}{R_L} \tag{2.73}$$

$$V_1 = \pi \left(1 + \frac{\pi^2}{2} \frac{r_{sat}}{R_L}\right)^{-1} V_{cc}. \tag{2.74}$$

The dc power $P_0$ and fundamental-frequency output power $P$ can be obtained, using Eqs (2.73) and (2.74) and taking into account that $V_1 = V_R$, by

$$P_0 = 2V_{cc}I_0 = \frac{\pi^2}{2} \frac{V_{cc}^2}{R_L} \frac{1}{1 + \frac{\pi^2}{2} \frac{r_{sat}}{R_L}} \tag{2.75}$$

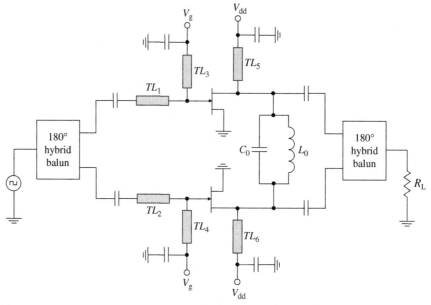

**FIGURE 2.11**

Microwave symmetrical current-switching configuration with parallel filter.

$$P = \frac{1}{2}\frac{V_{\text{R}}^2}{R_{\text{L}}} = \frac{\pi^2}{2}\frac{V_{\text{cc}}^2}{R_{\text{L}}}\frac{1}{\left(1 + \frac{\pi^2}{2}\frac{r_{\text{sat}}}{R_{\text{L}}}\right)^2}. \tag{2.76}$$

As a result, the collector efficiency $\eta$ of a symmetrical current-switching push–pull power amplifier with parallel filter can be written as

$$\eta = \frac{P}{P_0} = \frac{1}{1 + \frac{\pi^2}{2}\frac{r_{\text{sat}}}{R_{\text{L}}}}. \tag{2.77}$$

Microwave current-switching Class-D power amplifiers can be designed around 1 GHz by relying on the basic structure shown in Fig. 2.11, where the parallel $L_0C_0$ filter can be realized either by an air-wound inductor, a chip inductor, or by using high-impedance transmission lines together with a chip capacitor [11]. An advantage of this approach at microwaves is that the parasitic output capacitances of the devices can be partially absorbed into external tank capacitance $C_0$. This reduces the frequency-dependent capacitive losses typical for voltage-switching Class-D power amplifiers and makes the design less sensitive to component tolerance. For such a 900-MHz current-switching Class-D power amplifier, the required higher-order harmonic impedance termination can be provided by a second-order lattice lumped balun and a lumped-element parallel-tuned tank (air-coil inductor and ceramic capacitor) connected between the drain terminals of the two devices resulting in a peak drain efficiency of 75% at a saturated output power of 20 W using packaged GaN HEMT devices [12]. A drain efficiency of 71% can be achieved with an output power of 20.3 W at an operating frequency of 1 GHz using a narrow transmission line as a tank inductor between the two LDMOSFET devices [13]. However, the amplifier performance degrades as the operating frequency increases because of the bandwidth requirement of the balun, which should have absolute bandwidth of a few gigahertz in this case to cover at least the second-harmonic component. The other important problem is the implementation of a tank with lumped elements whose self-resonance frequency lies beyond the third harmonic of the operating frequency. Otherwise, uncontrolled impedance termination at the most relevant harmonic frequencies degrades the performance of a current-switching Class-D power amplifier, resulting in considerably lower efficiency than expected.

## 2.6 **Voltage-switching configuration with reactive load**

In practice, the load network of a switchmode Class-D power amplifier can have a nonzero reactance at the operating frequency caused by its mistuning or influence of the device parasitic reactive elements. Figure 2.12(a) shows the circuit schematic of a quasi-complementary voltage-switching configuration with reactive load, where $C_{c1}$ and $C_{c2}$ are the equivalent collector capacitances of each transistor and $X_L$ is the series load-network reactance. The series fundamentally tuned $L_0 C_0$ circuit reduces output currents at the harmonic frequencies flowing to the load to negligible level. The collector voltage waveforms of each transistor are unchanged by the load reactance. However, due to the effect of the series reactance $X_L$, the output current $i_R(\omega t)$ is shifted in phase relative to the collector voltage waveform, as shown in Fig. 2.12(b). Because of this phase shift, both collector currents $i_1(\omega t)$ and $i_2(\omega t)$ tend to be negative during a portion of each period. In this case, the series resonant circuit represents an inductive load. If both transistors $Q_1$ and $Q_2$ are MOSFET devices, the negative currents can be passed without any potential damage, since the MOSFET intrinsic body-drain p-n junction diode may be used as an antiparallel diode. However, bipolar devices in general do not conduct in reverse direction. As a result, the negative currents will charge their equivalent collector capacitances $C_{c1}$ and $C_{c2}$, producing large voltage spikes that can damage the transistors.

The series reactance reduces the amplitude of the output current and output power. If $Z_L = R_L + jX_L$, the load current $i_R(\omega t)$ of a complementary voltage-switching configuration with zero saturation resistance can be written using Eq. (2.36) as

$$i_R(\omega t) = \frac{4}{\pi} \frac{V_{cc}}{|Z_L|} \sin(\omega t + \varphi) \qquad (2.78)$$

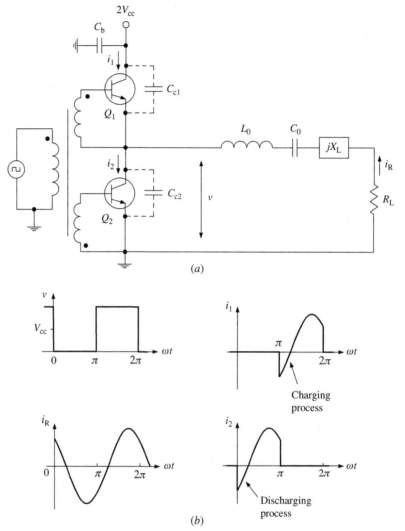

**FIGURE 2.12**

Voltage-switching Class-D power amplifier with reactive load.

where the phase shift $\varphi$ is defined by

$$\varphi = \tan^{-1}\frac{X_L}{R_L}. \qquad (2.79)$$

Consequently, the fundamental-frequency power $P$ delivered to the load $R_L$ will be reduced to

$$P = \frac{8}{\pi^2} \frac{V_{cc}^2}{R_L} \rho^2 \qquad (2.80)$$

where

$$\rho = \frac{R_L}{|Z_L|} < 1. \qquad (2.81)$$

A suitable reverse-direction current path is provided by diodes $D_1$ and $D_2$, as shown in Fig. 2.13(a). In this case, the sinusoidal phase-shifted load current always passes through one of the four devices ($Q_1$, $Q_2$, $D_1$, or $D_2$), preventing the collector voltage spikes, as shown in Fig. 2.13(b). Here, the shunt capacitance $C_c$ represents a sum of both collector capacitances $C_{c1}$ and $C_{c2}$, and may include any parasitic stray circuit capacitance associated with the method of practical implementation. The same protection approach can be used in a transformer-coupled voltage-switching configuration. Active devices used in the symmetrical and transformer-coupled current-switching Class-D power amplifiers may be protected from the negative collector voltages by placing diodes in series with the collectors. Note that the collector efficiency of a Class-D power amplifier is essentially unchanged by the load network reactance.

The operation with a capacitive load is not recommended because the antiparallel diodes generate high reverse-recovery current spikes [14]. These spikes occur in the switch current waveforms at both the switch turn-on and switch turn-off and may destroy the transistors. The transistors are turned on at high voltage equal to $2V_{cc}$, and the transistor output capacitance is short-circuited by a low transistor saturation resistance, dissipating the energy stored in that capacitance. Therefore, the turn-on switching loss is high, the effect of a feedback capacitance is significant, increasing the transistor input capacitance and the gate drive requirements, and the turn-on transition speed is reduced. Generally, the power amplifier can operate safely with an open circuit at the output. However, it is prone to catastrophic failure if the output is short-circuited at the operating frequency close to the resonant frequency of the output series resonant $L_0C_0$ circuit.

The presence of the parasitic collector capacitances $C_{c1}$ and $C_{c2}$ causes the power losses due to finite charge storage process. When the transistor $Q_1$ is turned on and the transistor $Q_2$ is turned off, the capacitor $C_{c1}$ is discharged through the $Q_1$ and the capacitor $C_{c2}$ is charged instantaneously to $2V_{cc}$. However, when the transistor $Q_2$ is turned on and the transistor $Q_1$ is turned off, the capacitor $C_{c2}$ is discharged instantaneously through $Q_2$ and the capacitor $C_{c1}$ is charged instantaneously to $2V_{cc}$. Since this occurs twice during each period and the power losses due to energy dissipated in the both transistors with charging and discharging processes are equal, the total power losses due to finite switching time for a complementary voltage-switching Class-D power amplifier with supply voltage of $2V_{cc}$ and zero saturation resistance can be written as

$$P_s = (C_{c1} + C_{c2})(2V_{cc})^2 f_0 = 4C_c V_{cc}^2 f_0 \qquad (2.82)$$

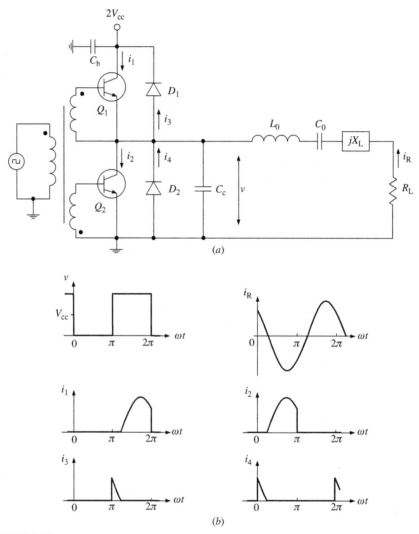

**FIGURE 2.13**

Voltage-switching Class-D power amplifier with reactive load and protection diodes.

where $C_c = C_{c1} + C_{c2}$ and $f_0$ is the operating frequency. The total switching power losses described by Eq. (2.82) can also characterize the transformer-coupled voltage-switching Class-D power amplifier with a supply voltage of $V_{cc}$, in which the parasitic collector capacitances $C_{c1}$ and $C_{c2}$ are charged and discharged between 0 and $2V_{cc}$.

In the current-switching Class-D power amplifier, the parasitic capacitances do not provide charging and discharging losses because the collector currents flowing through the transistor have fixed constant values (either zero or maximum $I_{dc}$). However, there is another mechanism of power losses associated with parasitic inductances $L_{c1}$ and $L_{c2}$ (due to finite lead length and leakage inductance in the transformer) in series with transistor collectors. In this case, the currents flowing through the active devices jump when switching occurs, since they must be changed from zero to $I_{dc}$ instantaneously, twice during each period. Therefore, the total power losses due to finite switching time for a current-switching Class-D power amplifier with zero saturation resistance can be written as

$$P_s = L_c I_{dc}^2 f_0 \tag{2.83}$$

where $L_c = L_{c1} + L_{c2}$.

## 2.7 Drive and transition time

The driving circuitry of a Class-D power amplifier must provide a driving signal sufficient to ensure that the active devices are alternately saturated or pinched off during the proper time period. Generally, if a current-switching Class-D power amplifier requires a rectangular current or voltage driving waveform, a voltage-switching Class-D power amplifier can be driven by either a rectangular or sinusoidal driving signal. Figure 2.14 shows the input part of a voltage-switching Class-D bipolar power amplifier and the waveforms associated with rectangular voltage and sinusoidal current driving waveforms.

The sinusoidal current driven through the primary winding of the input transformer causes the half-sinusoidal currents to be driven into each base from the transformer secondary winding. The current flowing into the corresponding base results in a rise of the base-emitter voltage to $V_{bmax}$, which is approximately equal to its threshold voltage of 0.7 V for a bipolar device and corresponds to the maximum value required to saturate the MOSFET device. At the same time, the current flowing into the base of the other transistor is shifted by 180°, causing the fall of its base-emitter voltage to $-V_{bmax}$, ensuring that the transistor is pinched off.

Hence, the rectangular voltage with levels $\pm (m/n)V_{bmax}$ appears across the primary winding. The impedance seen by the driver at the switching frequency represents a ratio of the fundamental-frequency voltage driving amplitude to the current amplitude written as

$$R_{dr} = \frac{4}{\pi} \left(\frac{n}{m}\right)^2 \frac{V_{bmax}}{I_{bmax}} \tag{2.84}$$

where $I_{bmax}$ is the peak base current amplitude, which must be large enough to sustain the collector current.

Then, the required driving power is given by

$$P_{dr} = \frac{2}{\pi} V_{bmax} I_{bmax}. \tag{2.85}$$

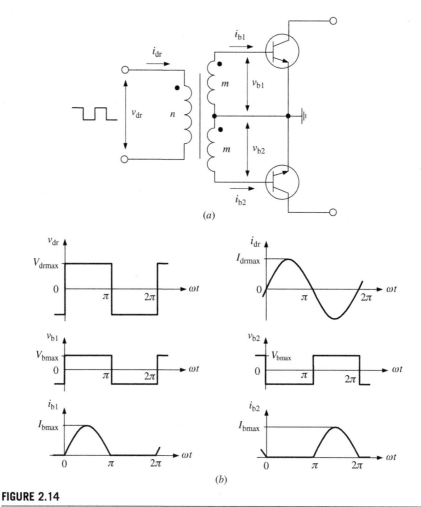

**FIGURE 2.14**

Driving voltage and current waveforms.

From Eq. (2.84), it follows that the driving resistance $R_{dr}$ is a function of the driving current amplitude and is not related to transistor parameters other than the voltage corresponding to a forward-biased base-emitter junction. For this reason, the driver output current must be limited to prevent transistor failure [5].

In a current-switching Class-D power amplifier, the rectangular collector (or drain) current requires the rectangular voltage and current driving waveforms. In this case, Eqs (2.84) and (2.85) can be rewritten as

$$R_{dr} = \left(\frac{n}{m}\right)^2 \frac{V_{bmax}}{I_{bmax}} \tag{2.86}$$

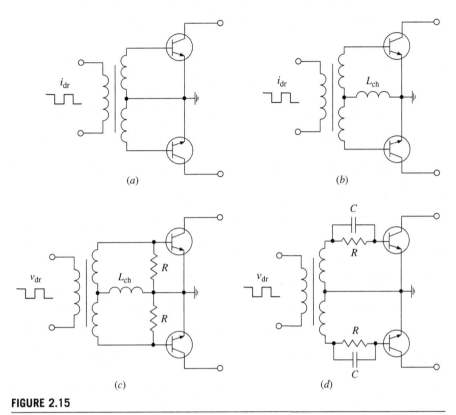

**FIGURE 2.15**

Input driving circuits.

$$P_{dr} = V_{bmax}I_{bmax} \qquad (2.87)$$

where the driver output current also must be limited.

Figure 2.15 shows the different configurations of the input driving circuits for a current-switching Class-D bipolar power amplifier. In the case of the driver with a rectangular current waveform, the active devices can be connected either in parallel, as shown in Fig. 2.15(a), or in series, as shown in Fig. 2.15(b), where $L_{ch}$ is the choke inductor. For a parallel connection, the driving current is distributed between bipolar transistors being inversely proportional to their input impedances. The input impedance of the transistor at saturation is significantly less than that in the active and pinch-off regions. Since the transistor cannot instantly change its saturation mode to pinch-off conditions under a negative driving signal, it absorbs the most part of the driving current, thus preventing the other transistor from going into saturation under a positive driving signal. Therefore, the series connection of the bipolar transistors is more effective because the driving current flows through both devices. However, the voltage across the reverse-biased base-emitter junction will be increased, which value must be controlled to prevent the device failure.

When the driver with a rectangular voltage waveform is used, the driving current flows through the forward-biased base-emitter junction of one transistor and the reverse-biased base-emitter junction of the other transistor when they are connected in series, as shown in Fig. 2.15(c). If a shunt resistor $R$ is connected in parallel to the transistor input, the effect of switching losses can be reduced because the device input capacitance is discharged faster through this resistor. A similar correction effect can be achieved by using a parallel $RC$ circuit connected in series to each transistor when both transistors are connected in parallel, as shown in Fig. 2.15(d). However, an increased current is required from the driver to saturate the transistor.

Generally, the input circuit of the transistor, either bipolar or MOSFET, can be represented as purely capacitive and only at low frequencies. In switching applications, however, the finite rise and fall times are the result of the effect of the much higher frequency components rather than the fundamental. For example, if the switching time of 4 ns can be tolerated at an amplitude of 80% with a 30-MHz carrier, it represents roughly a 100-MHz sine wave [15]. Besides, the input capacitance is usually nonlinear and varies over bias conditions. For example, the MOSFET gate-source and gate-drain capacitances vary with gate and drain voltages. At increased gate voltage, the gate-source capacitance goes down to its lowest value, just before reaching the threshold voltage, and then goes up to be constant at saturation. At the same time, when the MOSFET device begins to draw drain current, the drain voltage is lowered resulting in reduction of the depletion area and causing an overlap between the gate and bulk material. This in turn increases the value of the gate-drain capacitance, which takes maximum value at zero drain voltage and positive gate voltage corresponding to the maximum device transconductance. In bipolar transistors, the base-emitter nonlinear capacitance represents the large diffusion capacitance in the active and saturation regions and much lower junction capacitance in the pinch-off region.

However, at higher frequencies, the device equivalent input circuit must also include a series resistance, representing the base ohmic resistance for a bipolar transistor or effective gate resistance consisting of the distributed channel and gate electrode resistances for a MOSFET device. Let us consider the effect of a rectangular voltage drive with a duty ratio of 50% on the device input circuit shown in Fig. 2.16(a), where $R_{in}$ is the device input series resistance and $C_{in}$ is the device input capacitance. According to Kirchhoff's voltage law, the algebraic sum of all voltage drops taken around any closed path is zero, which gives

$$v_R + v_C - v_{in} = 0 \tag{2.88}$$

where

$$v_R = i_{dr}R = R_{in}C_{in}\frac{dv_C(t)}{dt} \tag{2.89}$$

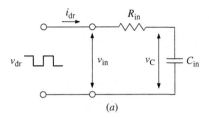

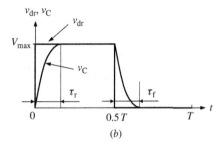

**FIGURE 2.16**

Rectangular driving of equivalent input transistor circuit.

and both the resistance $R_{in}$ and capacitance $C_{in}$ are assumed to be voltage-independent.

For a time period when $0 \leq t \leq 0.5T$,

$$v_{in} = V_{max} \tag{2.90}$$

where $V_{max}$ is the input peak voltage corresponding to the device saturation conditions.

Substituting Eqs (2.89) and (2.90) into Eq. (2.88) yields

$$R_{in}C_{in}\frac{dv_C(t)}{dt} + v_C(t) = V_{max}. \tag{2.91}$$

The general solution of the linear nonhomogeneous first-order differential equation under the initial condition of $v_C(t) = 0$ at $t = 0$, when there is no energy stored in the input capacitance, can be obtained by

$$\frac{v_C(t)}{V_{max}} = 1 - \exp\left(-\frac{t}{R_{in}C_{in}}\right) \tag{2.92}$$

which determines the length of a voltage angular rise time $\tau_r$ through the time constant $\tau_{in} = R_{in}C_{in}$ required for the input capacitance to charge up from 0 to $V_{max}$.

Similarly, for a time period when $0.5T \leq t \leq T$,

$$v_{in} = 0 \tag{2.93}$$

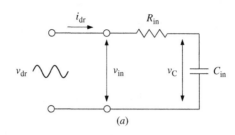

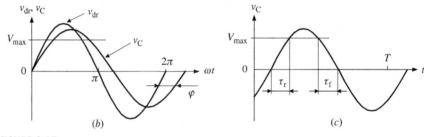

**FIGURE 2.17**

Sinusoidal driving of equivalent input transistor circuit.

resulting in the linear homogeneous first-order differential equation

$$R_{in}C_{in}\frac{dv_C(t)}{dt} + v_C(t) = 0. \tag{2.94}$$

Under the initial condition of $v_C(t) = V_{max}$ when $t = 0.5T$, the solution of Eq. (2.94) can be written as

$$\frac{v_C(t)}{V_{max}} = \exp\left(-\frac{t}{R_{in}C_{in}}\right) \tag{2.95}$$

which determines the length of a voltage angular fall time $\tau_f$ through the time constant $\tau_{in} = R_{in}C_{in}$ required for the input capacitance to discharge down from $V_{max}$ to zero.

From Eqs (2.92) and (2.95), it follows that the switching losses due to charging and discharging effects of the input capacitance are the same with voltage rise time $\tau_r$ and fall time $\tau_r$ equal both to $4\tau_{in}$ at the voltage level of 98% of $V_{max}$, which are shown in Fig. 2.16(b). Note that the power requirement for the input drive is independent of the switching speed, and the switching speed is ultimately limited by the input resistance. Hence, there is an intrinsic limit to how fast the input capacitance can be charged, which becomes a serious factor as the operating frequency increases.

Now let us consider a response of the same circuit to sinusoidal voltage driving and its effect on the power losses during the switching time. For the device input circuit shown in Fig. 2.17(a) with a sinusoidal drive $v_{dr}(\omega t) = V_{dr} \sin(\omega t)$, we can write

$$R_{in} C_{in} \frac{dv_C(t)}{dt} + v_C(t) = V_{dr} \sin(\omega t) \qquad (2.96)$$

where $V_{dr}$ is the voltage amplitude of the driving signal.

The general solution of the linear nonhomogeneous first-order differential equation can be written as

$$v_C(t) = A \exp\left(-\frac{t}{\tau_{in}}\right) + \frac{V_{dr}}{1 + (\omega \tau_{in})^2} [\sin(\omega t) - \tau_{in} \cos(\omega t)]. \qquad (2.97)$$

Let us assume that the driving voltage $v_{dr}(t)$ is zero at $t < 0$, so that the initial voltage across the input capacitance is zero. Then, under initial conditions of $v_C(t) = 0$ at $t = 0$ required to determine the unknown coefficient $A$, Eq. (2.97) can be rewritten in a normalized form as

$$\frac{v_C(t)}{V_{dr}} = \frac{1}{\sqrt{1 + (\omega \tau_{in})^2}} \left[ \sin \varphi \exp\left(-\frac{t}{\tau_{in}}\right) + \sin(\omega t - \varphi) \right] \qquad (2.98)$$

where $\varphi = \tan^{-1}(\omega \tau_{in})$ is the phase shift. From Eq. (2.98), it follows that, after a few time constants, the natural response corresponding to the capacitance-charging process becomes negligible, and only the sinusoidal response with phase shift of $\varphi$ remains, as shown Fig. 2.17(b). Then the circuit is operated in the sinusoidal steady-state mode with reduced amplitude.

The voltage rise time $\tau_r$ and fall time $\tau_f$ become longer with increasing input time constant $\tau_{in} = R_{in} C_{in}$, since the voltage amplitude across the capacitance reduces, becoming closer to the maximum amplitude $V_{max}$ required for the saturation mode of the transistor. Consequently, to minimize power losses, it is just necessary to increase the voltage amplitude of the driving signal. An advantage of the sinusoidal driving signal compared with rectangular drive is that there is no need to use a broadband input transformer, and all lead and leakage inductances in the gate-drive circuit can be absorbed into the input resonant circuit. With a sinusoidal driving technique, a quasi-complementary voltage-switching Class-D power amplifier was capable of providing an output power of 300 W at an operating frequency of 13.56 MHz with a drain efficiency of over 90% using a low-inductance half-bridge MOSFET die package on alumina substrate [2].

Generally, an exact analysis that includes transition waveforms and effects of elements of the complete device equivalent circuit is very complicated. This analysis of power losses can be substantially simplified by assuming that the resultant collector (or drain) voltage waveform of a complementary voltage-switching Class-D power amplifier is trapezoidal when the transitions produce ramp voltage waveforms, as shown in Fig. 2.18 [5]. From Fig. 2.17(c), we can see that, under sinusoidal drive with increased voltage amplitude, the shape of the rise time $\tau_r$ and fall time $\tau_f$ are close to ramp. The transition time required by a single

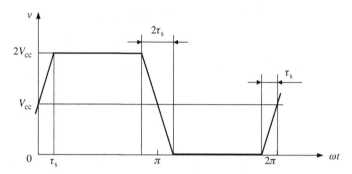

**FIGURE 2.18**

Transition time in Class-D power amplifier.

transistor to complete the entire switching process is shown in Fig. 2.18 as being converted to the angular time $\tau_s$. This time can include an effect of the output capacitance and other device parasitics. Both transistors are then assumed to have zero saturation resistances, and switching is completed within $2\tau_s$.

The fundamental-frequency collector voltage amplitude $V$ is obtained by a Fourier integral of the trapezoidal waveform taking into account the dc supply voltage of $2V_{cc}$ as

$$V = \frac{4}{\pi} V_{cc} \frac{\sin\tau_s}{\tau_s} \approx \frac{4}{\pi} V_{cc} \left(1 - \frac{\tau_s^2}{6}\right) \qquad (2.99)$$

where the linear approximation is valid only for small values of $\tau_s$.

For the fundamental-frequency output power $P = V^2/(2R_L)$ and dc current $I_0 = V/(\pi R_L)$, where $R_L$ is the load resistance, the collector efficiency can be calculated from

$$\eta = \frac{\pi}{4} \frac{V}{V_{cc}} = \frac{\sin\tau_s}{\tau_s}. \qquad (2.100)$$

The same results may also be obtained for the symmetrical current-switching and both voltage-switching and current-switching transformer-coupled Class-D power amplifier configurations [16].

## 2.8 Practical Class-D power amplifier implementation

Figure 2.19 shows the circuit schematic of a quasi-complementary MOSFET voltage-switching Class-D power amplifier where the dc power supply is connected to the drain of the top device and the source of the other device is connected to the ground [17]. The devices are driven via a three-winding transformer with appropriate polarities on the output so that the same drive is used for both

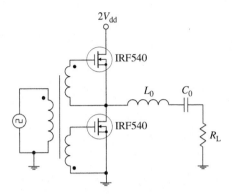

**FIGURE 2.19**

MOSFET voltage-switching Class-D power amplifier.

devices. The midpoint between the two devices is connected to the series resonant $L_0C_0$ circuit with a reasonable loaded quality factor of about 10, which can provide small impedance at the fundamental frequency and high impedances at the harmonic components. To design a 300-W Class-D power amplifier operating at 13.56 MHz from a 75-V dc supply voltage, it is necessary to provide the load resistance $R_L = 3.8\ \Omega$ and peak drain current $I_{max} = 12.56\ A$ in accordance with Eqs (2.36) and (2.38), assuming zero device saturation resistance. To satisfy these requirements, the MOSFET devices IRF540 were chosen, which can operate under dc voltage and current conditions of 100 V and 25 A, respectively. This MOSFET device has a saturation resistance equal to 0.085 $\Omega$, thus providing the conduction losses per device of 3.35 W. The output capacitance of the device is equal to 500 pF. Consequently, according to Eq. (2.82), the switching losses per device due to the capacitance discharging process are equal to 38.14 W. As a result, the maximum expected drain efficiency can reach a value of 80%. However, it also needs to take into account an effect of the device input circuit with a time constant $\tau_{in} = 39$ ns for a gate voltage of 10 V to rectangular drive that makes the overall transition time even longer.

Applying a current-switching Class-D configuration enables the elimination of switching losses due to the device output shunt capacitances. In this case, it is necessary to minimize the switching losses due to the parasitic series inductances which can be represented by the lead inductances of the device package. However, by on-chip integration of the parallel $LC$ resonator, it is possible to reduce the circuit complexity and eliminate parasitic reactance losses.

Figure 2.20 shows the circuit schematic of a bipolar current-switching InGaP/GaAs HBT Class-D power amplifier where active devices represent 80 emitter fingers with an emitter area of $2\ \mu m \times 20\ \mu m$ and peak current density of 0.11 mA/$\mu m^2$ [16]. A 180° input balun based on a 50-$\Omega$ coaxial line generates differential input signals, while a 180° output balun converts the balanced output with

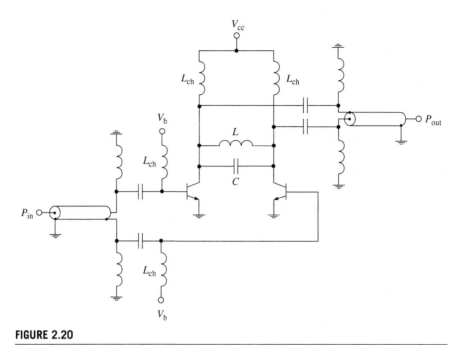

**FIGURE 2.20**

Schematic of bipolar current-switching Class-D power amplifier.

out-of-phase output signals to a single-ended output signal. To maximize power gain, the input and output matching circuits in the form of high-pass $L$-sections are applied to each transistor. The $LC$ resonator comprises a bond-wire inductor and a metal−insulator−metal (MIM) capacitor. Using a spiral inductor with a lower quality factor reduces collector efficiency by 5%. At a supply voltage of 3.4 V, the transistor saturation resistance $r_{sat}$ and transition time $\tau_s$ were equal to 0.58 $\Omega$ and 0.1 $\pi$, respectively. The bases of both HBTs are biased to a turn-on voltage of 1.2 V for operation as switches. As a result, the current-switching Class-D power amplifier achieved a collector efficiency of 78.5% at an output power of 29.5 dBm with a maximum power-added efficiency (*PAE*) of 68.5% at an operating frequency of 700 MHz. The best efficiency was achieved at a high drive level resulting in a power gain of about 9 dB. By using a similar approach with GaAs MESFETs devices, the power-added efficiency of 75.6% with an output power of 28.6 dBm and a power gain of 13.9 dB was achieved at the operating frequency of 900 MHz and supply voltage of 5 V [18].

Enhanced current-switching Class-D architecture for high-frequency operation can provide the required harmonic impedance conditions (open circuit for even-harmonic and short circuit for odd-harmonic components of the fundamental frequency) using a transmission-line multiharmonic output-load coupling network, which make the amplifier performance independent of the performance of the

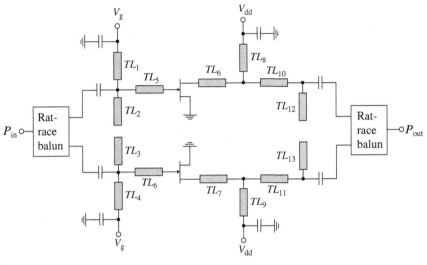

**FIGURE 2.21**

Microwave current-switching configuration with multiharmonic transformation networks.

balun and the parallel tank at higher harmonics. Practically, it is difficult to control more than the second- and third-harmonic components, which provide the most contribution to the output voltage and current waveforms. Therefore, the multiharmonic load-transformation network connected to the output terminal of each device simultaneously performs the fundamental load transformation and the correct impedance termination at the second and third harmonics of the fundamental frequency. Basically, this results in an inverse Class-F operation mode for each single amplification branch, which in turn is equivalent to the current-switching Class-D operation mode. In this case, harmonic components in the load are removed by the impedance transformation networks instead of by the combination of balun and parallel-tuned tank as in conventional architecture.

Figure 2.21 shows the schematic diagram of a high-frequency current-switching Class-D power amplifier with transmission-line multiharmonic transformation networks [11]. In this structure, the lumped-element tank is avoided and replaced by open-circuited quarterwave transmission lines ($TL_{12}$ and $TL_{13}$), which provide the low impedance at the drain of the active devices at this frequency, at the third harmonic. Should it be necessary to also present low impedance at higher-order odd harmonics, more stubs can be used. Instead of relying on the balun to provide high impedance at even harmonics, a short-circuit stub can be in the output network. Using this approach, the balun needs to be designed to operate at the fundamental frequency only, thus avoiding the difficulties associated with the implementation of a wideband balun at microwave frequencies. The drain bias for

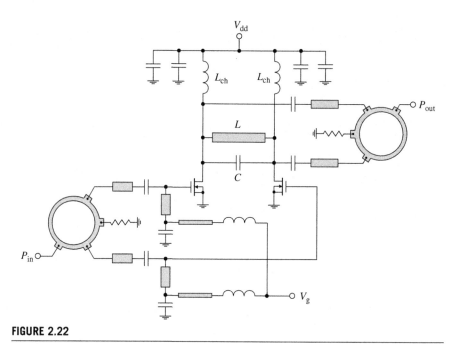

**FIGURE 2.22**

Schematic of LDMOSFET current-switching Class-D power amplifier.

each device is applied through the corresponding quarterwave transmission line ($TL_8$ or $TL_9$) shorted to ground by a large capacitor at one side. Each transmission line provides a high impedance at the fundamental frequency as well as at all odd harmonics, whereas it provides low input impedance at the second harmonic. This low impedance at the second harmonic is then transformed into a high impedance at the drain of the device by using an appropriate series transmission line ($TL_6$ or $TL_7$). As a result, a 39-dBm GaN HEMT current-switching Class-D power amplifier designed at 2.35 GHz achieved a drain efficiency of 68% and a *PAE* of 65% using a narrowband 180° rat-race balun. The baluns can also be constructed using low-loss quarterwave microstrip lines together with commercial 90° hybrids or microstrip-based coupled-line 180° hybrid structures [19,20].

Figure 2.22 shows a 1-GHz current-switching Class-D power amplifier using LDMOSFET devices [21]. The input and output baluns are represented by the 180° rat-race hybrids built on a high-dielectric substrate to minimize the physical size. The input matching circuits are added between the input of each transistor and hybrid to reduce the input return loss. The *LC* resonator was designed by converting the ideal inductor *L* and capacitor *C* to a real microstrip line and parallel capacitor. The power amplifier begins operating in a switching mode beyond 25-dBm input power. As a result, a maximum output power of 13 W was observed with a drain efficiency of 60% and power gain of 14 dB.

## 2.9 Class-D for digital pulse-modulation transmitters

Voltage-switching Class-D power amplifiers with a series filter are attractive candidates for digital RF transmitters based on a delta-sigma modulator (DSM) for linear and efficient amplification. In this case, the output switched voltage waveform is applied to a series fundamentally tuned filter, which exhibits a high impedance at all frequencies except for the resonant fundamental frequency, thus removing the out-of-band signals such as harmonics and quantization noise. Taking into account that no current flows outside of the desired frequency band, no power is dissipated at these frequencies. Since the two transistors are switched alternately, a voltage-switching Class-D amplifier can be approximated as a voltage-controlled voltage source, which operates effectively when feeding a series resonator. This high-efficiency level can be maintained even if it is driven by nonperiodic digital signals, as long as the reverse currents appearing in this condition can be provided by the transistors or parallel diodes when they are turned on. A simple series resonator is usually sufficient for periodic driving signals since the spectrum is discrete with widely spiced harmonics. However, for a broadband power spectrum such as bandpass DSM, a higher-order filter is required to properly attenuate quantization noise.

In CMOS voltage-switching Class-D power amplifiers, during the on-to-off transition of the devices, there is generally a short period of time when both pMOS and nMOS transistors are turned on, resulting in low resistance between the power supply and the ground. In this case, a large drain current (known as a shoot-through current) may be induced, which can cause significant energy loss. To minimize this loss, the pMOS and nMOS transistors are designed to have different driving circuits, as shown in Fig. 2.23(*a*) [22]. By modifying the pull-up and pull-down device size ratio of the drivers, the overlap of the turn-on time between the pMOS device and the nMOS device during the transition can be minimized.

A voltage-switching Class-D power amplifier is suitable for digital transmitters employing two-level DSM signals (levels 1 or 0). In a simplified case for a half-bridge voltage-switching Class-D power amplifier shown in Fig. 2.23(*b*), the driving signal states correspond to the two states of the Class-D amplifier operation. For example, level 1 corresponds to turn-on mode of the switch $S_1$ and turn-off mode of the switch $S_2$. At the same time, level 0 corresponds to turn-off mode of the switch $S_1$ and turn-on mode of the switch $S_2$. However, a single Class-D power amplifier is unable to differentiate the three driving states associated with three-level DSM signals. Therefore, the two voltage-switching Class-D power amplifiers can be configured in a full-bridge scheme shown in Fig. 2.23(*c*), with two pairs of switches to produce the three different driving conditions. For example, level 1 corresponds to conditions when switches $S_{11}$ and $S_{22}$ are turned on and switches $S_{12}$ and $S_{21}$ are turned off, level −1 corresponds to conditions when switches $S_{11}$ and $S_{22}$ are turned off and switches $S_{12}$ and $S_{21}$ are turned on, and

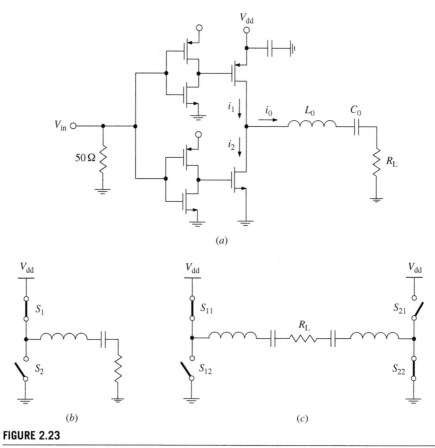

**FIGURE 2.23**

Schematics of CMOS voltage-switching Class-D power amplifiers.

level 0 corresponds to conditions when switches $S_{11}$ and $S_{21}$ are turned off and switches $S_{12}$ and $S_{22}$ are turned on. The output combining circuit represents the two quarter-wavelength transmission lines connected in series to the outputs of each Class-D amplifier and required for impedance transformation and a quarter-wavelength coaxial-line balun connected to the load. For CDMA IS-95 signals at 800 MHz, the drain efficiency can be increased from 31% to 33% at an output power of 15 dBm for three-level DSM with adjacent-channel power ratio (*ACPR*) of −43 dBc [22]. By reducing the capacitance associated with the transistors resulting in a shorter transition time, an amplifier efficiency can be improved significantly, especially in a low-power region. A full-bridge switching power amplifier suited as a driver for a high-power GaN HEMT amplifier in a Class-S transmission scheme, which can be operated with a pseudo-random digital pulse train up to 7.5 Gbit/s, can be implemented using a 0.25-μm SiGe BiCMOS technology [23].

Although out-of-band spectral components do not dissipate any power in an ideal Class-D circuit, a high level of out-of-band spectral power is wasteful and reduces the utility of the pulse train to generate load power. Significant out-of-band power is generated by one-bit quantization, and coding efficiency is introduced as a figure of merit to evaluate encoder performance. The coding efficiency is defined as the ratio of reconstructed load power relative to the total pulse-train power, and can be written as

$$\eta_{code} = \frac{8}{\pi^2} \sin^2(\pi D) \tag{2.101}$$

where $D$ is the duty ratio [24]. The coding efficiency is maximized with a duty ratio of 50% when equal to $8/\pi^2 \approx 81\%$, as in Eq. (2.29). In this case, the further efficiency degradation due to the device saturation resistance $r_{sat}$, which is equivalently connected in series to each ideal switch, can be determined by

$$\eta = \frac{1}{1 + \frac{4}{\eta_{code}} \frac{r_{sat}}{R_L}} \tag{2.102}$$

where $R_L$ is the load resistance [25].

Figure 2.24(a) shows a 0.5-μm CMOS amplifying path consisting of a multistage driver and a complementary voltage-switching Class-D power amplifier driven with a bandpass DSM signal [26]. The multistage driver consisting of five inverters of increasing gate width is necessary to provide enough current to drive the gate capacitances of the Class-D power amplifier. The Class-D amplifier stage consisting of the two complementary switches can deliver above 50 mW of output power with expected drain efficiency of about 32% for a WCDMA signal at 181 MHz with a dc supply voltage of 3.3 V. The switches on the Class-D power amplifier have a large gate width to minimize the channel resistance when the switches are turned on. As a result, the power dissipation in the Class-D switches consists of both capacitive switching power losses and ohmic conductive power losses for the load current. The output pulse train signal is filtered by a sixth-order Butterworth bandpass reconstruction filter with a 3-dB bandwidth of 10 MHz.

Figure 2.24(b) shows the complementary voltage-switching Class-D amplifier circuit with switches modeled by an ideal switch, a current source, a gate resistance $R_g$, and three intrinsic capacitances $C_{gs}$, $C_{gd}$, and $C_{ds}$ [24]. Both capacitances $C_{gs}$ and $C_{gd}$ are considered nonlinear capacitances which are highly nonlinear when the switch is turned on and relatively constant when the switch is turned off, while the capacitance $C_{ds}$ is modeled as a linear capacitance. An analysis of energy loss from parasitic capacitances shows that the gate-drain capacitance $C_{gd}$ is the dominant parasitic and has the largest change in stored energy. When the switch is turned on, the voltage across the capacitances is relatively

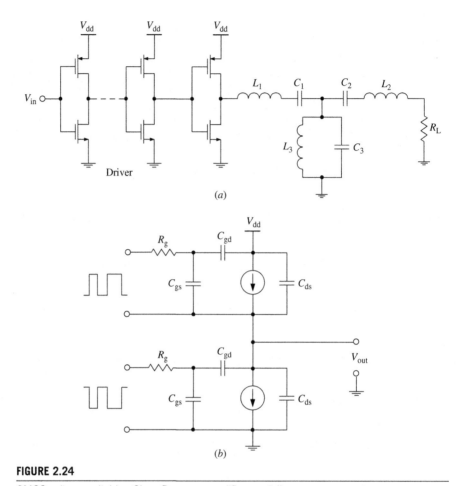

**FIGURE 2.24**

CMOS voltage-switching Class-D power amplifier modeling.

small and the stored energy is small, even though the nonlinear capacitances $C_{gs}$ and $C_{gd}$ are large in the on state. In the off state, the nonlinear capacitances are small, but the voltage is large, and the stored energy is much larger since the stored energy is proportional to the square of the voltage. As a result, it was shown for the driving pulse train with a modulator sample rate of 1.5 GHz at a carrier frequency of 500 MHz and the load signal flowing through a 15-MHz six-order reconstruction filter that the noise floor of the reconstructed signal with linear models for the capacitances of the 1-mm pHEMT switches is dependent only on the modulator noise floor. At the same time, when the switch has the nonlinear resistance characteristic of a pHEMT in the on-state, significant distortion is introduced by the power amplifier when the signal-to-noise ratio ($SNR$) is reduced by about 30 dB. The *SNR* is reduced a further 10 dB when the full nonlinear device circuit model with nonlinear capacitances is used. The drain efficiency including the conductive losses, switching losses, and filter insertion loss of 0.5 dB for a WCDMA signal with a peak-to-average ratio (*PAR*) of 7.1 dB reduces from approximately 60% at 500 MHz through 50% at 1 GHz to less than 40% at 2 GHz. Generally, improvements in efficiency could be obtained by providing better coding efficiency or using better devices with lower on-resistance, lower gate-drain capacitances, and high breakdown voltages.

## REFERENCES

1. Popov IA, ed. *Transistor Generators of Harmonic Oscillations in Switching Mode* (in Russian). Moskva: Radio i Svyaz; 1985.
2. Theodoridis MP, Mollov SV. Robust MOSFET driver for RF, Class-D inverters. *IEEE Trans. Industrial Electronics*. February 2008;IE-55:731−740.
3. Osborne MR. Design of tuned transistor power amplifiers. *Electronic Eng.* August 1968;40:436−443.
4. Baxandall PJ. Transistor sine wave LC oscillators, some general considerations and new developments. *IEE Proc. Electric Power Appl.* May 1959;106:748−758.
5. Krauss HL, Bostian CW, Raab FH. *Solid State Radio Engineering*. New York: John Wiley & Sons; 1980.
6. Artym AD. *Class D Power Amplifiers and Switching Generators in Radio Communication and Broadcasting* (in Russian). Moskva: Svyaz; 1980.
7. Kazimierczuk MK. *RF Power Amplifiers*. Chichester: John Wiley & Sons; 2008.
8. Chudobiak WJ, Page DF. Frequency limitations of Class-D transistor amplifiers. *IEEE J. Solid-State Circuits*. February 1969;SC-4:25−37.
9. Raab FH, Rupp DJ. A quasi-complementary Class-D HF power amplifier. *RF Design*. September 1992;15:103−110.
10. Kazimierczuk M, Modzelewski JS. Drive-transformerless Class-D voltage-switching tuned power amplifier. *Proc. IEEE*. June 1980;68:740−741.
11. Aflaki P, Negra R, Ghannouchi FM. Enhanced architecture for microwave current-mode Class-D amplifiers applied to the design of an S-Band GaN-Based power amplifier. *IEE Proc. Microwave Antennas Propag.* September 2009;3:997−1006.
12. Gustavsson U, Lejon T, Fager C, Zirath H. Design of highly efficient, high output power, L-Band Class D$^{-1}$ RF power amplifiers using GaN MESFET devices. *Proc. 37th Europ. Microwave Conf.* 2007;1089−1092.
13. Nemati HM, Fager C, Zirath H. High efficiency LDMOS current mode Class-D power amplifier at 1 GHz. *Proc. 36th Europ. Microwave Conf.* 2006;176−179.
14. Kazimierczuk MK. Class D voltage-switching MOSFET power amplifier. *IEE Proc. Electric Power Appl.* November 1991;138:285−296.
15. Granberg HO. Applying power MOSFETs in Class D/E RF power amplifier design. *RF Design*. June 1985;8:42−47.
16. Hung T-P, Metzger AG, Zampardi PJ, Iwamoto M, Asbeck PM. Design of high-efficiency current-mode Class-D amplifiers for wireless applications. *IEEE Trans. Microwave Theory Tech.* January 2005;MTT-53:144−151.

17. El-Hamamsy S-A. Design of high-efficiency RF Class-D power amplifier. *IEEE Trans. Power Electronics*. May 1994;PE-9:297–308.
18. Kobayashi H, Hinrichs JM, Asbeck PM. Current-mode Class-D power amplifiers for high-efficiency RF applications. *IEEE Trans. Microwave Theory Tech.* December 2001;MTT-49:2480–2485.
19. Kim J-Y, Han D-H, Kim J-H, Stapelton SP. A 50 W LDMOS current mode 1800 MHz Class-D power amplifier. *2005 IEEE MTT-S Int. Microwave Symp. Dig.* June 2005:1295–1298.
20. Tanany AA, Sayed A, Boeck G. A 2.14 GHz 50 Watt 60% power added efficiency GaN current mode class D power amplifier. *Proc. 38th Europ. Microwave Conf.* 2008;432–435.
21. Long A, Yao J, Long SI. A 13 W current mode class D high efficiency 1 GHz power amplifier. *45th Midwest Circuits and Systems Symp. Dig.* 2002;1:33–36.
22. Hung T-P, Rode J, Larson LE, Asbeck PM. Design of H-Bridge Class-D power amplifiers for digital pulse modulation transmitters. *IEEE Trans. Microwave Theory Tech.* December 2007;MTT-55:2845–2885.
23. Heck S, Schmidt M, Braeckle A. A switching-mode amplifier for Class-S transmitters for clock frequencies up to 7.5 GHz in 0.25 µm SiGe-BiCMOS. *2010 IEEE RFIC Symp. Dig.* May 2010:565–568.
24. Johnson T, Stapleton SP. RF Class-D power amplification with bandpass sigma-delta modulator drive signals. *IEEE Trans. Circuits and Systems − I: Regular Papers.* December 2006;CAS-I-53:2507–2520.
25. Braeckle A, Heck S, Berroth M, Dettmann I. Efficiency of current-mode Class-D amplifiers with delta-sigma modulated drive signals. *Proc. 2009 German Microwave Conf.* March 2009:1–4.
26. Johnson T, Sobot R, Stapleton S. CMOS RF Class-D power amplifier with bandpass sigma-delta modulation. *Microelectronics J.* March 2007;38:439–446.

# Class-F Power Amplifiers

## INTRODUCTION

Highly efficient operation of the power amplifier can be obtained by applying biharmonic or polyharmonic modes when an additional single-resonant or multi-resonant circuit tuned to the odd harmonics of the fundamental frequency is added into the load network. An infinite number of odd-harmonic resonators results in an idealized Class-F mode with a square voltage waveform and a half-sinusoidal current waveform at the device output terminal. In Class-F power amplifiers analyzed in the frequency domain, the fundamental and harmonic load impedances are optimized by short-circuit termination and open-circuit peaking to control the voltage and current waveforms at the device output in order to obtain maximum efficiency. In this chapter, different Class-F techniques using lumped and transmission-line elements including a quarterwave transmission line are analyzed. The effect of the saturation resistance and parasitic shunt capacitance is demonstrated. The design examples and practical RF and microwave Class-F power amplifiers are described and discussed.

## 3.1 Biharmonic and polyharmonic operation modes

The possibility of improving the efficiency of vacuum tube power amplifiers by a proper approximation of the anode voltage and current waveforms was discussed in the early 1920s [1]. It was then concluded that, theoretically, three-electrode tubes constitute converters of direct current to alternating current having generally unforeseen efficiencies. This means that, for different waveforms, anode efficiency will be different. For example, it may change from 50% for a purely sinusoidal anode current to 100% for rectangular pulses. Under operation with $\theta = \pi/2$ or $90°$, where $\theta$ corresponds to a half-conduction angle in a Class-B operation mode, the maximum theoretical anode efficiency achieves $\pi/4$ or 78.5%, characterized by the sinusoidal anode voltage waveform and half-sinusoidal current waveform.

In real practical power amplifier design, especially at high frequencies and low supply voltage, efficiency significantly degrades. Therefore, some design solutions contributing to efficiency improvement were published a long time ago with regard to the vacuum-tube power amplifiers and was based on the harmonic tuning at the anode of the active device. To understand the basic approach, let us evaluate the contribution of each harmonic component to an ideal half-sinusoidal current waveform and square voltage waveform by using a Fourier-series analysis. In this case, it is useful to calculate the partial Fourier series of current $i(\omega t)$ and voltage $v(\omega t)$ in normalized forms according to

$$\frac{i(\omega t)}{I_0} = 1 - \frac{\pi}{2}\sin \omega t - 2 \sum_{n=2,4,6,\ldots}^{N} \frac{\cos n \omega t}{n^2 - 1} \tag{3.1}$$

$$\frac{v(\omega t)}{V_0} = 1 + \frac{4}{\pi}\sin \omega t + \frac{4}{\pi} \sum_{n=3,5,7,\ldots}^{N} \frac{\sin n \omega t}{n} \tag{3.2}$$

where $I_0$ and $V_0$ are the dc current and voltage components, respectively.

Figure 3.1 shows that the shapes of the voltage and current waveforms can be significantly changed with increasing fundamental voltage amplitude by adding even one additional harmonic component with a proper phase. For example, the combination of the fundamental and third-harmonic components, being $180°$ out of phase at the center point, results in a flattened voltage waveform with depression in its center. It is clearly seen from Fig. 3.1($a$) that the proper ratio between the amplitudes of the fundamental and third-harmonic components can provide the flattened voltage waveform with minimum depression and maximum difference between its peak amplitude and the amplitude of the fundamental harmonic. Similarly, the combination of the fundamental and second-harmonic components, being in phase at the center point, flattens the current waveform corresponding to

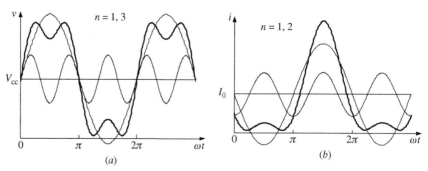

(a)          (b)

**FIGURE 3.1**

Fourier voltage and current waveforms with third and second harmonics.

the maximum values of the voltage waveform and sharpens the current waveform corresponding to the minimum values of the voltage waveform, as shown in Fig. 3.1(*b*). The optimum ratio between the amplitudes of the fundamental and second current harmonic components can maximize a peak value of the current waveform with its minimized value determined by the device saturation resistance in a practical circuit. Thus, power loss due to the active device can be minimized since the results of the integration over the period when minimum voltage corresponds to maximum current will give a small value compared with the power delivered to the load. In a common case, the same result can be achieved by adding the second harmonic into the voltage waveform and the third harmonic into the current waveform resulting in an inverse operation mode.

Ideally, the half-sinusoidal current waveform does not contain the third-harmonic component, as it follows from Eq. (3.1), because its third-harmonic Fourier current coefficient is equal to zero, that is $\gamma_3(\theta) = 0$. However, a load line analysis of a Class B power amplifier with sinusoidal output voltage waveform — under overdriven conditions when the device operates in pinch-off, active, and saturation modes during the period — shows that operation in the saturation mode is characterized by a depression in the output current waveform. From Fourier analysis it follows that such a current waveform includes the third harmonic component, which is 180° out of phase with the fundamental component at the point of symmetry of $\omega t = \pi/2$. Therefore, when an additional resonant circuit tuned to the third harmonic is included in the anode circuit operating in a saturation mode, the voltage drop with opposite phase will appear across this resonant circuit resulting in a similar depressed voltage waveform shown in Fig. 3.1(*a*) by the solid line. Hence, for the increased fundamental voltage amplitude, the output power at the fundamental frequency and anode efficiency can be increased for the same input drive. Physically, an efficiency improvement can be explained by the fact that fundamental voltage or current has negative values during some part of the period, corresponding to the negative power as an integration of a product of the instantaneous fundamental voltage and current. This means that the power loss on the active device is partly compensated by the reactive power provided by the harmonic resonator.

Adding the one or more higher-order harmonic components can further improve the voltage or current waveform. Figure 3.2(*a*) shows the voltage waveform with third- and fifth-harmonic peaking, which is close to an ideal rectangular waveform, while Fig. 3.2(*b*) shows the current waveform with second- and fourth-harmonic peaking, which is close to an ideal half-sinusoidal waveform.

Thus, based on the results of a Fourier analysis, it was concluded at that time that efficiency and output conditions may be changed materially if the load network is made to be responsive to certain harmonics, so that all input on these harmonics need not be absorbed as a tube loss. Figure 3.3(*a*) shows the type of load network with a third-harmonic trap which was discussed and first published in 1923 as a journal paper with regard to a vacuum-tube oscillator [2]. At the same time, it was experimentally found that applying the biharmonic driving

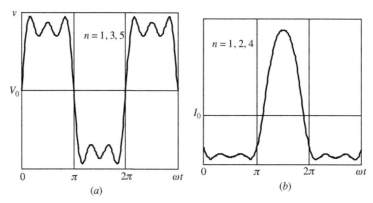

**FIGURE 3.2**

Fourier voltage and current voltage waveforms with three harmonics.

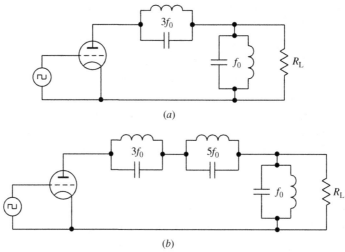

**FIGURE 3.3**

Biharmonic and polyharmonic power amplifiers with even-harmonic resonant tanks.

signal containing the fundamental and second-harmonic components produces the signal amplification more efficiently because of the much steeper driving waveform [3]. In this case, the resultant driving waveform consists of the fundamental and second-harmonic components being in phase at their maximum amplitudes, as shown in Fig. 3.1(*b*), with the amplitude of the second harmonic chosen as approximately one-quarter the amplitude of the fundamental. To maximize efficiency of the vacuum-tube amplifier, the use of a square voltage driving waveform and an additional resonator tuned to the fifth harmonic was also suggested, as shown in Fig. 3.3(b) [4]. The effect of the inclusion of the parallel resonant circuit tuned to the third-harmonic component and located in series to the anode was then described in a textbook [5]. Consequently, the basic theoretical background and potential circuit solutions to increase efficiency of vacuum tube power amplifiers and oscillators were generally understood in the 1920s.

Nevertheless, in next few decades, the basic operation mode for the vacuum tubes in power amplifiers of amplitude-modulated broadcasting radio transmitters was Class B with a resonant tank tuned to the fundamental. In this case, in order to improve efficiency, different transmitter architectures were proposed, based on Chireix and Doherty power amplifiers. However, during 1930s, some Russian papers stating an efficiency improvement of 25−30% in broadcasting radio transmitters by using a biharmonic mode for power amplifiers were published. It was shown that the symmetrical anode voltage waveform and minimum level of its depression for a biharmonic power amplifier shown in Fig. 3.3(a) can be provided with opposite phase conditions between the fundamental and third harmonic and optimum value of the ratio between their voltage amplitudes. Also, it was noted that high efficiency can be achieved even when impedance of the third-harmonic resonator is equal to or slightly greater than impedance of the fundamental tank circuit. In addition, such an approach can slightly improve the modulation properties of the power amplifier using either grid or anode modulation techniques because saturation for the fundamental frequency as a part of the flattened anode voltage waveform occurs later than that for the sinusoidal anode voltage [6].

In the comprehensive research given in [7], it was confirmed that the second- or third-harmonic voltages introduced into the anode or grid circuits in the proper phase improve the vacuum-tube performance. The proper phasing can be done by various means including an auxiliary tube, and the power output and overall efficiency of the main tube acting as a Class C power amplifier are increased. Fourth- and higher-order harmonics were, at best, of little value in improving the path of operation. Generally, however, in view of the parasitic anode−cathode capacitance comprising the interelectrode and case capacitances and series plate inductance, the entire anode circuit should be tuned to the third harmonic, not just a single resonator. For example, such an anode circuit may include a parallel third-harmonic resonator, which is slightly mistuned in this case, and an additional series $LC$ circuit connected in parallel to the tube, which has a capacitive reactance at the fundamental frequency and inductive reactance at the third-harmonic component tuned to the resonance with other elements in the anode circuit [8].

The load parallel resonator tuned to the third harmonic can be replaced by a low-pass filter with two series inductors and a shunt capacitor, as shown in Fig. 3.4(a), designed to pass the third harmonic of the fundamental frequency, terminating in a parallel resonant circuit tuned to the fundamental [9]. In this case, the ratio between the series inductor $L/2$ and shunt capacitor $C$ can be chosen to resonate the third-harmonic component $\omega = 3\omega_0 = \sqrt{2}/\sqrt{LC} = \omega_c/\sqrt{2}$, where $\omega_c = 2/\sqrt{LC}$ is the filter cutoff frequency, thus providing ideally infinite

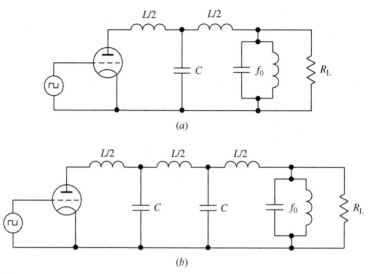

**FIGURE 3.4**

Biharmonic and polyharmonic vacuum-tube power amplifiers with low-pass filters.

impedance seen by the device anode at the third harmonic when it is assumed an infinite quality factor for the parallel fundamentally tuned resonant circuit. The parallel resonators tuned to the third and fifth harmonics can be replaced by a low-pass filter with the three sections, as shown in Fig. 3.4(b), whose elements are designed to pass the third and fifth harmonics of the fundamental frequency, terminating in a parallel resonant circuit tuned to the fundamental [9]. However, it is difficult to correctly specify the impedances at these harmonics seen by the anode circuit. This may result in a situation when a square-top anode voltage waveform cannot maintain its form in a circuit having inductive or capacitive reactance or both, even though, in the latter case, the reactive elements are so chosen that the circuit would be resonant for any one of the three frequencies including fundamental component. The reactances cause phase shifting of the component waves with consequent distortion of the resultant wave and loss of efficiency.

Figure 3.5 shows the biharmonic bipolar-transistor power amplifier with an additional resonant tank in the load network tuned to the third harmonic [10]. Since the third-harmonic current coefficient $\gamma_3(\theta)$ is positive for $\theta < 90°$, it is necessary to operate partly in a slightly saturated mode to obtain $\gamma_3(\theta) < 0$. Besides, higher efficiency can be achieved in a Class C with smaller angle $\theta$ that can easily be provided by the emitter $RC$ circuit. As a result, an output power of about 300 mW with a collector efficiency of 88% was achieved at the operating frequency of 106 kHz. It was then noted that an ideal solution requires the switch-mode amplifying circuit when the current flows through the device at minimum

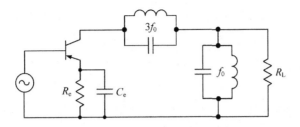

**FIGURE 3.5**

Biharmonic bipolar-transistor power amplifier.

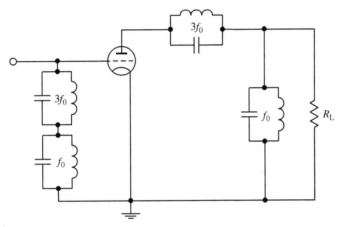

**FIGURE 3.6**

Biharmonic power amplifier with input harmonic control.

collector voltage (switch is turned on), but the current is zero at maximum collector voltage (switch is turned off).

In practice, the effective driving waveform may differ from the idealized square wave depending on what types of harmonic resonators are located in the load network [11]. For example, applying the biharmonic driving signal containing the fundamental and second-harmonic components produces the signal amplification more efficiently because of the much steeper driving waveform [3]. The detailed mathematical explanation of the effect of the biharmonic driving signal consisting of the fundamental and third-harmonic components is given in [12,13]. Figure 3.6 shows the simplified circuit of a vacuum-tube power amplifier containing the tank and third-harmonic resonant circuits, both located in grid and anode circuits.

Consider the effect of the biharmonic input signal on the MOSFET transistor, whose simplified equivalent circuit is represented by an ideal voltage-controlled current source with transconductance $g_m$ only. The biharmonic signal in the voltage form can be written as

$$v_g(\omega t) = V_g + V_{g1}(\cos \omega t - a_n \cos n \omega t) \tag{3.3}$$

where $V_g$ is the gate bias voltage and $a_n = V_{gn}/V_{g1}$ is the coefficient of the $n$th harmonic injection.

By substituting Eq. (3.3) into the idealized device piecewise-linear transfer characteristic, one can write for the output current

$$i(\omega t) = g_m(v_g - V_p) = I_0 + g_m V_{g1}(\cos \omega t - a_n \cos n \omega t) \tag{3.4}$$

where $I_0 = g_m(V_g - V_p)$ is the dc current and $V_p$ is the device pinch-off voltage. The output current $i(\omega t)$ takes a zero value when $\omega t = \theta$, where $\theta$ is one-half a conduction angle. Then,

$$I_0 = -g_m V_{g1}(\cos \theta - a_n \cos n\theta). \tag{3.5}$$

As a result, Eq. (3.4) can be rewritten in the form

$$i(\omega t) = g_m V_{g1}[\cos \omega t - \cos \theta - a_n(\cos n \omega t - \cos n\theta)]. \tag{3.6}$$

In this case, the Fourier current harmonic coefficients can be written as

$$\gamma_0'(\theta) = \frac{1}{\pi}\left[\sin \theta - \theta \cos \theta + a_n\left(\theta \cos n\theta - \frac{\sin n\theta}{n}\right)\right]$$

$$= \gamma_0(\theta) - \frac{a_n}{n}\gamma_0(n\theta) \tag{3.7}$$

$$\gamma_1'(\theta) = \frac{1}{\pi}\left[\theta - \frac{\sin 2\theta}{2} - na_n\left(\frac{\sin(n-1)\theta}{n-1} - \frac{\sin(n+1)\theta}{n+1}\right)\right]$$

$$= \gamma_1(\theta) - n^2 a_n \gamma_n(\theta) \tag{3.8}$$

$$\gamma_n'(\theta) = \frac{1}{\pi}\left[\frac{\sin(n-1)\theta}{n(n-1)} - \frac{\sin(n+1)\theta}{n(n+1)} - a_n\left(\theta - \frac{\sin 2n\theta}{2n}\right)\right]$$

$$= \gamma_n(\theta) - \frac{a_n}{n}\gamma_1(n\theta) \tag{3.9}$$

where $\gamma_n(\theta)$ are the current harmonic coefficients of an idealized Class-C power amplifier with a single-carrier driving signal.

To compare the operation modes with the output cosinusoidal voltage $v = V_{dd} - V\cos \omega t$ and biharmonic voltage $v = V_{dd} - V_1(\cos \omega t - a_n\cos n \omega t)$, where $a_n = V_n/V_1$ and $V_{dd}$ is the dc supply voltage, let us assume the conduction angles and maximum output current amplitudes are equal. The latter condition implies that the minimum values of the saturation voltages for both cases are equal too, appearing however at different time moments depending on the number of injected harmonic components. Then, the maximum drain voltage peak factor $\xi' = V_1/V_{dd}$ in a biharmonic mode can be obtained by [12]

$$\xi' = \frac{\xi}{\cos\dfrac{\pi}{2n}} \tag{3.10}$$

where $\xi = V/V_{dd}$ is the drain voltage peak factor corresponding to a cosinusoidal driving mode, when the voltage coefficient $a_n$ takes an optimum value

$$a_n^0 = \frac{1}{n}\sin\frac{\pi}{2n}. \tag{3.11}$$

Generally, the drain efficiency $\eta$ is written through the fundamental power $P_1$ and dc power $P_0$ in a single-carrier mode as

$$\eta = \frac{1}{2}\frac{P_1}{P_0} = \frac{1}{2}\frac{V}{V_{dd}}\frac{I_1}{I_0} = \frac{1}{2}\xi\frac{\gamma_1(\theta)}{\gamma_0(\theta)}. \tag{3.12}$$

By using Eqs (3.7) through (3.12), the drain efficiency $\eta'$ in a biharmonic mode can be calculated from

$$\eta' = \frac{1}{2}\frac{n^2}{\cos\dfrac{\pi}{2n}}\frac{\gamma_1(\theta) - n\,\gamma_n(\theta)\sin\dfrac{\pi}{2n}}{n^2\gamma_0(\theta) - \gamma_0(n\theta)\sin\dfrac{\pi}{2n}}\xi. \tag{3.13}$$

From Eq. (3.13), it follows that, in a Class-B bias condition with $\theta = 90°$, the drain efficiency in a biharmonic mode with a third-harmonic injection ($n = 3$) can achieve a value

$$\eta' = (0.85 \div 0.86)\xi \tag{3.14}$$

resulting in a maximum efficiency of $(85 \div 86)\%$ in an ideal case of zero saturation voltage when $\xi = 1$. Similar efficiency improvement of about 8% compared with a cosinusoidal driving signal was achieved for the case of third-harmonic injection with $a_3 = 0.14$ [12]. Moreover, the drain efficiency $\eta'$ can be further improved by optimizing the conduction angle and voltage coefficient. For example, Eq. (3.14) can be rewritten as $\eta' = (0.95 \div 0.96)\xi$ for optimum values $\theta_{opt} \approx 63°$ and $a_n^0 \approx 0.205$ [12].

However, it is difficult to achieve an optimum value of the voltage coefficient $a_n$ in real practical conditions because its value significantly depends on the ratio between resonant circuit equivalent resistance $R_n$ at the $n$th harmonic component and the equivalent resistance $R_1$ at fundamental frequency according to

$$a_n = \frac{V_n}{V_1} = \frac{I_nR_n}{I_1R_1} = \frac{\gamma_n'(\theta)}{\gamma_1'(\theta)}\frac{R_n}{R_1}. \tag{3.15}$$

From Eq. (3.15), it follows that, for a certain value of $n$ and an optimum value of $\theta$, the condition $a_n = a_n^0$ is satisfied with more accuracy for a greater ratio of $R_n/R_1$. As a rule of thumb, it is sufficient to choose their ratio equal or greater than 10. In this case, the worsening of the drain efficiency $\eta'$ will be equal to or less than 0.5%.

In previous input harmonic-control analysis, it was assumed that the active device represents an ideal voltage-controlled current source, which is a good approximation at sufficiently low frequencies for vacuum tubes or MOSFET devices. However, when using a high-frequency bipolar, MESFET, or HEMT transistor as an active element, it is necessary to take into account the significant nonlinearity of the device input capacitance. For example, the voltage-dependent input gate-source capacitance of a GaAs MESFET device can be modeled as a junction capacitance creating higher-order harmonics at the device input, each having a different phase. In this case, the proper relationships between amplitudes and phases of the harmonics at the transistor input are needed to approximate the required symmetrical drain voltage and current waveforms [14]. The pHEMT device, in contrast to a conventional MESFET device, shows a very steep gradient in its gate-source capacitance *versus* gate-bias voltage. As a result, the efficiency increase up to 81% can be achieved for a microwave power amplifier with ideal Class-F load network by input termination of all harmonics when the input signal is purely sinusoidal compared with a non-terminated case with severely distorted input voltage waveform [15]. This situation appears as the simplest and most effective alternative to multi-resonant input circuit, which is difficult to realize at very high frequencies, taking into account the active device parasitics.

Generally, the biharmonic driving signal for the final stage is formed in preceding amplifying stages by the proper harmonic tuning. However, if the driving signal source represents a sine-wave output voltage, the formation of the biharmonic driving signal can be done by subtracting the voltage of the harmonic component used in the anode circuit from the sinusoidal input voltage. Figure 3.7(*a*) shows the biharmonic vacuum-tube power amplifier, in which a low-reactance resonator compared with the anode resonator but tuned to the same third-harmonic frequency component is inserted into the cathode circuit [11]. The high degree of negative feedback developed makes it impossible for any significant third-harmonic current to develop, no matter what grid-voltage waveform is employed. In practice, with a sine wave input correctly related in amplitude with the grid bias, the relative voltages developed between grid and cathode are self-adjusting to provide the flattened voltage waveforms.

Unfortunately, it is not always easy to realize the proper amplitude conditions because of the finite losses associated with two third-harmonic resonators, since both resonators affect the voltage waveforms in the grid and anode circuits simultaneously. For example, the tubes with variable, weak, and small grid current characteristics cannot be used for such a biharmonic configuration. The effect of this problem can be significantly minimized if the resonant circuit in the cathode circuit tuned to the second harmonic together with a third-harmonic

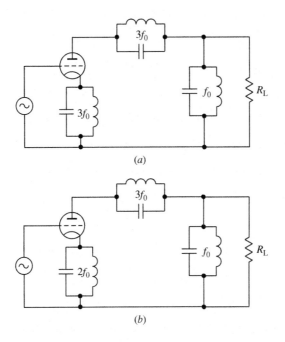

(a)

(b)

**FIGURE 3.7**

Biharmonic power amplifiers with cathode harmonic control.

resonant circuit in the anode circuit is used, as shown in Fig. 3.7(b), resulting in the input biharmonic voltage $v_g = V_{g1}\cos \omega t - V_2\cos 2\omega t$ and output voltage $v = V_1\cos \omega t + V_2\cos 2\omega t - V_3\cos 3\omega t$ [12]. In this case, the maximum available anode efficiency can reach values of 85–88%.

## 3.2 Idealized Class-F mode

Generally, an infinite number of odd-harmonic tank resonators can maintain a square voltage waveform with a half-sinusoidal current waveform at the device output terminal. Figure 3.8(a) shows such a Class-F power amplifier with a multiple-resonator output filter to control the harmonic content of its collector (anode or drain) voltage and/or current waveforms, thereby shaping them to reduce dissipation and to increase efficiency [16,17].

To simplify an analysis of a Class-F power amplifier, a simple equivalent circuit of which is shown in Fig. 3.8(b), the following several assumptions are introduced:

- The transistor has zero saturation voltage, zero saturation resistance, infinite off-resistance, and its switching action is instantaneous and lossless.
- The RF choke allows only a dc current and has no resistance.

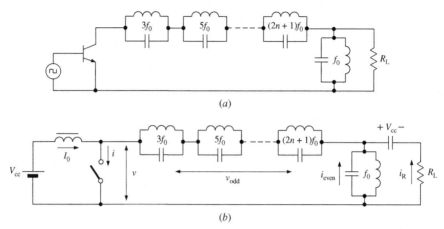

**FIGURE 3.8**

Basic circuits of Class-F power amplifier with parallel resonant circuits.

- Quality factors of all parallel resonant circuits have infinite impedance at the corresponding harmonic and zero impedance at other harmonics.
- There are no losses in the circuit except only into the load $R_L$.
- Operation mode with a 50% duty ratio.

To determine the idealized collector voltage and current waveforms, let us consider the distribution of voltages and currents in the load network assuming the sinusoidal fundamental current flowing into the load as $i_R(\omega t) = I_R \sin(\omega t)$, where $I_R$ is its amplitude. The voltage $v(\omega t)$ across the switch can be represented as a sum of the dc voltage $V_{cc}$, fundamental voltage $v_R = i_R R_L$ across the load resistor, and voltage $v_{odd}$ across the odd-harmonic resonators, written as

$$v(\omega t) = V_{cc} + v_{odd}[(2n + 1)\,\omega t] + v_R(\omega t). \tag{3.16}$$

Since the time moment $t$ was chosen arbitrarily, by introducing a phase shift of $\pi$, Eq. (3.16) can be rewritten for periodical sinusoidal functions as

$$v(\omega t + \pi) = V_{cc} - v_{odd}[(2n + 1)\,\omega t] - v_R(\omega t). \tag{3.17}$$

Then, the summation of Eq. (3.16) and Eq. (3.17) yields

$$v(\omega t) = 2V_{cc} - v(\omega t + \pi). \tag{3.18}$$

From Eq. (3.18), it follows that the maximum value of the collector voltage cannot exceed $2V_{cc}$ and the time duration with maximum voltage of $v = 2V_{cc}$ coincides with the time duration with minimum voltage of $v = 0$. Since the collector voltage is zero when the switch is turned on, the only possible waveform for the collector voltage is a square wave composed of only dc, fundamental-frequency, and odd-harmonic components.

During the interval $0 < \omega t \leq \pi$ when the switch is turned on, the current $i(\omega t)$ flowing through the switch can be written as

$$i(\omega t) = I_0 + i_{even}(2n\omega t) + i_R(\omega t) \tag{3.19}$$

whereas, during the interval $\pi < \omega t \leq 2\pi$ when the switch is turned off, the current $i(\omega t + \pi)$ is equal to zero resulting in

$$0 = I_0 + i_{even}(2n\omega t) - i_R(\omega t). \tag{3.20}$$

Then, by substituting Eq. (3.20) into Eq. (3.19), one can write

$$i(\omega t) = 2i_R(\omega t) = 2I_R \sin(\omega t) \tag{3.21}$$

from which it follows that the amplitude of the current flowing through the switch during interval $0 < \omega t \leq \pi$ is two times greater than the amplitude of the fundamental current. Thus, in a common case, Eq. (3.19) can be rewritten as

$$i(\omega t) = I_R(\sin \omega t + |\sin \omega t|) \tag{3.22}$$

which means that the switch current represents half-sinusoidal pulses with amplitude equal to double the load current amplitude.

Consequently, for a purely sinusoidal current flowing into the load shown in Fig. 3.9(a), the ideal collector voltage and current waveforms can be represented by the appropriate normalized waveforms shown in Fig. 3.9(b) and 3.9(c), respectively. Here, a sum of odd harmonics approximates a square voltage waveform, and a sum of the fundamental and even harmonics approximates a half-sinusoidal collector current waveform. As a result, the shapes of the collector current and voltage waveforms provide a condition when the current and voltage do not overlap simultaneously. Such a condition, with symmetrical collector voltage and current waveforms, corresponds to an idealized Class-F operation mode with 100% collector efficiency.

A Fourier series analysis of the current and voltage waveforms allows us to obtain the equations for the dc current and fundamental voltage and current components in the collector voltage and current waveforms: the dc current $I_0$ can be calculated from Eq. (3.22) as

$$I_0 = \frac{1}{2\pi} \int_0^\pi 2I_R \sin \omega t \, d\omega t = \frac{2I_R}{\pi} \tag{3.23}$$

the fundamental current component can be calculated from Eq. (3.22) as

$$I_1 = \frac{1}{\pi} \int_0^\pi 2I_R \sin^2 \omega t \, d\omega t = I_R \tag{3.24}$$

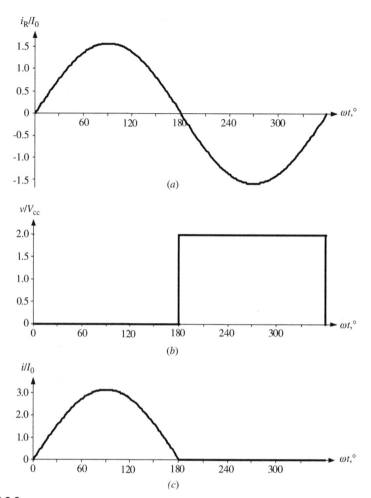

**FIGURE 3.9**

Ideal waveforms of Class-F power amplifiers.

and the fundamental voltage component can be calculated using Eq. (3.18) as

$$V_1 = V_R = \frac{1}{\pi} \int\limits_{\pi}^{2\pi} 2V_{cc} \sin(\omega t + \pi) d\omega t = \frac{4V_{cc}}{\pi}. \tag{3.25}$$

Then, the dc power and output power at the fundamental frequency are calculated from

$$P_0 = V_{cc}I_0 = \frac{2V_{cc}I_R}{\pi} \tag{3.26}$$

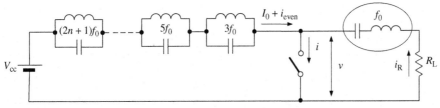

**FIGURE 3.10**

Equivalent circuit of Class-F power amplifier with series filter.

and

$$P_1 = \frac{V_1 I_1}{2} = \frac{2V_{cc}I_R}{\pi} \qquad (3.27)$$

resulting in a theoretical collector efficiency with maximum value

$$\eta = \frac{P_1}{P_0} = 100\%. \qquad (3.28)$$

The impedance conditions seen by the device collector for idealized Class-F mode are obtained by

$$Z_1 = R_1 = \frac{8}{\pi^2}\frac{V_{cc}}{I_0} \qquad (3.29)$$

$$Z_{2n} = 0 \quad \text{for even harmonics} \qquad (3.30)$$

$$Z_{2n+1} = \infty \quad \text{for odd harmonics} \qquad (3.31)$$

which are similar to that derived from the limiting case of the optimum efficiency Class-B mode [18].

The collector square-voltage and half-sinusoidal current waveforms can be similarly obtained by using a multiple-resonator circuit shown in Fig. 3.10, where the parallel resonators tuned to the third-, fifth-, and higher-order odd harmonics are located between the voltage supply and collector, while the series resonant circuit tuned to the fundamental frequency $f_0$ provides the sinusoidal current flowing into the load.

## 3.3 **Class-F with maximally flat waveforms**

Although it is impossible to realize the ideal harmonic-impedance conditions in real practice, the peaking of at least several current and voltage harmonic components should be provided to achieve high-efficiency operation of the power amplifier. The more the voltage waveform provided by higher-order harmonic components can be flattened, the less power dissipation due to flowing of the output current—when output voltage is extremely small—occurs. To understand common design principles and to numerically calculate power-amplifier efficiency according to the appropriate number of the frequency harmonic components of voltage and current waveforms, it is useful to introduce a design technique applied to Class-F approximation with maximally flat waveforms [19]. The output network is assumed ideal to deliver only the fundamental-frequency power to the load without loss. The active device represents an ideal current source with zero saturation voltage and output capacitance. Flattening of the voltage and current waveforms to realize Class-F operation can be accomplished by using odd-harmonic components to approximate a square voltage waveform, and even-harmonic components to approximate a half-sinusoidal current waveform given by

$$v(\omega t) = V_{cc} + V_1 \sin \omega t + \sum_{n=3,5,7,\dots}^{\infty} V_n \sin n \, \omega t \tag{3.32}$$

$$i(\omega t) = I_0 - I_1 \sin \omega t - \sum_{n=2,4,6,\dots}^{\infty} I_n \cos n \, \omega t. \tag{3.33}$$

For the symmetrical flattened voltage waveforms shown in Fig. 3.11, the medium points where the voltage waveform reaches its maximum and minimum values are at $\omega t = \pi/2$ and $\omega t = 3\pi/2$, respectively. Maximum flatness at the minimum voltage requires the even derivatives to be zero at $\omega t = 3\pi/2$. Since the odd-order derivatives are equal to zero because $\cos(n\pi/2) = 0$ for odd $n$, it is necessary to define the even-order derivatives of the voltage waveform given by Eq. (3.32). As a result, for a voltage spectrum including odd-frequency components up to seventh component, the second, fourth, and sixth derivatives are

$$\frac{d^2v}{d(\omega t)^2} = -V_1\sin \omega t - 9V_3\sin3\omega t - 25V_5\sin5\omega t - 49V_7\sin7\omega t \tag{3.34}$$

$$\frac{d^4v}{d(\omega t)^4} = V_1\sin \omega t + 81V_3\sin3\omega t + 625V_5\sin5\omega t + 2401V_7\sin7\omega t \tag{3.35}$$

$$\frac{d^6v}{d(\omega t)^6} = -V_1\sin \omega t - 729V_3\sin3\omega t - 15625V_5\sin5\omega t - 117649V_7\sin7\omega t. \tag{3.36}$$

At the minimum points, these derivatives must be equal to zero,

$$\left.\frac{d^2v}{d(\omega t)^2}\right|_{\omega t=\frac{3}{2}\pi} = \left.\frac{d^4v}{d(\omega t)^4}\right|_{\omega t=\frac{3}{2}\pi} = \left.\frac{d^6v}{d(\omega t)^6}\right|_{\omega t=\frac{3}{2}\pi} = 0. \tag{3.37}$$

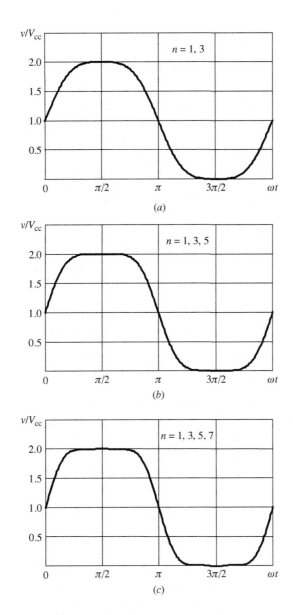

**FIGURE 3.11**

Voltage waveforms for $n$th-harmonic peaking.

Consequently, a system of three equations to calculate the odd-harmonic voltage amplitudes ($V_3$, $V_5$, and $V_7$) through the dc component $V_{cc}$ and fundamental voltage amplitude $V_1$ can be written as

$$V_1 - 9V_3 + 25V_5 - 49V_7 = 0 \qquad (3.38)$$

$$V_1 - 81V_3 + 625V_5 - 2401V_7 = 0 \tag{3.39}$$

$$V_1 - 729V_3 + 15625V_5 - 117649V_7 = 0. \tag{3.40}$$

For the third-harmonic peaking when only the third-harmonic component together with the fundamental one is present (assuming $V_5 = V_7 = 0$), Eq. (3.38) gives

$$V_3 = \frac{1}{9}V_1. \tag{3.41}$$

Then, from Eq. (3.32) at $\omega t = 3\pi/2$ when $v(\omega t) = 0$, the optimum amplitudes of the first and third harmonics are defined by

$$V_1 = \frac{9}{8}V_{cc} \qquad V_3 = \frac{1}{8}V_{cc}. \tag{3.42}$$

For the fifth-harmonic peaking with third- and fifth-harmonic components (assuming $V_7 = 0$), simultaneous solution of Eqs (3.38) and (3.39) yields

$$V_3 = \frac{1}{6}V_1 \qquad V_5 = \frac{1}{50}V_1 \tag{3.43}$$

and, consequently,

$$V_1 = \frac{75}{64}V_{cc} \qquad V_3 = \frac{25}{128}V_{cc} \qquad V_5 = \frac{3}{128}V_{cc}. \tag{3.44}$$

Similarly, for the seventh-harmonic peaking with third-, fifth-, and seventh-harmonic components, simultaneous solution of Eqs (3.38) to (3.40) yields

$$V_3 = \frac{1}{5}V_1 \qquad V_5 = \frac{1}{25}V_1 \qquad V_7 = \frac{1}{245}V_1 \tag{3.45}$$

resulting for zero collector voltages in

$$V_1 = \frac{1225}{1024}V_{cc} \qquad V_3 = \frac{245}{1024}V_{cc}$$
$$V_5 = \frac{49}{1024}V_{cc} \qquad V_7 = \frac{5}{1024}V_{cc} \tag{3.46}$$

The voltage waveforms for the third-harmonic peaking ($n = 1, 3$), fifth-harmonic peaking ($n = 1, 3, 5$), and seventh-harmonic peaking ($n = 1, 3, 5, 7$) are shown in Fig. 3.11.

For the symmetrical current waveforms shown in Fig. 3.12, the medium points where the current waveform reaches its minimum and maximum values are at $\omega t = \pi/2$ and $\omega t = 3\pi/2$, respectively. Since the odd-order derivatives are equal to zero because $\cos(\pi/2) = 0$ and $\sin(n\pi/2) = 0$ for even $n$, it is sufficient to determine the even-order derivatives of the current waveform given by Eq. (3.33). Maximum

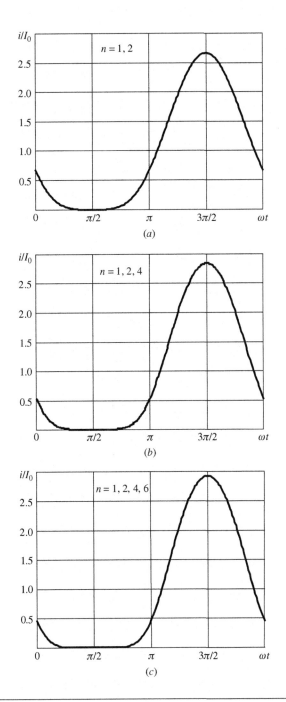

**FIGURE 3.12**

Current waveforms for $n$th-harmonic peaking.

flatness at the minimum current requires the even derivatives to be zero at $\omega t = \pi/2$. As a result, for a current spectrum including even-harmonic components up to the sixth component, the second, fourth, and sixth derivatives of the current waveform are

$$\frac{d^2 i}{d(\omega t)^2} = I_1 \sin \omega t + 4I_2 \cos 2\omega t + 16I_4 \cos 4\omega t + 36I_6 \cos 6\omega t \qquad (3.47)$$

$$\frac{d^4 i}{d(\omega t)^4} = -I_1 \sin \omega t - 16I_2 \cos 2\omega t - 256I_4 \cos 4\omega t - 1296I_6 \cos 6\omega t \qquad (3.48)$$

$$\frac{d^6 i}{d(\omega t)^6} = I_1 \sin \omega t + 64I_2 \cos 2\omega t + 4096I_4 \cos 4\omega t + 46656I_6 \cos 6\omega t. \qquad (3.49)$$

At the minimum points, these derivatives must be equal to zero,

$$\left. \frac{d^2 i}{d(\omega t)^2} \right|_{\omega t = \frac{\pi}{2}} = \left. \frac{d^4 i}{d(\omega t)^4} \right|_{\omega t = \frac{\pi}{2}} = \left. \frac{d^6 i}{d(\omega t)^6} \right|_{\omega t = \frac{\pi}{2}} = 0. \qquad (3.50)$$

Hence, a system of three equations to calculate the even-harmonic current amplitudes ($I_2$, $I_4$ and $I_6$) through the dc component $I_0$ and fundamental current amplitude $I_1$ can be written as

$$I_1 - 4I_2 + 16I_4 - 36I_6 = 0 \qquad (3.51)$$

$$I_1 - 16I_2 + 256I_4 - 1296I_6 = 0 \qquad (3.52)$$

$$I_1 - 64I_2 + 4096I_4 - 46656I_6 = 0. \qquad (3.53)$$

For the second-harmonic peaking when only the second harmonic component together with the fundamental one is present (assuming $I_4 = I_6 = 0$), Eq. (3.51) gives

$$I_2 = \frac{1}{4}I_1. \qquad (3.54)$$

Then, from Eq. (3.51) at $\omega t = \pi/2$ when $i(\omega t) = 0$, the optimum amplitudes are equal to

$$I_1 = \frac{4}{3}I_0 \qquad I_2 = \frac{1}{3}I_0. \qquad (3.55)$$

For the fourth-harmonic peaking with second and fourth harmonic components (assuming $I_6 = 0$), simultaneous solution of Eqs (3.51) and (3.52) yields

$$I_2 = \frac{5}{16}I_1 \qquad I_4 = \frac{1}{64}I_1 \qquad (3.56)$$

and, consequently,

$$I_1 = \frac{64}{45}I_0 \qquad I_2 = \frac{4}{9}I_0 \qquad I_4 = \frac{1}{45}I_0. \qquad (3.57)$$

Similarly, for the sixth-harmonic peaking with second, fourth, and sixth harmonic components, simultaneous solution of Eqs (3.51) to (3.53) yields

$$I_2 = \frac{175}{512} I_1 \qquad I_4 = \frac{7}{256} I_1 \qquad I_6 = \frac{1}{512} I_1 \qquad (3.58)$$

resulting for zero collector current in

$$I_1 = \frac{256}{175} I_0 \qquad I_2 = \frac{1}{2} I_0$$

$$I_4 = \frac{1}{25} I_0 \qquad I_6 = \frac{1}{350} I_0. \qquad (3.59)$$

The current waveforms for the second-harmonic peaking ($n = 1, 2$), fourth-harmonic peaking ($n = 1, 2, 4$), and sixth-harmonic peaking ($n = 1, 2, 4, 6$) are shown in Fig. 3.12.

The effectiveness of the operation modes with different voltage and current harmonic peaking can be compared by calculating the collector (drain) efficiency $\eta$ of each operation mode in accordance with

$$\eta = \frac{P_1}{P_0} = \frac{1}{2} \frac{V_1}{V_{cc}} \frac{I_1}{I_0}. \qquad (3.60)$$

The resultant efficiencies for various combinations of voltage and current harmonic components, as presented in Table 3.1, show that the efficiency increases with an increase in the number of voltage and current harmonic components. To increase efficiency, it is more desirable to provide harmonic peaking in consecutive numerical order − both for voltage and current harmonic components − than to increase the number of the harmonic components into only voltage or current waveforms. Class-F operation becomes mostly effective in comparison with Class-B operation if at least third voltage harmonic peaking and fourth current harmonic peaking are realized. An inclusion of fifth voltage harmonic component increases the efficiency to 83.3%. An additional inclusion of sixth-harmonic component into the current waveform and seventh-harmonic component into the voltage waveform leads to efficiencies up to 94%.

Maximum efficiency for a given set of harmonics can be additionally increased if first the fundamental-frequency amplitude is fixed and then the amplitude of the harmonics is adjusted to minimize the downward excursion of the overall waveform [20]. Fixing the waveform minimum at zero gives the minimum supply voltage needed for full output power, thus maximizing efficiency. As a result, the maximum efficiency can be generally improved by 6−8% compared to those for maximally flat waveforms. Analytically, the

**Table 3.1** Resultant Efficiencies for Various Combinations of Voltage and Current Harmonic Components

| Current Harmonic Components | Voltage Harmonic Components | | | | |
|---|---|---|---|---|---|
| | 1 | 1, 3 | 1, 3, 5 | 1, 3, 5, 7 | 1, 3, 5, …, ∞ |
| 1 | 1/2 = 0.500 | 9/16 = 0.563 | 75/128 = 0.586 | 1225/2048 = 0.598 | 2/π = 0.637 |
| 1, 2 | 2/3 = 0.667 | 3/4 = 0.750 | 25/32 = 0.781 | 1225/1536 = 0.798 | 8/3π = 0.849 |
| 1, 2, 4 | 32/45 = 0.711 | 4/5 = 0.800 | 5/6 = 0.833 | 245/288 = 0.851 | 128/45π = 0.905 |
| 1, 2, 4, 6 | 128/175 = 0.731 | 144/175 = 0.823 | 6/7 = 0.857 | 7/8 = 0.875 | 512/175π = 0.931 |
| 1, 2, 4, …, ∞ | π/4 = 0.785 | 9π/32 = 0.884 | 75π/256 = 0.920 | 1225π/4096 = 0.940 | 1 = 1.000 |

theoretical maximum efficiency depending only upon the terminating conditions can be written as

$$\eta = \frac{\pi}{2(m+1)} \cot\left[\frac{\pi}{2(m+1)}\right] \qquad (3.61)$$

where $m$ is the number of controlled harmonic components [21]. Moreover, when higher-order harmonic components are terminated in finite reactive impedances, the same maximum efficiency is obtained with the correct complex termination at the fundamental frequency.

Most Class-F power amplifiers employ a conduction angle of 180° when all of the odd harmonics of a half-sinusoidal current waveform are nulled, as are all of the even harmonics of a rectangular voltage waveform. However, conduction angles other than 180° generally cause all harmonics to be present. In this case, a given harmonic is nulled at a specific conduction angle, but all or most others remain. Consequently, flattening of the voltage waveform during the time of conduction must be accomplished by the addition of a single harmonic [22]. For example, for a conduction angle of approximately 130°, the use of the fourth harmonic increases efficiency by about 6%, compared with that for a conventional Class-C power amplifier.

## 3.4 Class-F with quarterwave transmission line

Ideally, a control of an infinite number of the harmonics maintaining a square voltage waveform and a half-sinusoidal current waveform at the drain can be provided using a serious quarterwave transmission line and a parallel-tuned resonant circuit, as shown in Fig. 3.13. This type of a Class-F power amplifier was initially proposed to be used at higher frequencies where implementation of the load networks with only lumped elements is difficult and the parasitic device output

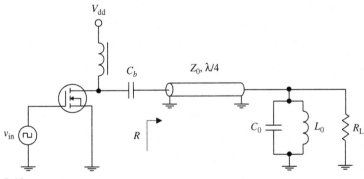

**FIGURE 3.13**

Class-F power amplifier with series quarterwave transmission line.

(bondwire or package lead) inductance is sufficiently small [16,17]. In this case, the quarterwave transmission line transforms the load impedance as

$$R = \frac{Z_0^2}{R_L} \tag{3.62}$$

where $Z_0$ is the characteristic impedance of a transmission line [23]. For even harmonics, the short circuit on the load side of the transmission line is repeated, thus producing a short circuit at the drain. However, the short circuit at the load produces an open circuit at the drain for odd harmonics with resistive load at the fundamental.

Generally, at low drive level, the active device acts as a current source (voltage-controlled in the case of the MOSFETs or MESFETs and current-controlled in the case of bipolar transistors). As input drive increases, the active device enters saturation resulting in a harmonic-generation process. Since the quarterwave transmission line presents the high impedance conditions to all odd harmonics, all odd harmonics provide a proper contribution to the output-voltage waveform. As a result, at high drive level, the output-voltage waveform becomes a complete square wave and the active device is saturated for a full half-cycle. In this case, the transistor acts as a switch rather than a saturating current source.

An alternative configuration of the Class-F power amplifier with a shunt quarterwave transmission line located between the dc power supply and device collector is shown in Fig. 3.14(a). In this case, there is no need to use an RF choke and a series blocking capacitor because a series fundamentally tuned resonant circuit is used instead of a parallel fundamentally tuned resonant circuit. However, unlike the case with a series quarterwave transmission line, such a Class-F load network configuration with a shunt quarterwave transmission line does not provide an impedance transformation. Therefore, the load resistance $R$, which is equal to the active device equivalent output resistance at the fundamental, must then be transformed to the standard load impedance. Let us now derive analytically some basic fundamental properties of a quarterwave transmission line. The transmission line in the time domain can be represented as an element with finite delay time depending on its electrical length. Consider a simplified load network of the Class-F power amplifier shown in Fig. 3.14(b), which consists of a parallel quarterwave transmission line grounded at the end through power supply, a series fundamentally tuned $L_0 C_0$ circuit, and a load resistance $R$. In an idealized case, the intrinsic device output capacitance is assumed to have negligible effect on the power amplifier RF performance. The loaded quality factor $Q_L$ of the series resonant $L_0 C_0$ circuit is high enough to provide the sinusoidal output current $i_R$ flowing into the load $R$.

To define the collector voltage and current waveforms, consider the electrical behavior of a homogeneous lossless quarterwave transmission line connected to the dc voltage supply with RF grounding [24]. In this case, the voltage $v(t, x)$ in any cross-section of such a transmission line can be represented as a sum of the

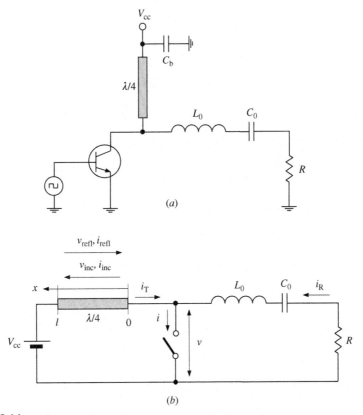

**FIGURE 3.14**

Class-F power amplifier with a shunt quarterwave transmission line.

incident and reflected voltages, $v_{inc}(\omega t - 2\pi x/\lambda)$ and $v_{refl}(\omega t + 2\pi x/\lambda)$, generally with an arbitrary waveform. When $x = 0$, the voltage $v(t, x)$ is equal to the collector voltage

$$v(\omega t) = v(t, 0) = v_{inc}(\omega t) + v_{refl}(\omega t).\qquad(3.63)$$

At the same time, at another end of the transmission line when $x = \lambda/4$, the voltage is constant and equal to

$$V_{cc} = v(t, \pi/2) = v_{inc}(\omega t - \pi/2) + v_{refl}(\omega t + \pi/2).\qquad(3.64)$$

Since the time moment $t$ was chosen arbitrarily, let us rewrite Eq. (3.64) using a phase shift of $\pi/2$ for each voltage by

$$v_{inc}(\omega t) = V_{cc} - v_{refl}(\omega t + \pi).\qquad(3.65)$$

Substituting Eq. (3.65) into Eq. (3.63) yields

$$v(\omega t) = v_{refl}(\omega t) - v_{refl}(\omega t + \pi) + V_{cc}. \tag{3.66}$$

Consequently, for the phase shift of $\pi$, the collector voltage can be obtained by

$$v(\omega t + \pi) = v_{refl}(\omega t + \pi) - v_{refl}(\omega t + 2\pi) + V_{cc}. \tag{3.67}$$

For an idealized operation condition with a 50% duty ratio (or cycle) when during half a period the transistor is turned on and during another half a period the transistor is turned off with an overall period of $2\pi$, the voltage $v_{refl}(\omega t)$ can be considered the periodical function with a period of $2\pi$,

$$v_{refl}(\omega t) = v_{refl}(\omega t + 2\pi). \tag{3.68}$$

As a result, the summation of Eqs (3.66) and (3.67) results in the expression for collector voltage in the form

$$v(\omega t) = 2V_{cc} - v(\omega t + \pi). \tag{3.69}$$

From Eq. (3.69), it follows that the maximum value of the collector voltage cannot exceed a value of $2V_{cc}$, and the time duration with maximum voltage of $v = 2V_{cc}$ coincides with the time duration with minimum voltage of $v = 0$.

Similarly, the equation for the current $i_T$ flowing into the quarterwave transmission line can be obtained by

$$i_T(\omega t) = i_T(\omega t + \pi) \tag{3.70}$$

which means that the period of a signal flowing into the quarterwave transmission line is equal to $\pi$ because it contains only even harmonics, since a quarterwave transmission line has an infinite impedance at odd harmonics.

Let the transistor operate as an ideal switch when it is turned on during the interval $0 < \omega t \leq \pi$ where $v = 0$ and turned off during the interval $\pi < \omega t \leq 2\pi$ where $v = 2V_{cc}$ according to Eq. (3.69). During the interval $\pi < \omega t \leq 2\pi$ when the switch is turned off, the load is connected directly to the transmission line and $i_T = -i_R = -I_R \sin \omega t$. Consequently, during the interval $0 < \omega t \leq \pi$ when the switch is turned on, $i_T = I_R \sin \omega t$ according to Eq. (3.70). Hence, the current flowing into the quarterwave transmission line at any $\omega t$ can be represented by

$$i_T(\omega t) = I_R|\sin \omega t| \tag{3.71}$$

where $I_R$ is the amplitude of current flowing into the load.

Since the collector current is defined as $i = i_T + i_R$, then

$$i(\omega t) = I_R(\sin \omega t + |\sin \omega t|) \tag{3.72}$$

which means that the collector current represents half-sinusoidal pulses with amplitude equal to double load-current amplitude.

Consequently, for a purely sinusoidal current flowing into the load due to the infinite loading quality factors of the series the fundamentally tuned $L_0C_0$ circuit

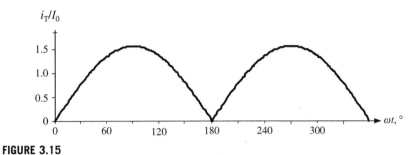

**FIGURE 3.15**

Ideal current waveform in quarterwave transmission line.

shown in Fig. 3.9(*a*), the ideal collector voltage and current waveforms can be represented by the corresponding normalized square and half-sinusoidal waveforms shown in Figs. 3.9(*b*) and 3.9(*c*), respectively, where $I_0$ is the dc current. Here, a sum of odd harmonics approximates a square voltage waveform, and a sum of the fundamental and even harmonics approximates a half-sinusoidal collector current waveform. The waveform corresponding to the normalized current flowing into the quarterwave transmission line shown in Fig. 3.15 represents a sum of even harmonics. As a result, the shapes of the collector current and voltage waveforms provide a condition where the current and voltage do not overlap simultaneously.

Figure 3.16 shows a practical example of the circuit schematic of a Class-F power amplifier with a shunt quarterwave line based on a 28-V 10-W Cree GaN HEMT power transistor CGH40010. To better illustrate the drain voltage and current wave-forms with minimum effect of the device parasitic output parameters, a sufficiently low operating frequency of 100 MHz was chosen. In this case, a simple lossy *RL* input shunt network is used to match the device input impedance to a 50-Ω source and to compensate for the device input gate-source capacitance $C_{in}$ of about 5 pF at the fundamental frequency that resulted in a small-signal $S_{11}$ better than −25 dB. The series 50-Ω resistor connected to the device gate is additionally included to provide unconditional operation stability. The simulated drain voltage close to a square wave-form and drain current close to a half-sinusoidal waveform are shown in Fig. 3.17(*a*), where small waveform ripples (minimized by load network parameter optimization) can be explained due to some effect of the device output drain-source capacitance $C_{out}$ of about 1.3 pF and package parasitics at higher-order harmonic components. As a result, a maximum drain efficiency of 84.7% with a power gain of 20 dB and an output power of 40 dBm at a supply voltage of 28 V were obtained with a sine-wave driving signal, as shown in Fig. 3.17(*b*). In this case, to better approximate the switchmode operation of a 10-W GaN HEMT device, it is necessary to slightly increase the input signal amplitude to operate in a saturated mode characterized by more than 3-dB gain compression point. Due to the diode-based nonlinearity of the device input circuit, the amplitude harmonic ratio between the fundamental-frequency, second-, and third-harmonic components at the gate terminal is 8.2:1.3:1.

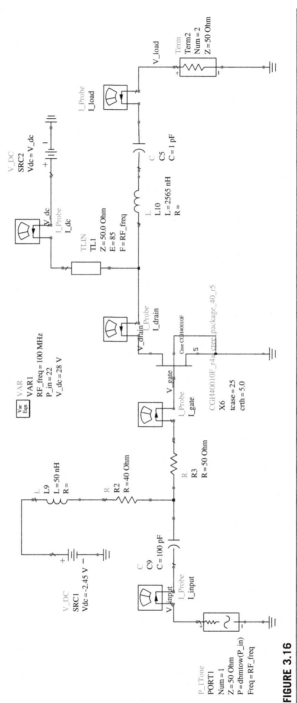

**FIGURE 3.16**

Circuit schematic of a Class-F GaN HEMT power amplifier with shunt quarterwave line.

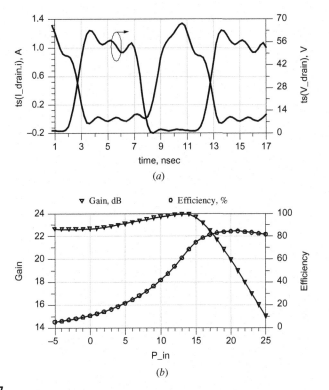

**FIGURE 3.17**

Simulated waveforms and performance of Class-F GaN HEMT power amplifier.

## 3.5 **Effect of saturation resistance and shunt capacitance**

It is useful to analytically estimate the effect of a saturation (or on-resistance) $r_{sat}$ which is not equal to zero in a real transistor, and transistor therefore dissipates some amount of power due to the collector current flowing through this resistance when the transistor is turned on. The simplified equivalent circuit of a Class-F power amplifier with a quarterwave transmission line where the transistor is represented by a non-ideal switch with saturation resistance $r_{sat}$ and parasitic output capacitance $C_{out}$ is shown in Fig. 3.18. During the interval $0 < \omega t \leq \pi$ when the switch is turned on, the saturation voltage $v_{sat}$ due to the current $i(\omega t)$ flowing through the switch can be written as

$$v_{sat}(\omega t) = V_{sat}\sin \omega t = 2I_R r_{sat}\sin \omega t \qquad (3.73)$$

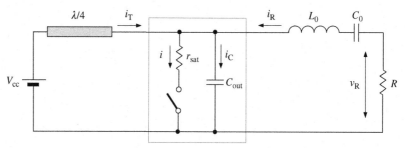

**FIGURE 3.18**

Effect of parasitic on-resistance and shunt capacitance.

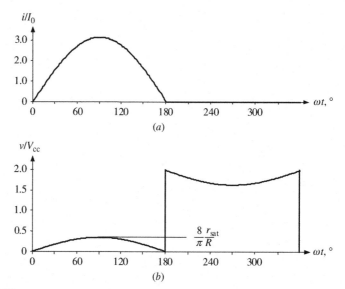

**FIGURE 3.19**

Idealized collector current and voltage waveforms with nonzero on-resistance.

where, by using Eq. (3.25), the saturation voltage amplitude $V_{sat}$ can be obtained by

$$V_{sat} = 2V_R \frac{r_{sat}}{R} = \frac{8V_{cc}}{\pi} \frac{r_{sat}}{R}. \tag{3.74}$$

The corresponding collector current and voltage waveforms are shown in Fig. 3.19 where the half-sinusoidal current flowing through the saturation resistance $r_{sat}$ causes the deviation of the voltage waveform from the ideal square waveform. In this case, the bottom part of the voltage waveform becomes sinusoidal with the amplitude $V_{sat}$ during the interval $0 < \omega t \leq \pi$. From Eq. (3.18) it follows that the same sinusoidal behavior will correspond to the top part of the voltage waveform during the interval $\pi < \omega t \leq 2\pi$.

The power losses and collector efficiency due to presence of the saturation resistance $r_{sat}$ can be evaluated using Eqs (3.21), (3.23), and (3.25) as

$$\frac{P_{sat}}{P_0} = \frac{1}{2\pi} \int_0^{2\pi} \frac{i^2(\omega t) r_{sat}}{I_0 V_{cc}} d\omega t = \frac{r_{sat}}{2\pi I_0 V_{cc}} \int_0^{2\pi} (2I_R)^2 \sin^2 \omega t \, d\omega t$$

$$= \frac{r_{sat} I_R}{V_{cc}} \frac{I_R}{I_0} = \frac{r_{sat}}{R} \frac{I_R}{I_0} \frac{V_R}{V_{cc}} = \frac{2r_{sat}}{R}. \quad (3.75)$$

Hence, the collector efficiency can be calculated from

$$\eta = 1 - \frac{P_{sat}}{P_0} = 1 - \frac{2r_{sat}}{R}. \quad (3.76)$$

In practice, the idealized collector voltage and current waveforms can be realized at low frequencies when the effect of the device collector capacitance is negligible. At higher frequencies, the effect of the collector capacitance contributes to a nonzero switching time resulting in time periods when the collector voltage and collector current exist at the same time when $v > 0$ and $i > 0$. Consequently, such a load network with shunt capacitance cannot provide the switchmode operation with an instantaneous transition from the device pinch-off to saturation mode. Therefore, during a nonzero time interval, the device operates in the active region as a current source with the reverse-biased collector-base junction, and the collector current is provided by this current source.

The current flowing through the collector capacitance can be determined by differentiating both parts of Eq. (3.18), taking into account that voltage $v$ is the voltage across the capacitance $C$, as

$$i_C(\omega t) = -i_C(\omega t + \pi) \quad (3.77)$$

which means that the current due to the capacitance charging process is equal to the current due to the capacitance-discharging process with opposite sign, and the durations of the charging and discharging periods are equal.

From Fig. 3.18, it follows that the current flowing through the collector capacitance at the arbitrary time moment $t$ can be written as

$$i_C(\omega t) = i_T(\omega t) + i_R(\omega t) - i(\omega t) \quad (3.78)$$

whereas, at the time moment $(t + \pi/\omega)$, it can be obtained by

$$i_C(\omega t + \pi) = i_T(\omega t + \pi) + i_R(\omega t + \pi) - i(\omega t + \pi). \quad (3.79)$$

The output current flowing into the load is written as sinusoidal

$$i_R(\omega t) = I_R \sin(\omega t + \varphi) \tag{3.80}$$

where $\varphi$ is the initial phase shift due to the finite value of the collector capacitance. Then, by taking into account Eqs (3.70) and (3.77), from Eq. (3.79) it follows that

$$-i_C(\omega t) = i_T(\omega t) - i_R(\omega t) - i(\omega t + \pi). \tag{3.81}$$

Adding Eq. (3.78) and Eq. (3.81) yields

$$i(\omega t) + i(\omega t + \pi) = 2i_T(\omega t) \tag{3.82}$$

that specify the relationship in the time domain between the collector current and current flowing into the transmission line.

The collector current and voltage waveforms are shown in Fig. 3.20 where the phase angle $\varphi_1$ corresponds to the beginning of the transistor saturation mode,

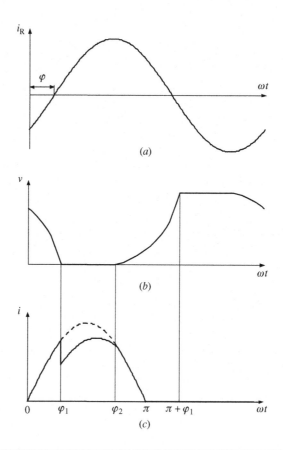

**FIGURE 3.20**

Effect of shunt capacitance on voltage and current waveforms.

and the phase angle $\varphi_2$ corresponds to the beginning of the active mode and collector capacitance charging process start-up. During the saturation interval when $\varphi_1 < \omega t < \varphi_2$, the collector current $i(\omega t)$ can be defined using Eqs (3.78), (3.80), and (3.81) by

$$i(\omega t) = i_T(\omega t) + i_R(\omega t) = 2i_R(\omega t) = 2I_R\sin(\omega t + \varphi). \qquad (3.83)$$

In the active region when $0 \le \omega t \le \varphi_1$ and $\varphi_2 \le \omega t \le \pi$, the collector current flowing into the load network is defined by the input driving signal and, for the conduction angle of 180°, represents the periodic half-sinusoidal pulses written as

$$i(\omega t) = I_{\text{active}}(\sin \omega t + |\sin \omega t|) \qquad (3.84)$$

where the collector current amplitude $I_{\text{active}}$ in the active region is characterized by a higher value than the one in saturation mode when $I_{\text{sat}} = 2I_R$, as follows from Eq. (3.83), due to the shunting effect of the forward-biased collector-base diode junction when device is saturated. The moment of the opening of the collector-base junction corresponds to the time moment $\varphi_1$ with instantaneous reduction in the collector current waveform, as shown in Fig. 3.20(c). Physically, this effect can be explained by the carrier injection from the device collector to its base region as a result of the forward-biasing of the collector-base junction. The saturation period is characterized by the diffusion capacitance of the forward-biased collector-base junction whereas, in active or pinch-off regions, the reverse-biased collector-base junction is described by the junction capacitance, the value of which is substantially smaller. The saturation period is ended at the moment $\varphi_2$ corresponds to the beginning of the active mode, and the process of charging the collector junction capacitance is started up.

By using Eqs (3.71), (3.83), and (3.84), the current flowing into the transmission line $i_T$ and current flowing through the collector capacitance $i_C$ can be obtained by

$$i_T(\omega t) = I_{\text{active}}|\sin \omega t| \qquad (3.85)$$

$$i_C(\omega t) = 2[I_R\sin(\omega t + \varphi) - I_{\text{active}}\sin \omega t]. \qquad (3.86)$$

Power losses due to the charging and discharging processes of the device collector capacitance can be calculated from

$$P_{\text{loss}} = \frac{1}{2\pi} \int_0^{\varphi_1} v(\omega t)i(\omega t)d\omega t + \frac{1}{2\pi} \int_{\varphi_2}^{\pi} v(\omega t)i(\omega t)d\omega t \qquad (3.87)$$

where the collector voltage $v$ coincides with the voltage across the capacitance $C$. From Eq. (3.87), it follows that the longer the active region, due to the larger collector capacitance, the more power is lost and this reduces the efficient operation of the power amplifier. From the results of numerical calculations, the maximum operating frequency where the collector efficiency of a Class-F power amplifier with the effect of the output capacitance $C_{out}$ is higher than the collector efficiency of a conventional Class-B power amplifier can be evaluated from

$$f_{max} \cong \frac{0.47}{RC_{out}} \tag{3.88}$$

where the output capacitance $C_{out}$ is assumed to be a total collector capacitance (including the capacitances of the passive and active parts of the collector-base junction) corresponding to the dc bias operation point [24].

## 3.6 Load networks with lumped elements

Theoretical results show that the proper control of only second and third harmonics can significantly increase the collector efficiency of the power amplifier by flattening the output voltage waveform. Since practical realization of a multi-element high-order $LC$ resonant circuit can cause a serious implementation problem, especially at higher frequencies, it is sufficient to be confined to a three- or four-element resonant circuit composing the load network of the power amplifier. In addition, it is necessary to take into account that, in practice, both extrinsic and intrinsic transistor parasitic elements such as output shunt capacitance or serious inductance have a substantial effect on the efficiency. The output capacitance $C_{out}$ can represent the collector capacitance $C_c$ in the case of the bipolar transistor or the sum of the drain-source and gate-drain capacitances, $C_{ds} + C_{gd}$, in the case of the FET device. The output inductance $L_{out}$ is generally composed of the bondwire and lead inductances for a packaged transistor, the effect of which becomes significant at higher frequencies. The typical two-terminal reactive networks with series and parallel resonators used in a practical design procedure, which provide ideally infinite impedances at the fundamental and third harmonics and zero impedance at the second harmonic, are shown in Fig. 3.21 [25−27].

At microwave and millimeter-wave frequencies, it is required to minimize a number of elements to reduce the effect of any possible circuit parasitics. Figure 3.22 shows the simple Class-F load network with second- and third-harmonic control, where $C_{ds}$ is the drain-source capacitance of the MESFET device tuned together with inductance $L_1$ and capacitance $C_1$ for a parallel resonance at the third harmonic, while the series combination of $L_1$ and $C_1$ provides a series resonance at the second harmonic operating as a second-harmonic trap [28]. The load network input impedance $Z_{net}$ can be written as

$$Z_{net} = j\frac{\omega^2 L_1 C_1 - 1}{\omega C_1 - \omega C_{ds}(\omega^2 L_1 C_1 - 1)}. \tag{3.89}$$

As a result, applying two harmonic-impedance conditions, open-circuited for the third harmonic $Z_{net}(3\omega_0) = \infty$ and short-circuited for the second harmonic

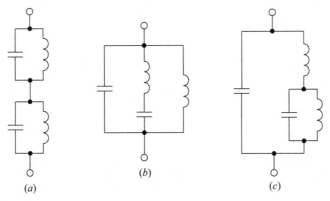

**FIGURE 3.21**

Two-terminal reactive networks with series and parallel resonators.

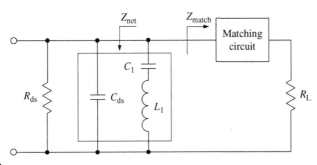

**FIGURE 3.22**

Class-F load network with second- and third-harmonic control.

$Z_{\text{net}}(2\omega_0) = 0$, where $\omega_0$ is the fundamental angular frequency, the values of load-network elements as a function of $C_{\text{ds}}$ can be calculated from

$$C_1 = \frac{5}{4} C_{\text{ds}} \tag{3.90}$$

$$L_1 = \frac{1}{5\omega_0^2 C_{\text{ds}}}. \tag{3.91}$$

However, since the resulting impedance at the fundamental frequency is capacitive, it is necessary to compensate for the capacitive reactance by choosing the parameters of the matching circuit having an inductive reactance of its input impedance $Z_{\text{match}}$ at the fundamental and high impedance conditions at third- and higher-order harmonic components. The output matching circuit based on an $L$-type lumped transformer with a series inductor and a shunt capacitor implemented in a

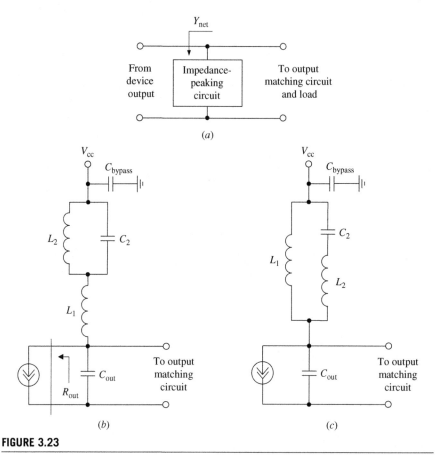

**FIGURE 3.23**

Load networks with parallel and series resonant circuits.

monolithic microwave design can be used to provide an impedance matching at the fundamental frequency and a high-impedance condition at the third harmonic together with a second-harmonic trap [29].

By adding an additional element to the load network shown in Fig. 3.22, it is possible to compensate for the capacitive reactance at the fundamental frequency, thus providing both high impedance at the fundamental and third harmonics and zero impedance at the second harmonic. Examples of such load networks with additional parallel and series resonant circuits located between the dc power supply and device output are shown in Fig. 3.23 [27,30]. Here, the output circuit of the active device is represented by a multiharmonic current source, and $R_{out}$ is the equivalent output resistance at the fundamental frequency defined as a ratio of the fundamental voltage at the device output to the fundamental current flowing into the device.

The reactive part of the output admittance (or susceptance) $B_{net} = \text{Im}Y_{net}$ of the load network with a parallel resonant tank shown in Fig. 3.23($b$), including the device output capacitance $C_{out}$, can be written as

$$B_{net} = \omega C_{out} - \frac{1 - \omega^2 L_2 C_2}{\omega L_1 (1 - \omega^2 L_2 C_2) + \omega L_2}. \tag{3.92}$$

By applying three harmonic-impedance conditions $B_{net}(\omega_0) = B_{net}(3\omega_0) = 0$ and $B_{net}(2\omega_0) = \infty$ at the device output (collector or drain), determining an open circuit for the fundamental and third harmonic components and a short circuit for the second harmonic component, Eq. (3.92) can be rewritten in the form of three equations as

$$(1 - \omega_0^2 L_1 C_{out})(1 - \omega_0^2 L_2 C_2) - \omega_0^2 L_2 C_{out} = 0 \tag{3.93}$$

$$L_1(1 - 4\omega_0^2 L_2 C_2) + L_2 = 0 \tag{3.94}$$

$$(1 - 9\omega_0^2 L_1 C_{out})(1 - 9\omega_0^2 L_2 C_2) - 9\omega_0^2 L_2 C_{out} = 0. \tag{3.95}$$

As a result, the ratios between elements of this impedance-peaking load network are

$$L_1 = \frac{1}{6\omega_0^2 C_{out}} \tag{3.96}$$

$$L_2 = \frac{5}{3} L_1 \tag{3.97}$$

$$C_2 = \frac{12}{5} C_{out} \tag{3.98}$$

where the sum of the reactances of the parallel resonant tank, consisting of an inductor $L_2$ and a capacitor $C_2$, and inductor $L_1$ creates open-circuit parallel resonance conditions at the fundamental and third harmonic components, while the series capacitive reactance of the tank circuit in series with an inductance $L_1$ creates a short-circuit series resonance condition at the second harmonic component [30,31].

Applying the same conditions for the load network with a series-resonant circuit shown in Fig. 3.23($c$) results in the ratios between elements given by

$$L_1 = \frac{4}{9\omega_0^2 C_{out}} \tag{3.99}$$

$$L_2 = \frac{9}{15} L_1 \tag{3.100}$$

$$C_2 = \frac{15}{16} C_{out} \tag{3.101}$$

where an inductance $L_2$ and a capacitance $C_2$ create a short-circuit condition at the second harmonic, and all elements create the parallel-resonant tanks for the fundamental and third harmonic components [27].

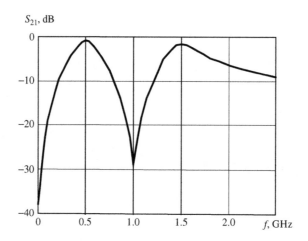

**FIGURE 3.24**

Frequency response of load network with parallel resonant circuit.

To determine the transfer performance of the impedance-peaking load network in the frequency domain, it is best to represent such a load network as shown in Fig. 3.21(a), then simulate the small-signal $S$-parameters, and finally plot a magnitude of $S_{21}$ in decibels over wide frequency range. As an example, the frequency–response characteristic of the load network with a parallel resonant circuit, whose parameters are calculated based on the chosen fundamental frequency $f_0 = 500$ MHz, is shown in Fig. 3.24. In this case, the load-network parameters are $C_{out} = 2.2$ pF, $R_{out} = R_L = 200$ $\Omega$, $C_2 = 5.3$ pF, $L_1 = 7.7$ nH and $L_2 = 12.8$ nH with an inductor quality factor $Q_{ind} = 20$. It should be noted that the power amplifier efficiency can be even higher if the first element of the output matching circuit adjacent to the transistor output is in series and inductive to provide high-impedance conditions at higher-order harmonics.

Careful design must be provided at higher frequencies or in the case of high-power mode when the transistor equivalent output resistance at the fundamental frequency is sufficiently small. In this case, an effect of the output series inductance, including the bondwire and lead inductances for a packaged active device, becomes significant. The equivalent circuit of such an impedance-peaking load network is shown in Fig. 3.25. Here, the series circuit consisting of an inductor $L_1$ and a capacitor $C_1$ creates a short-circuit condition at the third harmonic. Since the output inductor $L_{out}$ and capacitor $C_{out}$ are tuned to create an open-circuit condition at the third harmonic, the device collector sees the resultant high impedance at the third harmonic.

To achieve the third-harmonic high impedance, an external series inductor may be added to interconnect the device output inductance $L_{out}$ directly at the output terminal (collector or drain) if its value is smaller than required. In addition,

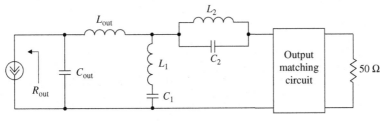

**FIGURE 3.25**

Impedance-peaking load network including device output inductance.

the parallel resonant tank $(L_2, C_2)$ is tuned to the second harmonic, while the series network $(L_1 + L_{out}, C_1)$ provides a short-circuit condition at the second harmonic [31]. As a result, the ratios between the load-network parameters are

$$L_{out} = \frac{1}{9\omega_0^2 C_{out}} \tag{3.102}$$

$$L_1 = \frac{4}{5}L_{out} \tag{3.103}$$

$$C_1 = \frac{5}{4}C_{out} \tag{3.104}$$

$$L_2 = \frac{1}{4\omega_0^2 C_2}. \tag{3.105}$$

As a first approximation for comparison between different operation modes, the output device resistance $R_{out}$ at the fundamental frequency required to realize a Class-F operation mode with third-harmonic peaking can be estimated as the equivalent resistance determined at the fundamental frequency for an ideal Class-F operation and written as $R_{out} = R_1^{(F)} = V_1/I_1$, where $V_1$ and $I_1$ are the fundamental-frequency voltage and current amplitudes at the device output. Assuming zero saturation voltage and using Eq. (3.25) yields

$$R_1^{(F)} = \frac{4}{\pi}\frac{V_{cc}}{I_1} = \frac{4}{\pi}R_1^{(B)} \tag{3.106}$$

where $R_1^{(B)} = V_{cc}/I_1$ is the device equivalent output resistance at the fundamental frequency in an ideal Class B mode.

The load network can represent a low-pass $LC$ ladder network commencing with a shunt capacitor and terminated with a narrowband shunt resonant in parallel with the resistive load [21]. In this case, the device output capacitance can be a part of the external load network. This $LC$ ladder-type load network should provide the correct complex impedance at the fundamental frequency, a short circuit at the first $(2m-2)$ even harmonics, an open circuit at the first $(2m-1)$ odd harmonics, and reactive terminations at all harmonics $2m$ and above. The

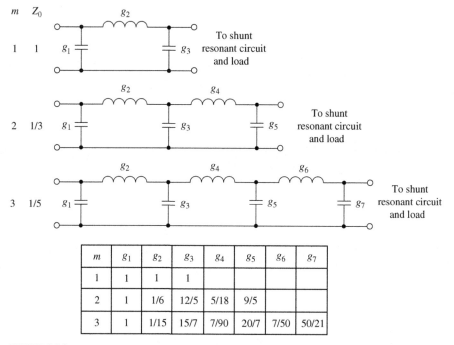

**FIGURE 3.26**

Ideal ladder low-pass Class-F load-networks and their parameters.

configurations of these networks for values of $m$ up to the third degree with corresponding normalized coefficients $g_i$ are shown in Fig. 3.26.

The general formulas for explicit values of the elements in the low-pass ladder load network with arbitrary degree based on a cutoff frequency $\omega_c = 1$ and a reference impedance of $1\ \Omega$ are derived from

$$g_1 g_2 = \frac{1}{m(2m-1)}$$

$$g_i g_{i+1} = \frac{4}{(2m-1+i)(2m-i)} \tag{3.107}$$

$$g_{2m} g_{2m+1} = \frac{1}{m}$$

where $g_1 = 1$, and the characteristic impedance of the network operating as an impedance inverter which is obtained by

$$Z_0 = \frac{1}{2m-1} \tag{3.108}$$

where $2 \leq i \leq 2m - 1$. By using such a low-order low-pass $LC$ ladder network for input and output matching with second- and third-harmonic control at the device input and output terminals simultaneously, a drain efficiency of 76% for a Class-F power amplifier implemented in a 0.5-$\mu$m pHEMT process was achieved at an operating frequency of 1.84 GHz with 16 W of output power [32].

## 3.7 **Load networks with transmission lines**

Generally, the design of a transmission-line or distributed matching circuit can be based on approximate equivalence between lumped and distributed elements that can be established by applying a Richards's transformation which provides a sequence of equal-length open- and short-circuited transmission lines representing redundant transmission-line sections as unit elements of $\lambda/8$ long at cutoff frequency $\omega_c$ [33]. The transmission-line impedances and electrical lengths of the load network can then be optimized and tuned to simultaneously provide the correct impedance at the fundamental frequency and also the required complex terminations [34]. The shunt fundamentally tuned resonant circuit is realized using a quarter-wavelength transmission line, which operates as a harmonic short at even harmonics.

It is problematic to realize the ladder $LC$ load network by using commercially available lumped element chip capacitors and inductors, due to their self-resonance frequency and insertion loss limitations. Therefore, at higher frequencies it is useful to transform a lumped ladder circuit to its distributed equivalent based on a multisection stepped-impedance transmission line. As an example, Fig. 3.27 shows that the two-section lumped ladder $LC$ circuit, including the device parasitic shunt output capacitance $C_p$ and series bondwire inductance $L_p$, can be equivalently transformed to a corresponding Class-F load network with a series two-section stepped-impedance transmission line, whose transmission-line sections are

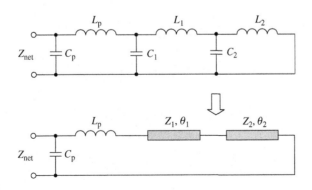

**FIGURE 3.27**

Transformation from lumped to distributed Class-F load network.

characterized by the characteristic impedances $Z_i = \sqrt{L_i/C_i}$ and electrical lengths $\theta_i = \sin^{-1}(\omega_0\sqrt{L_iC_i})$, where $i = 1, 2$. [35]. Since poles of the input impedance of the stepped-impedance transmission line can vary depending on the ratio of the characteristic impedances and electrical lengths of the transmission-line sections, the whole circuit with parasitic elements can be optimized to provide a short-circuit condition at even harmonics and open-circuit condition at odd harmonics. The experimental results of the fabricated 2-W 25-V microstrip GaN HEMT Class-F power amplifier show a maximum drain efficiency of about 80% at an operating frequency of 5.86 GHz.

Generally, the optimum impedance matching at the fundamental frequency with the second- and third-harmonic tuning can be provided by using the series transmission-line sections and shunt open-circuit or short-circuit stubs to obtain open-circuit peaking and short-circuit termination seen by the device output at corresponding harmonic component [36]. Such an approach is very simple in terms of the load-network design and very convenient for practical implementation. Figure 3.28(a) shows the idealized basic configuration of the transmission-line Class-F load network where a shunt quarter-wavelength transmission line

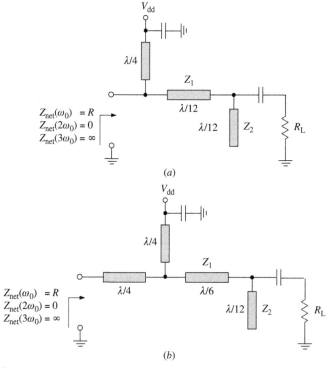

**FIGURE 3.28**

Idealized transmission-line Class-F load networks.

providing a short-circuit termination for even voltage harmonic at the device output is connected to the dc power supply with a bypass capacitor [37]. The series transmission-line section and open-circuit stub, both having an electrical length of 30° at the fundamental frequency, provide an open-circuit mode at the third harmonic because an open-circuit stub has a quarter wavelength at the third harmonic to realize a short-circuit condition at the right-hand side of the series transmission line having a quarter wavelength at the third harmonic as well. The ideal Class-F power amplifier with all even harmonic short-circuit termination and third harmonic peaking achieves a maximum drain efficiency of 88.4%. Such a Class-F load network implemented using coplanar lines into a 24-GHz MMIC GaAs pHEMT power amplifier had contributed to a drain efficiency of 59% at an output power of 20 dBm [38].

The load-network impedance $Z_{net}$ seen by the device output at the fundamental frequency can be written as

$$Z_{net} = Z_1 \frac{R_L(Z_2 - Z_1 \tan^2\theta) + jZ_1Z_2 \tan\theta}{Z_1Z_2 + j(Z_1 + Z_2)R_L \tan\theta} \tag{3.109}$$

where $\theta = \theta_1 = \theta_2 = 30°$, $Z_1$ and $\theta_1$ are the characteristic impedance and electrical length of the series transmission line, and $Z_2$ and $\theta_2$ are the characteristic impedance and electrical length of the open-circuit stub. Hence, the impedance matching with the load at the fundamental can be provided by proper choice of the characteristic impedances $Z_1$ and $Z_2$.

Separating Eq. (3.109) into real and imaginary parts and taking into account that $\text{Re}Z_{net} = R$ and $\text{Im}Z_{net} = 0$, the system of two equations with two unknown parameters is obtained as

$$Z_1^2 Z_2^2 - R_L^2(Z_1 + Z_2)(Z_2 - Z_1 \tan^2\theta) = 0 \tag{3.110}$$

$$(Z_1 + Z_2)^2 R_L^2 R \tan^2\theta - Z_1^2 Z_2^2 [R_L(1 + \tan^2\theta) - R] = 0 \tag{3.111}$$

which enables the characteristic impedances $Z_1$ and $Z_2$ to be properly calculated. This system of two equations can be explicitly solved as a function of the parameter $r = R_L/R$ resulting in

$$\frac{Z_1}{R_L} = \frac{\sqrt{4r - 3}}{r} \tag{3.112}$$

$$\frac{Z_1}{Z_2} = 3\left(\frac{r - 1}{r}\right). \tag{3.113}$$

Consequently, for the specified value of the parameter $r$ with the required Class-F optimum fundamental-frequency load resistance $R$ and standard load resistance $R_L = 50\ \Omega$, the characteristic impedance $Z_1$ is calculated from Eq. (3.112) and then the characteristic impedance $Z_2$ is calculated from Eq. (3.113). For example, if the required Class-F optimum load resistance $R$ is equal to 12.5 $\Omega$ resulting in $r = 4$, the characteristic impedance of the series transmission line $Z_1$ is equal to 45 $\Omega$ and the characteristic impedance of the open-circuit stub $Z_2$ is equal to 20 $\Omega$.

Figure 3.28(*b*) shows the simplified Class-F load network with the second- and third-harmonic tuning using a series quarter-wavelength transmission line which provides zero impedance at the second harmonic and infinite impedance at the third harmonic components [39]. The short-circuit impedances at the right-hand side of the series quarter-wavelength transmission line are provided by the short-circuited shunt quarter-wavelength transmission line at the second harmonic and an open-circuit $\lambda/12$ stub together with a series $\lambda/6$ transmission line at the third harmonic. In this case, an impedance matching at the fundamental frequency can be provided either using a separate matching circuit or by optimizing the characteristic impedances of the load-network transmission lines.

The close approximation to an ideal Class-F mode in the microwave region can be achieved by using a load network shown in Fig. 3.29, where each open-circuit stub has a quarter wavelength at each higher-order harmonic of the fundamental frequency providing zero impedance at the higher harmonic at point $A$ [40]. The transmission line $TL_{11}$ having a quarter wavelength at the fundamental frequency transforms these zero impedances into open-circuit impedance at odd harmonics and short-circuit impedance at even harmonics seen from the device output. The quarterwave transmission line $TL_{12}$ is necessary for impedance matching at the fundamental frequency of the output impedance $Z_{out}$ with the standard load $R_L$. To compensate for the reactive impedance at the fundamental frequency due to the open-circuit stubs $TL_2, \ldots, TL_7$, the reactance compensation open-circuit stubs $TL_2^*, \ldots, TL_7^*$ are added in series to form an overall half wavelength for each stub. However, such a Class-F load network can be effectively used at low power levels when the device parasitic output capacitance $C_{out}$ is sufficiently small.

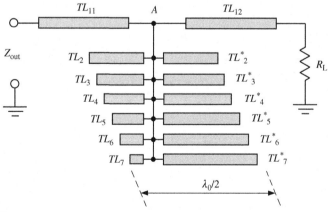

**FIGURE 3.29**

Class-F load network with reactance compensation circuits.

However, the efficiency of the power amplifier can be limited by the transistor output capacitance $C_{out}$ (mostly represented by drain-source capacitance $C_{ds}$ or collector capacitance $C_c$) if this capacitance is not absorbed into a multiharmonic load network without compromising the ability to properly terminate the second and third harmonic components. Figure 3.30(a) shows the simplified circuit schematic of the transmission-line transistor power amplifier, whereas the equivalent load network corresponding to Class-F mode with a shunt short-circuited quarter-wavelength transmission line which provides a short circuit at even harmonics is shown in Fig. 3.30(b) [30,31].

In Class-F mode with a transmission-line load network shown in Fig. 3.30(b), the electrical length of an open-circuit stub $TL_3$ is chosen to have a quarter wavelength at the third harmonic to realize short-circuit condition at the right-hand side of the series transmission line $TL_2$, whose electrical length $\theta_2$ should provide an inductive reactance to resonate with the device output capacitance $C_{out}$ at the third harmonic. In this case, a quarterwave transmission line $TL_1$ located between the dc power supply and drain terminal provides short-circuit termination for even voltage harmonics. Such a load network is very practical when the dc drain voltage is very high to increase the device output impedance, or the operating frequency is

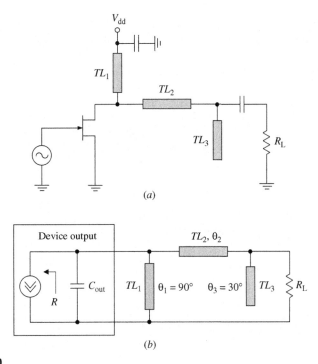

(a)

(b)

**FIGURE 3.30**

Transmission-line Class-F load network with shunt capacitance.

sufficiently low, and bare die is used instead of packaged transistor to easily absorb bondwire inductances by the load network [41]. As a result, the electrical lengths of the transmission lines at the fundamental frequency can be obtained as

$$\theta_1 = \frac{\pi}{2} \tag{3.114}$$

$$\theta_2 = \frac{1}{3}\tan^{-1}\left(\frac{1}{3Z_0\omega_0 C_{out}}\right) \tag{3.115}$$

$$\theta_3 = \frac{\pi}{6} \tag{3.116}$$

where $Z_0$ is the characteristic impedance of the series transmission line $TL_2$ and $\omega_0$ is the fundamental radian frequency.

Figure 3.31 shows an example of the frequency−response characteristic of the microstrip impedance-peaking circuit shown in Fig. 3.30(b), assuming an alumina substrate for microstrip lines and the device equivalent output resistance $R = 50\ \Omega$ and output capacitance $C_{out} = 2.2$ pF, microstrip-line characteristic impedance $Z_0 = 50\ \Omega$, and electrical length $\theta_2 = 15°$ according to Eq. (3.115). From Fig. 3.31, it follows that the corresponding short-circuited conditions for all even harmonics and third-harmonic peaking have been provided. However, an additional output impedance matching to compensate for the reactive part and to match the real part of the equivalent output impedance with the standard 50-$\Omega$ load impedance at the fundamental frequency $f_0 = 500$ MHz is required by optimizing the characteristic impedances of the series microstrip line and open-circuit stub.

When the Class-F power amplifier is fabricated as a hybrid integrated circuit, it needs to take into account influence of the parasitic series bondwire inductance. In this case, it is possible to use either an additional series compensation line connected to the device output [42] or a load network with optimized parameters

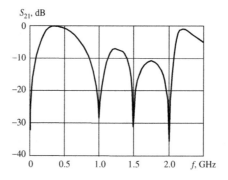

**FIGURE 3.31**

Frequency−response of microstrip impedance-peaking circuit.

shown in Fig. 3.32(a) [43]. In the latter case, the open-circuit condition for the third harmonic can be satisfied by modifying Eq. (3.115) as

$$\theta_2 = \frac{1}{3}\tan^{-1}\left(\frac{1}{3Z_0\omega_0 C_{\text{out}}} - \frac{3\omega_0 L_{\text{out}}}{Z_0}\right) \quad (3.117)$$

where $\theta_2$ is the electrical length of the series transmission line $TL_2$ (at the operating frequency $\omega_0$) with the characteristic impedance $Z_0$. However, the short-circuit condition for the second- and higher-order even harmonics cannot be met, and the performance of the Class-F power amplifier is adversely affected as the operating frequency increases. Therefore, the transistor drain voltage and current waveforms are not perfectly symmetrical, which are characterized by significant transition times from a saturation region to a pinch-off region and *vice versa*, mainly due to the effect of the second harmonic having the certain phase shift provided by the series bondwire inductance.

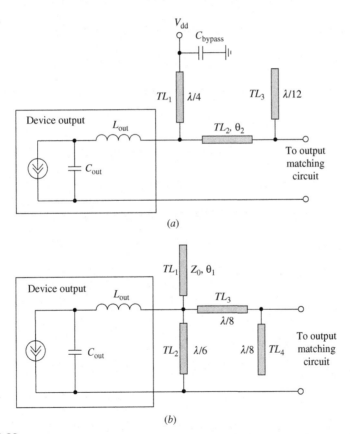

(a)

(b)

**FIGURE 3.32**

Transmission-line Class-F load networks including device bondwire inductance.

The more complicated impedance-peaking circuit to improve efficiency, including the output series parasitic inductance $L_{out}$, which can generally represent the bondwire and package lead inductances, is shown in Fig. 3.32(b). Here, to create the second-harmonic short-circuit condition and third-harmonic peaking, it is convenient to use a combination of the open-circuit and short-circuit transmission-line stubs in the load network. The device output elements $L_{out}$ and $C_{out}$ must create a parallel resonance at the third harmonic of the fundamental frequency since the short-circuit stub $TL_2$ has a half wavelength at the third harmonic with short-circuit conditions at both its ends. It should be mentioned that the open-circuit stub having a quarter wavelength at the third harmonic of the fundamental frequency can also be used [44]. The electrical length $\theta_1$ of an open-circuit stub $TL_1$ is chosen to have less than a quarter wavelength at the second harmonic to realize an overall capacitive reactance together with the short-circuited transmission line $TL_2$ to be resonant with output inductance $L_{out}$ at the second harmonic of the fundamental frequency. The transmission lines $TL_3$ and $TL_4$ must be of quarter wavelengths at the second harmonic to provide the second-harmonic high impedance condition at the input of $TL_3$. As a result, the ratios between the load-network parameters are [31]

$$L_{out} = \frac{1}{9\omega_0^2 C_{out}} \tag{3.118}$$

$$\theta_1 = \frac{1}{2}\tan^{-1}\left(\frac{Z_0}{2\omega_0 L_{out}} - \frac{1}{\sqrt{3}}\right). \tag{3.119}$$

## 3.8 LDMOSFET power amplifier design examples

The effectiveness of the Class-F load-network design technique can be demonstrated based on the example of high-power LDMOSFET amplifiers. The small-signal equivalent circuit of the LDMOSFET cell with a gate length of 1.25 μm and a gate width of 1.44 mm is shown in Fig. 3.33(a) [33]. The device model parameters were extracted from pulsed current–voltage (I–V) and small-signal S-parameter measurements. The parameters of the small-signal device equivalent circuit are given at a bias voltage for Class AB with a quiescent current $I_q = 15$ mA at a supply voltage $V_{dd} = 28$ V. The measured and modeled output $I_{ds} - V_{ds}$ characteristics of the high-power device with a total gate width of $28 \times 1.44$ mm are shown in Fig. 3.33(b). Based on these characteristics, it is easy to choose the peak drain current, which allows us to maximize the drain efficiency by minimizing the saturation voltage. For example, choosing a peak current of 3.5 A results in a dc current of approximately $3.5/\pi \approx 1.1$ A, according to Eqs (3.21) and (3.23), that leads to a saturation voltage of about 4 V only. As a result, the maximum drain efficiency of about 80% providing a delivery of the

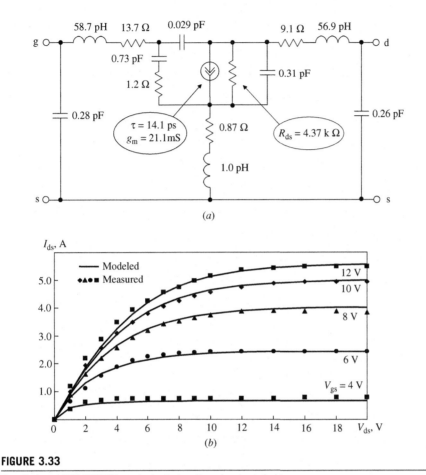

**FIGURE 3.33**

Small-signal LDMOSFET equivalent circuit and output I–V curves.

output power of more than 20 W into the load can be achieved using a supply voltage of 24 V.

The circuit schematic of the simulated 500-MHz single-stage lumped LDMOSFET power amplifier is shown in Fig. 3.34. In this particular case, the total gate width of a high-voltage LDMOSFET device is $7 \times 1.44$ mm to achieve 8 W of output power. The drain efficiency and power gain of the amplifier versus input power $P_{in}$ for the case of ideal inductors are given in Fig. 3.35(a). The drain efficiency over 75% (curve 2) is obtained due to a short-circuit condition at the second harmonic and open-circuit condition at the third harmonic. Generally, it is important to provide high-impedance conditions at higher-order harmonics that can be readily done by using an output matching circuit with the series inductor

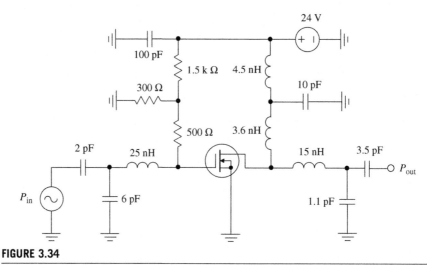

**FIGURE 3.34**

Simulated lumped LDMOSFET Class-F power amplifier.

as a first element. This shortens the switching time from the pinch-off region to the voltage-saturation region by better approximating the idealized drain voltage square waveform, as shown in Fig. 3.35(c). It should be noted that the drain current waveform differs from a half-sinusoidal waveform because it includes higher-order odd harmonic components together with the current flowing through the device intrinsic equivalent circuit capacitors.

As follows from Eq. (3.32) for a symmetrical voltage waveform, the initial phases for the fundamental frequency and its harmonics should be equal, which is easy to realize by short-circuit and open-circuit conditions. However, according to Eq. (3.33) for a half-sinusoidal current waveform, the phases for any harmonic component should differ from the phase for the fundamental frequency by 90°. This condition is easily realized in a Class-B load network where the fundamental component of the drain voltage is in phase with the fundamental component of the drain current, but, for all higher-order current harmonics, the impedance of the resonant circuit will be capacitive since the drain current harmonics mostly flow through the shunt capacitor. Therefore, the accurate harmonic phasing is very important to improve effectiveness of a Class-F load network. The amplifier drain efficiency and power gain can be significantly reduced when the values of the quality factor of the load-network inductors are sufficiently small. For example, the maximum value of the drain efficiency can reach only 71% when an inductor quality factor at the fundamental frequency is $Q_{ind} = 30$, as shown in Fig. 3.35(b).

Therefore, at a high power level, it is preferred to use the load networks that employ microstrip lines. The equivalent circuit of a simulated 500-MHz

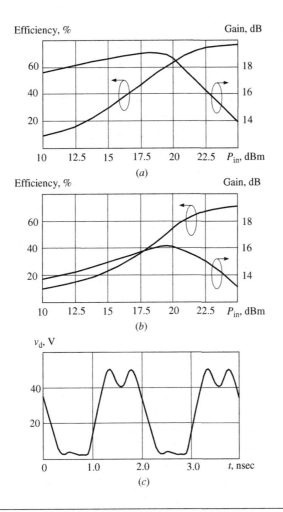

**FIGURE 3.35**

Drain efficiency, power gain, and voltage waveform.

single-stage microstrip **LDMOSFET** power amplifier using an active device with the same geometry is shown in Fig. 3.36. The input and output matching circuits represent a *T*-type matching circuit each, consisting of a series microstrip line, a parallel open-circuit stub, and a series capacitor. To provide even-harmonic short-circuit termination and third-harmonic peaking for a Class-F mode, an RF grounded quarter-wavelength microstrip line and a combination of the series short-length microstrip line and open-circuit stub with electrical length of 30° at the fundamental frequency are used. Such an output-circuit configuration approximates the square drain voltage waveform accurately, as shown in Fig. 3.37(*a*), and provides a drain efficiency over 75% with a maximum output

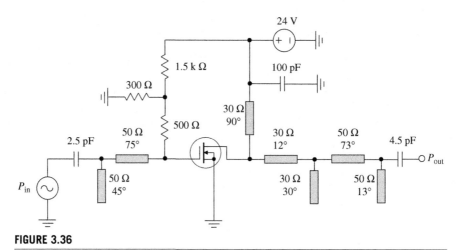

**FIGURE 3.36**

Simulated microstrip LDMOSFET Class-F power amplifier.

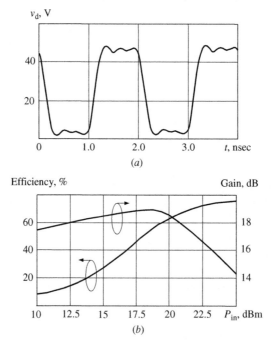

**FIGURE 3.37**

Drain voltage waveform, efficiency, and power gain.

power of 8 W, as shown in Fig. 3.37(b). The smaller value of the drain efficiency compared to the theoretically achievable value can be explained by the non-optimized impedances at higher-order harmonics since, unlike a lumped inductor, the transmission line exhibits an equidistant impedance performance in the frequency domain with consecutive poles and zeros at the characteristic frequencies. This means that using a simple T-type transmission-line transformer does not provide high impedance conditions at all higher-order harmonics simultaneously.

It should be mentioned that, at frequencies close to the device transition frequency $f_T$, the drain (or collector) waveform becomes stretched due to a delay effect of the transistor input circuit with different and significant phase shifts $\omega\tau_{in} = \omega R_{in} C_{in}$ for higher-order harmonic components, where $R_{in}$ is the series input resistance (gate resistance for field-effect transistor or base resistance for bipolar transistor) and $C_{in}$ is the shunt input capacitance (gate-source capacitance for field-effect transistor or diffusion base-emitter capacitance for bipolar transistor). In this case, the concept of the low-frequency conduction angle is not valid anymore when the fundamental-frequency and third-harmonic components of the output voltage are out-of-phase for conduction angles above 180°. In other words, the low-frequency or external conduction angle considered at the input of the transistor is different from the high-frequency or internal conduction angle considered at the transistor junction directly, due to a transient effect of the device input circuit. Thus, to achieve a 50% duty ratio of the collector voltage, the low-frequency conductance angles at the input may be significantly less than 180° [45]. In addition, it is necessary to introduce an additional phase shift at the third-harmonic component by tuning a third-harmonic component to make a collector voltage waveform more symmetrical.

## 3.9 Broadband capability of Class-F power amplifiers

Generally, the high-efficiency Class-F performance can be realized across the relatively narrow frequency bandwidth due to high-$Q$ resonant conditions at the fundamental-frequency and higher-order harmonic components. However, for a sufficiently low-power device when the real part of its input and load impedances is close to 50 $\Omega$, it becomes possible to simplify the design of the broadband input matching circuit and load network. In this case, it is enough to use a simple lossy input matching circuit composed of a single series inductive element, which can be realized using a high-impedance microstrip line, and a series small-value resistor, with a shunt resistor added to the gate-bias circuit, whose value is sufficiently small as well. It is also important to include both resistors to provide unconditionally stable operating conditions. The load network was designed primarily to achieve quasi-optimal Class-F loads at the fundamental-frequency and second-harmonic components. A nonresonant structure composed only of a single low-impedance microstrip line and a dc blocking capacitor was used in conjunction with the drain-bias quarter-wavelength line. Such a design with broadband capability achieved 21-dBm output power with a 0.5-dB flatness and a *PAE* of better than 65% from 575 to 915 MHz, corresponding to a relative bandwidth of 45% [46]. Another possibility to extend high-efficiency performance of a Class-F power amplifier over a wider frequency range is to provide for its operation in a deep saturation mode in the desired frequency range. For a two-stage 2-GHz Class-F GaN HEMT power amplifier with a final stage operating in saturation mode with a 3-dB gain compression point or more, a *PAE* greater than 60% is reached over 20% frequency bandwidth from 1.8 to 2.2 GHz with an output power varying between 40 and 42 dBm [47].

In handset applications where monolithic implementation for high-efficiency power amplifiers is required, the microstrip lines are too bulky to be employed in a broadband Class-F load network. Figure 3.38 shows the two-stage broadband monolithic Class-F power amplifier for handset application, where the input and interstage matching circuits are based on two-section *L*-type *LC* matching circuits (high-pass and low-pass sections for input matching circuit and two high-pass sections for interstage matching circuit), and a load network includes a few matching sections with the second- and third-harmonic resonators [48,49]. By using a two-section *L*-type matching section where each section is characterized by the same quality factor, it is convenient to match a low device input impedance of about 2 Ω with an input terminal impedance of 50 Ω through the intermediate impedance of 10 Ω when the matching circuit loaded quality factor $Q_L$ is reduced to $Q_L = \sqrt{(50/10) - 1} = \sqrt{(10/2) - 1} = 2$ (instead of $Q_L = \sqrt{(50/2) - 1} = 4.9$ for a single-section *LC* matching circuit) that ideally provides more than two times wider frequency bandwidth (BW) according to BW $= f_0/Q_L$, where $f_0$ is the center bandwidth frequency [33]. The interstage matching circuit comprises the bias-line inductance at the collector of the driver stage, and the series resistance is

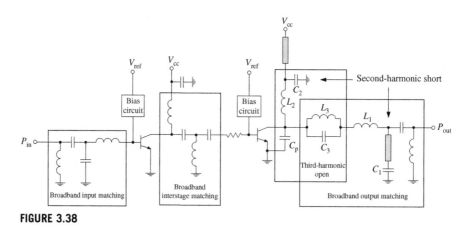

**FIGURE 3.38**

Schematic of broadband MMIC Class-F power amplifier.

connected to the base of the final stage to increase the real part of the device input impedance and to improve the operating stability of the power amplifier.

The load network comprises a broadband impedance matching at the fundamental frequency, the second-harmonic short circuits ($L_2C_2$ has a near zero impedance at the upper band of the second harmonic and $C_1$ with a short microstrip line has a near zero impedance at the lower band of the second harmonic), and the third-harmonic open circuit. The $L_3C_3$ tank resonator, which has an inductive reactance at the fundamental frequency, provides the high impedance at the third-harmonic frequency, whereas the device output (collector) capacitance $C_p$ is resonated out at the third-harmonic frequency by the inductance at the collector-bias line. The $L_1$ represents the series bondwire inductance, which is a part of the broadband matching circuit at the fundamental frequency. As a result, the proper second- and third-harmonic control could enhance the *PAE* up to 48% across 300-MHz bandwidth from 1.8 to 2.1 GHz for a handset power amplifier with an output power of 30 dBm with less than 1-dB variations at a supply voltage of 3.4 V implemented in the InGaP/GaAs HBT process.

## 3.10 **Practical Class-F power amplifiers and applications**

A typical VHF high-efficiency lumped-element bipolar power amplifier, which can provide a 10-W output power with a *PAE* of about 60% in a zero-bias Class C operation, is shown in Fig. 3.39(*a*). Using a *T*-type output-matching *LC* transformer with a series inductor creates high-impedance conditions for the second- and higher-order harmonics at the collector terminal, thereby improving the collector efficiency. In this case, the collector current waveform is close to the sinusoidal waveform, while the collector voltage waveform is characterized by a high value of its peak factor. To provide a reliable transistor operation when maximum collector voltage amplitude should be less than the collector-emitter breakdown voltage, it is necessary to reduce the collector supply voltage. Due to the small value of the transistor input impedance of about 1 $\Omega$, the frequency bandwidth of such a bipolar power amplifier is sufficiently narrow and normally does not exceed several percentage points. The inductor $L_3$ is required to provide a zero base-emitter bias voltage, while the inductor $L_1$ and bypass capacitor $C_b$ are necessary to isolate dc power supply from RF signal. Their values are sufficiently large to influence the amplifier matching conditions. Such an RF power amplifier in a slightly overdriven Class-B operation mode can provide a 10-W output power with a power gain of 8 dB and a *PAE* close to 70% at an operating frequency of 250 MHz [18].

However, to improve the power amplifier reliability by reducing a peak factor to a theoretical maximum value of 2, it is sufficient to use an RF grounded quarter-wavelength transmission line instead of an RF choke. Such an approach provides short-circuit conditions for collector voltage even harmonics, thus

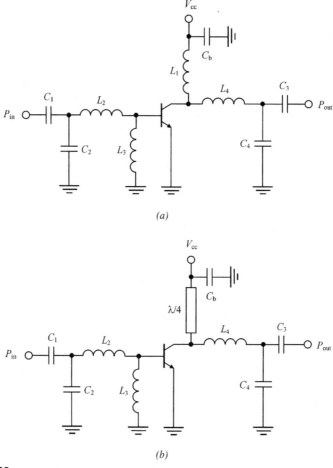

(a)

(b)

**FIGURE 3.39**

Typical VHF high-efficiency bipolar power amplifiers.

resulting in the square voltage and half-sinusoidal current waveform approximations typical for Class-F operation mode. To increase the impedance conditions
for higher-order harmonic components, it is necessary to use the series high-$Q$
resonant circuit tuned to the fundamental frequency and followed by the output
matching circuit. The typical schematic of such a high-efficiency Class-F VHF
power amplifier with a shunt quarterwave transmission line is shown in Fig. 3.39
($b$). As a result, a collector efficiency approaching 90% at an operating frequency
of 250 MHz for a 10-W hybrid bipolar power amplifier can be achieved [33]. At
higher frequencies when the output matching circuit is fabricated using the transmission-line technology replacing a lumped inductor by a series microstrip line
and a shunt capacitor by an open-circuit stub, its input impedance is optimized to

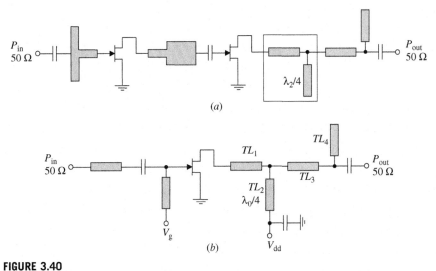

**FIGURE 3.40**

Simplified schematics of microstrip Class-F power amplifier.

realize high impedance condition at the third harmonic, also taking into account the active device parasitics [50].

The simplified circuit topology of a microstrip two-stage 900-MHz GaAs MESFET power amplifier is shown in Fig. 3.40($a$) [51]. The microstrip line between the MESFET device and open-circuit stub with an electrical length $\lambda_2/4$, where $\lambda_2$ is the second-harmonic wavelength, is a compensation line to compensate for the equivalent device output reactance. The $T$-type transmission-line impedance transformer, which consists of a series microstrip line, an open-circuit microstrip stub, and a series capacitor, provides an output impedance matching with a 50-$\Omega$ load. The input and interstage matching circuits at the fundamental frequency were designed based on the microstrip lines as well. As a result, with the second-harmonic controlled by a series microstrip line and an open-circuit microstrip stub (shown in a marked box), such a power amplifier demonstrates a drain efficiency of more than 80%, a $PAE$ of 71%, and an output power of 2 W at a supply voltage of 6 V.

A similar load-network configuration of a single-stage 1.75-GHz MESFET power amplifier is shown in Fig. 3.40($b$) [42,52]. It consists of a short-circuited quarterwave microstrip line $TL_2$, having high-impedance conditions at the fundamental and third harmonic components at its input, and an open-circuit stub $TL_4$ of a quarter wavelength at the third harmonic, having low impedance at the third harmonic component at its input. To realize close to a square-wave voltage and half-sinusoidal current waveforms at the drain terminal, these short- and open-circuit conditions are transformed to the device output by two series microstrip lines $TL_1$ and $TL_3$. This resulted in a drain efficiency of 75% and a power gain of 11 dB at an output power of 24.5 dBm with a drain bias voltage of 3 V. To better optimize impedance at the second harmonic, an additional short-circuited quarter-wave line can be included further into the output matching circuit of the microwave $X$-band power amplifier [53]. In a monolithic integrated circuits design, the second-harmonic short at the drain can be realized also by using a series resonant circuit connected to the ground, while the high-impedance condition at the third harmonic is provided by a series spiral inductor being a part of the output matching circuit [54].

One of the most important factors for high-efficiency Class-F operation mode is the value of the device saturation resistance $r_{sat}$ (or on-resistance $r_{on}$), especially at a low supply voltage. Here, $r_{sat}$ is the ratio of the drain-source voltage at the saturated drain current to the saturated drain current. It is difficult to improve the efficiency of a small-scale MESFET with a narrow gate width when, in a low-voltage operation, the ratio $r_{sat}/R_{out}$ (where $R_{out}$ is the real part of the device equivalent output impedance) is not small enough. To provide a high-efficiency operation mode of the power amplifier, it is required to increase this ratio as much as possible. For example, by decreasing $r_{sat}$ by half, the drain efficiency can be improved by about 10%. A MESFET device, which has a saturation resistance $r_{sat}$ of about 1 $\Omega$, can demonstrate a drain efficiency of 90% at a supply voltage of 6 V in a 900-MHz Class-F power amplifier [51].

The circuit schematic of a Class-F power amplifier implemented in a deep submicron 0.2-$\mu$m CMOS technology is shown in Fig. 3.41 [55]. Applying a Class-F operation mode has the advantage of substantially less drain voltage peak factor compared with a Class-E mode. This helps to overcome the problem of a low oxide breakdown voltage, which limits the maximum output power and efficiency of the CMOS power amplifier because of the lower supply voltage

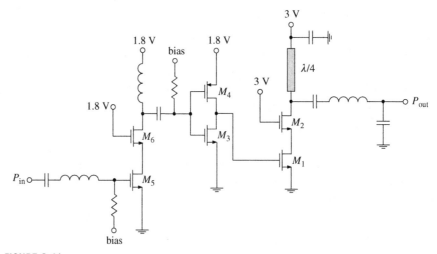

**FIGURE 3.41**

Schematic of Class-F power amplifier with quarterwave transmission line.

required for the device protection. The Class-F operation mode is achieved by using an external quarter-wavelength transmission line together with a series on-chip resonant circuit in the load network that provides high impedance at the second- and higher-order harmonics. In a cascode configuration of the final stage, the thin gate device $M_1$ is protected by a thick oxide (80 Å) device $M_2$ with no threat to oxide breakdown under supply voltage of 3 V. The driver stage based on a complementary nMOS and pMOS pair eliminates the problem of negative voltage swing across the gate of the cascode device $M_1$, which is normally the case for a single-ended nMOS device, and provides the driving signal waveform closer to a square wave. Such a CMOS Class-F power amplifier operating at 900 MHz can deliver a maximum output power of 1.5 W with a *PAE* of 43%.

Figure 3.42 shows the simplified schematic of a parallel-amplifier architecture implemented in a 0.25-μm CMOS technology and intended to provide high efficiency at backoff output power levels [56]. This architecture employs three binary-weighted Class-F power amplifiers, the output powers of which are combined in a power-combining network based on the parallel quarterwave transmission lines loaded by the parallel resonant tank tuned to the fundamental frequency. The capability to completely turn off each individual power amplifier without interfering with the operation of other individual power amplifiers is

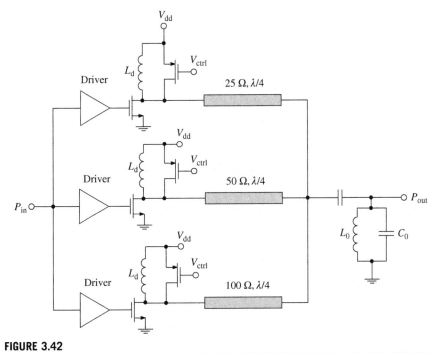

**FIGURE 3.42**

Class-F parallel-power amplifier architecture.

provided by the addition of pMOS shorting switches, resulting in high impedance at the end of the corresponding transmission line. The power-amplifier architecture operating at 1.4 GHz from a 1.5-V power supply occupies an active die size of 0.43 mm$^2$ and achieves a *PAE* of 49% at a maximum output power of 300 mW, while maintaining a *PAE* of greater than 43% over a lower output power range down to 100 mW. The transmission lines are implemented using the printed circuit board (PCB) microstrip lines. On-chip transmission-line fabrication in CMOS technology by using *LC* ladders makes it significantly shorter. For example, the frequency response of 10 sections containing a series inductor and a shunt capacitor each can approximate that of the transmission line within 5% occupying the area of about 14 times shorter. In this case, it is enough to use a spiral inductor to implement both series inductor and shunt capacitor which can be obtained with the bottom-plate parasitic capacitance of the spiral. However, the maximum *PAE* of the parallel on-chip power amplifier architecture degrades by 10−15%.

An efficient Class-F operation mode can also be applied to a distributed power amplifier. Figure 3.43 shows the simplified schematic of a modified Class-F single-ended dual-fed distributed MESFET power amplifier with a device

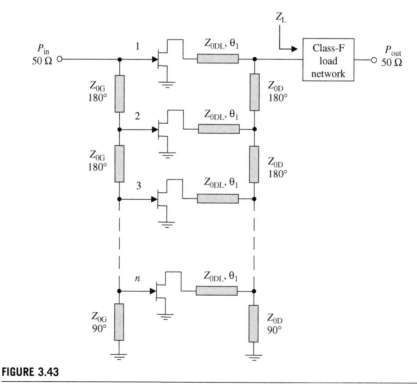

**FIGURE 3.43**

Simplified schematic of microstrip Class-F distributed amplifier.

electrical spacing of 180° at the center bandwidth frequency for optimum operation [57]. The two-port Class-F load network simultaneously provides impedance
matching at the fundamental frequency, high impedance at the third and fifth
harmonics, and low impedance at even harmonics. Namely, $Z_L$ is equal to $Z_{0D}$ at
the fundamental frequency but infinite at the third and fifth harmonic components. The open-circuit and short-circuit terminations of the output line are
applied directly to the drain of each device through the half-wave transmission
lines. For an idealized Class-F optimum operation, the impedance $Z_L(\omega_0)$ seen
from the MESFET drain at the fundamental frequency considering both forward
and reverse traveling waves can be obtained by

$$Z_L(\omega_0) = Z_{0D} = \frac{8}{n\pi^2} \frac{V_{dd}}{I_0} \tag{3.120}$$

where $V_{dd}$ is the drain bias voltage, $I_0$ is the total dc current, and $n$ is the number
of MESFETs.

To compensate for the device parasitics, an extra transmission line is cascaded
with each device drain port. The characteristic impedance $Z_{0DL}$ and electrical
length $\theta_1$ of this transmission line are

$$Z_{0DL} = \sqrt{\frac{L_d}{C_{ds}}} \tag{3.121}$$

$$\theta_1 = \pi - \omega_0\sqrt{L_dC_{ds}} \tag{3.122}$$

where $\omega_0$ is the fundamental angular frequency, $L_d$ is the series drain inductance, and
$C_{ds}$ is the shunt drain-source capacitance. The combination of $L_d$ and $C_{ds}$, forming a
simple low-pass filter section, and extra transmission line, behaves as a half-wave
transmission-line transformer if $\omega\sqrt{L_dC_{ds}}$ is less than 36° or one-tenth of the
wavelength at least up to the third harmonic. In a hybrid technology, such a 2-FET
($n = 2$) Class-F distributed power amplifier is able to achieve a drain efficiency of
71% and an output power of 22 dBm at an operating frequency of 1.75 GHz.

Modern wireless communication systems require feeding the signal with a
non-constant envelope through the power amplifier. In this case, there is a tradeoff
between power-amplifier efficiency and linearity with an improvement in one
coming at the expense of the other. In a classical analog envelope elimination and
restoration (EER) Kahn approach where special devices are required to separate
amplitude (envelope) and RF phase-modulated signals, one type of a power amplifier is responsible for envelope signal amplification, while another type of a power
amplifier is fed by a constant-envelope RF signal, as shown in Fig. 3.44. The constant-envelope RF signal can be amplified efficiently by a nonlinear power amplifier (PA) using a Class-F operation mode. For example, at the operating frequency
of 8.4 GHz, a Class-F MESFET power amplifier can provide the maximum instantaneous efficiency of 55% with an output power of 610 mW [58]. Higher drain
efficiency of about 73% at an average output power of 31 dBm can be achieved
for a transmission-line Class-F GaN HEMT power amplifier optimized to operate

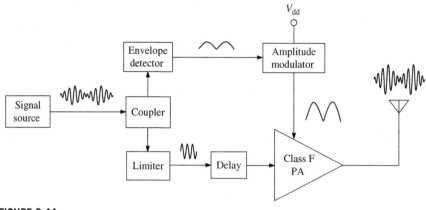

**FIGURE 3.44**

Block diagrams of Kahn EER transmitter with a Class-F power amplifier.

in the EER transmitter at 2.14 GHz [39]. Amplitude modulation of the final stage of the power amplifier based on a Class-S modulator restores the envelope to the phase-modulated carrier signal creating an amplitude replica of the input signal [16]. In contrast to linear power amplifiers, a Kahn EER transmitter is operated with high efficiency over a wide dynamic range of backoff output power levels and, therefore, produces an average efficiency that is typically three to five times higher. To minimize misalignment between phase and amplitude, the delay line located in phase-modulated or envelope amplifying path is required.

In modern radio transmitters intended for wireless applications, both the envelope and phase-modulated signals can be easily generated separately using a digital signal processing (DSP) technique [33]. The average efficiency of 26.4% for a multi-carrier signal and 43.8% for a quadrature amplitude modulation (QAM) with better linearity can be achieved in an X-band Class-F power amplifier using the Kahn technique compared with 9.5% and 28.7% for a linear power-amplifier mode, respectively [58]. Average efficiency can even be increased by applying a modified Kahn technique using an additional drive modulation when dynamic RF input amplitude varies proportionally to the signal envelope.

# REFERENCES

1. Latour M, Chireix H. The efficiency of three-electrode tubes used for the production of continuous waves in radio telegraphy, that is, for the conversion of direct current into alternating current. *Proc. IRE*. September 1923;11:551−558.
2. Prince DC. Vacuum tubes as power oscillators, part III. *Proc. IRE*. September 1923;11:527−550.
3. Shelleng JC. Amplifying system. U.S. Patent 1,484,967, February 1924.

4. Round HJ. Wireless telegraph and telephone transmission. U.S. Patent 1,564,627, December 1925.

5. Zenneck J, Rukop H. *Lehrbuch der Drahtlosen Telegraphie*. Stuttgart: Ferdinand Enke; 1925.

6. Fomichev IN. A new method to increase efficiency of the radio broadcasting station (in Russian). *Elektrosvyaz*. June 1938;58−66.

7. Sarbacher RI. Power-tube performance in Class C amplifiers and frequency multipliers as influenced by harmonic voltage. *Proc. IRE*. November 1943;31:607−625.

8. Model ZI, Ivanov BI, Person SV, Soloviev GF. Increasing of the efficiency of a high power HF vacuum-tube oscillator by separating the third harmonic (in Russian). *Radiotekhnika*. April 1947;2:15−23.

9. Royden GT. High-frequency amplifier. U.S. Patent 2,498,711, February 1950.

10. Berman LS. Increasing of useful power of the resonant semiconductor power amplifier by increasing its efficiency, Part I (in Russian). *Radiotekhnika*. November 1957;12:62−65.

11. Tyler VJ. A new high-efficiency high-power amplifier. *Marconi Review*. Fall 1958;21: 96−109.

12. Fuzik NS. Biharmonic modes of a tuned RF power amplifier. *Telecommun. Radio Eng*. July 1970;25(Part 2):117−124.

13. Zivkovic Z, Marcovic A. Third harmonic injection increasing the efficiency of high-power HF amplifiers. *IEEE Trans. Broadcasting*. June 1985;BC-31:34−39.

14. Rudyakova AN, Krizhanovski VG. Driving waveforms for Class-F power amplifiers. *2000 IEEE MTT-S Int. Microwave Symp. Dig*. 473−476.

15. White PM. Effect of input harmonic terminations on high efficiency Class-B and Class-F operation of PHEMT devices. *1998 IEEE MTT-S Int. Microwave Symp. Dig*. 3:1611−1614.

16. Krauss HL, Bostian CW, Raab FH. *Solid State Radio Engineering*. New York: John Wiley & Sons; 1980.

17. Raab FH. An introduction to Class-F power amplifiers. *RF Design*. May 1996;19: 79−84; July 1996, 14.

18. Snider DM. A theoretical analysis and experimental confirmation of the optimally loaded and overdriven RF power amplifier. *IEEE Trans. Electron Devices*. December 1967;ED-14:851−857.

19. Raab FH. Class-F power amplifiers with maximally flat waveforms. *IEEE Trans. Microwave Theory Tech*. November 1997;MTT-45:2007−2012.

20. Raab FH. Maximum efficiency and output of Class-F power amplifiers. *IEEE Trans. Microwave Theory Tech*. June 2001;MTT-49:1162−1166.

21. Rhodes JD. Output universality in maximum efficiency linear power amplifiers. *Int. J. Circuit Theory Appl*. July−August 2003;31:385−405.

22. Raab FH. Class-F power amplifiers with reduced conduction angles. *IEEE Trans. Broadcasting*. December 1998;BC-44:455−459.

23. Raab FH. FET power amplifier boosts transmitter efficiency. *Electronics*. June 1976;49:122−126.

24. Borisov VA, Voronovich VV. Analysis of switchmode transistor amplifier with parallel forming transmission line (in Russian). *Radiotekhnika i Elektronika*. August 1986;31:1590−1597.

25. Fuzik NS, Sadykov EA, Serguchev VI. Electrical design of the oscillatory circuits of the final stage of a radio transmitter operating in a biharmonic mode. *Telecommun Radio Eng*. January 1970;25(Part 2):141−145.

26. Voronovich VV, Galakh VP. Investigation of a polyharmonic bipolar transistor oscillator by monoharmonic excitation. *J. Commun. Technol. Electronics*. January 1997;42:102−108.

27. Trask C. Class-F amplifier loading networks: a unified design approach. *1999 IEEE MTT-S Int. Microwave Symp. Dig*. 351−354.

28. Kopp WS, Pritchett SD. High efficiency power amplification for microwave and millimeter frequencies. *1989 IEEE MTT-S Int. Microwave Symp. Dig*. 857−858.

29. Gao S, Xu H, Mishra UK, York RA. MMIC Class-F power amplifiers using field-plated AlGaN/GaN HEMTs. *2006 IEEE Compound Semiconductor Integrated Circuit Symp. Dig*. 81−84.

30. Grebennikov AV. Effective circuit design techniques to increase MOSFET power amplifier efficiency. *Microwave J*. July 2000;43:64−72.

31. Grebennikov AV. Circuit design technique for high efficiency Class F amplifiers. *2000 IEEE MTT-S Int. Microwave Symp. Dig*. 771−774.

32. Akkul M, Roberts M, Walker V, Bosch W. High efficiency power amplifier input/output topologies for base station and WLAN applications. *2004 IEEE MTT-S Int. Microwave Symp. Dig*. 843−846.

33. Grebennikov A. *RF and Microwave Power Amplifier Design*. New York: McGraw−Hill; 2004.

34. Wren M, Brazil TJ. Experimental Class-F power amplifier design using computationally efficient and accurate large-signal pHEMT model. *IEEE Trans. Microwave Theory Tech*. May 2005;MTT-53:1723−1731.

35. Kuroda K, Ishikawa R, Honjo K. Parasitic compensation design technique for a C-Band GaN HEMT Class-F amplifier. *IEEE Trans. Microwave Theory Tech*. November 2010;MTT-58:2741−2750.

36. Giannini F, Scucchia L. A complete class of harmonic matching networks: synthesis and application. *IEEE Trans. Microwave Theory Tech*. March 2009;MTT-57:612−619.

37. Grebennikov A. Load network design technique for Class F and inverse Class F PAs. *High Frequency Electronics*. May 2011;10:58−76.

38. Negra R, Ghannouchi FM, Baechtold W. Study and design optimization of multiharmonic transmission-line load networks for Class-E and Class-F *K*-Band MMIC power amplifiers. *IEEE Trans. Microwave Theory Tech*. June 2007;MTT-55:1390−1397.

39. Hong S, Woo YY, Kim I, et al. High efficiency GaN HEMT power amplifier optimized for OFDM EER transmitter. *2007 IEEE MTT-S Int. Microwave Symp. Dig*. 1247−1250.

40. Honjo K. A simple circuit synthesis method for microwave Class-F ultra-high-efficiency amplifiers with reactance-compensation circuits. *Solid-State Electronics*. August 2000;44:1477−1482.

41. Schmelzer D, Long SI. A GaN HEMT Class F amplifier at 2 GHz with >80% PAE. *IEEE J. Solid-State Circuits*. October 2007;SC-42:2130−2136.

42. Dietsche S, Duvanaud C, Pataut G, Obregon J. Design of high power-added efficiency FET amplifiers operating with very low drain bias voltages for use in mobile telephones at 1.7 GHz. *Proc. 23rd Europ. Microwave Conf*. 1993;252−254.

43. Ko S, Wu W, Lin J, et al. A high efficiency Class-F power amplifier using AlGaN/GaN HEMT. *Microwave and Optical Technology Lett*. October 2006;48:1955−1957.

44. Mitzlaff JE. High efficiency RF power amplifier. U.S. Patent 4,717,884, January 1988.

45. Rudiakova AN. BJT Class-F power amplifier near transition frequency. *IEEE Trans. Microwave Theory Tech*. September 2005;MTT-53:3045−3050.

46. Butterworth P, Gao S, Ooi SF, Sampbell A. High-efficiency Class-F power amplifier with broadband performance. *Microwave and Optical Technology Lett.* February 2005;44:243−247.

47. Ramadan A, Reveyrand T, Martin A, et al. Two-stage GaN HEMT amplifier with gate-source voltage shaping for efficiency versus bandwidth ehnacements. *IEEE Trans. Microwave Theory Tech.* March 2011;MTT-59:699−706.

48. Kang D, Choi J, Jun M, et al. Broadband Class-F power amplifiers for handset applications. *Proc. 39th Europ. Microwave Conf.* 2009;484−487.

49. Kang D, Kim D, Choi J, Kim J, Cho Y, Kim B. A multimode/multiband power amplifier with a boosted supply modulator. *IEEE Trans. Microwave Theory Tech.* October 2010;MTT-58:2598−2608.

50. Ooi SF, Gao S, Sambell A, Smith D, Butterworth P. A high efficiency Class-F power amplifier design technology. *Microwave J.* November 2004;47:110−122.

51. Chiba K, Kanmuri N. GaAs FET power amplifier module with high efficiency. *Electronics Lett.* November 1983;19:1025−1026.

52. Duvanaud C, Dietsche S, Pataut G, Obregon J. High-efficient Class F GaAs FET amplifiers operating with very low bias voltages for use in mobile telephones at 1.75 GHz. *IEEE Microwave and Guided Wave Lett.* August 1993;3:268−270.

53. Boesch RD, Thompson JA. X-Band 0.5, 1, and 2 Watt power amplifiers with marked improvement in power-added efficiency. *IEEE Trans. Microwave Theory Tech.* June 1990;MTT-38:707−711.

54. Maeda M, Nishijima M, Takehara H, Adachi C, Fujimoto H, Ishikawa O. A 3.5 V, 1.3 W GaAs power multi-chip IC for cellular phones. *IEEE J. Solid-State Circuits.* October 1994;SC-29:1250−1256.

55. Kuo TC, Lusignan BB. A 1.5-W Class-F RF power amplifier in 0.2-μm CMOS technology. *2001 IEEE Int. Solid-State Circuits Conf. Dig.* 154−155.

56. Shirvani A, Su DK, Wolley B. A CMOS RF power amplifier with parallel amplification for efficient power control. *IEEE J. Solid-State Circuits.* June 2002;SC-37:684−693.

57. Eccleston KW. Modified Class-F distributed amplifier. *IEEE Microwave and Wireless Components Lett.* October 2004;14:461−483.

58. Weiss MD, Raab FH, Popovic Z. Linearity of X-Band Class-F power amplifiers in high-efficiency transmitters. *IEEE Trans. Microwave Theory Tech.* June 2001; MTT-49:1174−1179.

# Inverse Class-F

## INTRODUCTION

Highly efficient operation of the power amplifier can also be obtained by applying biharmonic or polyharmonic modes when an additional single-resonant or multi-resonant circuit tuned to the even harmonics of the fundamental frequency is added into the load network. An infinite number of even-harmonic resonators results in an idealized inverse Class-F mode with a half-sinusoidal voltage waveform and a square current waveform at the device output terminal. In inverse Class-F power amplifiers analyzed in the frequency domain, the fundamental and harmonic load impedances are optimized by short-circuit termination and open-circuit peaking to control the voltage and current waveforms at the device output in order to obtain maximum efficiency. In this chapter, different inverse Class-F techniques using lumped and transmission-line elements including a quarterwave transmission line are analyzed. Design examples and practical RF and microwave inverse Class-F power amplifiers for different wireless applications are described and discussed.

## 4.1 Biharmonic and polyharmonic operation modes

Highly efficient biharmonic operation mode can be realized using a second-harmonic peaking when an additional parallel resonant circuit tuned to the second harmonic of the fundamental frequency is included in series into the load network. Similarly to the load network with the third-harmonic peaking, the additional resonator creates a high impedance at the second harmonic resulting in an efficiency improvement. In this case, by limiting to a biharmonic operation condition, the partial Fourier series of current $i(t)$ and voltage $v(t)$ in normalized form can be respectively written as

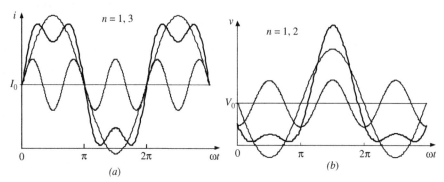

**FIGURE 4.1**

Fourier current and voltage waveforms with third and second harmonics.

$$\frac{v(\omega t)}{V_0} = 1 - \frac{\pi}{2}\sin \omega t - \frac{2}{3}\cos 2\omega t \tag{4.1}$$

$$\frac{i(\omega t)}{I_0} = 1 + \frac{4}{\pi}\sin \omega t + \frac{4}{3\pi}\sin 3\omega t \tag{4.2}$$

where $V_0$ and $I_0$ are the dc voltage and current components, respectively. Note that an infinite number of the voltage and current harmonics presented in a Fourier-series expansion results in the ideal half-sinusoidal voltage and square current waveforms.

Figure 4.1 shows that the shapes of the voltage and current waveforms can be significantly transformed by adding even one additional harmonic component with a proper phase. For example, the combination of the fundamental-frequency and third-harmonic components with 180° out-of-phase shift at the center of symmetry results in a flattened current waveform with depression in its center, as shown in Fig. 4.1(a), which can be minimized by using a proper ratio between the amplitudes of the fundamental and third harmonics. Similarly, the combination of the fundamental-frequency and second-harmonic components, which are in phase at the center of symmetry, sharpens the voltage waveform corresponding to minimum values of the voltage waveform, as shown in Fig. 4.1(b). The optimum ratio between the amplitudes of the fundamental and second current harmonics can maximize the current waveform in one-half of the period and minimize the current waveform during the other half of the period determined by the device saturation resistance in a practical circuit. This means that the power loss due to the active device can be minimized since the results of the integration over the period when minimum current corresponds to maximum voltage will give a small value compared with the power delivered to the load.

The effect of the inclusion of the parallel resonant circuit tuned to the second harmonic and located in series at the anode, as shown in Fig. 4.2(a), was first described and analyzed in early 1940s [1,2]. It was shown that the symmetrical

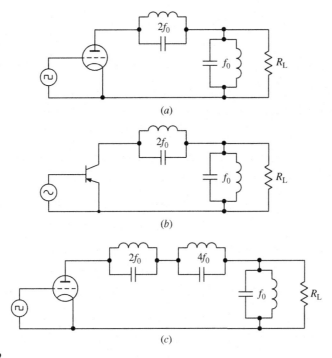

**FIGURE 4.2**

Biharmonic and polyharmonic power amplifiers.

anode current waveform and level of its depression can be provided with the opposite phase conditions between the fundamental-frequency and second-harmonic components ($\theta > 100°$ for anode current) and an optimum value of the ratio between their voltage amplitudes. It was also noted that high operation efficiency can be achieved even when impedance of the parallel circuit to second harmonic is equal to or slightly greater than impedance of the tank circuit to fundamental frequency. In practical vacuum-tube power amplifiers intended for operation at very high frequencies, the peak output power and anode efficiency can therefore be increased by 1.15–1.2 times [3]. In addition, such an approach can improve the modulation properties of the power amplifier when the phase of the second voltage harmonic becomes negative compared to that of the fundamental frequency [1]. However, generally, in view of the parasitic capacitance realized between anode and cathode, the entire anode circuit should be tuned to the second harmonic, not only a single resonator.

Figure 4.2(b) shows the biharmonic bipolar-transistor power amplifier with an additional resonant tank in the load network tuned to the second harmonic which achieved a collector efficiency of 92% and an output power of about 400 mW at an operating frequency of 106 kHz [4]. The required phase shift between the fundamental and second voltage harmonics at the collector is achieved due to the device inertia provided by its input capacitance and operation in a slightly saturated mode. However, in this case, the collector voltage peak factor is higher compared to that for a power amplifier with the third-harmonic resonator. Also, an additional resonator tuned to the fourth harmonic can be connected in series with the second-harmonic resonator, as shown in Fig. 4.2(c), to maximize the anode efficiency of the vacuum-tube amplifier with a square voltage driving waveform [5].

An efficiency improvement can also be significant in a Class-C power amplifier by using a single fourth-harmonic resonator in an anode circuit. In this case, the grid driving voltage can represent a simple sinusoidal waveform, since the current coefficients for the fundamental-frequency and fourth-harmonic components have opposite signs providing their 180° out-of-phase voltage conditions at the anode at the center of waveform symmetry. For an idealized active device whose output is represented by the voltage-controlled current source only, the flattening of the anode voltage waveform can take place when the biharmonic signal, including the fundamental-frequency and fourth-harmonic components at the anode, can be written as

$$v = V_a - V_{a1} \cos \omega t + V_{a4} \cos 4\omega t \tag{4.3}$$

where $V_a$ is the anode dc voltage, $V_{a1}$ is the fundamental-frequency amplitude, and $V_{a4}$ is the fourth-harmonic amplitude. For a piecewise-linear approximation of the active device transfer current–voltage characteristic, the anode current will represent a sequence of pulses, the duration of which is defined by a conduction angle $2\theta$ determined by the grid dc-bias conditions. The opposite signs of the voltage amplitudes $V_{a1}$ and $V_{a4}$ cause the anode voltage waveform to be bottom flattened, thus reducing the power losses during the time interval when the anode current is high. Figure 4.3 shows the simplified schematics of both (a) single-ended and (b) push–pull biharmonic vacuum-tube high power amplifiers, each having an additional harmonic resonator tuned to the fourth harmonic [6].

The ratio of the fundamental-frequency and fourth-harmonic amplitudes expressed through the anode current coefficients $\gamma_1(\theta)$ and $\gamma_4(\theta)$ is defined by

$$\frac{V_{a1}}{V_{a4}} = \frac{I_{a1}}{I_{a4}} \frac{R_L}{R_{L4}} = \frac{\gamma_1(\theta) R_L}{\gamma_4(\theta) R_{L4}} \tag{4.4}$$

where $I_{a1}$ is the amplitude of the fundamental current, $I_{a4}$ is the amplitude of the fourth-harmonic current, $R_{L4}$ is the load resistance for the fourth-harmonic current due to a finite value of the quality factor of the fourth-harmonic resonator,

$$\gamma_1(\theta) = \frac{1}{\pi} \left( \theta - \frac{\sin 2\theta}{2} \right) \tag{4.5}$$

$$\gamma_4(\theta) = \frac{\sin 4\theta \cos \theta - 4 \cos 4\theta \sin \theta}{30\pi}$$

$$= -\frac{2}{15\pi} \sin \theta \, (6 \cos^4 \theta - 7 \cos^2 \theta + 1). \tag{4.6}$$

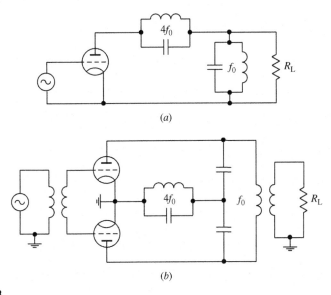

(a)

(b)

**FIGURE 4.3**

Single-ended and push—pull power amplifiers with fourth harmonic peaking.

From Eqs (4.5) and (4.6), it can be seen that the signs of the fundamental-frequency current coefficient $\gamma_1(\theta)$ and fourth-harmonic current coefficient $\gamma_4(\theta)$ are opposite when

$$\cos^{-1}\frac{1}{\sqrt{6}} < \theta < \pi - \cos^{-1}\frac{1}{\sqrt{6}} \tag{4.7}$$

which results in $65.9° < \theta < 114.1°$, thus satisfying the required relationship between $V_{a1}$ and $V_{a4}$ in Eq. (4.3). This means that the bottom flattening of the anode voltage waveform can be achieved in Class-B or Class-C power amplifiers with a sinusoidal driving signal. Figure 4.4(a) shows the anode voltage waveform where the amplitude of the fourth-harmonic component is one-eighth of the amplitude of the fundamental component, while the anode pulsed current waveform corresponding to a Class-C operation mode is shown in Fig. 4.4(b). Higher anode efficiency can be achieved in a deep Class-C mode when $\theta$ is significantly smaller than 90°, since the bottom flattening of the voltage waveform occurs during the time interval shorter than half a period.

To further increase efficiency of the biharmonic power amplifier, it is advisable to provide a biharmonic driving signal consisting of the fundamental frequency and its second harmonic component [7,8]. In this case, the fundamental-frequency and second-harmonic components must be 180° out of phase at their maximum amplitudes, and the amplitude of the second harmonic is preferably chosen to have approximately three-eighths the amplitude of the fundamental. Figure 4.5(a) shows

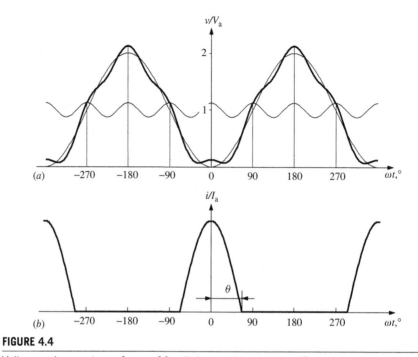

**FIGURE 4.4**

Voltage and current waveforms of fourth-harmonic power amplifier.

the simplified circuit schematic of a vacuum-tube power amplifier containing the fundamental tank and second-harmonic resonant circuits, both in grid and anode circuits. From Eq. (3.13) given in Chapter 3, it follows that, for optimum values $\theta_{opt} \approx 80°$ and $a_n^0 \approx 0.41$, the anode efficiency in a biharmonic mode with second-harmonic injection when $n = 2$ can be increased up to $\eta' = (0.95 \div 0.96)\xi$ [7].

The simple solution to realize 180° out-of-phase conditions between the voltage fundamental-frequency and second-harmonic components at the device output is to use a second-harmonic tank resonator connected in series to the device input, as shown in Fig. 4.5(*b*) [1]. Such an approach makes it possible to flatten the anode voltage waveform in the active region avoiding the device saturation mode. Due to the diode-type input of the vacuum tube, the resultant grid current pulse will contain a strong second-harmonic component resulting in a second harmonic voltage component across the input resonator. The loaded quality factor of the second-harmonic resonator must be high enough to neglect the voltage drop at the fundamental frequency. As a result, the second-harmonic resonator has no effect on the voltage fundamental-frequency component. However, it provides a phase shift of 180° for the second-harmonic component, since increasing a voltage drop across the resonator results in the decreasing of the voltage drop across the grid-cathode (base-emitter or gate-source) terminals.

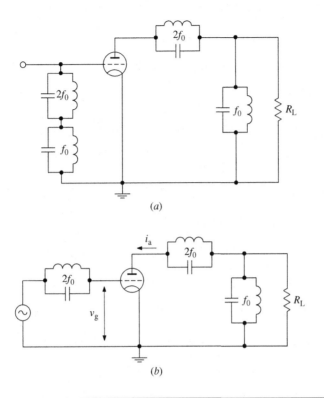

(a)

(b)

**FIGURE 4.5**

Biharmonic power amplifier with input harmonic control.

Figure 4.6(a) shows the combination of the fundamental-frequency and second-harmonic components shifted by 180°. In comparison with Fig. 4.1(b) where the fundamental-frequency and second-harmonic components are in phase at maximum point of the fundamental-frequency component resulting in a waveform bottom flattening, this voltage waveform at the device input is characterized by its top flattening when the second harmonic has minimum value at maximum point of the fundamental-frequency component. Then, choosing the bias point $V_g$ equal to the device pinch-off voltage $V_p$, the selection of which corresponds to Class-B mode with the conduction angle of 180° for monoharmonic operation, will result in the anode biharmonic current pulses with conduction angle $2\theta > 180°$, as shown in Fig. 4.6(b). At the same time, using a second-harmonic resonator in the load network contributes to the anode voltage waveform, as shown Fig. 4.1(b).

Similar biharmonic approach to affect individually the fundamental-frequency and second-harmonic conditions by using a transmission-line technique can be applied to the two-stage transistor power amplifiers intended to operate at ultra-high and microwave frequencies. In this case, the proper amplitude and phase

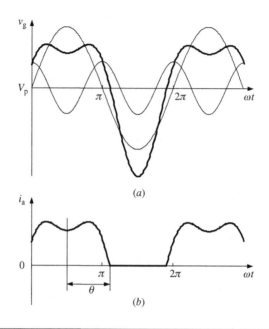

**FIGURE 4.6**

Input and output voltage and current waveforms with second harmonic.

conditions for the fundamental-frequency and second-harmonic components at the input of the final power stage separating from the driver stage output can be realized using the quarter-wavelength transmission lines in the form of the open-circuit and short-circuit stubs [9]. As a result, a drain efficiency of 77% can be achieved for a microstrip biharmonic GaAs MESFET power amplifier operated at the carrier frequency of 1.62 GHz with an output power of 27.9 dBm.

## 4.2 Idealized inverse Class-F mode

Generally, an infinite number of even-harmonic tank resonators can maintain a square current waveform, also providing a half-sinusoidal voltage waveform at the anode (collector or drain). Figure 4.7 shows the basic schematics of an inverse Class-F power amplifier with a multiple-resonator output filter to control the harmonic content of its collector voltage and current waveforms, thereby shaping them to reduce dissipation and to increase efficiency.

The term "inverse" means that the collector voltage and current waveforms are interchanged compared to a conventional case under the same idealized assumptions. Consequently, for a purely sinusoidal current flowing into the load, as shown in Fig. 4.8(a), the ideal collector current waveform is composed by a

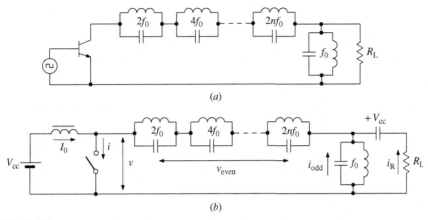

**FIGURE 4.7**

Basic circuits of inverse Class-F power amplifier with parallel resonant circuits.

fundamental-frequency and odd-harmonic components approximating a square waveform, as shown in Fig. 4.8(b). At the same time, the collector voltage waveform is composed by the fundamental-frequency and even-harmonic components approximating a half-sinusoidal waveform, as shown in Fig. 4.8(c). As a result, the shapes of the collector current and voltage waveforms provide a condition when the current and voltage do not overlap simultaneously, similarly to a conventional Class-F mode. Such a condition, with symmetrical collector voltage and current waveforms, corresponds to an idealized inverse Class-F operation mode with a 100% collector efficiency.

By using Eqs (3.18) and (3.22) given in Chapter 3 for a conventional Class-F mode, similar analysis of the distribution of voltages and currents in the inverse Class-F load network results in equations for the collector current and voltage waveforms as

$$i(\omega t) = 2I_0 - i(\omega t + \pi) \tag{4.8}$$

where $I_0$ is the dc current, and

$$v(\omega t) = V_R(\sin \omega t + |\sin \omega t|) \tag{4.9}$$

where $V_R$ is the fundamental-frequency amplitude. From Eq. (4.8), it follows that maximum value of the collector current cannot exceed a value of $2I_0$, and the time duration with maximum amplitude when $i = 2I_0$ coincides with the time duration with minimum amplitude when $i = 0$. Since the collector current is zero when the switch is turned off, the only possible waveform for the collector current is a square wave composing of only dc, fundamental-frequency, and odd-harmonic components.

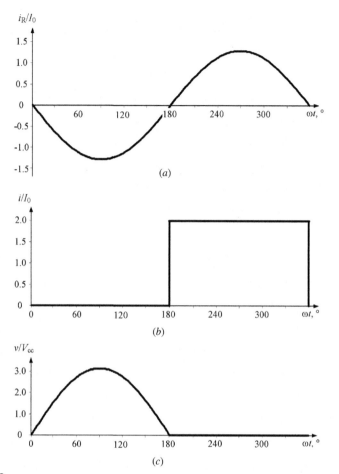

**FIGURE 4.8**

Ideal waveforms of inverse Class-F power amplifier.

By using a Fourier analysis of the current and voltage waveforms, the following equations for the dc voltage, fundamental voltage and current components in the collector voltage and current waveforms can be obtained:

the fundamental current component can be calculated using Eq. (4.8) as

$$I_1 = I_R = \frac{1}{\pi} \int_{\pi}^{2\pi} 2I_0 \sin(\omega t + \pi) \, d\omega t = \frac{4I_0}{\pi} \tag{4.10}$$

the dc voltage $V_{cc}$ can be calculated from Eq. (4.9) as

$$V_{cc} = \frac{1}{2\pi} \int_{0}^{\pi} 2V_R \sin \omega t \, d\omega t = \frac{2V_R}{\pi} \tag{4.11}$$

the fundamental voltage component can be calculated from Eq. (4.9) as

$$V_1 = \frac{1}{\pi} \int_0^\pi 2V_R \sin^2 \omega t \, d\omega t = V_R \tag{4.12}$$

Then, the ratio between the dc power $P_0$ and the output power at the fundamental frequency $P_1$ can be given by

$$P_1 = \frac{V_1 I_1}{2} = \frac{1}{2} \frac{\pi V_{cc}}{2} \frac{4I_0}{\pi} = P_0 \tag{4.13}$$

resulting in a theoretical collector efficiency of 100%.

The impedance conditions seen by the device collector for an idealized inverse Class-F mode must be equal to

$$Z_1 = R_1 = \frac{\pi^2}{8} \frac{V_{cc}}{I_0} \tag{4.14}$$

$$Z_{2n+1} = 0 \quad \text{for odd harmonics} \tag{4.15}$$

$$Z_{2n} = \infty \quad \text{for even harmonics} \tag{4.16}$$

## 4.3 Inverse Class-F with quarterwave transmission line

An idealized inverse Class-F operation mode can also be represented by using a sequence of the series resonant circuits tuned to the fundamental and odd harmonics, as shown in Fig. 4.9(a). In this case, it is assumed that each resonant circuit has zero impedance at the corresponding fundamental frequency $f_0$ and its odd-harmonic components $(2n + 1)f_0$ and infinite impedance at even harmonics $2nf_0$ realizing the idealized inverse Class-F square current and half-sinusoidal voltage waveforms at the device output terminal. As a result, the active device which is driven to operate as a switch sees the load resistance $R_L$ at the fundamental frequency, while the odd harmonics are shorted by the series resonant circuits.

An infinite set of the series resonant circuits tuned to the odd harmonics can be effectively replaced by a quarter-wavelength transmission line with the same operating capability. Such a circuit representation of an inverse Class-F power amplifier with a series quarterwave transmission line loaded by the series resonant circuit tuned to the fundamental frequency is shown in Fig. 4.9(b) [10,11]. The series-tuned output circuit presents a load resistance at the frequency of operation to the transmission line. At the same time, the quarter-wavelength transmission line transforms the load impedance according to

$$R = \frac{Z_0^2}{R_L} \tag{4.17}$$

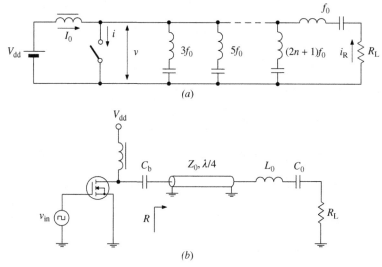

**FIGURE 4.9**

Inverse Class-F power amplifier with series quarterwave transmission line.

where $Z_0$ is the characteristic impedance of a transmission line. For even harmonics, the open circuit on the load side of the transmission line is repeated, thus producing an open circuit at the drain. However, the quarter-wavelength transmission line converts the open circuit at the load to a short circuit at the drain for odd harmonics with resistive load at the fundamental frequency.

Consequently, for a purely sinusoidal current flowing into the load due to infinite loaded quality factor of the series fundamentally tuned circuit, the ideal drain current and voltage waveforms can be represented by the corresponding normalized square and half-sinusoidal waveforms shown in Figs. 4.8(b) and 4.8(c), respectively. Here, a sum of the fundamental and odd harmonics approximates a square current waveform and a sum of the fundamental and even harmonics approximates a half-sinusoidal drain voltage waveform. As a result, the shapes of the drain current and voltage waveforms provide a condition when the current and voltage do not overlap simultaneously. The quarter-wavelength transmission line causes the output voltage across the load resistor $R_L$ to be phase-shifted by 90° relative to the fundamental-frequency components of the drain voltage and current.

Figure 4.10 shows a practical example of the circuit schematic of an inverse Class-F power amplifier with a series quarterwave line based on a 28-V, 10-W Cree GaN HEMT power transistor CGH40010. Similarly to the conventional Class-F GaN HEMT power amplifier, a simple lossy $RL$ input shunt network is also used to match the device input impedance to a 50-Ω source and to compensate for the device input gate-source capacitance $C_{gs}$ of about 5 pF at the fundamental frequency, providing a small-signal $S_{11}$ better than −20 dB at an operating frequency

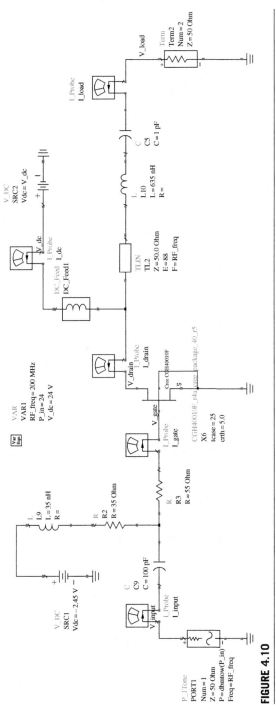

**FIGURE 4.10**

Schematic of inverse Class-F GaN HEMT power amplifier with series quarterwave line.

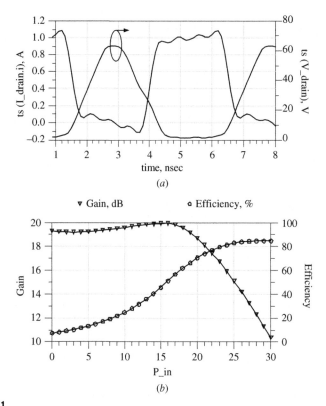

**FIGURE 4.11**

Simulated waveforms, gain, and efficiency of inverse Class-F GaN HEMT power amplifier.

of 200 MHz. The series 55-Ω resistor is necessary to provide unconditional operational stability. The simulated drain voltage (close to half-sinusoidal) and current
(close to square) waveforms are shown in Fig. 4.11(a) where small deviations from
the ideal waveforms (with optimized load-network parameters) can be explained due
to effect of the device output drain-source capacitance $C_{ds}$ and package parasitics
In this case, a maximum drain efficiency of 84.8% with a power gain of 12.3 dB
and an output power of 40.3 dBm at a supply voltage of 24 V were obtained with
a sine-wave driving signal in a deep saturation condition at an input power
$P_{in} = 28$ dBm, as shown in Fig. 4.11(b).

## 4.4 Load networks with lumped elements

Theoretical results show that the proper control of the second harmonic can
significantly increase the collector efficiency of the power amplifier by flattening
the output current waveform and minimizing the product of integration of the voltage

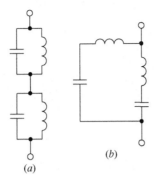

**FIGURE 4.12**

Two-terminal reactive network with series and parallel resonators.

and current waveforms. Practical realization of a multi-element high-order $LC$ resonant circuit can cause a serious implementation problem, especially at higher frequencies and in monolithic integrated circuits, when only three harmonic components can be effectively controlled. Therefore, it is sufficient to be confined to the three- or four-element resonant circuit comprising the load network of the power amplifier. In this case, the operation with the second-harmonic open circuit and third-harmonic short circuit is a promising concept for low-voltage power amplifiers [12].

In addition, it is necessary to take into account that, in practice, both extrinsic and intrinsic transistor parasitic elements such as output shunt capacitance or serious inductance have a substantial effect on the efficiency. The output capacitance $C_{out}$ can represent the collector capacitance $C_c$ in the case of the bipolar transistor or the sum of drain-source capacitance and gate-drain capacitance $C_{ds} + C_{gd}$ in the case of the FET device. The output inductance $L_{out}$ is generally composed of the bondwire and lead inductances for a packaged transistor, the effect of which becomes significant at higher frequencies. Figure 4.12 shows the typical two-terminal lumped reactive networks with series and parallel resonators used in a practical design procedure, which ideally provide infinite impedances at the fundamental-frequency and second-harmonic components.

The equivalent circuit of the second-harmonic impedance-peaking circuit is shown in Fig. 4.13($a$). Here, the series circuit consisting of an inductor $L_1$ and a capacitor $C_1$ creates a resonance at the second harmonic. Since the device output inductance $L_{out}$ and capacitance $C_{out}$ are tuned to create an open-circuited condition at the second harmonic, the device collector sees the resultant high impedance at the second harmonic. To achieve the second-harmonic high impedance, an external inductance can be added to interconnect the device output inductance $L_{out}$ directly at the output terminal (collector or drain) if its value is not sufficient. As a result, the values of the network parameters are

$$L_{out} = \frac{1}{4\omega_0^2 C_{out}} \tag{4.18}$$

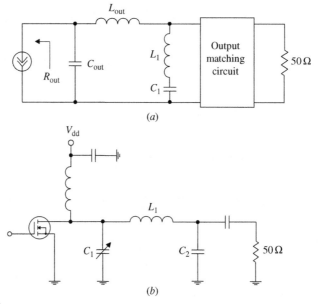

**FIGURE 4.13**

Second-harmonic lumped impedance-peaking circuits.

$$L_1 = \frac{1}{4\omega_0^2 C_1}. \tag{4.19}$$

The lumped load network with second-harmonic peaking, which was used in an inverse Class-F LDMOSFET power amplifier operated at 1 GHz with an output power of 13 W, is shown in Fig. 4.13(b) [13]. In this case, the capacitances $C_1$ and $C_2$ associated to inductance $L_1$ were optimized to present high impedance to second harmonic and an optimal load resistance seen by the device output. The capacitance $C_1$ was chosen to be adjustable in order to take into account the technology implementation accuracy. With accurate tuning of the load-network parameters, the power-added efficiency can be increased by more than 10% compared to the conventional Class-B mode.

The open- and short-circuit conditions for an inverse Class-F power amplifier can be obtained by using a lumped-element harmonic-control circuit which consists of two two-port reactance networks, one connected in series to the transistor output and the other connected in parallel with the transistor or load. The shunt reactance circuit is used to maintain zero impedances to the transistor at odd harmonics except at fundamental level, The real part of the transformed load

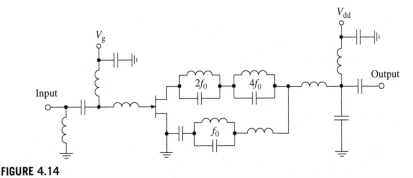

**FIGURE 4.14**

Schematic of inverse Class-F power amplifier with up to fourth-order harmonic control.

impedance is provided for the transistor at zero points of the circuit. Figure 4.14 shows the circuit configuration of a GaN HEMT inverse Class-F power amplifier considering up to fourth-order harmonic control [14]. Here, the series reactance circuit has pole points at $2f_0$ and $4f_0$, and zero point at $3f_0$, whereas the shunt reactance circuit has pole point at $f_0$ and zero point at $3f_0$, where $f_0$ is the fundamental frequency. The output matching circuit is added between the inverse Class-F harmonic-control circuit and 50-$\Omega$ load. Since the upper limit for the self-resonant frequency of the commercially available chip capacitors and inductors is around 8 GHz, the operating frequency $f_0 = 860$ MHz was chosen to demonstrate the drain efficiency over 75% at about 1-W output power.

As a first approximation for comparison between different operation modes, the output device resistance $R_{\text{out}}$ at the fundamental frequency required to realize an inverse Class-F operation mode with second-harmonic peaking can be estimated as an equivalent resistance $R_{\text{out}} = R_1^{(\text{invF})} = V_1/I_1$ determined at the fundamental frequency for an ideal inverse Class-F operation. Assuming zero saturation voltage and using Eq. (4.11) yields

$$R_1^{(\text{invF})} = \frac{\pi \, V_{\text{cc}}}{2 \, I_1} = \frac{\pi^2}{8} R_1^{(F)} = \frac{\pi}{2} R_1^{(B)} \qquad (4.20)$$

where $R_1^{(F)}$ is the equivalent fundamental output resistance in a conventional Class-F mode and $R_1^{(B)} = V_{\text{cc}}/I_1$ is the equivalent fundamental output resistance in an ideal Class B. As follows from Eq. (4.20), the equivalent output resistance for an ideal inverse Class-F mode is higher by more than 1.5 times compared to a conventional Class-B operation. Therefore, using an inverse Class-F operation mode simplifies the output matching circuit design by minimizing the impedance transformation ratio. This is very important for high output power level when the output resistance is sufficiently small. However, the maximum amplitude of the output voltage waveform can exceed the supply voltage by about three times.

## 4.5 Load networks with transmission lines

The ideal inverse Class-F power amplifier cannot provide all the voltage required by third and higher-order odd-harmonic short-circuit termination by the use of a single parallel transmission line, as can be easily realized by a quarter-wavelength transmission line for even-harmonic shorting in the conventional Class-F power amplifier. In this case, with a sufficiently simple circuit schematic convenient for practical realization, applying the current second-harmonic peaking and voltage third-harmonic shorting can result in a maximum drain efficiency of more than 80% [15,16]. For example, by providing a proper high-impedance second-harmonic peaking and load matching at the fundamental, the collector efficiency of 83% for a 425-MHz microstrip bipolar power amplifier and the drain efficiency of 68% for a 4-GHz MMIC GaN HEMT power amplifier have been achieved [17,18]. Generally, to control the second and third harmonics at microwave frequencies, it is preferable to use short-circuit and open-circuit stubs instead of lumped capacitors in the load network for better performance predictability and tuning accuracy [19,20].

Figure 4.15(*a*) shows the idealized transmission-line inverse Class-F load-network schematic where a shunt $\lambda/6$ transmission line grounded through the bypass capacitor provides a low impedance for the third harmonic and certain impedances for the fundamental and second harmonics at the device drain terminal [16,21]. The second-harmonic peaking is achieved by using an open-circuit $\lambda/8$ stub which creates a low impedance at the right-hand side of the series $\lambda/12$ transmission line at the second harmonic. As a result, this series $\lambda/12$ transmission line with electrical length of 60° at the second harmonic (combined in parallel with the shunt $\lambda/6$ transmission line with electrical length of 120° at the second harmonic) represents a second-harmonic tank.

It is also useful and very practical for inverse Class-F power amplifiers to consider an alternative transmission-line load network with a series $\lambda/8$ transmission line as a first element, as shown in Fig. 4.15(*b*) [22,23]. In this case, the series $\lambda/8$ transmission line short-circuited at its right-hand side by the shunt quarter-wavelength transmission line at the second harmonic provides an open circuit at the second harmonic seen by the device output, while the combined $(\lambda/8 + \lambda/24 = \lambda/6)$ series transmission line short-circuited at its right-hand side by the open-circuit $\lambda/12$ stub provides a short circuit at the third harmonic seen by the device output. By proper choice of the characteristic impedances of the transmission line, the impedance matching conditions at the fundamental frequency can be realized.

The load-network impedance $Z_{net}$ seen by the device output at the fundamental can be written as

$$Z_{net} = Z_1 \frac{R_L(Z_2 - Z_1 \tan 30° \tan 60°) + jZ_1Z_2 \tan 60°}{Z_1Z_2 + j(Z_1 \tan 30° + Z_2 \tan 60°)R_L} \tag{4.21}$$

where $Z_1$ is the characteristic impedance of the combined series transmission line and $Z_2$ is the characteristic impedance of an open-circuit $\lambda/12$ stub. Separating Eq. (4.21)

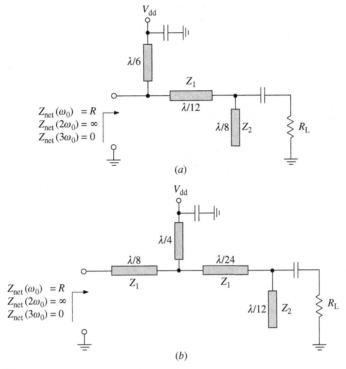

**FIGURE 4.15**

Idealized transmission-line inverse Class-F load networks.

into real and imaginary parts and taking into account that $\mathrm{Re}Z_{\mathrm{net}} = R$ and $\mathrm{Im}Z_{\mathrm{net}} = 0$, the system of two equations with two unknown parameters is obtained by

$$(Z_1 + 3Z_2)^2 R_{\mathrm{L}}^2 R - 3Z_1^2 Z_2^2 (4R_{\mathrm{L}} - 3R) = 0 \tag{4.22}$$

$$3Z_1^2 Z_2^2 - R_{\mathrm{L}}^2 (Z_2 - Z_1)(Z_1 + 3Z_2) = 0 \tag{4.23}$$

which allows direct calculation of the characteristic impedances $Z_1$ and $Z_2$. This system of two equations can be explicitly solved as a function of the parameter $r = R_{\mathrm{L}}/R$ resulting in

$$\frac{Z_1}{R_{\mathrm{L}}} = \frac{\sqrt{4r - 1}}{\sqrt{3}\,r} \tag{4.24}$$

$$\frac{Z_1}{Z_2} = \frac{r - 1}{r}. \tag{4.25}$$

Consequently, if the required optimum load resistance and standard load resistance are equal to $R = 20\ \Omega$ and $R_L = 50\ \Omega$, respectively, resulting in $r = 2.5$, then the characteristic impedance of the series transmission line calculated from Eq. (4.22) is equal to $Z_1 = 35\ \Omega$ and the characteristic impedance of the open-circuit stub calculated from Eq. (4.23) is equal to $Z_2 = 58\ \Omega$.

Figure 4.16 shows the simplified inverse Class-F load network with the second- and third-harmonic control using a series stepped-impedance transmission line, where $\theta_1$ and $\theta_2$ are the electrical lengths of the stepped-impedance transmission-line sections and $M$ is the characteristic impedance ratio. In this case, a simple combination of two loaded transmission lines with high and low characteristic impedances can provide the fundamental impedance matching and second-harmonic tuning simultaneously to a high impedance nearby open-circuit condition [18]. However, by optimizing the characteristic impedances of the transmission-line sections, infinite impedance at the second harmonic and zero impedance at the third harmonic can be provided simultaneously. For the simplicity of analytical calculations, the transmission-line lengths can be set equal as $\theta = \theta_1 = \theta_2$.

The input network impedance $Z_{net}$ as a function of electrical length $\theta$ of equal sections of the stepped-impedance transmission line is written as

$$Z_{net} = Z_0 \frac{Z_L(1 - M \tan^2 \theta) + jZ_0 \left( \dfrac{1}{M} + 1 \right) \tan \theta}{Z_0 \left( 1 - \dfrac{\tan^2 \theta}{M} \right) + jZ_L(M + 1) \tan \theta} \tag{4.26}$$

where $Z_L$ is the load impedance including the short-circuit and open-circuit stubs.

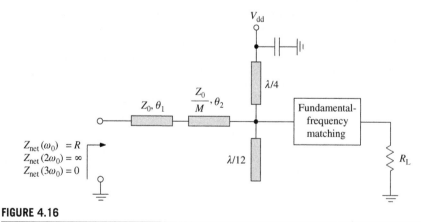

**FIGURE 4.16**

Transmission-line inverse Class-F load network with stepped-impedance line.

For $Z_L = 0$,

$$Z_{net} = jZ_0 \frac{\dfrac{1}{M} + 1}{1 - \dfrac{\tan^2 \theta}{M}} \tan \theta \tag{4.27}$$

whose zeroes correspond to $\theta_{0k} = (\pi/2)k$ and poles correspond to

$$\theta_{pk} = k\pi \pm \tan^{-1}\sqrt{M} \tag{4.28}$$

where $k = 0, 1, 2, \ldots$ . Thus, by choosing the proper characteristic impedance ratio $M$, the first pole $\theta_{p1}$ can be moved from initial value of $45°$ for uniform line towards the second zero at $\theta_{02} = 90°$ and set at $\theta_{p1} = 60°$ for the stepped-impedance transmission line with $M = 3$. As a result, by using such a stepped-impedance transmission line, an open circuit at the second harmonic and short circuit at the third harmonic of the fundamental frequency can be provided if the fundamental frequency corresponds to $\theta = 30°$.

Figure 4.17(*a*) shows the design example of a stepped-impedance transmission-line inverse Class-F load network with harmonic tuning which provides an

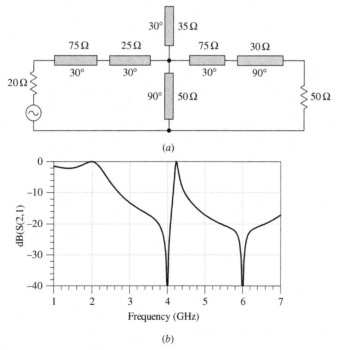

(*a*)

(*b*)

**FIGURE 4.17**

Frequency response of load network with second- and third-harmonic control.

impedance matching at the fundamental frequency between the input impedance of 20 Ω and load impedance of 50 Ω at $f_0 = 2$ GHz [20,24]. In this case, the stepped-impedance transmission line with equal section lengths of 30° loaded by a short-circuited quarter-wavelength line and an open-circuit $\lambda/12$ stub creates an open circuit at the second harmonic and short circuit at the third harmonic seen by the 20-Ω source at its input. Using a subsequent short-length transmission line as an element of the output fundamental-frequency matching circuit is required to compensate for the reactive part of the modified source impedance (any source impedance can be transformed to a real load impedance using a $\lambda/8$ transformer whose characteristic impedance is equal to the magnitude of the source impedance [25]) followed by the quarter-wavelength transmission line which provides the transformation of the real part of the modified source impedance to the standard load of 50 Ω. The frequency response of such a stepped-impedance transmission-line inverse Class-F load network with harmonic tuning is shown in Fig. 4.17(*b*), which demonstrates the fundamental-frequency impedance matching at 2 GHz, second-harmonic open circuit at 4 GHz, and third-harmonic short circuit at 6 GHz.

However, the efficiency of the power amplifier can be limited by the transistor output capacitance $C_{out}$ (mostly represented by drain-source capacitance $C_{ds}$ or collector capacitance $C_c$) if this capacitance is not absorbed into multi-harmonic load network without compromising the ability to properly terminate the second- and third-harmonic components. Figure 4.18(*a*) shows the simplified circuit schematic of a transmission-line transistor power amplifier, whereas the equivalent load network corresponding to inverse Class-F mode with a shunt short-circuited $\lambda/3$ transmission line which provides a short circuit at the third harmonic at the device output is shown in Fig. 4.18(*b*) [16]. This circuit configuration is similar to that one used to provide a conventional Class-F operation mode.

For such an inverse Class-F microstrip power amplifier, it is necessary to provide the following electrical lengths for the transmission lines at the fundamental frequency:

$$\theta_1 = \frac{\pi}{3} \tag{4.29}$$

$$\theta_2 = \frac{1}{2}\tan^{-1}\left[\left(2Z_0\,\omega_0 C_{out} + \frac{1}{\sqrt{3}}\right)^{-1}\right] \tag{4.30}$$

$$\theta_3 = \frac{\pi}{4} \tag{4.31}$$

where $Z_0$ is the characteristic impedance of the microstrip lines. The transmission line $TL_1$ with electrical length $\theta_1 = 60°$ provides a short-circuited condition for the third harmonic, whereas the remaining two transmission lines, together with the device output capacitance, form a parallel resonant circuit to realize an

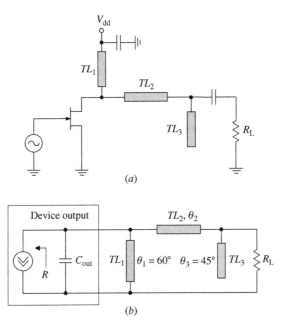

**FIGURE 4.18**

Transmission-line inverse Class-F load network with shunt capacitance.

open-circuit condition for the second harmonic at the drain terminal. The open-circuit stub $TL_3$ with electrical length $\theta_3 = 45°$ creates a short-circuited condition at the end of the transmission line $TL_2$ at the second harmonic. Thus, the series short-circuited transmission line $TL_2$ having an inductive reactance is tuned to the parallel resonance condition with the device output capacitance $C_{out}$ and short-circuited transmission line $TL_1$.

As an example, Fig. 4.19 shows the frequency–response characteristic of the microstrip impedance-peaking circuit using an alumina substrate for the device equivalent output resistance $R_{out} = 50\ \Omega$ and output capacitance $C_{out} = 2.2$ pF, characteristic impedance of microstrip lines $Z_0 = 50\ \Omega$ and electrical length $\theta_2 = 19°$. From Fig. 4.19, it follows that, for the second-harmonic peaking and third-harmonic short-circuit termination, an additional output matching at the fundamental frequency $f_0 = 500$ MHz is required, taking into account the reactance introduced by the impedance-peaking load network.

Generally, to ensure that the fundamental matching did not affect the harmonic control, the load network can represent a separate fundamental matching circuit and the harmonic-control circuit, as shown in Fig. 4.20 [26,27]. In this case, the harmonic-control circuit contains two shunt open-circuit stubs and two series transmission lines which provide the second- and third-harmonic control. The electrical length of each open-circuit stub is chosen to present a short circuit

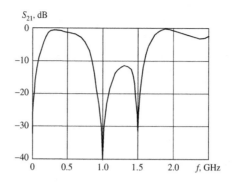

**FIGURE 4.19**

Frequency response of the microstrip impedance-peaking circuit.

for the frequency of the corresponding harmonic at the junction with the series transmission line, in order not to be affected by the tuning of the fundamental matching network. The series transmission lines with electrical lengths $\theta_1$ and $\theta_2$ are used to provide the right phase of the optimum harmonic reflection coefficient $\Gamma_{\mathrm{opt}}$ to the drain of the transistor. Figure 4.20($a$) shows the load-network configuration where the second-harmonic stub is closer to the drain then the third-harmonic stub. Since its length is $\lambda/8$, which corresponds to $\lambda/4$ at the second harmonic, the second-harmonic stub transforms the open end to a short circuit, allowing for the achievement of a reflection coefficient independent of the fundamental matching and with a magnitude very close to unity at $2f_0$. Similarly, the third-harmonic stub with a length of $\lambda/12$ at $f_0$ is used to tune the third harmonic independently from the fundamental matching. The load-network configuration with a third-harmonic stub placed closer to the transistor drain is shown in Fig. 4.20($b$). Even though both configurations present the same reflection coefficient to the transistor at the fundamental, second, and third harmonic frequencies, their resulting efficiencies can be different depending on the difference in the conduction loss introduced by each topology. Despite high drain efficiencies provided by these load-networks in implemented 10-W GaN HEMT inverse Class-F power amplifiers, more than 80% at 1 GHz and about 75% at 2.45 GHz, making the entire tuning procedure properly match the impedances at the fundamental frequency and provide the second- and third-harmonic control is quite complicated and each load network requires a significant substrate space to be implemented.

At higher frequencies, especially at microwaves, the effect of the output series inductance $L_{\mathrm{out}}$, including a bondwire inductance and a lead inductance for the packaged device, becomes significant. Therefore, it is useful to describe in a simple analytical way the relationship between the internal device parameters and parameters of the external harmonic-control circuit. The circuit schematic of a simple transmission-line impedance-peaking circuit is shown in Fig. 4.21($a$), where a combination of the series transmission line $TL_1$ and open-circuit stub $TL_2$

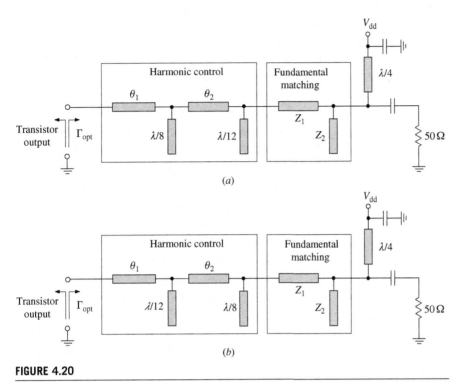

**FIGURE 4.20**

Idealized transmission-line inverse Class-F load networks.

together with the output capacitance $C_{out}$ and inductance $L_{out}$ is used to provide an open-circuited condition seen by the device multiharmonic current source at the second harmonic. Since the open-circuit stub at the second harmonic has zero impedance, the network admittance at the second harmonic can be written as

$$Y_{net}(2\omega_0) = j2\omega_0 C_{out} + \frac{1}{j2\omega_0 L_{out} + jZ_0 \tan 2\theta}. \tag{4.32}$$

Hence, when $\text{Im}Y_{net}(2\omega_0) = 0$, the ratios between the parameters of the impedance-peaking circuit at the second harmonic are defined by

$$\theta = \frac{1}{2}\tan^{-1}\left(\frac{1 - 4\omega_0^2 L_{out} C_{out}}{2\omega_0 C_{out}Z_0}\right) \tag{4.33}$$

resulting in a parallel-resonant tank consisting of the shunt capacitance $C_{out}$ and parallel inductance composed by the output inductance $L_{out}$ and transmission line $TL_1$.

Another configuration of the impedance-peaking circuit including the device output series inductance $L_{out}$ is shown in Fig. 4.21(b), where the open-circuit and

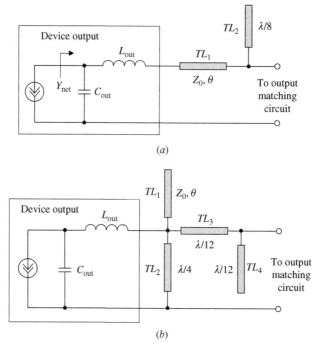

**FIGURE 4.21**

Transmission-line inverse Class-F load networks with bondwire inductance.

short-circuit transmission-line stubs are used to create both the third-harmonic short-circuit termination and second-harmonic peaking. The device output elements $C_{out}$ and $L_{out}$ (if necessary an additional series inductor should be added) must create a parallel resonance at the second harmonic since a short-circuit stub $TL_2$ has a half wavelength at the second harmonic with short-circuited conditions at its both ends. The electrical length $\theta$ of an open-circuit stub $TL_1$ is chosen to have less than a quarter wavelength at the third harmonic to realize a capacitive reactance to be resonant with an output inductance $L_{out}$ at the third harmonic of the fundamental frequency. The transmission lines $TL_3$ and $TL_4$ must be of quarter wavelengths at the third harmonic to provide a third-harmonic high-impedance condition at the input of $TL_3$. As a result, the ratios between the elements are

$$L_{out} = \frac{1}{4\omega_0^2 C_{out}} \tag{4.34}$$

$$\theta = \frac{1}{3}\tan^{-1}\left(\frac{Z_0}{3\omega_0 L_{out}}\right). \tag{4.35}$$

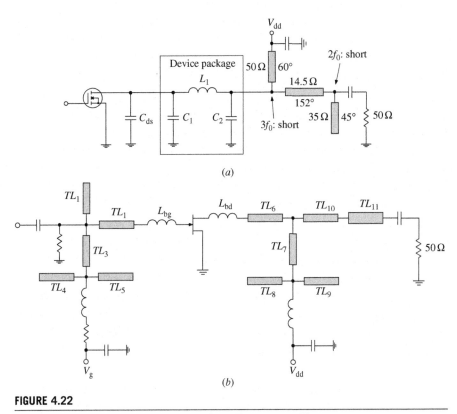

**FIGURE 4.22**

Schematics of transmission-line inverse Class-F power amplifier.

As an alternative to the simple analytical approach, a complicated and time-consuming load-pull optimization procedure can be used to optimize impedance at the fundamental-frequency, second, and third harmonics. Figure 4.22($a$) shows the transmission-line inverse Class-F load network which includes the device drain-source capacitance $C_{ds}$ and package parasitics represented by a $\pi$-type low-pass $C_1 L_1 C_2$ network [28]. By using load-pull optimization, the load network was designed to provide an open circuit at the second harmonic and a short circuit at the third harmonic at the intrinsic drain port. Moreover, the matching conditions between the 50-$\Omega$ standard load and 35-$\Omega$ equivalent load required to realize an optimum inverse Class-F load at the fundamental frequency are provided by the load network as well. As a result, a drain efficiency of 66% and an output power of 13.8 W were obtained for a 2.14-GHz 28-V LDMOSFET inverse Class-F power amplifier.

Figure 4.22($b$) shows the circuit schematic of a 3.5-GHz, 28-V GaN HEMT inverse Class-F power amplifier where a transistor bare-die is used to reduce parasitics of the package and facilitate harmonic impedance optimization at the transistor output reference plane [29]. Here, $L_{bg}$ and $L_{bd}$ are used to model the input and output bondwire inductances, respectively. The general optimization procedure consists of two steps: a source-pull/load-pull simulation to find the optimum source and load impedances at the fundamental frequency for efficiency and output power and then a harmonic load-pull simulation to define the optimum second- and third-harmonic load and source impedances for high-efficiency operation. The input matching circuit consists of the transmission lines from $TL_1$ to $TL_5$ and the output load network consists of the transmission lines from $TL_6$ to $TL_{11}$, which are all optimized to provide the optimum matching at the fundamental frequency and the corresponding second- and third-harmonic control. Finally, a maximum drain efficiency of 82% with a power gain of 12 dB was achieved at an output power of 40.4 dBm using a RO5870 substrate for board fabrication.

## 4.6 LDMOSFET power amplifier design examples

The effectiveness of the transmission-line load-network design technique for inverse Class-F application can be demonstrated by the example of a 28-V high-power LDMOSFET amplifier with the device of the same geometry as for a conventional Class-F mode, the small-signal equivalent circuit and output current−voltage characteristics of which are shown in Fig. 3.33 in Chapter 3. The circuit schematic of the simulated 500-MHz single-stage microstrip inverse Class-F power amplifier is shown in Fig. 4.23. In this case, both input and output matching circuits represent the $T$-type low-pass matching circuits with a series microstrip line, a shunt open-circuit stub having a capacitive reactance, and a series capacitor. To provide the third-harmonic short-circuit termination and second-harmonic peaking corresponding to

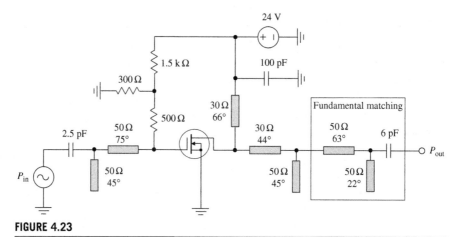

**FIGURE 4.23**

Simulated 500-MHz single-stage microstrip power amplifier with $T$-transformer.

an inverse Class-F operation mode, the short-circuited microstrip line with electrical length of 60° and combination of a series microstrip line and an open-circuit stub with electrical length of 45° for the second-harmonic termination are used.

Figure 4.24(a) shows the drain voltage waveform which is only slightly different from the half-sinusoidal shape due to the effect of the fourth- and higher-order harmonics that are not properly controlled. Nevertheless, a drain efficiency of up to 71% with a maximum output power of 8 W was achieved, as shown in Fig. 4.24(b). In this case, the peak value of the drain voltage is more than two times greater than the drain supply voltage of 24 V, resulting in a peak factor approximately equal to 58/24 = 2.4.

To minimize the number of the circuit elements, the output fundamental matching circuit of such a transmission-line power amplifier can also be realized in the form of a high-pass L-transformer with a series capacitor and a shunt transmission line grounded at its end, as shown in Fig. 4.25(a). Simulation results indicate that, for this particular case, the length of a short-circuited transmission line in the output matching circuit is close to a quarter wavelength, corresponding to a significant inductive reactance at the fundamental frequency. Consequently, in this case, the load network designed to provide the second- and third-harmonic

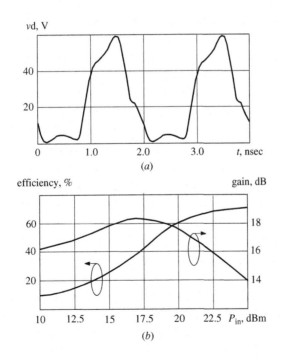

(a)

(b)

**FIGURE 4.24**

Drain voltage waveform, efficiency and power gain.

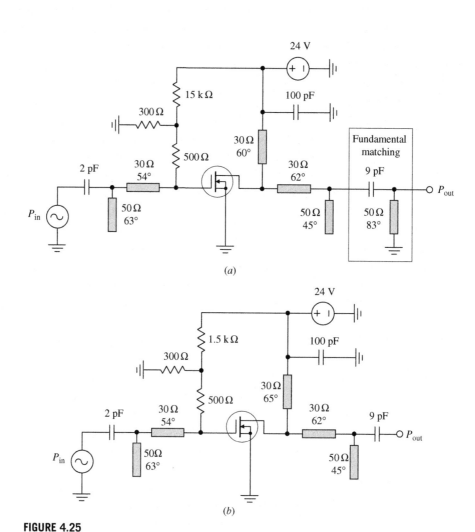

**FIGURE 4.25**

Schematics of simulated 500-MHz microstrip power amplifier.

control can perform a function of the matching circuit as well, together with the series capacitance required for dc blocking. The simplified and slightly modified circuit schematic of the simulated 500-MHz single-stage microstrip inverse Class-F high-power amplifier with total LDMOSFET channel width of $28 \times 1.44$ mm is shown in Fig. 4.25(b).

As a result, the drain efficiency up to 78% for an output power of about 25 W with a power gain of 14 dB can be achieved, as shown in Fig. 4.26(a). An analysis of the drain voltage and current waveforms plotted in Fig. 4.26(b) indicates

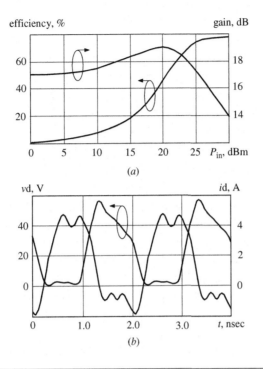

**FIGURE 4.26**

Drain efficiency, power gain, voltage and current waveforms.

that the operation mode obtained is close to an inverse Class-F mode where the drain current waveform approximates a square wave, while the drain voltage waveform is close to a half-sinusoidal waveform. It should be noted that the negative current values are due to the current flowing through the intrinsic drain-source capacitance when the device multiharmonic voltage-controlled current source is pinched off. Besides, there is a small phase shift between voltage and current waveforms due to uncompensated phases at the harmonics using such a simple load network which provides both fundamental matching and second- and third-harmonic control. In this case, the maximum drain voltage amplitude does not even reach a value of 60 V.

Figure 4.27 shows (*a*) the harmonic spectrum corresponding to the drain current waveform and (*b*) the harmonic spectrum corresponding to the drain voltage waveform up to fifth harmonic components. It can be seen that both the second harmonic of a drain current waveform and the third harmonic of a drain voltage waveform are slightly more increased than expected from the ideal waveforms. In addition, the higher-order harmonic components also make their contribution to the shapes of the drain voltage and current waveforms. In practical design, it is possible to use chip capacitors instead of the open-circuit microstrip

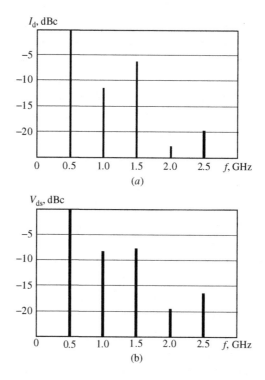

**FIGURE 4.27**

Drain current and voltage waveform spectra.

stubs to minimize the overall size of a power-amplifier board. In this case, the presence of a parasitic series inductance of the chip capacitor together with a parasitic via inductance can be useful to tune the capacitor's series self-resonance condition to the second harmonic.

## 4.7 Examples of practical implementation

The inverse Class-F concept can be successfully used in the design of low-voltage monolithic CMOS power amplifiers for wireless applications where lower current consumption is preferable due to on-chip inductors having low-quality factors. Figure 4.28 shows the inverse Class-F CMOS power amplifier with second-harmonic peaking implemented in a 0.6-$\mu$m CMOS double-poly double-metal technology [30]. The MOSFET device was characterized by a gate length of 0.6 $\mu$m and a gate width of 1200 $\mu$m. The load network consists of a shunt on-chip capacitor $C_1$, a series bondwire inductor $L_0$, and a shunt off-ship chip

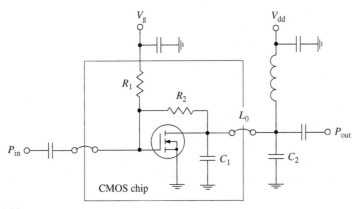

**FIGURE 4.28**

CMOS inverse Class-F power amplifier schematic.

capacitor $C_2$. At the second harmonic, the capacitor $C_2$ acts as a short circuit because of the self-resonance condition with its series parasitic inductance. At the same time, the capacitor $C_1$ and bondwire inductor $L_0$ are tuned on the second harmonic providing a second-harmonic peaking at the device drain terminal. To stabilize the power amplifier operation, the feedback resistance $R_2$ is connected between the drain and the gate. The transistor was biased at half the maximum expected drain current. As a result, a small-signal power gain of 10.5 dB, a saturated output power of 22.8 dBm, and a maximum *PAE* of 42% were achieved at an operating frequency of 1.9 GHz and a supply voltage of 3 V.

However, the level of the third-order intermodulation component $IM_3$ at high output powers is sufficiently high (approximately −20 dBc at $P_{out} = 18$ dBm) that this amplifier could not be directly used in an application that requires linear response to a varying-envelope signal. The theoretical analysis and measurements of the intermodulation distortions for Class-AB operation show that, for the small-signal conditions, $IM_3$ follows a well-known 3 dB per dB slope. As a result of the different contribution of the device transfer function components, there are two sweet spots where $IM_3$ is minimal [31]. The first sweet spot appears because of the turn-on knee region contribution, whereas the second sweet spot close to the output-power compression point is due to the combined effect of the quadratic-to-linear and compression transitions of the device transfer function.

Figure 4.29 shows the inverse Class-F pulsed bipolar power amplifier operated at 425 MHz with a peak output power of 50 W [17]. The transmission lines in the load network serve two purposes, one is to match the low impedance of the device collector to the standard 50-$\Omega$ output, and the other is to present the second-harmonic peaking required to produce the desired waveforms at the device collector for inverse Class-F operation. The electrical length $\theta$ of the first series microstrip line is calculated from Eq. (4.33) based on the value of the output

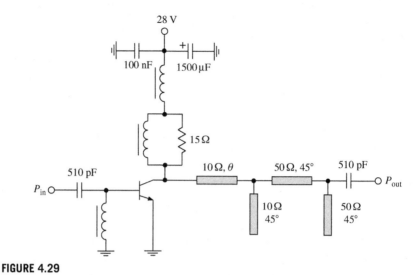

**FIGURE 4.29**

Schematic of bipolar high power amplifier.

device series inductance $L_{out}$. The first open-circuit microstrip stub having zero impedance at the second harmonic is necessary to provide high impedance seen by the device internal multiharmonic current source at the second harmonic. The second open-circuit microstrip stub is necessary to maximize the suppression of the even-harmonic components. Both open-circuit stubs operate as capacitors at the fundamental frequency realizing a two-section low-pass output matching circuit. The inductor–resistor parallel circuit contains a ferrite core wire-wound inductor and 15-$\Omega$ resistor to provide a supply path while presenting an open circuit to the RF power. The right-hand capacitor on the supply branch is a 1500-$\mu$F charge storage capacitor to preserve the squareness of the current pulse at the collector by preventing pulse slump and poor rise time. The left-hand capacitor is a 100-nF parallel-plate bypass capacitor to bring RF ground to that node to prevent any RF power from entering the power supply.

In modern communication systems operating in frequency ranges from 225 MHz to 2.5 GHz, the silicon LDMOSFET devices are widely used in high-power transmitters due to their efficient and linear operation, simple bias-circuit realization, and cost-per-watt performance. Figure 4.30 shows the schematic of a 1-GHz LDMOSFET power amplifier using lumped resonators [32]. Such an inverse Class-F schematic provides high impedance at the second harmonic of the fundamental frequency, low impedance at the third harmonic, and conjugate matching of the output equivalent device impedance at the fundamental frequency to a standard 50-$\Omega$ load. By using an LDMOSFET device with a gate length of 0.8 $\mu$m and a total gate width of 31.9 mm, a maximum drain efficiency of 77.8% with an output power of 12.4 W were achieved at a supply voltage of 28 V.

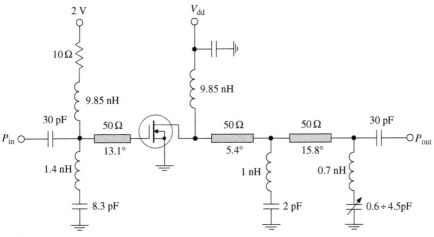

**FIGURE 4.30**

Schematic of 1-GHz power amplifier with second-harmonic resonator.

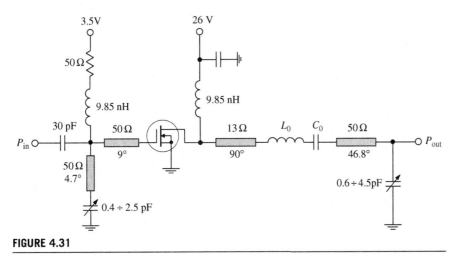

**FIGURE 4.31**

Schematic of 1.78-GHz power amplifier with quarterwave transmission line.

Figure 4.31 shows the schematic of a 1.78-GHz LDMOSFET power amplifier using the same device which is based on a quarterwave transmission-line topology [32]. At the fundamental frequency, a quarter-wavelength transmission line operates as an impedance transformer followed by a series $L_0 C_0$ filter tuned to the fundamental frequency and an $L$-type output matching circuit required to provide conjugate matching and additional harmonic suppression. For the second and

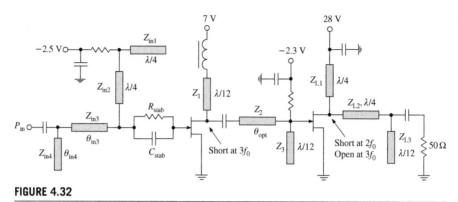

**FIGURE 4.32**

Schematic of high-efficiency two-stage transmission-line power amplifier.

higher-order harmonic components, the series filter presents an open circuit at the output of a quarterwave transmission line. Hence, an open circuit is presented at the device output for any even harmonic and a short circuit is presented for any odd harmonic of the fundamental frequency due to the equidistant frequency properties of the input impedance of a loaded quarterwave transmission line. The proper phasing of the second harmonic can be easily realized in practice when the active device is driven into a saturation mode. As a result, a drain efficiency of 60% with an output power of 13 W and a power gain of 10 dB were achieved at a supply voltage of 26 V.

To reduce the drain voltage and current overlapping resulting in an improved efficiency, a technique providing the gate driving with a quasi-half-sine-wave voltage by using a second-harmonic injection can be applied to a Class-F power amplifier. Figure 4.32 shows the circuit schematic of a two-stage 2-GHz GaN HEMT power amplifier with an output power of 15 W, a power gain of 28 dB, and a *PAE* of 70% [33]. The driver stage is designed to operate in inverse Class-F mode and feeds the power stage with both fundamental and second-harmonic components having an optimized ratio of their amplitudes, equal to approximately 7:1, and zero phase shift between them. In this case, the driver transistor operating in inverse Class-F mode sees high impedance at the second harmonic and short impedance at the third harmonic due to the shunt $\lambda/12$ open-circuit stub and series microstrip line with optimum electrical length $\theta_{\mathrm{opt}}$ connected to its drain port, thus providing a proper half-sine wave that directly feeds the gate port of the power stage. At the same time, low impedance at the second harmonic, due to the shunt $\lambda/4$ short-circuit stub, and high impedance at the third harmonic, due to the series $\lambda/4$ microstrip line and shunt $\lambda/12$ open-circuit stub, are provided at the drain port of the power transistor operating in a Class-F mode. The load network is synthesized to achieve impedance transformation at the fundamental frequency by optimizing the characteristic impedances $Z_{\mathrm{L}2}$ and $Z_{\mathrm{L}3}$. A parallel *RC* network

($R_{stab}$ and $C_{stab}$) is added in series at the input of the driver stage to ensure stability. It was observed that the active gate-source voltage-shaping technique offers significant improvement of *PAE versus* frequency that resulted in a *PAE* higher than 60% over a 400-MHz bandwidth around 2-GHz center bandwidth frequency at about 4-dB gain compression point.

## 4.8 Inverse Class-F GaN HEMT power amplifiers for WCDMA systems

In modern wireless communication systems, it is required that the power amplifier is able to operate with high efficiency, high linearity, and low-harmonic output levels simultaneously. To increase the efficiency of the power amplifier, it is possible to apply a switchmode inverse Class-F mode technique which can ideally provide a drain efficiency of 100% with a rectangular drive. This kind of power amplifier requires operation in deep saturation mode resulting in a poor linearity, and therefore is not suitable to directly replace linear power amplifiers in conventional WCDMA, CDMA2000, or OFDM transmitters with nonconstant envelope signals. However, to obtain both high efficiency and good linearity, such kinds of nonlinear high-efficiency power amplifiers operating in an inverse Class-F mode can be used in advanced transmitter architectures such as Doherty, LINC (linear amplification using nonlinear components), or EER (envelope elimination and restoration) with digital predistortion [23,34].

In a power amplifier using the packaged device at high frequencies or with high output power, the presence of a transistor output series bondwire and lead inductance $L_{out}$ creates some problems in providing an acceptable second- or third-harmonic open- or short-circuit termination. In this case, it is convenient to use a series transmission line as a first element of the load network connected to the device output, as shown in Fig. 4.33(*a*), where the transmission line $TL_1$ is placed between the device drain and shunt short-circuited quarter-wavelength transmission line $TL_3$. However, if the length of combined series transmission line $TL_1 + TL_2$ becomes very long in Class-F mode with a short circuit at the second harmonic and an open circuit at the third harmonic and additional fundamental-frequency matching circuit is required, then such a load network in inverse Class-F mode is compact, convenient for harmonic tuning, and very practical. Figure 4.33(*b*) shows the equivalent circuit of a transmission-line inverse Class-F load network, where the complex-conjugate load matching is provided at the fundamental and both high reactance at the second harmonic and low reactance at the third harmonic are created at the device output by using two series transmission lines $TL_1$ and $TL_2$, the electrical lengths of which depend on the values of the device output shunt capacitance $C_{out}$ and series inductance $L_{out}$, a quarter-wavelength short-circuit stub $TL_3$, and an open-circuit stub $TL_4$ with electrical length of 30° [20,35]. The output shunt capacitance $C_{out}$ can represent both

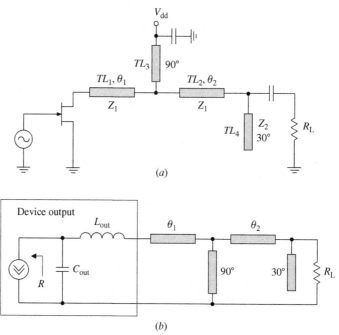

**FIGURE 4.33**

Transmission-line inverse Class-F power amplifier and its equivalent circuit.

intrinsic bias-dependent drain-source capacitance $C_{ds}$ and extrinsic bias-independent drain pad-contact capacitance $C_{dp}$ of the nonlinear large-signal equivalent circuit for a GaN HEMT device, whereas the series output inductance $L_{out}$ model combines the effect of metallization, bond wire, and package inductances [36].

The harmonic conditions for an inverse Class-F load network seen by the device multi-harmonic current source derived from Eqs (4.14) to (4.16) for the first three harmonic components including fundamental are

$$\text{Re } Z_{net}(\omega_0) = R \tag{4.36}$$

$$\text{Im } Z_{net}(2\omega_0) = \infty \tag{4.37}$$

$$\text{Im } Z_{net}(\omega_0) = \text{Im } Z_{net}(3\omega_0) = 0 \tag{4.38}$$

where the load resistance (or equivalent output resistance) $R$ seen by the device output at the fundamental frequency can be written in inverse Class-F mode as

$$R = \frac{\pi}{2} R^{(B)} = \frac{\pi}{4} \frac{(V_{dd} - V_{sat})^2}{P_{out}} \tag{4.39}$$

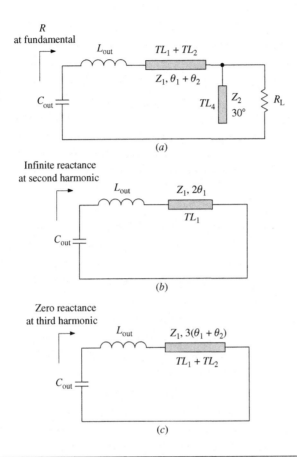

**FIGURE 4.34**

Load networks seen by the device output at corresponding harmonics.

where $R^{(B)} = (V_{dd} - V_{sat})^2/2P_{out}$ is the equivalent output resistance obtained in a Class-B mode, $V_{dd}$ is the dc supply voltage, $V_{sat}$ is the saturation voltage, and $P_{out}$ is the fundamental output power.

Figure 4.34(a) shows the transmission-line load network seen by the device multi-harmonic current source at the fundamental frequency, where the combined series transmission line $TL_1 + TL_2$ (together with an open-circuit capacitive stub $TL_4$ with electrical length of $30°$) provides an impedance matching between the optimum equivalent output device resistance $R$ and the standard load resistance $R_L$ by proper choice of the transmission-line characteristic impedances $Z_1$ and $Z_2$, where $C_{out}$ and $L_{out}$ are the elements of the matching circuit. For simplicity of calculation, the characteristic impedances of the transmission lines $TL_1$ and $TL_2$ are set to be equal to $Z_1$.

The load network seen by the device current source at the second harmonic (taking into account the short-circuit effect of the grounded quarter-wavelength transmission line $TL_3$) is shown in Fig. 4.34(b), where the transmission line $TL_1$ provides an open-circuit condition for the second harmonic at the device output by forming a second-harmonic tank together with the shunt capacitor $C_{out}$ and series inductance $L_{out}$. Similar load network at the third harmonic is shown in Fig. 4.34(c), where the open-circuit effect of the grounded quarter-wavelength transmission line $TL_3$ and short-circuit effect of the open-circuit stub $TL_4$ at the third harmonic are used. In this case, the combined transmission line $TL_1 + TL_2$ short-circuited at its right-hand side and connected in series with an inductance $L_{out}$ provides a short-circuit condition for the third harmonic at the device output. Depending on the actual physical length of the device output electrode, the on-board adjusting of the transmission lines $TL_1$ and $TL_2$ can easily provide the required open-circuit and short-circuit conditions, as well as an impedance matching at the fundamental frequency, due to their series connection to the device output.

By using Eqs (4.37) and (4.38), the electrical lengths of the transmission lines $TL_1$ and $TL_2$, assuming the same characteristic impedance $Z_1$ for both series transmission-line sections, can be defined from

$$2\omega_0 C_{out} - \frac{1}{2\omega_0 L_{out} + Z_1 \tan 2\theta_1} = 0 \tag{4.40}$$

$$3\omega_0 L_{out} + Z_1 \tan 3(\theta_1 + \theta_2) = 0 \tag{4.41}$$

with the maximum total electrical length $\theta_1 + \theta_2 = \pi/3$ or $60°$ at the fundamental frequency or $180°$ at the third harmonic component when $L_{out} = 0$.

As a result, the electrical lengths of the transmission lines $TL_1$ and $TL_2$ as analytical functions of the device output series inductance $L_{out}$ and shunt capacitance $C_{out}$ are obtained as

$$\theta_1 = \frac{1}{2} \tan^{-1} \frac{1 - (2\omega_0)^2 L_{out} C_{out}}{2Z_1 \omega_0 C_{out}} \tag{4.42}$$

$$\theta_2 = \frac{\pi}{3} - \frac{1}{3} \tan^{-1} \frac{3\omega_0 L_{out}}{Z_1} - \theta_1 \tag{4.43}$$

where the transmission-line characteristic impedance $Z_1$ can be set in advance.

In order to omit an additional matching section at the fundamental, the inverse Class-F load network can also be used to match the equivalent device fundamental-frequency impedance $R$ with the standard load impedance $R_L$ (usually equal to 50 $\Omega$). In this case, it is necessary to properly optimize both characteristic impedances $Z_1$ and $Z_2$. Figure 4.35 shows the equivalent representation of an inverse Class-F load network (including the device output parameters $L_{out}$ and $C_{out}$) by a lumped low-pass $\pi$-type matching circuit where $C = \tan 30°/\omega_0 Z_2$ and $L \approx (Z_1/\omega_0)$ $\sin(\theta_1 + \theta_2) + L_{out}$ due to the sufficiently short length of the combined transmission

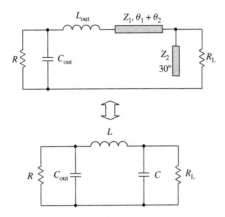

**FIGURE 4.35**

Equivalent representations of load network at fundamental frequency.

line $TL_1 + TL_2$, typically much less than $60°$ at the fundamental frequency depending on the device output parameters. As a result,

$$Z_1 \cong \frac{\omega_0(L - L_{\text{out}})}{\sin(\theta_1 + \theta_2)} \tag{4.44}$$

$$Z_2 = \frac{1}{\omega_0 C \sqrt{3}} \tag{4.45}$$

For a lumped low-pass $\pi$-type matching circuit with $R_L > R$ and $Q = \omega_0 C_{\text{out}} R$,

$$Q_L = \sqrt{\frac{R_L}{R}(1 + Q^2) - 1} \tag{4.46}$$

$$C = \frac{Q_L}{\omega_0 R_L} \tag{4.47}$$

$$L = \frac{Q + Q_L}{1 + Q_L^2} \frac{R_L}{\omega_0} \tag{4.48}$$

where the parameters $L_{\text{out}}$, $C_{\text{out}}$, $R$, and $R_L$ are fixed [16]. By subsequent calculation of the load quality factor $Q_L$ from Eq. (4.46) and then the capacitance $C$ from Eq. (4.47), the characteristic impedance $Z_2$ can be directly obtained from Eq. (4.45). After calculating the inductance $L$ from Eq. (4.48), the remaining parameters including the electrical lengths $\theta_1$ and $\theta_2$ and characteristic impedance $Z_1$ can be finally obtained from a system of three equations given by Eqs (4.42) to (4.44).

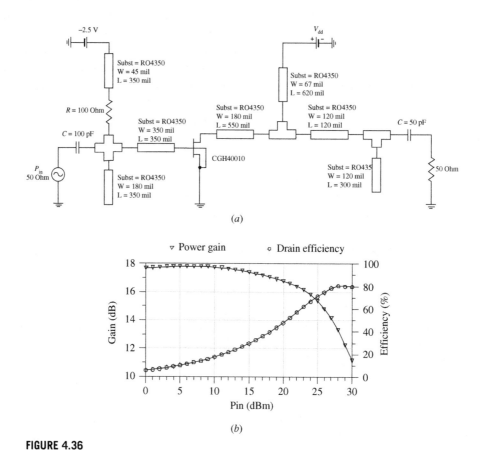

**FIGURE 4.36**

Transmission-line 10-W inverse Class-F GaN HEMT power amplifier.

For most practical cases when $\omega_0 L_{out}/Z_1 < 0.1$, from Eqs (4.43) and (4.44) it follows that the characteristic impedance $Z_1$ can be easily calculated from

$$Z_1 = \frac{\omega_0(2L - L_{out})}{\sqrt{3}}. \tag{4.49}$$

Figure 4.36(a) shows the simulated circuit schematic which approximates a transmission-line inverse Class-F power amplifier based on a 28-V, 10-W Cree GaN HEMT power transistor CGH40010 and transmission-line load network with the second- and third-harmonic tuning, as shown in Fig. 4.33(a). The input matching circuit provides a fundamental-frequency complex-conjugate matching with the standard 50-$\Omega$ source. The load network was slightly modified by optimizing the parameters of the series transmission line for implementation convenience. In this case, the device input and output package leads as external elements were properly modeled to take into account the effect of their inductances, and their models were then added to the simulation setup. The simulation results of a trans-mission-line inverse Class-F GaN HEMT power amplifier shown in Fig. 4.36(b) are based on a nonlinear device model supplied by Cree and technical parameters for a 30-mil RO4350 substrate. The maximum output power of 41.3 dBm, power gain of 13.3 dB (linear gain of about 18 dB), drain efficiency of 80.3%, and *PAE* of 76.5% are achieved at an operating frequency of 2.14 GHz with a supply volt-age of 28 V and a quiescent current of 40 mA. The experimental results of the test board of this transmission-line inverse Class-F GaN HEMT power amplifier using a 28-V, 10-W Cree GaN HEMT power transistor CGH40010P in a metal-ceramic pill package were very close to the simulated results with a maximum output power of 41.0 dBm, a drain efficiency of 76.0%, a *PAE* of 72.2%, and a power gain of 13.0 dB at an operating frequency of 2.14 GHz (gate bias voltage $V_g = -2.8$ V and drain supply voltage $V_{dd} = 28$ V), achieved without any tuning of the input matching circuit and load network [20].

Figure 4.37(a) shows the simulated circuit schematic of a transmission-line inverse Class-F power amplifier based on a 25-W Cree GaN HEMT power

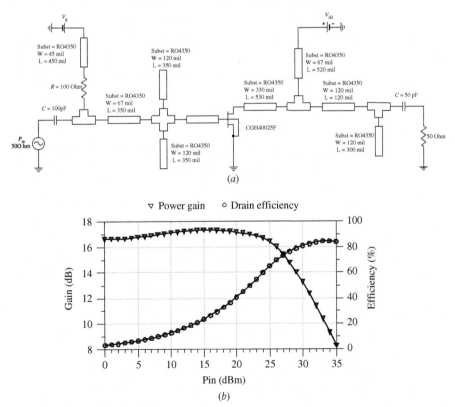

(a)

(b)

**FIGURE 4.37**

Transmission-line 25-W inverse Class-F GaN HEMT power amplifier.

transistor CGH40025F. The input matching circuit provides a complex-conjugate matching with the standard 50-$\Omega$ source. Similarly, the load network approximates the transmission-line structure with the second- and third-harmonic control shown in Fig. 4.33(a). In this case, the characteristic impedances of the shunt and series transmission lines were optimized for better performance and convenience of practical implementation. Special care was taken for modeling of the device input and output package leads to account for finite values of their inductances. Figure 4.37(b) shows the simulated results of an inverse Class-F power amplifier using a 30-mil RO4350 substrate. The maximum output power of 43.4 dBm, drain efficiency of 84.6%, and *PAE* of 78.5% with a power gain of 11.4 dB (linear gain of about 17 dB) at $V_g = -2.5$ V and $V_{dd} = 28$ V were achieved at an operating frequency of 2.14 GHz. The measured results of such an inverse Class-F GaN HEMT power amplifier with a maximum output power of 43.5 dBm, a drain efficiency of 82.1%, and a power gain of 12.2 dB at 2.14 GHz (quiescent current $I_q = 100$ mA and drain supply voltage $V_{dd} = 32$ V) were achieved without any tuning of the input matching circuit and load network [37].

For particular values of the parameters $C_{out}$ and $L_{out}$ when $C_{out}$ is sufficiently small and $L_{out}$ is enough large, the electrical length $\theta_2$ can be set to zero, thus resulting in a special simplified case of the inverse Class-F load network shown in Fig. 4.38(a). In a common case, Eqs (4.40) and (4.41) can be rewritten in a generalized form of

$$4\,pq + 2q\,\tan 2\theta_1 - 1 = 0 \tag{4.50}$$

$$3\,p + \tan 3\theta_1 = 0 \tag{4.51}$$

where $q = Z_1 \omega_0 C_{out}$ and $p = \omega_0 L_{out}/Z_1$. Substituting Eq. (4.51) into Eq. (4.50) results in a transcendental equation to define implicitly the required electrical length $\theta_1$ for proper $q$, which is written as

$$2\,q(2\,\tan 3\theta_1 - 3\,\tan 2\theta_1) + 3 = 0. \tag{4.52}$$

Then, the characteristic impedance $Z_1$ can be obtained from Eq. (4.51) which provides explicit calculation of the load-network parameter $p$ for given electrical length $\theta_1$.

Figure 4.38(b) shows the frequency behavior of the normalized parameters $p$ and $q$ which characterize a transmission-line inverse Class-F load network with a uniform series transmission line. In this case, a transmission-line network solution can be found for $p$ varying from 0.34 to 1.02 when $q < 0.1$, resulting in values of electrical length $\theta_1$ within 35° to 45°. For example, for $q = 0.09$ which gives $p = 0.74$ and $\theta_1 = 38°$, as follows from Fig. 4.38(b), the characteristic impedance $Z_1$ and series inductance $L_{out}$ are calculated for $C_{out} = 0.3$ pF as $Z_1 = 22\ \Omega$ and $L_{out} = 1.2$ nH at an operating frequency of 2.14 GHz.

Figure 4.39(a) shows the simulated circuit schematic of an inverse Class-F GaN HEMT power amplifier based on a 28-V, 5-W Nitronex RF power transistor NPTB00004 where the input matching circuit provides a complex-conjugate

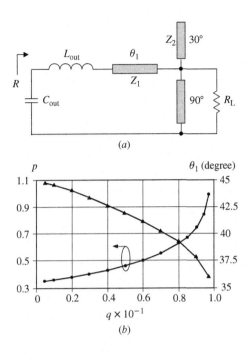

**FIGURE 4.38**

Special case of inverse Class-F load network and its parameters.

matching with a 50-$\Omega$ source and the load network represents the transmission-line circuit shown in Fig. 4.38(a). The simulated results of an inverse Class-F power amplifier using a 30-mil RO4350 substrate are shown in Fig. 4.39(b). The maximum output power of 37.1 dBm, drain efficiency of 73.1%, and *PAE* of 70.7% at an operating frequency of 2.14 GHz are achieved with a power gain of 15.1 dB (linear gain of around 20 dB) and a supply voltage of 25 V. To take into account the parasitic effect of the device output lead, the length of adjacent series microstrip line can be slightly reduced. As a result of practical implementation, a drain efficiency of 74.6% and a *PAE* of 70.5% with a power gain of 12.5 dB were achieved at 2.14 GHz (gate bias voltage $V_g = -1.4$ V, quiescent current $I_q = 20$ mA, and drain supply voltage $V_{dd} = 25$ V) [35]. In this case, deeper satu-ration mode with increased input power results in lower dc supply current that contributes to increased drain efficiency (78.5% and higher) with almost constant fundamental output power. The fundamental output power is varied almost line-arly from 33 dBm at $V_{dd} = 20$ V up to 37.5 dBm at $V_{dd} = 35$ V, and a maximum drain efficiency of about 82% is achieved at a dc supply voltage of 32.5 V. Due to the short-circuit and open-circuit effects of the corresponding quarterwave transmission-line stubs, the second- and third-harmonic components are sup-pressed by more than 25 dB.

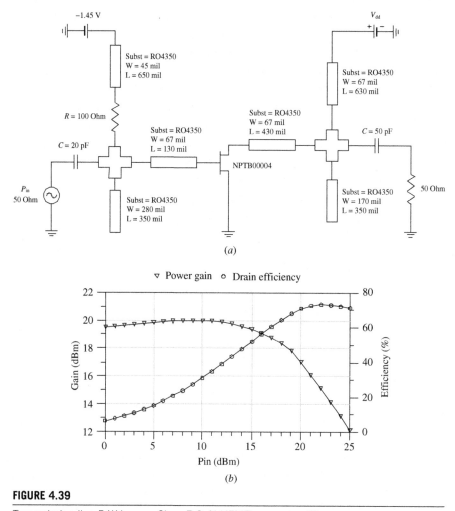

**FIGURE 4.39**

Transmission-line 5-W inverse Class-F GaN HEMT power amplifier.

Figure 4.40(a) shows the simulated circuit schematic of an inverse Class-F GaN HEMT power amplifier based on a 28-V, 60-W Cree RF power transistor CGH27060F. The input matching circuit provides a complex-conjugate matching with a 50-$\Omega$ source at the fundamental frequency. The load network represents the corresponding microstrip-line implementation of the transmission-line load circuit shown in Fig. 4.38(a), where a uniform series transmission line is replaced by the stepped-impedance series transmission line which is required for convenience of practical implementation and results in shorter overall length [38]. Special stripline models in air substrate for the device input and output package

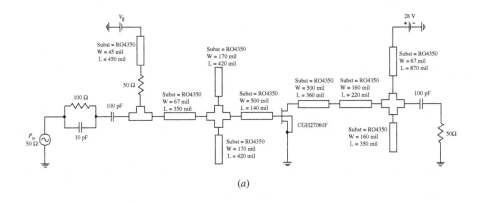

(a)

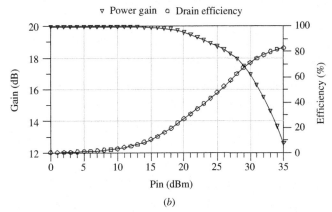

(b)

**FIGURE 4.40**

Transmission-line 50-W inverse Class-F GaN HEMT power amplifier.

leads are added to the simulation setup to take into account their inductances. For stability reasons, a parallel connection of a 10-pF capacitor and a 100-$\Omega$ resistor are inserted in series into the input circuit. Figure 4.40(b) shows the simulated results of a transmission-line 50-W inverse Class-F GaN HEMT power amplifier implemented on a 30-mil RO4350 substrate. The maximum output power of 47.6 dBm, drain efficiency of 82.4%, and *PAE* of 78.8% at an operating frequency of 2.14 GHz are achieved with a power gain of 13.6 dB (linear gain of around 20 dB) at a supply voltage of 32 V. High drain efficiency of more than 80% is achieved in a strong saturation mode when the power gain is compressed by more than 5 dB. In this case, the measured results demonstrate a saturated output power of 47.3 dBm, a drain efficiency of 82.3%, and a power gain of 14.3 dB, resulting in a *PAE* of 79.8% (gate bias voltage $V_g = -3.35$ V, quiescent current $I_q = 230$ mA, and drain supply voltage $V_{dd} = 32$ V), that was achieved without any tuning of the load network [38].

# REFERENCES

1. Kolesnikov AI. A new method to improve efficiency and to increase power of the transmitter (in Russian). *Master Svyazi*. June 1940:27−41.
2. Sarbacher RI. Power-tube performance in class C amplifiers and frequency multipliers as influenced by harmonic voltage. *Proc. IRE*. November 1943;31:607−625.
3. Glazman ES, Kalinin LB, Mikhailov YI. Improving VHF transmitter efficiency by using the biharmonic mode. *Telecommunications and Radio Engineering*. July 1975;30(Part 1):46−51.
4. Berman LS. Increasing of useful power of the resonant semiconductor power amplifier by increasing its efficiency, (Part II in Russian). *Radiotekhnika*. March 1958;13:70−73.
5. Tyler VJ. A new high-efficiency high-power amplifier. *Marconi Review*. Fall 1958;21: 96−109.
6. Lu X. An alternative approach to improving the efficiency of high power radio frequency amplifiers. *IEEE Trans. Broadcasting*. June 1992;BC-38:85−89.
7. Fuzik NS. Biharmonic modes of a tuned RF power amplifier. *Telecommunications and Radio Engineering*. July 1970;25(Part 2):117−124.
8. Zivkovic Z, Marcovic A. Increasing the efficiency of the high-power triode HF amplifier − why not with the second harmonic? *IEEE Trans. Broadcasting*. March 1986;BC-32:5−10.
9. Ingruber B, Pritzl W, Smely D, Wachutka M, Magerl G. A high-efficiency harmonic-control amplifier. *IEEE Trans. Microwave Theory Tech*. June 1998;MTT-46:857−862.
10. Kazimierczuk MK. A new concept of class f tuned power amplifier. Proc. 27th Midwest Circuits and Systems Symp. 1984;425−428.
11. Raab FH. An introduction to Class-f power amplifiers. *RF Design*. May 1996;vol. 19:79−84:p. 14, July 1996.
12. Heymann P, Doerner R, Rudolph M. Harmonic tuning of power transistors by active load-pull measurement. *Microwave J*. June 2000;43:22−37.
13. Quyahia A, Duperrier C, Tolant C, Temcamani F, Eudeni D, A 71.9% power-added-efficiency inverse Class-F LDMOS. *2006 IEEE MTT-S Int. Microwave Symp. Dig.* 1542−1545.
14. Abe Y, Ishikawa R, Honjo K. Inverse Class-F AlGaN/GaN HEMT microwave amplifier based on lumped element circuit synthesis method. *IEEE Trans. Microwave Theory Tech*. December 2008;MTT-56:2748−2753.
15. Wei CJ, DiCarlo P, Tkachenko YA, McMorrow R, Bartle D. Analysis and experimental waveform study on inverse Class-f mode of microwave power FETs. *2000 IEEE MTT-S Int. Microwave Symp. Dig.* 1:525−528.
16. Grebennikov A. *RF and Microwave Power Amplifier Design*. New York: McGraw-Hill; 2004.
17. McCalpin W. High efficiency power amplification with optimally loaded harmonic waveshaping. *Proc. RF Tech. Expo'86 (Anaheim, CA)*. 1986:119−124.
18. Yamasaki T, Kittaka Y, Minamide H, et al. A 68% efficiency, C-band 100 W GaN power amplifier for space applications. *2010 IEEE MTT-S Int. Microwave Symp. Dig.* 1384−1386.
19. Giannini F, Scucchia L. A complete class of harmonic matching networks: synthesis and application. *IEEE Trans. Microwave Theory Tech*. March 2009;MTT-57:612−619.

20. Grebennikov A. Load network design technique for Class F and inverse Class F PAs. *High Frequency Electronics*. May 2011;10:58−76.

21. Heima T, Inoue A, Ohta A, Tanino N, Sato K. A new practical harmonics tune for high efficiency power amplifier. *Proc. 29th Europ. Microwave Conf.* 1999:271−274.

22. Aflaki P, Negra R, Ghannouchi FM. Design and implementation of an inverse Class-F power amplifier with 79% efficiency by using a switch-based active device model. *Proc. 2008 IEEE Radio and Wireless Symp.* 423−426.

23. Kim I, Woo YY, Hong S, Kim B. High efficiency hybrid EER transmitter for WCDMA applications using optimized power amplifier. *Proc. 37th Europ. Microwave Conf.* 2007:182−185.

24. Grebennikov A. A simple stepped-impedance transmission-line load network for inverse Class F power amplifiers. *Microwave and Optical Technology Lett.* May 2011;53:1157−1160.

25. Steinbrecher DH. An interesting impedance matching network. *IEEE Trans. Microwave Theory Tech.* June 1967;MTT-22:382.

26. Ebrahimi MM, Helaoui M, Ghannouchi FM. Efficiency enhancement of a WiMAX switching mode GaN power amplifier through layout optimization of distributed harmonic matching networks. *Proc. 39th Europ. Microwave Conf.* 2009:1732−11735.

27. Helaoui M, Ghannouchi FM. Optimizing losses in distributed multiharmonic matching networks applied to the design of an RF GaN power amplifier with higher than 80% power-added efficiency. *IEEE Trans. Microwave Theory Tech.* February 2009;MTT-57:314−322.

28. Gerhard W, Knochel R. A 2.14 GHz Inverse Class F Si-LDMOS power amplifier with voltage second harmonic peaking. *2006 German Microwave Conf. Guide* 21.

29. Saad P, Nemati HM, Thorsell M, Andersson K, Fager C. An inverse Class-F GaN HEMT power amplifier with 78% PAE at 3.5 GHz. *Proc. 39th Europ. Microwave Conf.* 2009:496−499.

30. Fortes F, Rosario MJ. A second harmonic Class-F power amplifier in standard CMOS technology. *IEEE Trans. Microwave Theory Tech.* June 2001;MTT-49:1216−1220.

31. Fager C, Pedro JC, Carvalho NB, Zirath H, Fortes F, Rosario MJ. A comprehensive analysis of IMD behavior in RF CMOS power amplifiers. *IEEE J. Solid-State Circuits*. January 2004;SC-39:24−34.

32. Lepine F, Adahl A, Zirath H. *L*-Band LDMOS power amplifiers based on an inverse Class-F architecture. *IEEE Trans. Microwave Theory Tech.* June 2005;MTT-53:2007−2012.

33. Ramadan A, Reveyrand T, Martin A, et al. Two-stage GaN HEMT amplifier with gate-source voltage shaping for efficiency versus bandwidth enhancement. *IEEE Trans. Microwave Theory Tech.* March 2011;MTT-59:699−706.

34. Kim J, Kim B, Woo YY. Advanced design of linear doherty amplifier for high efficiency using saturation amplifier. *2007 IEEE MTT-S Int. Microwave Symp. Dig.*, 1573−1576.

35. Grebennikov A. High-efficiency transmission-line GaN HEMT inverse Class F power amplifier for active antenna arrays. *Proc. 2009 Asia−Pacific Microwave Conf.*, 317−320.

36. Jarndal A, Aflaki P, Negra R, Kouki AK, Ghannouchi F. Large-signal modeling methodology for GaN HEMTs for RF switching-mode power amplifiers design. *Int J RF and Microwave Computer-Aided Eng.* January 2011;21:45−51.

37. Grebennikov A. A high-efficiency 100-W four-stage Doherty GaN HEMT power amplifier module for WCDMA systems. *2011 IEEE MTT-S Int. Microwave Symp. Dig.* 1−4.

38. Grebennikov A. High-efficiency transmission-line inverse Class F power amplifiers for 2-GHz WCDMA systems. *Int. J. RF and Microwave Computer-Aided Eng.* July 2011;21:446−456.

# Class-E with Shunt Capacitance

5

## INTRODUCTION

The switchmode Class-E power amplifiers with shunt capacitance have found widespread application due to their design simplicity and high-efficiency operation. Their load-network configuration consists of a shunt capacitor, a series inductor, and a series filter tuned to the fundamental frequency to provide a high level of harmonic suppression. In the Class-E power amplifier, the transistor operates as an on-to-off switch, and the shapes of the current and voltage waveforms provide a condition in which the high current and the high voltage do not occur simultaneously. That minimizes the power dissipation and maximizes the power-amplifier efficiency. In this chapter, the historical aspect and modern trends of a Class-E power-amplifier design are presented. Different circuit configurations and load-network techniques using the push−pull mode, lumped elements, and transmission lines are analyzed. The effects of the power-transistor saturation resistance, finite switching time, and nonlinear shunt capacitance are described. The practical RF and microwave Class-E power amplifiers and their applications are given and discussed.

## 5.1 Effect of a detuned resonant circuit

Using resonant circuits in the load network tuned to the odd and/or even harmonics of the fundamental frequency can generate biharmonic or polyharmonic operation modes of vacuum-tube power amplifiers, which is very effective to increase their operating efficiency. This implies ideally the in-phase or out-of-phase harmonic conditions when symmetrical flattened voltage or current waveforms can be formed. However, as it turned out, this is not the only way to improve the power-amplifier efficiency. Figure 5.1 shows the circuit schematic

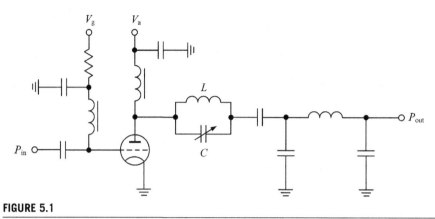

**FIGURE 5.1**

Class-C power amplifier with detuned resonant circuit.

of a vacuum-tube power amplifier with a parallel-tuned *LC* circuit inserted between the anode and the output matching circuit, which has a resonant frequency equal to about 1.5 times the carrier frequency of the signal to be amplified [1]. In other words, if the carrier signal is transmitting at a fundamental frequency $f_0$, the parallel resonant circuit will have a resonant frequency of about $1.5f_0$ followed by a filter or output matching circuit to suppress the harmonics of the fundamental frequency and to maximize the output power at the fundamental frequency, delivered to the standard load. As a result, an efficiency of 89% was achieved for a 3.2-MHz vacuum-tube high-power amplifier. Although it was assumed that such a parallel resonant circuit introduces considerable impedance to its own second harmonic, which is the third harmonic $3f_0$ of the carrier frequency and can result in a flattened anode voltage waveform, another interesting conclusion can be derived from this circuit topology. In this case, provided the output low-pass $\pi$-type matching circuit has purely resistive impedance at the fundamental frequency and capacitive reactances at the harmonic components, the device anode sees inductive impedance at the fundamental frequency and capacitive reactances at the second and higher-order harmonic components. This means that the voltage and current waveforms are not symmetrical any more, thus representing an alternative result of the efficiency improvement. Such an effect of increasing efficiency when the output resonant circuit of the vacuum-tube Class-C power amplifier is detuned from the carrier frequency was first described in early 1960s [2]. Anode efficiencies of about 92–93% were achieved for the phase angles of the output load network within 30° to 40°, resulting in an inductive reactance at the fundamental frequency and capacitive reactances at other harmonic components seen by the anode of the active device.

A few years later, it was discovered that very high efficiencies can be obtained with a series-tuned *LC* circuit connected to a transistor, as shown in Fig. 5.2(*a*) [3]. The reasons for this high efficiency are that the transistor operates in a pure

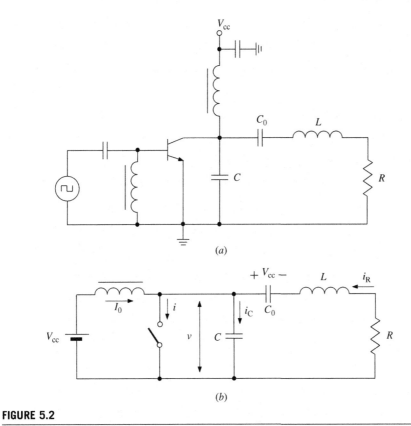

(a)

(b)

**FIGURE 5.2**

Basic circuits of switchmode power amplifier with nonsinusoidal output voltage.

switching mode, and the voltage across the transistor and the current flowing through it can both be made equal to zero during the switching transient interval when a proper choice of the transistor and circuit parameters are provided. To satisfy such a high-efficiency condition, the transistor current and voltage should be near zero at the time just prior to the conduction interval when the transistor goes into the saturation mode. The series-tuned circuit, consisting of a capacitor $C_0$ and an inductor $L$, must appear inductive at the operating frequency. In this case, a loaded quality factor $Q_L$ of the series-tuned circuit of about 10 will give a good sinusoidal shape to the load current. As a result, a 20-W, 500-kHz bipolar power amplifier was built having a collector efficiency of 94% with a conduction angle of 180°.

To realize high operational efficiency of the power amplifier, the ideal switch should represent the transistor, and the impedance seen from the collector into the matched circuit should not correspond to a short circuit for the second harmonic over the whole frequency range [4]. The simplified collector circuit of a switchmode power amplifier is shown in Fig. 5.2(b), where $C$ is the collector-emitter capacitance, assumed to be independent of voltage, $C_0$ is the dc-blocking capacitor, $L$ is the series inductor, and $R$ is the load resistor. When the voltage across the capacitor $C$ crosses zero, the switch will be closed. The basic switching operation can be described by two sets of linear first-order differential equations, one set for the on-state and another set for the off-state, with an approximate solution. It was assumed that high collector efficiency is a result of a zero collector-emitter voltage at the end of the off-state cycle due to transient without need for any current when only a small power is dissipated in the switching slopes. Collector efficiencies up to 85% were achieved in the frequency range of 48–70 MHz.

The exact theoretical analysis of the operational conditions of a single-ended switchmode power amplifier using its simplified equivalent circuit shown in Fig. 5.2 (*b*) with the calculation of the circuit parameters was firstly first given in [5,6]. Here, the active device is considered an ideal switch that is driven so as to provide the device switching between its on-state and off-state operation conditions. As a result, the collector voltage waveform is determined by the switch when it is turned on and by the transient response of the load network when the switch is turned off. To simplify a theoretical analysis, the following assumptions were introduced:

- Transistor has zero saturation voltage, zero saturation resistance, infinite off-resistance, and its switching action is instantaneous and lossless (except when discharging the shunt capacitance).
- Total shunt capacitance is independent of the collector and is assumed linear.
- RF choke allows only a constant dc current and has no resistance.
- There are no losses in the circuit except into the load $R$.

For a lossless operation mode, it is necessary to provide the following optimum conditions for voltage across the switch at turn-on time moment $t = T$, when the transistor is saturated:

$$v(t)|_{t=T} = 0 \tag{5.1}$$

$$\left.\frac{dv(t)}{dt}\right|_{t=T} = 0 \tag{5.2}$$

where $T$ is the period of input driving signal and $v(t)$ is the voltage across the switch. The second condition given by Eq. (5.2) means that the collector current $i(t)$ has no jump and is equal to zero at this moment since the collector capacitor $C$ is completely discharged.

The derivation of the load-network parameters is based on the consideration of the processes during on-state and off-state transistor operation modes separately [6,7]. When the switch is turned on, the equivalent amplifier system is described by a system of the first-order differential equations in the form

$$V_{cc} = L\frac{di_R(t)}{dt} + i_R(t)R \tag{5.3}$$

$$i(t) = I_0 + i_R(t) \tag{5.4}$$

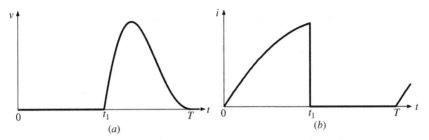

**FIGURE 5.3**

Idealized collector voltage and current waveforms of switchmode power amplifier with nonsinusoidal output voltage.

where the voltage $V_{cc}$ is applied to the plates of the blocking capacitor $C_0$. By taking into account the initial condition $i(0) = 0$, the current $i(t)$ flowing through the switch can be obtained by

$$i_s(t) = \frac{V_{cc} + I_0 R}{R}\left[1 - \exp\left(-\frac{R}{L}t\right)\right].$$  (5.5)

When the switch is turned off at the time moment $t = t_1$, another system of the first-order differential equations can be written by

$$V_{cc} = v(t) + L\frac{di_R(t)}{dt} + i_R(t)R$$  (5.6)

$$C\frac{dv(t)}{dt} = I_0 + i_R(t)$$  (5.7)

with the initial conditions $i_R(t_1) = i(t_1) - I_0$ and $v(t_1) = 0$.

Hence, the voltage $v(t)$ across the switch can be obtained by

$$v(t) = (V_{cc} + I_0 R)\left[1 + \frac{1}{\omega}\sqrt{d^2 + \omega^2}\,\exp(-\delta t)\sin(\omega t + \varphi)\right]$$  (5.8)

where

$$\omega_0 = \frac{1}{\sqrt{LC}} \qquad \delta = \frac{R}{2L} = \frac{\omega_0}{2Q_L}$$

$$d = \frac{i(t_1)}{(V_{cc} + I_0 R)\,C}\frac{1}{} - \delta$$

$$\varphi = \tan^{-1}\left(\frac{\omega}{d}\right) \quad \omega = \sqrt{\omega_0^2 - \delta^2}.$$

In an optimum operation mode when the collector (*a*) voltage and (*b*) current waveforms shown in Fig. 5.3 do not overlap each other, resulting in a zero power loss on the transistor, the voltage $v(t)$ in Eq. (5.8) should satisfy the optimum

**Table 5.1** Load-Network Parameters of Switchmode Power Amplifier

| Parameters | $\tau_{sat}$, **Degree** | | | | |
|---|---|---|---|---|---|
| | **108°** | **144°** | **180°** | **216°** | **252°** |
| $\dfrac{\omega L}{R}$ | 3.4872 | 2.4083 | 1.7879 | 1.3494 | 0.9887 |
| $\omega CR$ | 0.2063 | 0.2280 | 0.2177 | 0.1865 | 0.1437 |
| $\dfrac{P_{out}R}{V_{cc}^2}$ | 0.0732 | 0.1788 | 0.3587 | 0.6622 | 1.1953 |

zero-voltage and zero voltage-derivative conditions given by Eqs (5.1) and (5.2). As a result, the analytical relationships between the load-network components can be properly obtained, whose values for different duty cycle or saturation time $\tau_{sat} = \omega t_1$ are given in Table 5.1, where $P_{out} = I_0 V_{cc}$ is the output power corresponding to the idealized lossless operation conditions [8].

However, to provide such an idealized switching operation mode, the loaded quality factor of this $L$-type circuit should be sufficiently small, for example, $Q_L = 2.8656$ for a 50% duty cycle, increasing at smaller duty cycles. This leads to the variation of the harmonic coefficient from 15% at $\tau_{sat} = 216°$ to 3% at $\tau_{sat} = 144°$. As a result, such a switchmode bipolar power amplifier designed to operate at 1 MHz and supply voltage of 15 V with a total output power of 3 W could provide a collector efficiency of 95.9% under optimum conditions with a 50% duty cycle [8].

## 5.2 Load network with shunt capacitor and series filter

For additional harmonic suppression, it is necessary to connect the load through the series filter tuned to the fundamental frequency, for example, a simple $L_0 C_0$ filter shown in Fig. 5.4(a) [5,9]. The loaded quality factor $Q_L$ of the series resonant circuit consisting of an inductor $L_0$ and a capacitor $C_0$ tuned to the fundamental frequency $\omega_0 = 1/\sqrt{L_0 C_0}$ should be sufficiently high in order for the output current to be sinusoidal. The single-ended switchmode power amplifier with a shunt capacitor was introduced by the Sokals in 1975 as a Class-E power amplifier and has found widespread application due to their design simplicity and high operational efficiency [10,11]. This type of high-efficiency power amplifier was then widely used in different frequency ranges and output power levels ranging from several kilowatts at low RF frequencies up to about 1 W at microwaves [12].

The characteristics of a Class-E power amplifier can be determined by finding its steady-state collector voltage and current waveforms. The simplified equivalent circuit of a Class-E power amplifier with shunt capacitance is shown in Fig. 5.4(b), where the load network consists of a capacitor $C$ shunting the

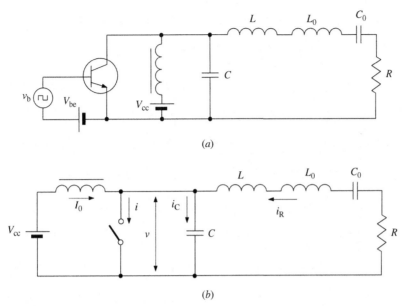

**FIGURE 5.4**

Basic circuits of Class-E power amplifier with shunt capacitance.

transistor output, a series inductor $L$, a series fundamentally tuned $L_0C_0$ circuit, and a load resistor $R$. In a common case, the shunt capacitor $C$ represents the transistor output capacitance and an external circuit capacitor. The collector of the transistor is connected to the supply voltage by an RF choke with high reactance at the fundamental frequency. The transistor is considered an ideal switch that is driven in such a way as to provide the device switching between its on-state and off-state operation conditions. As a result, the collector voltage waveform is determined by the switch when it is turned on and by the transient response of the load network when the switch is turned off.

To simplify the analysis of a Class-E power amplifier, the following assumptions are introduced:

- The transistor has zero saturation voltage, zero saturation resistance, infinite off-resistance, and its switching action is instantaneous and lossless.
- The total shunt capacitance is independent of the collector and is assumed to be linear.
- The RF choke allows only a constant dc current and has no resistance.
- The loaded quality factor $Q_L = \omega L_0/R = 1/\omega C_0 R$ of the series resonant $L_0C_0$ circuit tuned to the fundamental frequency is high enough for the output current to be sinusoidal at the switching frequency.
- There are no losses in the circuit except only in the load $R$.
- For simplicity, a 50% duty ratio is used.

For a lossless operation mode, it is necessary to provide the following optimum conditions for voltage across the switch (just prior to the start of switch on) at the moment $\omega t = 2\pi$, when transistor is saturated:

$$v(\omega t)|_{\omega t=2\pi} = 0 \tag{5.9}$$

$$\left.\frac{dv(\omega t)}{d\omega t}\right|_{\omega t=2\pi} = 0 \tag{5.10}$$

where $v(\omega t)$ is the voltage across the switch.

A detailed theoretical analysis of a Class-E power amplifier with shunt capacitance, for any duty ratio, is given in [13], where the load current is assumed to be sinusoidal,

$$i_R(\omega t) = I_R \sin(\omega t + \varphi) \tag{5.11}$$

where $\varphi$ is the initial phase shift.

When the switch is turned on for $0 \le \omega t < \pi$, the current through the capacitance

$$i_C(\omega t) = \omega C \frac{dv(\omega t)}{d\omega t} = 0 \tag{5.12}$$

and, consequently,

$$i(\omega t) = I_0 + I_R \sin(\omega t + \varphi) \tag{5.13}$$

under the initial on-state condition $i(0) = 0$. Hence, the dc current can be defined as

$$I_0 = -I_R \sin \varphi \tag{5.14}$$

and the current through the switch can be rewritten by

$$i(\omega t) = I_R[\sin(\omega t + \varphi) - \sin \varphi]. \tag{5.15}$$

When the switch is turned off for $\pi \le \omega t < 2\pi$, the current through the switch $i(\omega t) = 0$, and the current flowing through the capacitor $C$ can be written as

$$i_C(\omega t) = I_0 + I_R \sin(\omega t + \varphi) \tag{5.16}$$

producing the voltage across the switch by the charging of this capacitor according to

$$v(\omega t) = \frac{1}{\omega C} \int_{\pi}^{\omega t} i_C(\omega t) d\omega t = -\frac{I_R}{\omega C}[\cos(\omega t + \varphi) + \cos \varphi + (\omega t - \pi)\sin \varphi]. \tag{5.17}$$

Applying the first optimum condition given by Eq. (5.9) enables the phase angle $\varphi$ to be determined as

$$\varphi = \tan^{-1}\left(-\frac{2}{\pi}\right) = -32.482°. \tag{5.18}$$

Consideration of trigonometric relationships shows that

$$\sin\varphi = \frac{-2}{\sqrt{\pi^2+4}} \text{ and } \cos\varphi = \frac{\pi}{\sqrt{\pi^2+4}}. \qquad (5.19)$$

Then, the steady-state voltage waveform across the switch using Eqs (5.14) and (5.19) can be obtained by

$$v(\omega t) = \frac{I_0}{\omega C}\left(\omega t - \frac{3\pi}{2} - \frac{\pi}{2}\cos\omega t - \sin\omega t\right). \qquad (5.20)$$

Using Fourier series expansion, the expression to determine the supply voltage $V_{cc}$ can be written as

$$V_{cc} = \frac{1}{2\pi}\int_0^{2\pi} v(\omega t)\,d\omega t = \frac{I_0}{\pi\omega C}. \qquad (5.21)$$

As a result, the normalized steady-state collector voltage waveform for $\pi \le \omega t < 2\pi$ and current waveform for period of $0 \le \omega t < \pi$ are

$$\frac{v(\omega t)}{V_{cc}} = \pi\left(\omega t - \frac{3\pi}{2} - \frac{\pi}{2}\cos\omega t - \sin\omega t\right) \qquad (5.22)$$

$$\frac{i(\omega t)}{I_0} = \frac{\pi}{2}\sin\omega t - \cos\omega t + 1. \qquad (5.23)$$

Figure 5.5 shows the normalized (*a*) load current, (*b*) collector voltage, and (*c*) collector current waveforms for an idealized nominal Class-E mode with shunt capacitance. From collector voltage and current waveforms it follows that, when the transistor is turned on, there is no voltage across the switch and current $i(\omega t)$, consisting of the load sinusoidal current and dc current, flows through the transistor. However, when the transistor is turned off, this current flows through the shunt capacitor $C$. The jump in the collector current waveform at the instant of switching off is necessary to obtain nonzero output power at the fundamental frequency delivered to the load, which can be defined as an integration of the product of the collector voltage and current derivatives over the entire period [14]. Therefore, $dv/dt$ and $di/dt$ must both be nonzero during at least one of the switching transitions.

As a result, there is no nonzero voltage and current simultaneously, which means a lack of the power losses and gives an idealized collector efficiency of 100%. This implies that the dc power and fundamental-frequency output power delivered to the load are equal,

$$I_0 V_{cc} = \frac{I_R^2}{2}R. \qquad (5.24)$$

Consequently, the value of the dc supply current $I_0$ using Eqs (5.14) and (5.19) can be determined by

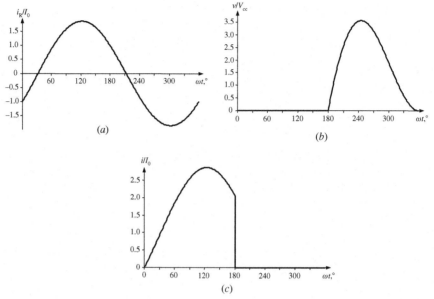

**FIGURE 5.5**

Normalized load current, collector voltage, and collector current waveforms for nominal Class-E with shunt capacitance.

$$I_0 = \frac{V_{cc}}{R} \frac{8}{\pi^2 + 4} = 0.577 \frac{V_{cc}}{R}. \tag{5.25}$$

Then, the amplitude of the output voltage $V_R = I_R R$ can be obtained from

$$V_R = \frac{4 V_{cc}}{\sqrt{\pi^2 + 4}} = 1.074 V_{cc}. \tag{5.26}$$

The peak collector voltage $V_{max}$ and current $I_{max}$ can be determined by differentiating the appropriate waveforms given by Eqs (5.22) and (5.23), respectively, and setting the results equal to zero, which gives

$$V_{max} = -2\pi\varphi V_{cc} = 3.562 V_{cc} \tag{5.27}$$

and

$$I_{max} = \left( \frac{\sqrt{\pi^2 + 4}}{2} + 1 \right) I_0 = 2.8621 I_0. \tag{5.28}$$

The fundamental-frequency voltage $v_1(\omega t)$ across the switch consists of the two quadrature components shown in Fig. 5.6, whose amplitudes can be found using Fourier formulas and Eq. (5.22) as

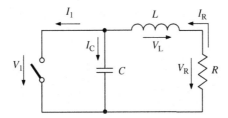

**FIGURE 5.6**

Equivalent Class-E load network at fundamental frequency.

$$V_R = -\frac{1}{\pi} \int_0^{2\pi} v(\omega t) \sin(\omega t + \varphi) d\omega t = \frac{I_R}{\pi \omega C} \left( \frac{\pi}{2} \sin 2\varphi + 2 \cos 2\varphi \right) \qquad (5.29)$$

$$V_L = -\frac{1}{\pi} \int_0^{2\pi} v(\omega t) \cos(\omega t + \varphi) d\omega t = -\frac{I_R}{\pi \omega C} \left( \frac{\pi}{2} + \pi \sin^2 \varphi + 2 \sin 2\varphi \right). \qquad (5.30)$$

As a result, the optimum series inductance $L$ and shunt capacitance $C$ can be calculated from

$$\frac{\omega L}{R} = \frac{V_L}{V_R} = 1.1525 \qquad (5.31)$$

$$\omega CR = \frac{\omega C}{I_R} V_R = 0.1836. \qquad (5.32)$$

The optimum load resistance $R$ for the supply voltage $V_{cc}$ and fundamental-frequency output power $P_{out}$ delivered to the load using Eqs (5.24) and (5.26) can be obtained by

$$R = \frac{8}{\pi^2 + 4} \frac{V_{cc}^2}{P_{out}} = 0.5768 \frac{V_{cc}^2}{P_{out}}. \qquad (5.33)$$

Finally, the phase angle of the load network seen by the switch and required for an idealized nominal Class-E mode with shunt capacitance can be determined through the load-network parameters using Eqs (5.31) and (5.32) as

$$\phi = \tan^{-1} \left( \frac{\omega L}{R} \right) - \tan^{-1} \left( \frac{\omega CR}{1 - \frac{\omega L}{R} \omega CR} \right) = 35.945°. \qquad (5.34)$$

When realizing a nominal Class-E operation mode, it is very important to know up to which maximum frequency such an idealized efficient operation mode can be extended. In this case, it is important to establish the relationship between a maximum frequency $f_{max}$, a parallel shunt capacitance $C$, and a dc supply voltage $V_{cc}$. As a result, substituting Eq. (5.32) into Eq. (5.25) gives

$$I_0 = \pi \omega C \, V_{cc}. \tag{5.35}$$

Then, by taking into account the relationship between $I_0$ and $I_{max}$ given in Eq. (5.28), the maximum frequency of a nominal Class-E power amplifier with shunt capacitance can be evaluated from

$$f_{max} = \frac{1}{\pi^2} \frac{1}{\sqrt{\pi^2 + 4} + 2} \frac{I_{max}}{C_{out} V_{cc}} = \frac{I_{max}}{56.5 \, C_{out} V_{cc}} \tag{5.36}$$

where $C = C_{out}$ is the transistor output capacitance limiting the maximum operation frequency of an ideal Class-E power-amplifier circuit [15].

The high-$Q_L$ assumption for the series-resonant $L_0 C_0$ circuit can lead to considerable errors if its value is substantially small in real circuits [16]. For example, for a 50% duty ratio, the values of the load-network parameters for the loaded quality factor $Q_L$ less than unity can differ by several tens of percentages. At the same time, for $Q_L \geq 7$, the errors are found to be less than 10% and they become less than 5% for $Q_L \geq 10$. A detailed overview of Class-E power amplifiers with shunt capacitance, including explicit design equations, applicable frequency ranges, optimization principles, and experimental results is given in [17].

## 5.3 Matching with a standard load

For most practical applications, it is necessary to match the required Class-E optimum load resistance $R$ with a standard load resistance $R_L$, usually equal to 50 $\Omega$. Figure 5.7($a$) shows the equivalent circuit of a Class-E power amplifier with shunt capacitance where the series $L_0 C_0$ filter is followed by an $L$-type low-pass matching circuit consisting of a series inductor $L_1$ and a shunt capacitor $C_1$ [18]. Such a connection of this matching circuit when its shunt capacitor $C_1$ is connected in parallel to the load resistor $R_L$ assumes that $R < R_L$. This is normally the case for the high-power or low-voltage power amplifiers. The main goal of the matching circuit is to provide a maximum delivery of the output power at the fundamental frequency $\omega_0$ to the standard load of 50 $\Omega$, since it is assumed that a series $L_0 C_0$ filter has sufficiently high $Q$-factor to suppress the harmonic components of the fundamental frequency.

In this case, the Class-E optimum resistance $R$ can generally be defined through the parameters of the $L$-type low-pass impedance-matching circuit and the load resistance $R_L$ as

$$R = \frac{R_L}{1 + (\omega_0 R_L C_1)^2} + j \left[ \omega L_1 - \frac{\omega_0 C_1 R_L^2}{1 + (\omega_0 R_L C_1)^2} \right]. \tag{5.37}$$

Since it is necessary to provide the complex-conjugate matching conditions to maximize the output power in the load, the imaginary part of Eq. (5.37) must

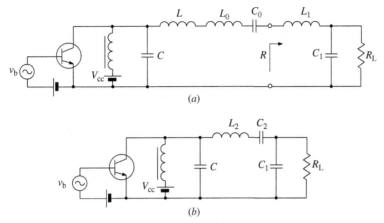

**FIGURE 5.7**

Schematics of Class-E power amplifiers with impedance-matching circuits.

be equated to zero. Then, Eq. (5.37) can be resolved separately for the real and imaginary parts as

$$R_L = R(1 + Q_L^2) \tag{5.38}$$

$$\frac{1}{\omega_0 C_1} = \omega_0 L_1 (1 + Q_L^{-2}) \tag{5.39}$$

where $Q_L = \omega_0 C_1 R_L = \omega_0 L_1/R$ is the loaded quality factor which is equal for both the series and parallel circuits at the fundamental frequency.

As a result, the loaded quality factor $Q_L$ can be expressed through the load resistances $R$ and $R_L$ as

$$Q_L = \sqrt{\frac{R_L}{R} - 1} \tag{5.40}$$

while the matching circuit parameters can be calculated from

$$L_1 = \frac{R Q_L}{\omega_0} \tag{5.41}$$

$$C_1 = \frac{Q_L}{\omega_0 R_L}. \tag{5.42}$$

The series high-$Q_L$ filter and $L$-type matching circuit can be combined into a $\pi$-type matching circuit with two shunt capacitors and series $L_2 C_2$ circuit, where $L_2 = L + L_0 + L_1$ and $C_2 = C_0$, as shown in Fig. 5.7($b$). Generally, a tandem connection of any type of low-pass or band-pass matching sections can be used for different ratios between a required optimum Class-E load resistance $R$ and a

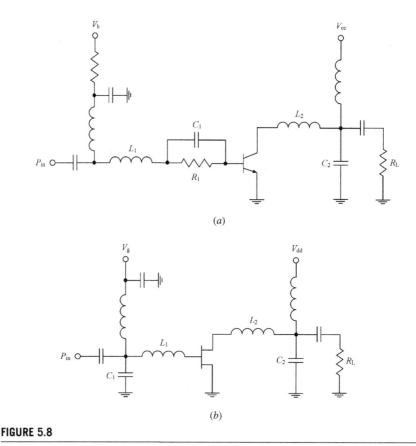

(a)

(b)

**FIGURE 5.8**

Schematics of a lumped-elements Class-E power amplifiers with shunt capacitance.

standard load resistance $R_L$. In this case, to simplify the matching design procedure, it is best to cascade the low-pass $L$-type matching sections with equal values
of their $Q_L$-factors [19,20].

Figure 5.8(a) shows the circuit schematic of a lumped-elements Class-E HBT
power amplifier operated in a frequency bandwidth of 5−6 GHz and based on a
single-section low-pass output matching circuit [21]. Here, the capacitor $C_2$ and
inductor $L_2$ are determined to obtain appropriate Class-E load impedance at the
fundamental frequency. Due to its low-pass characteristic, the impedance seen at
the collector terminal tends to approximate an open circuit with increasing frequency. The input network was designed to match the transistor input impedance
to the 50-$\Omega$ source impedance, where the ballasting resistor $R_1$ and large capacitance $C_1$ are used to improve low-frequency stability. In a monolithic power
amplifier with the circuit schematic shown in Fig. 5.8(b) based on a AlGaN/GaN
HEMT technology with $f_T = 18$ GHz, a shunt capacitor $C_2$ is fabricated using

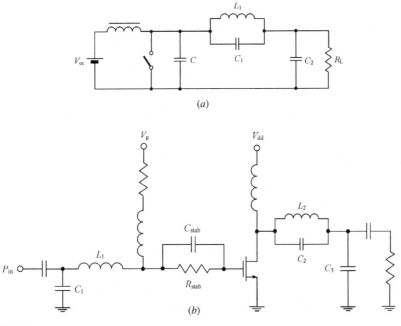

**FIGURE 5.9**

Schematics of a lumped-elements Class-E power amplifier with second-harmonic peaking.

parallel-plate technology with silicon nitride (SiN) passivation dielectric layer and a series multiturn inductor $L_2$ is implemented using air-bridges and a 3-μm-thick gold interconnected layer, resulting in a *PAE* of 61% and an output power of 33.8 dBm at an operating frequency of 4 GHz [22].

In the case of monolithic implementation of the Class-E power amplifier, the large value of an inductor $L_0$ shown in Fig. 5.7(a) results in the significant ohmic losses and low self-resonant frequency that is crucial at microwave frequencies. Therefore, to minimize the inductance value of the monolithically implemented spiral inductor, the Class-E load network can represent the alternative topology shown in Fig. 5.9(a) with optimized second-harmonic peaking and simultaneous impedance transformation at the fundamental frequency [21]. Here, the parallel tank formed by capacitor $C_2$ and inductor $L_2$ is designed to resonate at $2f_0$. The net impedance of the second-harmonic resonator must represent an inductive reactance at the fundamental frequency $f_0$, similar to the contribution of inductance $L$ in the Class-E load network shown in Fig. 5.7(a). In this case, it is assumed that the capacitive reactances of the parallel tank at the third and higher-order harmonic components are relatively high compared with that provided by the shunt capacitance $C$.

Consequently, the design equations for the load-network elements $L_2$, $C_1$, and $C_2$ in Fig. 5.9(a) can be derived by using Eq. (5.31) as

$$L_2 = \frac{3R}{4\omega_0}(Q_L + 1.1525) \tag{5.43}$$

$$C_2 = \frac{1}{4\omega_0^2 L_2} \tag{5.44}$$

$$C_1 = \frac{Q_L}{\omega_0 R_L}. \tag{5.45}$$

where $Q_L$ is obtained from Eq. (5.40).

Figure 5.9(b) shows the circuit schematic of a lumped 2.55-GHz Class-E power amplifier with a tank resonant circuit in the load network tuned to the second harmonic. Based on a 90-nm CMOS process, this power amplifier can achieve a simulated drain efficiency of 57.4% at an output power of 22.7 dBm when operated from a supply voltage of 2.5 V [23].

## 5.4 Effect of saturation resistance

In practical power amplifier design, especially when a value of the supply voltage is sufficiently small, it is very important to predict the overall degradation of power amplifier efficiency due to the finite value of the transistor saturation resistance. Figure 5.10 shows the simplified equivalent circuit of a Class-E power amplifier with shunt capacitance, including the saturation resistance (on-resistance) $r_{sat}$ connected in series to the ideal switch. To obtain a quantitative estimate of the losses due to the contribution of $r_{sat}$, the saturated output power $P_{sat}$ can be obtained with a simple approximation when the current $i(\omega t)$ flowing through the saturation resistance $r_{sat}$ is determined in an ideal case by Eq. (5.23).

An analytical expression to calculate the power losses due to the saturation resistance $r_{sat}$, whose value is assumed constant, can be represented in the normalized form as

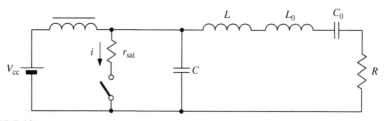

**FIGURE 5.10**

Equivalent Class-E load network with transistor saturation resistance.

$$\frac{P_{sat}}{P_0} = \frac{r_{sat}}{2\pi\, I_0 V_{cc}} \int\limits_0^\pi i^2(\omega t)d\omega t \qquad (5.46)$$

where $P_0 = I_0 V_{cc}$ is the dc power. As a result, by using a linear approximation of the collector current waveform for the Class-E load network with one inductor and one capacitor whose equivalent circuit without saturation resistance $r_{sat}$ is shown in Fig. 5.2(b), the averaged dissipated power $P_{sat}$ normalized to the dc power $P_0$ can be evaluated by

$$\frac{P_{sat}}{P_0} = \frac{8}{3}\frac{r_{sat}P_0}{V_{cc}^2} \qquad (5.47)$$

for the required dc power and supply voltage [18].

For a Class-E power amplifier with shunt capacitance represented by equivalent circuit shown in Fig. 5.10, by taking into account that

$$\int\limits_0^\pi \left(\frac{\pi}{2}\sin \omega t - \cos \omega t + 1\right)^2 d\omega t = \frac{\pi}{8}(\pi^2 + 28)\, I_0^2 \qquad (5.48)$$

Equation (5.46) can be finally rewritten using Eq. (5.25) by

$$\frac{P_{sat}}{P_0} = \frac{r_{sat}}{2\pi}\frac{I_0}{V_{cc}}\frac{\pi}{8}(\pi^2 + 28) = \frac{r_{sat}}{2R}\frac{\pi^2 + 28}{\pi^2 + 4} = 1.365\frac{r_{sat}}{R}. \qquad (5.49)$$

The collector efficiency $\eta$ can be calculated from

$$\eta = \frac{P_{out}}{P_0} = \frac{P_0 - P_{sat}}{P_0} = 1 - \frac{P_{sat}}{P_0}. \qquad (5.50)$$

The presence of the saturation resistance results in the finite value of the saturation voltage $V_{sat}$ which can be defined from

$$\frac{V_{sat}}{V_{cc}} = 1 - \frac{1}{1 + 1.365\dfrac{r_{sat}}{R}} \qquad (5.51)$$

where $V_{sat}$ is normalized to the dc collector voltage $V_{cc}$ [24].

In an optimum switching mode when both Class-E conditions given by Eqs (5.9) and (5.10) are satisfied, the second optimum condition is equivalent to

$$\left.\frac{di(\omega t)}{d\omega t}\right|_{\omega t=2\pi} = 0 \qquad (5.52)$$

when the collector current at the end of each period must start with zero derivative, since $v(\omega t) = r_{sat}i(\omega t)$. Figure 5.11 shows the collector voltage and current waveforms by solid lines corresponding to an optimum operation mode [9]. Here, the shape of the saturation voltage shown in Fig. 5.11(a) is determined by the collector current waveform. However, if the first optimum condition is not satisfied

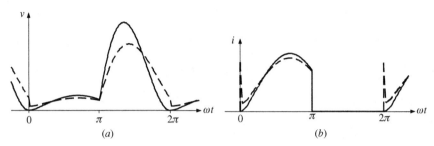

**FIGURE 5.11**

Collector voltage and current waveforms for idealized Class-E with saturation resistance.

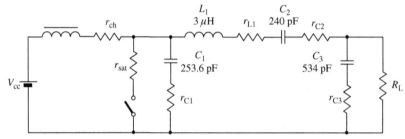

**FIGURE 5.12**

Equivalent Class-E load network with lossy elements.

when, for example, some positive value of the collector voltage at the time when the transistor is turned on is assumed, this results in the switching losses accompanied by the proper transient response of the current waveform shown in Fig. 5.11(*b*) by dashed lines in the impulse form with a finite amplitude whose duration is determined by the time constant $\tau_s = r_{sat}C$ [25]. In this case, the peak collector voltage reduces compared to the optimal case.

In addition to the saturation resistance $r_{sat}$, the power losses in the elements of the load network are also very important for predicting efficiency of the Class-E power amplifier. The contribution of each circuit element to the overall power loss and drain efficiency was experimentally validated on the example of a 300-W MOSFET Class-E power amplifier developed to operate at a switching frequency of 7.29 MHz, and a dc supply voltage of 130 V with a sinusoidal driving signal [26]. Being calculated at 42.5% duty ratio, the individual component losses of the elements of the equivalent Class-E load network shown in Fig. 5.12 are listed in Table 5.2. A power MOSFET transistor IRFP440 was used as a switch and its saturation resistance $r_{sat}$ was obtained from the manufacturer's data sheet. As a result, the total calculated power loss was 29.3 W, resulting in a drain

**Table 5.2** Component Resistance and Predicted Power Loss

| Component | Resistance, $\Omega$ | Predicted Power Loss, W |
|---|---|---|
| $r_{sat}$ | 0.85 | 12.75 |
| $r_{c2}$ | 0.42 | 6.29 |
| $r_{L1}$ | 0.28 | 4.19 |
| $r_{C3}$ | 0.42 | 3.77 |
| $r_{C1}$ | 0.23 | 1.22 |
| $r_{ch}$ | 0.2 | 1.07 |

efficiency of 91.1%, which is very close to the measured drain efficiency of 90.1%. As expected, more than 40% of the total power loss is contributed by the power transistor due to its nonzero value of the saturation resistance.

## 5.5 Driving signal and finite switching time

An analysis of the idealized operation of the Class-E power amplifier is based on a preliminary assumption of instant transistor switching of the saturation mode with zero on-resistance to the pinch-off mode with zero current. However, real transistors are characterized by nonzero transition times when switching time may constitute a significant part of the period, especially at high frequencies. In this case, there are different contributions of the nonzero switching conditions during the on-to-off and off-to-on transitions. If the power loss during the off-to-on transition is negligible because the collector voltage drops to zero at the end of pinch-off mode with collector or drain current starting from zero value, the power loss becomes significant during the on-to-off transition since the collector or drain current must decrease instantly to zero from a significant value which it achieves at the end of saturation mode. Physically, the finite switching time can be explained by the device inertia when the base charge reduces to zero value with some finite time delay $\tau_s$. Due to the device inertia mostly owing to the time delay in its input circuit, there is an active state where the collector or drain current is determined by the base or channel charge process, but not by the load network.

The input-port characteristics of bipolar, MOSFET, or MESFET transistors are so different that generally a different driver circuit should be used for each type of transistor. In this case, the best gate-voltage drive should represent a trapezoidal waveform, with the falling transition occupying 30% or less of the period. Shorter transition time is difficult to achieve because of the effect of the transistor equivalent circuit parameters, among which the most important are the gate resistance $R_g$, the gate-source capacitance $C_{gs}$, and the gate-drain capacitance $C_{gd}$ providing an

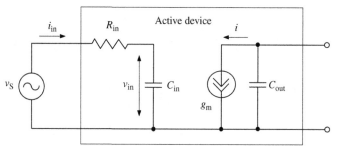

**FIGURE 5.13**

Simplified equivalent transistor model.

input time delay $\tau_{in} = R_g(C_{gs} + C_{gd})$. For both MOSFET and MESFET transistors, the optimum drive minimizes the sum of the output-stage power dissipation and the driver-stage power consumption. The peak of the drive waveform should be safely below the transistor maximum gate-source voltage rating. For MESFET devices, it should be less than the gate-source voltage, at which the gate-source diode conducts enough current to cause either of two undesired effects:

- Metal migration of the gate metallization at an undesirably rapid rate, making the transistor operating lifetime shorter than desired.
- Enough power dissipation to reduce the overall efficiency more than the efficiency is increased by the lower dissipation in the lower on-resistance that results from a higher upper level of the drive waveform.

The lower level of the trapezoid should be low enough to result in a satisfactorily small current during the off-state of the transistor operation. A sine wave is a usable approximation to the trapezoidal waveform, though not optimum. To obtain the transistor duty ratio of 50%, that is usually the best choice, the zero level of the sine wave should be positioned slightly above the transistor turn-on threshold voltage. It is a better approximation to remove the part of the sine waveform that goes below the value of the gate-source bias voltage that ensures fully off-state operation, replacing it with a constant voltage at that gate-source bias voltage. This reduces the input drive power by slightly less than 50%, almost doubling the power gain.

Figure 5.13 shows the simplified equivalent transistor model where both capacitances $C_{in}$ and $C_{out}$ include the feedback base-collector (or gate-drain) capacitance. Despite the simplification, such an equivalent model can describe with sufficient accuracy the behavior of the bipolar, MOSFET, or MESFET transistor up to approximately $0.1f_T$, where $f_T$ is the transition frequency. Moreover, for a MOSFET transistor, the gate-source capacitance varies insignificantly over a wide range of the gate-bias voltages, whereas the feedback gate-drain capacitance is sufficiently small, normally by an order of less than the gate-source capacitance.

Let us assume that the transistor is driven from the signal source with a sinusoidal voltage,

$$v_S(\omega t) = V_0 + V_S \sin(\omega t + \psi_0) \tag{5.53}$$

where $V_0$ is the base-emitter (or gate-source) bias voltage and $\psi_0$ is the initial phase shift.

For the input transistor circuit, the linear first-order differential equation can be given by

$$\omega C_{in} \frac{dv_{in}}{dt} = \frac{v_S - v_{in}}{R_{in}}. \tag{5.54}$$

By taking into account that $i = g_m v_{in}$, where $g_m$ is the device transconductance (assuming it is constant when the device is turned on and is equal to zero when the device is turned off), Eq. (5.54) can be rewritten as

$$\frac{di}{d\omega t} + \frac{i}{\omega \tau_{in}} - \frac{V_0 + V_S \sin(\omega t + \psi_0)}{\omega \tau_{in}} = 0 \tag{5.55}$$

whose general solution can be obtained by

$$i(\omega t) = C_0 \exp\left(-\frac{\omega t}{\omega \tau_{in}}\right) + A_1 \sin(\omega t + \psi_0) + A_2 \cos(\omega t + \psi_0) + A_3 \tag{5.56}$$

where $\tau_{in} = R_{in} C_{in}$.

The coefficients $A_1$, $A_2$, and $A_3$ are defined by substituting Eq. (5.56) into Eq. (5.55) and equating the corresponding components. Then, by setting the boundary conditions $i(0) = i(\pi) = 0$ to determine the remaining unknown coefficient $C_0$ and bias voltage $V_0$, we can finally write

$$i(\omega t) = \frac{g_m V_S}{\sqrt{1 + (\omega \tau_{in})^2}} \left[ \frac{1 - 2\exp\left(-\dfrac{\omega t}{\omega \tau_{in}}\right) + \exp\left(-\dfrac{\pi}{\omega \tau_{in}}\right)}{\exp\left(-\dfrac{\pi}{\omega \tau_{in}}\right) - 1} \right.$$
$$\left. \times \sin(\psi - \psi_0) + \sin(\omega t - \psi + \psi_0) \vphantom{\frac{1}{1}} \right] \tag{5.57}$$

where $\psi = \tan^{-1}(\omega \tau_{in})$ is the input phase angle. It should be noted that the collector (or drain) current waveforms can have different characteristics, depending on the boundary conditions, initial phase shift, and base-emitter (or gate-source) bias voltage. For example, a 50% duty ratio of the input signal is realized with the bias voltage calculated from

$$V_0 = -\frac{V_S}{\sqrt{1 + (\omega \tau_{in})^2}} \frac{1 + \exp\left(-\dfrac{\pi}{\omega \tau_{in}}\right)}{1 - \exp\left(-\dfrac{\pi}{\omega \tau_{in}}\right)} \sin(\psi - \psi_0) \tag{5.58}$$

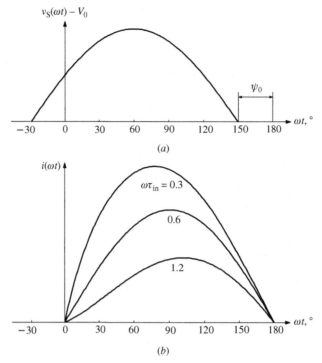

(a)

(b)

**FIGURE 5.14**

Collector current waveforms versus input parameter $\omega\tau_{\text{in}}$.

which can take positive, zero, and negative values when the transistor can be biased for Class-AB, Class-B, or Class-C operation modes, respectively, depending on its input circuit parameters.

For the sinusoidal input voltage with initial phase $\psi_0 = 30°$ shown in Fig. 5.14(a), the resulting collector (or drain) current waveforms for several values of the parameter $\omega\tau_{\text{in}}$ are plotted in Fig. 5.14(b), from which it follows that generally the collector current waveforms are asymmetrical, and the degree of this asymmetry depends on the input circuit phase angle $\psi$. In this case, close to symmetrical waveform can be obtained with $\omega\tau_{\text{in}} = 0.6$. At higher frequencies, the waveform asymmetry is shifted to the right-hand side, thus contributing to shorter switching time of the transistor. However, larger value of $\omega\tau_{\text{in}}$, which is the time delay due to the input $RC$ circuit, contributes to the smaller output current amplitude with an appropriately smaller fundamental-frequency component, resulting in a power-gain reduction. The input voltage amplitude $V_{\text{in}}$ across the capacitance $C_{\text{in}}$ can be defined as a frequency-dependent function by

$$\frac{V_{\text{in}}}{V_S} = \frac{1}{\sqrt{1 + (\omega\tau_{\text{in}})^2}}. \tag{5.59}$$

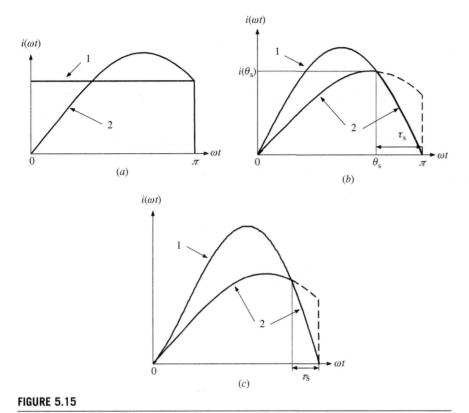

**FIGURE 5.15**

Collector current waveforms due to nonzero values of switching time.

From Eq. (5.59), it follows that, for the same input capacitance $C_{in}$, the voltage across the capacitance $C_{in}$ reduces with an increase of $R_{in}$ or $\tau_{in}$, thus resulting in a smaller output fundamental current amplitude. For zero input resistance $R_{in}$, when there is no power loss in the transistor input circuit, the power gain becomes infinite. The input power for a nonzero value of $R_{in}$ can be calculated from

$$P_{in} = \frac{I_{in}^2 R_{in}}{2} = \frac{(\omega \tau_{in})^2}{1 + (\omega \tau_{in})^2} \frac{V_S^2}{2R_{in}} \tag{5.60}$$

where $I_{in}$ is the input current amplitude.

Generally, the switching time is sufficiently small. Certainly, for an ideal active device without any memory effects due to intrinsic phase delays, the switching time is equal to zero when the rectangular input drive results in a rectangular output response with the required amplitude, as shown in Fig. 5.15(a) by rectangular curve 1, where curve 2 corresponds to an ideal Class-E collector-current waveform. Such an ideal case assumes zero feedback capacitance and zero input resistance $R_{in}$. Otherwise, if $R_{in}$ is not equal to zero, the input low-pass $R_{in}C_{in}$ filter section provides the suppression of higher-order harmonics of the fundamental frequency, resulting in the finite rise and fall times of the driving rectangular pulse in the time domain. This means that, during this finite on-time and off-time operation conditions, an active device cannot be instantly switched from the saturation mode to the pinch-off mode and operates in the active region when simultaneously output current and output voltage are positive corresponding to the output power dissipation inside the transistor.

However, in a real situation, especially at higher frequencies, it is very difficult to realize the driving signal close to the rectangular form as it leads to significant circuit complexity and requires the minimum device input delay ($R_{in}$ should be as small as possible) and transition frequency $f_T = g_m/2\pi C_{in}$ to be as high as possible ($C_{in}$ should be as much smaller as possible). Fortunately, to realize high-efficiency operation conditions, it is sufficient to drive the power amplifier simply with a sinusoidal signal. The finite-time transition from the saturation mode to the pinch-off mode through the device active mode takes place at the point of the intersection of the curve corresponding to the base (or channel) charge process (curve 1) and the curve corresponding to the required ideal collector-current waveform provided by the load network (curve 2), as shown in Fig. 5.15(b). To minimize the switching time interval, it is sufficient to slightly overdrive the active device with signal amplitude by 20–30% higher than is required for a conventional Class-B power amplifier, as shown in Fig. 5.15(c). As an alternative, the second harmonic component (approximation of a half-sinusoidal waveform) or third harmonic component (approximation of a rectangular waveform with close to trapezoidal waveform) with proper phasing can be added to the input driving signal. In both cases, the overall driving signal amplitude will be increased compared with simply sinusoidal driving signal, thus resulting in a faster switching operation time.

The power dissipated during this on-to-off transition can be calculated assuming zero on-resistance as

$$P_s = \frac{1}{2\pi} \int_{\theta_s}^{\pi} i(\omega t)\, v(\omega t) d\omega t \qquad (5.61)$$

where the collector voltage during the transition time $\tau_s = \pi - \theta_s$ is defined by

$$v(\theta_s) = \frac{1}{\omega C} \int_{\theta_s}^{\pi} i_C(\omega t)\, d\omega t. \qquad (5.62)$$

The short duration of the switching time and the proper behavior of the resulting collector (or drain) waveform allows us to make an additional assumption of a linearly decreasing collector current during fall time $\tau_s = \pi - \theta_s$, starting at $i(\theta_s)$ at time $\theta_s$ and decaying to zero at time $\pi$, which can be written as

$$i(\omega t) = i(\theta_s)\left(1 - \frac{\omega t - \theta_s}{\tau_s}\right) \tag{5.63}$$

where $i(\theta_s)$ corresponds to the peak collector current shown in Fig. 5.15(b) [18,24]. In this case, the capacitor-charging current $i_C(\omega t) = i(\theta_s) - i(\omega t)$, being zero during saturation mode, varies linearly between zero and $i(\theta_s)$ during on-to-off transition according to

$$i_C(\omega t) = i(\theta_s)\frac{\omega t - \theta_s}{\tau_s}. \tag{5.64}$$

The collector voltage produces a parabolic voltage waveform during the switching interval according to Eq. (5.62) as

$$v(\omega t) = \frac{i(\theta_s)}{2\omega C\tau_s}(\omega t - \theta_s)^2. \tag{5.65}$$

As a result, the power dissipated during transition according to Eq. (5.61) is then

$$P_s = \frac{i^2(\theta_s)\tau_s^2}{48\pi\omega C}. \tag{5.66}$$

For an optimum Class-E power amplifier by assuming in view of a short transition time that $i(\theta_s) = i(\pi)$, from Eq. (5.23) it follows that $i(\pi) = 2I_0$, hence

$$P_s = \frac{I_0^2\tau_s^2}{12\pi\omega C}. \tag{5.67}$$

By taking into account Eq. (5.35), the switching loss power $P_s$ normalized to the dc power $P_0$ can be obtained by

$$\frac{P_s}{P_0} = \frac{I_0\tau_s^2}{12\pi\omega CV_{cc}} = \frac{\tau_s^2}{12}. \tag{5.68}$$

As a result, the collector efficiency $\eta$ can be estimated as

$$\eta = 1 - \frac{P_s}{P_0} = 1 - \frac{\tau_s^2}{12}. \tag{5.69}$$

As follows from Eq. (5.68), the power losses due to the nonzero switching time are sufficiently small and, for example, for $\tau_s = 0.35$ or $20°$, they are only about 1%, whereas they are approximately equal to 10% for $\tau_s = 60°$. A more exact analysis assuming a linear variation of the collector current during on-to-off transition results in similar results when efficiency degrades to 97.72% for $\tau_s = 30°$ and to 90.76% for $\tau_s = 60°$ [27]. Considering an exponential collector current decay rather than linear during the fall time shows the similar result for $\tau_s = 30°$ when $\eta = 96.8\%$, but the collector efficiency degrades more significantly at longer fall times when, for example, $\eta = 86.6\%$ for $\tau_s = 60°$ [28].

It should be noted that all these results were obtained for a particular case of the infinite $Q_L$ factor of the series-tuned resonant circuit. However, for the loaded quality factors $Q_L$ having finite and sufficiently small values with the assumption of harmonic distortion in the output signal, the collector efficiency increases by several percents [29]. As a result, in terms of efficiency and linearity, the most desirable range of $Q_L$ is from 5 to 10. In terms of practical implementation, the smaller the gate length of the transistor, the lower the fall-time angles that can be expected, and the loaded quality factor becomes of less importance for collector efficiency. For example, by using a 0.6-μm CMOS device, the lower decay time with the drain efficiency greater by about 10% than that when using a 0.8-μm CMOS device can be achieved [29]. Similarly, using a 1.0-μm GaN HEMT device results in an output power of 93 mW and a drain efficiency of 72% at an operating frequency of 1 GHz, whereas a drain efficiency of 82% can be achieved with a 0.12-μm GaN HEMT device [30].

## 5.6 Effect of nonlinear shunt capacitance

At high operating frequencies close to the maximum switching frequency $f_{max}$, the shunt capacitance $C$ required for an ideal switchmode Class-E operation will be fully represented by the output capacitance of the active device. However, the intrinsic output transistor capacitance (collector capacitance of a bipolar device or drain-source capacitance of a field-effect transistor) is generally nonlinear. If its contribution to the overall shunt capacitance is significant, it is necessary to take into account the nonlinear nature of this capacitance. Figure 5.16 shows the equivalent Class-E load network with a nonlinear shunt capacitor $C$.

The nonlinear capacitance can be represented as a junction capacitance by

$$C(v) = \frac{C_0}{\left(1 + \dfrac{v}{V_{bi}}\right)^{\gamma}} \tag{5.70}$$

where $v$ is the reverse voltage over the diode junction, $V_{bi}$ is the built-in potential, $\gamma$ is the junction sensitivity or gradual coefficient ($\gamma = 1/3$ for gradient junction and $\gamma = 1/2$ for abrupt junction), and $C_0 = C(0)$ is the collector capacitance at $v = 0$, which can be defined through the supply voltage $V_{cc}$ as

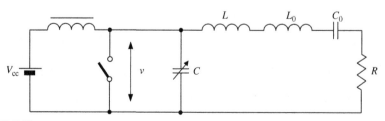

**FIGURE 5.16**

Equivalent Class-E load network with nonlinear shunt capacitor.

$$C(0) = C(V_{cc})\left(1 + \frac{V_{cc}}{V_{bi}}\right)^{\gamma}. \tag{5.71}$$

The nonlinear capacitance has an effect only during the time period when the transistor is turned off. Therefore, the collector current during the time period when the transistor is turned on can be determined by Eq. (5.23). Then, the collector voltage due to the current flowing through the nonlinear capacitance

$$i_C(\omega t) = \omega C(v)\frac{dv(\omega t)}{d\omega t} \tag{5.72}$$

can be calculated by substituting Eqs (5.70) and (5.16) into Eq. (5.72) and integrating of its both parts as

$$\int_0^v \frac{C_0}{\left(1 + \frac{v}{V_{bi}}\right)^{\gamma}}\,dv = \frac{1}{\omega}\int_\pi^{\omega t} i_C(\omega t)\,d\omega t. \tag{5.73}$$

As a result,

$$\frac{V_{bi}}{1-\gamma}\left[\left(1 + \frac{v}{V_{bi}}\right)^{1-\gamma} - 1\right] = \frac{I_0}{\omega C_0}\left(\omega t - \frac{3\pi}{2} - \frac{\pi}{2}\cos\omega t - \sin\omega t\right) \tag{5.74}$$

or

$$v = V_{bi}\left\{\left[\frac{I_0}{\omega C_0}\frac{1-\gamma}{V_{bi}}\left(\omega t - \frac{3\pi}{2} - \frac{\pi}{2}\cos\omega t - \sin\omega t\right) + 1\right]^{\frac{1}{1-\gamma}} - 1\right\}. \tag{5.75}$$

Since the optimum load phase angle, dc current, and output voltage are not affected by the capacitance nonlinearity, Eq. (5.74) can be rewritten using Eq. (5.35) as

$$\frac{V_{bi}}{1-\gamma}\left[1 + \frac{v}{V_{bi}} - \left(1 + \frac{v}{V_{bi}}\right)^{\gamma}\right] = V_{cc}\pi\left(\omega t - \frac{3\pi}{2} - \frac{\pi}{2}\cos\omega t - \sin\omega t\right) \tag{5.76}$$

which represents a nonlinear algebraic equation with respect to the collector voltage $v(\omega t)$ for any type of the device junction [31]. For abrupt diode junction with $\gamma = 0.5$, it can be simplified to the quadratic equation and solved analytically. The optimum small-signal value of the shunt capacitance $C_0$ can then be calculated from Eq. (5.70) by taking into account Eqs (5.32) and (5.33).

Figure 5.17 shows the difference between the collector voltage waveforms corresponding to the circuits with nonlinear capacitance (curve 1) and linear capacitance (curve 3). The nonlinear nature of this capacitance must be taken into account when specifying the breakdown voltage of the transistor. For example, the collector voltage waveform will rise in the case of the shunt capacitance described by abrupt diode junction in comparison with the linear capacitance, and its maximum voltage can be greater by about 20% for a 50% duty ratio [9,31].

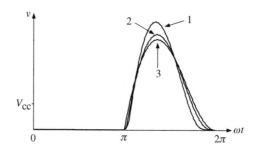

**FIGURE 5.17**

Collector voltage waveforms due to nonlinear shunt capacitance.

However, stronger nonlinearity of the shunt capacitance causes the peak voltages to be higher [32]. At the same time, the deviations of the optimum load network parameters are insignificant, less than 5% in a wide range of supply voltages. Since the nonlinear capacitance is largest at zero voltage, the collector waveform will rise more slowly than in the linear case. As the collector voltage increases, the capacitance will decrease, and hence the voltage should begin to rise more quickly than in the linear case. If the shunt capacitance consists of both nonlinear and linear capacitances, the collector voltage waveform is intermediate (curve 2) and located between the two extreme cases of entirely nonlinear or entirely linear capacitance [33]. In some cases, when the manufacturer provides the device intrinsic capacitances at specified bias values, the output capacitance can be considered as an equivalent linear shunt capacitance specified at the operating bias point and expressed as a function of the supply voltage [34,35]. It was found that the Class-E power amplifier with nonlinear shunt capacitance does not achieve zero-voltage-switching operation when the supply voltage is lower than the value designed for an idealized optimum Class-E operation mode, thus resulting in switching power loss and lower power efficiency [36]. The highest power efficiency can be obtained at a lower dc supply voltage than the designed dc supply voltage, even though the circuit was designed to achieve idealized optimum Class-E operation at the designed dc supply voltage [37].

## 5.7 Optimum, nominal, and off-nominal Class-E operation

More detailed theoretical analysis of the time-dependent behavior of the collector voltage and current waveforms shows that, for a finite value of the saturation resistance $r_{sat}$, the optimum conditions for an idealized operation mode given by Eqs (5.9) and (5.10) no longer correspond to the minimum dissipated power losses, and there are optimum nonzero values of the collector voltage and its derivative at switching time instant corresponding to minimum overall power losses [38]. For example, even for small losses with the normalized loss parameter $\omega C r_{sat} = 0.1$ for a duty ratio of 50%, the optimum series inductance $L$ is almost two times greater and the optimum shunt capacitance $C$ is about 20% greater than those obtained under optimum conditions given by Eqs (5.9) and (5.10). However, for collector efficiencies of 90% and greater, both optimum inductance and optimum capacitance differ by less than 20% from their optimum values for $r_{sat} = 0$ [25]. It should be noted that, if the first condition is satisfied, then the power losses close to minimum can be achieved with zero second condition, since the positive voltage derivative results in positive current jump, which requires greater driving amplitude, while the negative voltage derivative with negative current jump demonstrates reduction in power loss of only 3% [39]. Thus, generally the switching conditions given by Eqs (5.9) and (5.10) can be considered optimum only for an idealized case of the Class-E load network with zero saturation resistance providing the switchmode transistor operation when it is operated in only pinch-off and saturation regions. However, they can be considered as a sufficiently accurate initial assumption for further design and optimization of the practical high-efficiency power amplifier circuits.

The term "optimum" for the Class-E conditions given by Eqs (5.9) and (5.10) means that the voltage across the switch should be equal to zero and there are no current jumps across the capacitor $C$ (capacitor must be discharged) at the moments when the switch is turned on with further instant transitions from off-state to on-state modes, thus resulting in a maximum achievable collector efficiency. Otherwise, if current starting to flow through the switch at this moment is not equal to zero, the device cannot be considered an ideal switch because of the appearance of the active operation mode of the device between its pinch-off and saturation modes. In this case, the collector current and voltage waveforms overlap each other reducing the collector efficiency because of the power dissipation in the device. Therefore, it is more proper to call the *optimum conditions* given by Eqs (5.9) and (5.10) for a lossless operation mode with ideal switch as the *nominal (or idealized optimum) conditions*.

*Optimum* electrical operation is defined as the operating condition that gives the highest possible drain or collector efficiency at a specified output power and peak switch voltage for a given set of parasitic parameters such as the transistor saturation resistance, turn-off and turn-on switching transition times, and finite quality factors of the load-network inductors and capacitors. In this case, each set of components results in a specific "optimum" off-nominal design. The smaller the parasitic series resistances, and the larger the parasitic parallel resistances, the closer is the optimum switch-voltage waveform to the nominal waveform. The designer can trade-off among the power dissipations, so as to minimize the total power dissipation at the specified output power and peak switch voltage.

Typically, manual or automatic optimization of the Class-E load-network parameters can reduce the power dissipation of a nominal design by about 30%. As an example, Fig. 5.18 shows the switch (*a*) voltage and (*b*) current waveforms for a nominal 13.56-MHz 101-W switchmode Class-E design, providing a drain

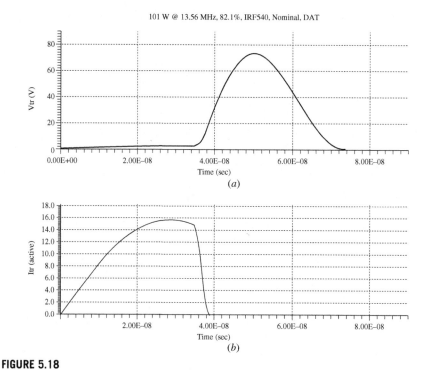

**FIGURE 5.18**

Switching voltage and current waveforms for nominal Class-E operation.

efficiency of 82.1%. At the same time, Fig. 5.19 shows the switch (a) voltage and (b) current waveforms for an optimized version of that design, achieving a drain efficiency of 86.6% at 102-W output. In that example, the 23-W power dissipation of the nominal design was reduced by 6.4 W (to 16.6 W) in the optimized design. As a result, the 6.4-W reduction of dissipated power represents 28% of the 23-W power dissipated in the nominal design. In this optimized design, as seen from Fig. 5.19(a), the drain voltage waveform at the switch turn-on time is slightly positive.

In a nominal (or idealized optimum) Class-E operation, the load network discharges the device output capacitance prior to the turning on of the device, producing ideally 100% efficiency. Below the maximum frequency, at which the shunt susceptance required for optimum operation is provided by only the device output capacitance, it is generally possible to adjust the series load reactance to achieve almost 100% efficiency. For example, optimum operation can be achieved by adding external shunt capacitance to the device output. However, above the maximum frequency, it is impossible to achieve an ideal 100% efficiency by varying the series load network parameters. As frequency increases, the collector voltage waveform approaches the ramp produced by dc charging of the

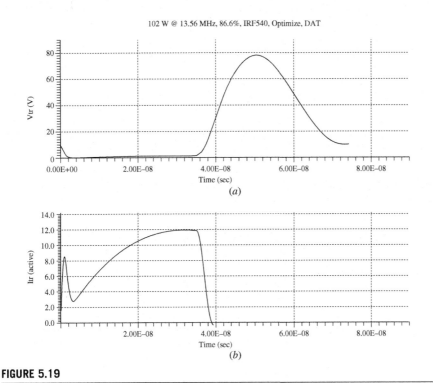

**FIGURE 5.19**

Switching voltage and current waveforms for optimized Class-E operation.

shunt capacitor. Consequently, the maximum achievable collector (or drain) efficiency decreases as the frequency of operation is increased above the maximum frequency. The maximum possible efficiency $\eta_{max}$ and the normalized circuit parameter $\omega L/R$ required to produce it, along with the normalized peak voltage $V_{max}/V_{cc}$, for a fixed supply voltage $V_{cc}$, are shown in Table 5.3 [40].

From Table 5.3 it follows that, above the maximum frequency, efficiency can be maximized by proper selection of the series inductance and load resistance when it looks reasonable. For example, at operating frequency $f = 2.512f_{max}$, the collector efficiency of an ideal Class-E mode remains still high, being even higher than that for a Class-F mode with control of three collector voltage and three collector current harmonic components. The collector efficiency of an ideal Class-E power amplifier drops at $f = 3.162f_{max}$ to 77.87%, which approximately corresponds to the maximum collector efficiency of an ideal Class-B power amplifier of 78.5%.

For off-nominal Class-E operation when only the zero-voltage switching condition is satisfied, the designer can use a higher shunt capacitance than for nominal operation at the same switching frequency and load resistance [41]. In this case, both peak voltage and current values are higher for off-nominal operation which occurs for $-0.5\pi < \varphi < 0$ for phase angles $\varphi > \varphi_{opt}$, where $\varphi_{opt}$ is the optimum phase angle determined by Eq. (5.18) for a nominal Class-E mode at 50% duty ratio. Generally, the peak switch voltage and current values vary with the turn-on switch voltage slope and the duty ratio [42].

The small load variation has no significant effect on the efficiency of an ideal Class-E power amplifier operated in nominal switching conditions. For example, efficiency varies gradually, remaining at 95% or more for variations in the load resistance $R$ of +55% to $-37$% relative to its nominal value [43]. Generally, the minimum efficiency for a given output $VSWR$ (voltage standing wave ratio) decreases almost linearly, as shown in Table 5.4 [44]. For the usual design requirement of operation with $VSWR \leq 2$, efficiency is no lower than 89%. However, for some load impedance of any specified $VSWR$, it is possible to achieve the maximum collector efficiency of 100%, which is realized at phase angles of the complex reflection coefficient $\Gamma$ equal to $+65°$ and $-115°$. Consequently, the contour of $\eta = 100$% on a Smith chart is a straight line. The contours for lower values of efficiency are curved, but symmetrical about the contour of $\eta = 100$%. These results are very helpful in practical implementation. Since 100% efficiency can be achieved along a line that goes through the center of the Smith chart, a single tuning element is generally sufficient to transform any specified load impedance into an impedance on the line corresponding to

**Table 5.3** Suboptimum Operation Above Maximum Frequency

| $f/f_{max}$ | $\omega L/R$ | $V_{max}/V_{cc}$ | $\eta_{max}$, % |
|---|---|---|---|
| 1.000 | 1.152 | 3.562 | 100 |
| 1.259 | 1.330 | 3.198 | 99.59 |
| 1.585 | 1.053 | 2.981 | 96.96 |
| 1.995 | 0.852 | 2.789 | 92.16 |
| 2.512 | 0.691 | 2.632 | 85.62 |
| 3.162 | 0.561 | 2.519 | 77.87 |

**Table 5.4** Minimum Efficiency for Different Values of $VSWR$

| $VSWR$ | $V_{max}/V_{cc}$ | $\eta_{min}$, % |
|---|---|---|
| 1.0 | 3.5621 | 100 |
| 1.5 | 3.0133 | 96.00 |
| 2.0 | 2.6814 | 88.88 |
| 2.5 | 2.4449 | 81.63 |
| 3.0 | 2.2615 | 74.99 |

$\eta = 100\%$. The specified output power can then be obtained by adjusting the dc supply voltage $V_{cc}$. When adjusting the dc supply voltage, take care not to exceed the safe $V_{cc}$.

## 5.8 Push–pull operation mode

The push–pull power-amplifier configuration offers a possibility of combining powers from two linear power amplifiers to obtain a larger overall output power. The switchmode power amplifiers can also be configured to operate in a push–pull mode. The simplest way is to use a push–pull power amplifier with a resistive load is by adding either a series or a parallel $LC$ circuit. In a push–pull power amplifier shown in Fig. 5.20(a), the rectangular pulses of voltage $V_{cc} \pm V_{cc}$ drive the series resonant $LC$ circuit, resulting in a switchmode power amplifier with switching voltage. In a push–pull power amplifier shown in Fig. 5.20(b), the rectangular pulses of current $I_0 \pm I_0$ drive the parallel resonant $LC$ circuit, resulting in a switchmode power amplifier with a switching current.

With the high-$Q_L$ resonant circuits tuned to the fundamental frequency, these power amplifiers operate as the switchmode Class-D power amplifiers with a sinusoidal load current. However, when the low-$Q_L$ resonant $LC$ circuit shown in

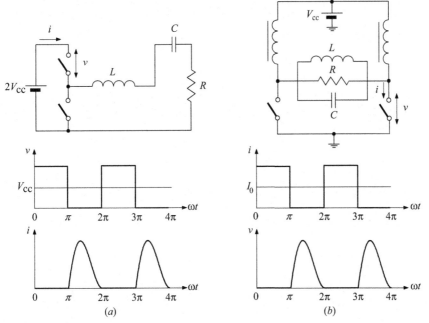

**FIGURE 5.20**

Basic circuits of push–pull switchmode power amplifiers.

Fig. 5.20(b) is mistuned to the frequency of about $1.5f_0$, where $f_0$ is the fundamental frequency, the collector voltage waveform will represent the typical Class-E collector waveform with zero voltage and zero voltage-derivative nominal conditions [25]. In this case, the load current will not contain the even harmonics of the fundamental frequency. However, to provide a purely sinusoidal load current, the remaining third and higher-order harmonic components can be eliminated by using an additional high-$Q_L$ filter tuned to the fundamental frequency.

Figure 5.21 shows the push−pull configuration of an optimally tuned Class-E power amplifier with a series-resonant filter representing an open-circuit for harmonic components of the fundamental frequency, where the input transformer provides the device driving signals with opposite phases [13,45]. Each device, however, operates as if it were a single-ended Class-E power amplifier. When a given transistor is driven on, it provides a ground connection on the primary winding of the output transformer, causing the dc current and transformed sinusoidal output current to charge the capacitor shunting the other transistor. The dc power is supplied to the switching devices through the center tap of the primary winding of the output transformer. The voltage appearing on the secondary winding of the output transformer contains both positive and negative Class-E collector voltage waveforms. Consequently, it has a fundamental-frequency component that has twice the amplitude of the fundamental-frequency component of either collector waveform. The resulting impedance seen by either half of the power amplifier looking into primary winding of the output transformer depends on the squared ratio between the number of turns of the primary winding and the secondary winding. The sinusoidal load voltage will be shifted relative to the collector voltage by the same phase as for the case of a single-ended Class-E power amplifier.

The push−pull Class-E power amplifier can be designed based on a balanced circuit topology with symmetrical Class-E load networks with shunt capacitances, connected to a common load. Figure 5.22(a) shows the bipolar push−pull Class-E power amplifier with a series load resistor [46]. Here, both bipolar transistors operate as switches with opposite 180° phases to each other when the even harmonics can be significantly suppressed in the load resistor. To provide the transition from a balanced load to an unbalanced load, an additional transformer or balun can be used. Being designed to operate at a carrier frequency of 100 kHz with a 50% duty ratio, such a push−pull bipolar Class-E power amplifier achieved a *PAE* of 82.2% with an output power of 1.89 W at a supply voltage of 5 V.

A symmetrically driven push−pull Class-E power amplifier using high-power MOSFET devices is shown in Fig. 5.22(b) [47]. In this circuit, the load resistor is connected in parallel to the series capacitor. For a symmetrically driven Class-E power amplifier, the switches are driven on and off within each of the half-operating period. The duty ratio of the switches is adjusted to realize zero-voltage and zero voltage-derivative nominal conditions. However, the nominal Class-E operation mode can be maintained only for relatively low loaded

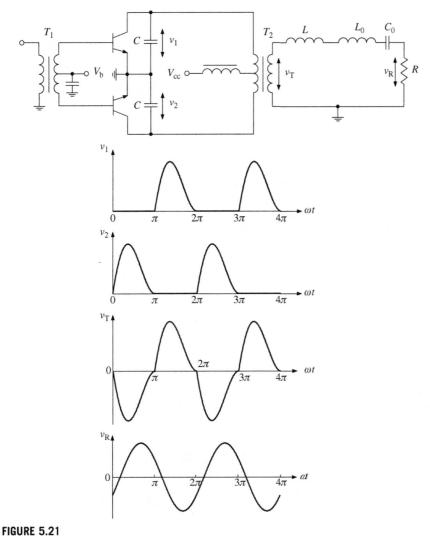

**FIGURE 5.21**

Basic concept of Class-E push–pull operation.

quality factors below 2, which is a design constraint for this symmetrically driven push–pull Class-E power amplifier. Although the use of low $Q_L$-factors has an advantage of low component stress, the harmonic content of the output signal may be significant. As a result, the measured drain efficiency of about 85% with a $Q_L$-factor of 1.9 was achieved within the ranges of output power from 30 to 150 W and supply voltages from 15 to 32 V at a switching frequency of about 1 MHz. Note that, in view of the nonzero device saturation voltage, the

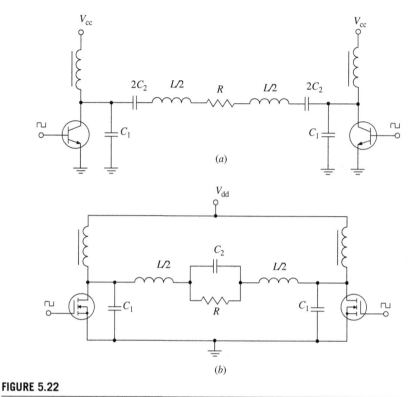

**FIGURE 5.22**

Symmetrically driven push—pull Class-E power amplifiers.

zero-voltage and zero voltage-derivative conditions are no longer optimum to provide maximum operating efficiency.

At microwave frequencies, push—pull Class-E power amplifiers can be designed by using transmission-line power dividers and combiners. Figure 5.23(a) shows the circuit schematic of a monolithic X-band push—pull pHEMT power amplifier, including an input power divider and an output power combiner based on slotlines and coplanar waveguides [48]. The input balun separates the input signal into two differential signals that are 180° out-of-phase, while the output balun is reverse-oriented to the input one. The input and output matching circuits consist of the series coplanar waveguides providing inductive impedances and shunt coplanar waveguides giving capacitive reactances at the fundamental frequency. The Class-E load network, which is followed by the output matching circuit, includes the internal device shunt capacitance and part of the series coplanar waveguide. The input transmission-line balun shown in Fig. 5.23(b) provides first the balanced-to-unbalanced signal transformation by transition from the waveguide coplanar with left open-circuit end to slotline, then followed by the

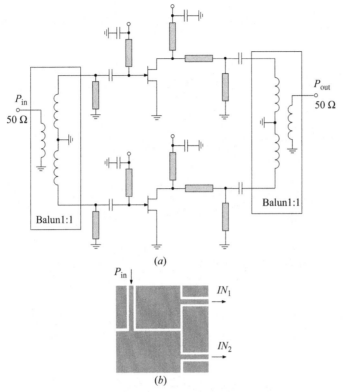

*(a)*

$P_{in}$

$IN_1$

$IN_2$

*(b)*

**FIGURE 5.23**

Microwave push–pull Class-E power amplifier.

slotline *T*-junction, and finally connected to the two differential paths based on the coplanar waveguides with open-circuit terminations at each of their ends. An *X*-band power amplifier based on these baluns with broadband performance and compact size and pHEMT transistors with gate geometry of 0.3 μm × 600 μm can achieve a *PAE* close to 60% in a frequency band of 9–10 GHz with output powers of several hundred miliwatts.

## 5.9 **Load networks with transmission lines**

At ultrahigh and microwave frequencies, different types of transmission lines are often preferred over lumped inductors because of the convenience of their practical implementation, more predictable performance, less insertion loss, and less effect of the parasitic elements. For example, the matching circuit can be composed with any types of transmission lines, including open-circuit or short-circuit

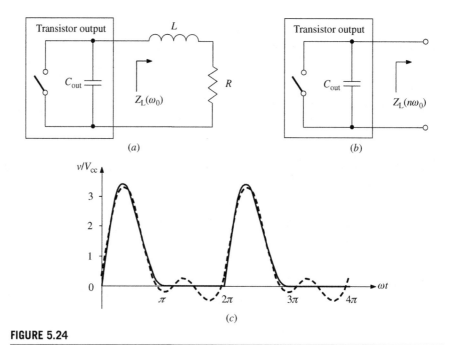

**FIGURE 5.24**

Optimum load impedance and two-harmonic Class-E voltage waveform.

stubs to provide the required matching and harmonic suppression conditions. In this case, to approximate the idealized Class-E operation mode of the microwave power amplifier, it is necessary to design the transmission-line load network satisfying the required idealized optimum impedances at the fundamental-frequency and harmonic components. Generally, the device output capacitance can represent the required shunt capacitance whose nominal or idealized optimum Class-E value is defined by Eq. (5.32). Consequently, the main challenge is to simultaneously satisfy the idealized optimum requirements at the fundamental-frequency impedance $Z_L(\omega_0)$ shown in Fig. 5.24(a) and harmonic component impedances $Z_L(n\omega_0)$ shown in Figs. 5.24(b), which are expressed by using Eq. (5.31) at the fundamental frequency $f_0$ as

$$Z_L(\omega_0) = R + j\omega L = R\left(1 + j\frac{\omega L}{R}\right) = R(1 + j\tan 49.052°) \tag{5.77}$$

and at the harmonic components $nf_0$, where $n = 2, 3, \ldots, \infty$, as

$$Z_L(n\omega_0) = \infty. \tag{5.78}$$

Generally, it is practically impossible to realize these conditions for an infinite number of harmonic components by using only transmission lines. However, as it turned out from the Fourier-series analysis, a good approximation to Class-E

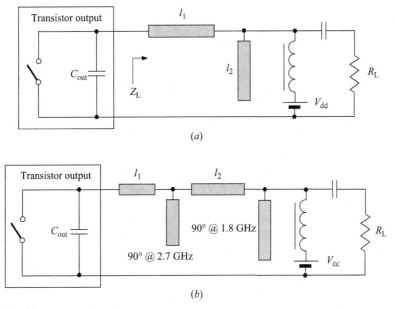

**FIGURE 5.25**

Equivalent circuits of Class-E power amplifiers with transmission lines.

mode may be obtained with the dc, fundamental-frequency, and second harmonic components of the voltage waveform across the switch [15,49]. In this case, the drain efficiency will be the same as in a maximum-efficiency Class-F mode with second harmonic control [50]. Figure 5.24(c) shows the collector (or drain) voltage waveform containing these two harmonic components (dashed line) plotted along with an ideal voltage waveform (solid line). In practical implementation, the two-harmonic Class-E load network designed for microwave applications will include a series microstrip line $l_1$ and an open-circuit stub $l_2$, as shown in Fig. 5.25(a). The electrical lengths of microstrip lines $l_1$ and $l_2$ are chosen to be of about 45° at the fundamental frequency to provide an open-circuit condition seen by the device output at the second harmonic, according to Eq. (5.78). Their characteristic impedances are calculated to satisfy the required inductive-impedance condition at the fundamental frequency given by Eq. (5.77). In the case of a packaged active device, its output bondwire and lead inductance can be accounted for by shortening the length $l_1$.

In some cases, a value of the device output capacitance exceeds the required nominal value for a Class-E mode with shunt capacitance. In this situation, it is possible to approximate Class-E mode with high efficiency by setting a properly optimized load at the fundamental frequency and strongly reactive load at the second-harmonic and third-harmonic components [51]. Such a harmonic-control network consists of open-circuit quarterwave stubs at the second- and third-harmonic components separately, as shown in Fig. 5.25(*b*), where the third-harmonic quarter-wave stub is located before the second-harmonic quarterwave stub. As a result, very high collector efficiency can be achieved even with values of the device output capacitance higher than conventionally required, at the expense of lower output power, keeping the load at the second and third harmonics strictly inductive. Maximum collector efficiency over 90% with an output power of 1.5 W for the test power amplifier using a commercial bipolar transistor MRF557 was measured at a carrier frequency of 900 MHz. The second-harmonic open-circuit stub can be replaced by a $\lambda/4$-long shortened (by the decoupling capacitor) stub, which is a half-wave long at $2f_0$, used to bias the transistor [52].

An analysis of a nominal Class-E mode in the frequency domain shows that the combined effect of the saturation resistance $r_{sat}$ and second-harmonic loading is characterized by a shift of the idealized optimum-load fundamental-frequency impedance [53]. Besides, parasitic effects due to packaging and microstrip junction discontinuities may upset the transmission-line open-circuit requirement. Therefore, it is particularly important to predict the power amplifier performance due to nonideal harmonic terminations. For example, the voltage peak factor increases when the second-harmonic load varies from capacitive to inductive, and excessive current flow can be realized under capacitive load reactance. In this case, to achieve an optimum switchmode Class-E operation mode, the maximum collector efficiency with lower output power is obtained for the negative collector voltage-derivative condition [54]. Generally, higher efficiency is achieved for open-circuit terminations of the second- through fifth-harmonic components, with up to 5% efficiency degradation when the loading configurations for odd harmonics are modified from open-circuit to short-circuit termination across 9–11 GHz [55]. The variation of the second-harmonic load reflection coefficient by 10% in magnitude from 1 to 0.9 and $\pm 20°$ in phase angle results in an insignificant efficiency variation of only 1%.

The resultant transmission-line topology of the Class-E load network also including the matching properties can be designed based on a lumped circuit prototype [56]. The design procedure includes the following steps:

1. Calculation of the optimum Class-E load network parameters (shunt capacitance $C$, series inductance $L$, and load resistance $R$) based on the specified supply voltage $V_{cc}$ and power delivered to the load $P_L$.
2. Design of the lumped-element circuit to match the nominal (or optimized) load resistance $R$ with standard load resistance of 50 $\Omega$ at the fundamental frequency, with corresponding harmonic suppression.
3. Transformation of the lumped-element circuit into its transmission-line circuit equivalent carefully observing the impedances at the higher harmonics.
4. Circuit simulation to validate its operation and component parameters to obtain Class-E approximation.
5. Transformation of the transmission-line equivalent circuit into a distributed microstrip layout.

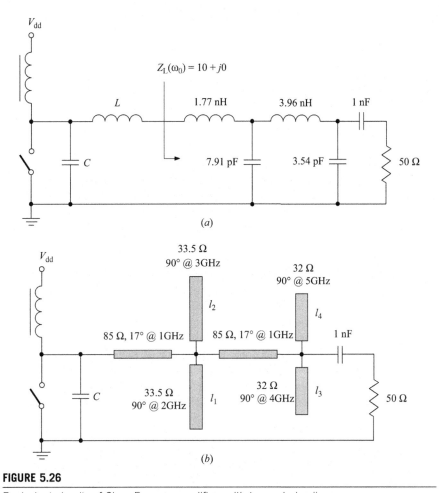

**FIGURE 5.26**

Equivalent circuits of Class-E power amplifiers with transmission lines.

Figure 5.26(a) shows the lumped-element Class-E load network with impedance transformation for a 1-GHz power amplifier delivering 1 W into a 50-$\Omega$ load. It is assumed that the MESFET device has a saturation resistance of 1 $\Omega$ and output shunt capacitance of around 2 pF. Taking into account the drain breakdown voltage of 15 V, the maximum supply voltage would be limited to $V_{max} = 15/3.562 = 4.2$ V according to Eq. (5.27). In this case, to achieve an output power of 1 W, an ideal value of $R = 0.5768 \times (4.2)^2 = 10\ \Omega$ is required according to Eq. (5.33). The inductive reactance $X = \omega L = 11.5\ \Omega$ is chosen to provide a load phase angle of 49.052° [13]. The required load-network shunt capacitance is $C = 0.1836/(2\pi \times 10) = 2.92$ pF according to Eq. (5.32). To achieve the required impedance transformation from 10 $\Omega$ to 50 $\Omega$ and to provide

**Table 5.5** First Three Frequencies Suppressed by Each of Four Harmonic Stubs

| Harmonic Stub | Suppressed Frequency | | |
|---|---|---|---|
| $l_1$ | $2f_0$ | $6f_0$ | $10f_0$ |
| $l_2$ | $3f_0$ | $9f_0$ | $15f_0$ |
| $l_3$ | $4f_0$ | $12f_0$ | $20f_0$ |
| $l_4$ | $5f_0$ | $15f_0$ | $25f_0$ |

a sufficient harmonic suppression in the load, a two-stage cascaded low-pass $L$-type transformer was chosen, where the first $L$-section transforms from 10 $\Omega$ to 22 $\Omega$ and the second $L$-section transforms from 22 $\Omega$ to 50 $\Omega$.

The conversion of the lumped-element circuit involves replacing the series inductors and shunt capacitors with equivalent transmission-line elements. The series inductors can be realized by high-impedance transmission-line short sections with electrical length of $\theta \leq \lambda/8$, calculated from equivalent equation $\omega L = Z_0 \tan \theta$, where $Z_0$ is the characteristic impedance of the line, while shunt capacitances are replaced by low-impedance open-circuit stubs with electrical length of $\theta \leq \lambda/8$, calculated from equivalent equation $\omega C = \tan \theta/Z_0$ [20]. The electrical parameters of the capacitive stubs (characteristic impedance and electrical length) can be optimized to simultaneously provide the correct reactance at the fundamental frequency $f_0$ and enhanced suppression of harmonics, reaching the load by arranging the stub lengths to have low input impedance at selected harmonics, i.e. short-circuiting these harmonics to ground. This is achieved by choosing the stub length such that its electrical length is exactly 90° at the particular harmonic that one would like to suppress. The characteristic impedance of the stub is then chosen to provide the desired capacitive reactance at the fundamental frequency.

Figure 5.26(*b*) shows the resultant transmission-line topology of a 1-GHz Class-E power amplifier where each shunt capacitor is replaced by two stubs, providing the harmonic suppression at four frequencies [56]. The capacitive stubs were chosen to have an electrical length of 90° at frequencies $2f_0$, $3f_0$, $4f_0$, and $5f_0$. Table 5.5 lists the first three frequencies suppressed by each of the four harmonic stubs. By using the printed circuit board (PCB) with a relative dielectric permittivity of 2.2, a loss tangent of 0.009 at 10 GHz, and substrate thickness of 0.381 mm for microstrip realization of this Class-E load network, a drain efficiency of 72% at 950 MHz with an output power of 458 mW and a harmonic suppression of more than 40 dB at a drain supply voltage of 3.53 V were achieved. The equivalent output capacitance of the used MESFET device ATF8140 was approximately 2.5 pF, which is 0.6 pF less than the required switch capacitance of 3.1 pF. An additional 0.6 pF capacitance was made up of two short capacitive stubs attached at the point where the device drain is connected to the load network. To simplify the overall design procedure of a highly efficient MMIC power amplifier at $K$-band, the load network may contain only two capacitive stubs to provide the

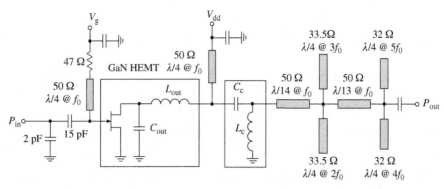

**FIGURE 5.27**

Equivalent circuits of Class-E power amplifiers with transmission lines.

reactance required for the fundamental load transformation and low impedances at selected second and third harmonics by making the electrical lengths of these stubs exactly one quarter wavelength at a particular harmonic [57].

For a packaged high-power transistor with large output capacitance, additionally a series capacitor can be used to compensate for the package parasitic inductance followed by the shunt short-circuited transmission line, which is required to compensate for excessive device output capacitance [58]. In this case, a *PAE* of 73.6% and a power gain of 14.8 dB were measured at an output power of 39.1 dBm for a 1-GHz LDMOSFET Class-E power amplifier using MRF282 device biased with a gate voltage of 3.2 V and a drain voltage of 21.4 V. The peak *PAE* of 79.2% and drain efficiency of 80.4% with a power gain of 18.1 dB are achieved at an output power of more than 41.1 dBm for a 1-GHz GaN HEMT Class-E power amplifier using a compensating circuit with series capacitor and shunt inductor followed by a harmonic-control network with quarterwave open-circuit stubs for the second to fifth harmonics [59].

Figure 5.27 shows the circuit schematic of a 25-V GaN HEMT Class-E power amplifier designed for WCDMA applications at 2.14 GHz [60]. Here, the series capacitor $C_c$ and shunt inductor $L_c$ are used to compensate for parasitic components of bonding inductance $L_{out}$ and output capacitance $C_{out}$. The harmonic-control load network suppresses all harmonic power levels below −60 dBc for the whole output power range. As a result, the peak *PAE* of 70% with a power gain of 13 dB was achieved at an output power of 43 dBm. The broadband performance with a power gain over 12 dB and a *PAE* over 60% is maintained through 200 MHz. Using the LDMOSFET device in this Class-E design results in a *PAE* of 62.5% with a power gain of 13.8 dB at an output power of 39.8 dBm [61].

The transmission-line Class-E load network shown in Fig. 5.25(*b*) can be modified in order to obtain simple analytical equations to explicitly define the transmission-line parameters [62]. Such a modified transmission-line Class-E load

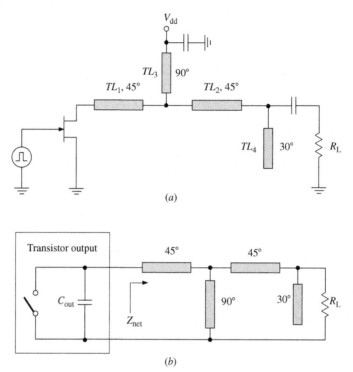

**FIGURE 5.28**

Modified transmission-line Class-E power amplifier.

network is shown in Fig. 5.28, where the combined series quarterwave transmission line provides an impedance transformation at the fundamental frequency, and the open-circuit stubs with electrical lengths of 90° and 30° create the open-circuit conditions seen from the device output at the second and third harmonics, respectively.

Figure 5.29($a$) shows the load network seen from the transistor output at the fundamental frequency. Here, the combined quarter-wavelength series transmission line $TL_1 + TL_2$, together with an open-circuit capacitive stub $TL_4$ having an electrical length of 30°, simultaneously provide the required inductive reactance and impedance transformation of the nominal or optimized Class-E load resistance $R$ to the standard load resistance $R_L$ by proper choice of the transmission-line characteristic impedances $Z_1$ and $Z_2$.

The capacitive load impedance $Z_L$ provided by the load resistance $R_L$ and capacitive stub $TL_4$ can be written at the fundamental frequency as

$$Z_L = \frac{Z_2 R_L}{Z_2 + jR_L \tan 30°} \tag{5.79}$$

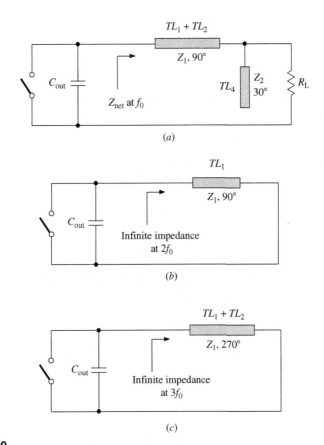

Load networks seen by the device output at harmonics.

where $Z_2$ is the characteristic impedance of a 30° open-circuit stub. Generally, the input impedance of the loaded transmission line can be written as

$$Z_{net} = Z_1 \frac{Z_L + jZ_1 \tan \theta}{Z_1 + jZ_L \tan \theta} \tag{5.80}$$

where $\theta$ is the electrical length of the transmission line. Then, substituting Eq. (5.79) into Eq. (5.80) for $\theta = 90°$ results in an inductive input impedance

$$Z_{net} = \frac{Z_1^2}{Z_L} = \frac{Z_1^2}{Z_2 R_L} (Z_2 + jR_L \tan 30°) \tag{5.81}$$

when the required nominal or optimized Class-E resistance can be provided by proper choice of the characteristic impedance $Z_1$, while the required optimum

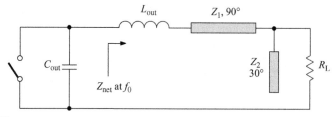

**FIGURE 5.30**

Load network with series inductance at fundamental.

Class-E inductive reactance can be achieved with the corresponding value of the characteristic impedance $Z_2$.

Separating Eq. (5.81) into real and imaginary parts results in the following system of two equations with two unknown parameters:

$$\text{Re } Z_{\text{net}} = \frac{Z_1^2}{R_{\text{L}}} \tag{5.82}$$

$$\text{Im } Z_{\text{net}} = \frac{Z_1^2}{\sqrt{3}Z_2} \tag{5.83}$$

which allows direct calculation of the characteristic impedances $Z_1$ and $Z_2$. As a result, by using Eq. (5.77),

$$Z_1 = \sqrt{R_{\text{L}}R} \tag{5.84}$$

$$Z_2 = 0.5\,R_{\text{L}} \tag{5.85}$$

where $R_{\text{L}} = 50\,\Omega$ and $R$ is calculated from Eq. (5.33).

The transmission-line Class-E load network seen from the device output at the second harmonic is shown in Fig. 5.29(b), taking into account the shorting effect of the quarterwave short-circuit stub $TL_3$, where the transmission line $TL_1$ provides an open-circuit condition for the second harmonic. At the third harmonic, the transmission-line Class-E load network can similarly be represented, as shown in Fig. 5.29(c), due to the open-circuit effect of the short-circuited quarterwave line $TL_3$ and short-circuit effect of the open-circuited harmonic stub $TL_4$ at the third harmonic. In this case, the combined transmission line $TL_1 + TL_2$ provides an open-circuit condition for the third harmonic at the device output being shorted at its right-hand side.

However, in a common case, it is necessary to take into account the transistor output parasitic series bondwire and lead inductance $L_{\text{out}}$ shown in Fig. 5.30, which provides an additional inductive reactance at the fundamental frequency and does not affect the open-circuit conditions at the second and third harmonics. The inductive effect at the input of the series quarter-wavelength transmission line can be reduced by proper changing of the characteristic impedance $Z_2$. In this case, Eq. (5.83) can be rewritten as

$$\text{Im } Z_{\text{net}} = \frac{Z_1^2}{\sqrt{3}Z_2} + \omega_0 L_{\text{out}}. \tag{5.86}$$

Hence, by using Eqs (5.77) and (5.84), the characteristic impedance $Z_2$ can now be calculated from

$$Z_2 = \frac{R_L}{2} \frac{1}{1 - \dfrac{\omega_0 L_{\text{out}}}{1.1586R}} \tag{5.87}$$

resulting in a higher characteristic impedance of the open-circuit stub for greater values of the series inductance $L_{\text{out}}$.

Figure 5.31 shows the simulated circuit schematic of a transmission-line Class-E power amplifier based on a commercial 28-V, 5-W GaN HEMT power transistor NPTB00004 in a plastic package [62]. The input matching circuit with an open-circuit stub and a series transmission line provides a complex-conjugate matching with the standard 50-$\Omega$ source. The load network represents the optimized transmission-line Class-E load network shown in Fig. 5.28($a$). Using a 30-mil RO4350 substrate, an output power of 37 dBm, a drain efficiency of 73%, and a *PAE* of 71% at an operating frequency of 2.14 GHz are achieved with a power gain of 14 dB (linear gain of 19 dB) and a supply voltage of 25 V.

## 5.10 Practical Class-E power amplifiers and applications

A high level of output power with very high operational efficiency can be easily achieved in Class-E mode by using high-voltage power MOSFET devices at sufficiently low frequencies. Figure 5.32($a$) shows the circuit schematic of a 13.56-MHz, 400-W Class-E power amplifier providing a drain efficiency of 82% with an input sinusoidal drive of 12 W at a supply voltage of 120 V [63]. The series inductor $L_s$ and capacitor $C_s$ form the resonant network that produces the rising and falling voltage waveform required for a Class-E operation. The series tank circuit at the load composed of a capacitor $C_1$ and an inductor $L_1$ is a trap for the second-harmonic component, which contributes to the overall level of harmonic suppression of more than 40 dB below the carrier. Since at the fundamental frequency this second-harmonic resonant circuit represents a capacitive reactance, it transforms together with a part of the series inductance $L_s$ the standard load of 50 $\Omega$ to around 13 $\Omega$ required for a nominal Class-E mode. The impedance of the gate is small with a real part of about 3 $\Omega$ and an inductive reactance of about 4 $\Omega$. The input transformer with a voltage transformation ratio of 6:1 is used to step up from the input gate impedance to the driving source impedance of 50 $\Omega$. This transformer also sets the dc gate bias to 0 V and ensures the transistor is turned off when it is not driven, as this is far below the threshold voltage of 4 V.

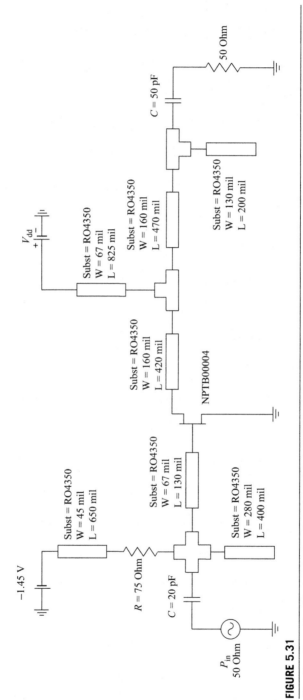

**FIGURE 5.31**

Circuit schematic of transmission-line Class-E GaN HEMT power amplifier.

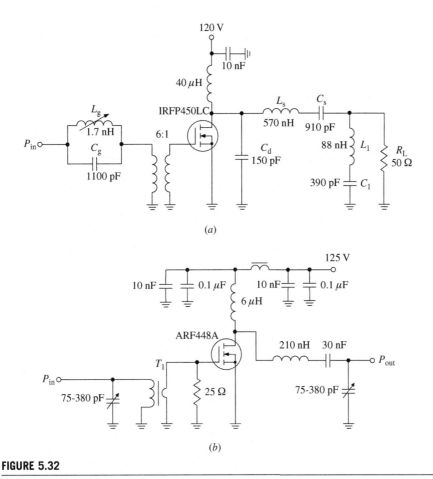

**FIGURE 5.32**

High-power high-frequency Class-E MOSFET power amplifiers.

The capacitor $C_g$ with the variable inductor $L_g$ is used to compensate for the input inductive reactance of the transistor providing the input *VSWR* of 1.6:1.

Figure 5.32(*b*) shows the circuit schematic of a 27.12-MHz 500-W Class-E MOSFET power amplifier with a drain efficiency of 83% at a supply voltage of 125 V [64]. The input ferrite transformer provides the 2:1 transformation voltage ratio to match the gate impedance, which is represented by the parallel equivalent circuit with a capacitance of 2200 pF and a resistance of 210 $\Omega$. Use of the external parallel resistor of 25 $\Omega$ simplifies the matching procedure and improves the amplifier stability conditions. The transformer secondary winding provides an inductance of 19 nH, which is required to compensate for the device input capacitance at the operating frequency. High-quality passive components are necessary to use in the low-pass *L*-type output network, where the quality factor of the bare copper wire inductor was equal to 375. The series blocking capacitor consists of

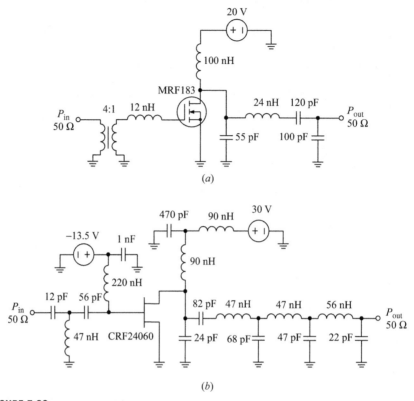

**FIGURE 5.33**

High-power VHF Class-E power amplifier.

three parallel disc ceramic capacitors. To realize a Class-E operation with shunt capacitance, it is sufficient to be limited to only the output device capacitance with a value of 125 pF. This is just slightly larger than that required to obtain the idealized optimum drain voltage and current Class-E waveforms.

Silicon LDMOSFET devices made it possible to achieve a high output power level in Class-E operation with sufficiently high efficiency at higher frequencies. Figure 5.33(a) shows the simplified circuit schematic of a high-power VHF LDMOSFET Class-E power amplifier achieving a drain efficiency of 70% for maximum output power of 54 W at an operating frequency of 144 MHz with an input drive of 5 W [65]. The drain efficiency can be increased to 88% if the output power level is reduced to 14 W by an appropriate increase of the series inductance in the load network. The input device impedance is sufficiently low; therefore, a ferrite transformer and a series inductor are used at the input. At the output, the standard load impedance of 50 $\Omega$ is transformed to the load resistance of 1.5 $\Omega$ required for an idealized optimum Class-E mode by a lumped $L$-transformer with a series inductor whose value is included with the inductance of 24 nH and a shunt 100-pF capacitor. The required value of a shunt switching capacitance is provided by the values of the intrinsic device output capacitance of 38 pF and external capacitance of 55 pF. The loaded quality factor of the resonant circuit was chosen at a sufficiently low value of 5 that allows some frequency bandwidth operation to be provided and sensitivity of amplifier performance to the resonant-circuit parameters to be reduced. To decrease the loss in the load network, the inductor was fabricated by using a 5-mm-wide copper ribbon that provides the inductor quality factor of 150−250, depending on the distance to the ground plane. By inserting a spacer between the ribbon and the ground plane, the inductance can be tuned at least a factor of two.

Figure 5.33(*b*) shows the simplified circuit schematic of a silicon carbide (SiC) MESFET Class-E power amplifier which provides a maximum drain efficiency of 86.8% at an output power of 20.5 W at 145 MHz reached at a drain voltage of 30 V, with an input drive power level of 27 dBm [66]. The output nominal Class-E impedance of approximately 18 $\Omega$ was matched to a 50-$\Omega$ load with a low-pass three-section *L*-type matching network to suppress harmonics by at least 60 dB below the carrier. The input of the active device was matched to 50-$\Omega$ source by means of a high-pass filter network to prevent the attenuation of the high-frequency harmonic components of the drive signal. Since this power amplifier was designed to provide linear amplification by restoring the input signal envelope with drain amplitude modulation, the drain bias network was built with a low-pass filter that allows drain modulating frequencies of up to a few megahertz to pass through it with minimum attenuation, while at the same time achieving acceptable isolation at the carrier frequency and its harmonics.

The modified Class-E power amplifier with a tuned series-parallel resonance load network is shown in Fig. 5.34 which is characterized by higher maximum operating frequency and maximum drain peak voltage compared to the conventional Class-E power amplifier with shunt capacitance [67]. With a high-power LDMOSFET device, a drain efficiency of 81.77%, a *PAE* of 79.58%, and an output power of 32.71 dBm at 250 MHz were measured. This power amplifier can retain its high efficiency over 78% and output power over 32.7 dBm within approximately 21% relative bandwidth from 240 to 290 MHz.

The transmission-line Class-E power amplifier topology is shown in Fig. 5.35(*a*), where the electrical lengths of microstrip lines $l_3$ and $l_4$ in the load network must be close to 45° so that an approximate open circuit at the second harmonic will be presented to the switch shunt capacitor, which is an equivalent output device capacitance. Microstrip lines with a characteristic impedance of 50 $\Omega$ each were fabricated using a substrate with thickness of 2.54 mm and effective dielectric permittivity $\varepsilon_r = 10.5$. For a MESFET device having a drain-source capacitance $C_{ds} = 2.4$ pF, a *PAE* of 80% was achieved at 0.5 GHz with an output power of 0.55 W [49]. In this case, the electrical lengths of microstrip lines are $l_1 = 73°$, $l_2 = 79°$, $l_3 = 58°$, and $l_4 = 46°$. The power-added efficiency remains above 75% over a 10% bandwidth and above 50% over a 26% bandwidth. Moreover, a *PAE* of 73% can be realized at 1.0 GHz with an

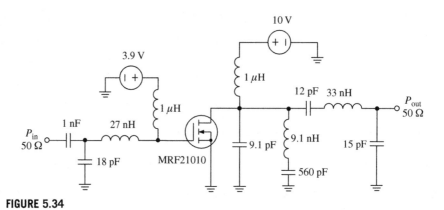

**FIGURE 5.34**

Schematic of a VHF LDMOSFET Class-E power amplifier with series-parallel network.

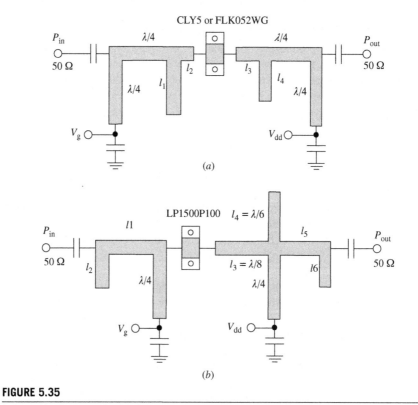

**FIGURE 5.35**

Schematic diagram of a 1.8 GHz pHEMT Class-E power amplifier.

output power of 0.94 W. By using a MESFET device with a sufficiently higher transition frequency $f_T$ in such a microstrip Class-E power amplifier, an output power of 0.61 W, a 1-dB compressed power gain of 7.6 dB, a drain efficiency of 81%, and a *PAE* of 72% were achieved at 5 GHz [15]. The power amplifier was fabricated on a substrate with a thickness of 0.508 mm and $\varepsilon_r = 2.2$. As a result, the lengths of 50-$\Omega$ microstrip lines (1.6 mm wide) are 9 mm for $l_1$, 1.8 mm for $l_2$, 5.3 mm for $l_3$, and 6.2 mm for $l_4$, respectively. The power-added efficiency is greater than 70% over a 5% bandwidth and greater than 60% over a 10% bandwidth. A similar design approach can be used to design a monolithic Class-E power amplifier in *X*-band. As a result, the measured performance of a monolithic power amplifier that employs a pHEMT device with a geometry of 0.3 $\mu$m $\times$ 600 $\mu$m showed a peak *PAE* of 63% at 10.6 GHz and a constant output power of greater than 24 dBm together with a power gain of 10 dB over a frequency bandwidth of 9−11 GHz [68].

Figure 5.35(*b*) shows the schematic diagram of a 1.8-GHz transmission-line Class-E power amplifier using a commercially available pHEMT device which can offer a power gain over 12 dB for up to 15 GHz at a supply voltage of 3 V [53]. The gate and drain biasing are accomplished through the quarter-wavelength stubs printed on a substrate with thickness of 0.79 mm and $\varepsilon_r = 2.33$. In this case, the input quarterwave stub provides a short-circuit termination of the second harmonic at the device input. The load impedance at the second harmonic is made close to an open circuit by using microstrip line $l_3$ having an electrical wavelength of 45° at the fundamental frequency. Fundamental input and output matching is achieved through microstrip lines $l_1$, $l_2$, $l_4$, $l_5$, and $l_6$, which were made adjustable by using high-quality copper foil. Initially, the device output capacitance $C_{out}$ of 0.3−0.5 pF and saturation resistance $r_{sat}$ of 1−2 $\Omega$ were assumed, and $l_4$ was set to 60°. The transistor was biased near pinch-off with a quiescent current of 5 mA, which is about 5% of the maximum dc drain current of about 100 mA. The amplifier exhibited a maximum saturated power of 22 dBm, a maximum drain efficiency of 93%, and a maximum *PAE* of 88%.

Significant progress in CMOS technology has shown a promising future for CMOS RF power amplification. Much progress has been achieved at the research level, and the obvious possibility to minimize cost and size of the integrated circuits for RF handset transmitters, especially power-amplifier MMICs, makes CMOS technology very feasible and brings considerable economical benefits. However, realizing high-efficiency operation of power amplifiers is limited by technology issues such as high value of the device saturation resistance, low value of breakdown voltage, and lossy silicon substrate. Therefore, it is vital to apply high-efficiency techniques in the design of CMOS power amplifiers. Figure 5.36(*a*) shows the design structure of a two-stage CMOS Class-E power amplifier, where biasing inductors are used to perform the function of dc-level shifting, since the device threshold voltage is positive for the nMOS devices [69]. Note that these inductors should have extremely low parasitic resistance in order to reduce the metal power loss. To further reduce the power loss in passive components, bondwire

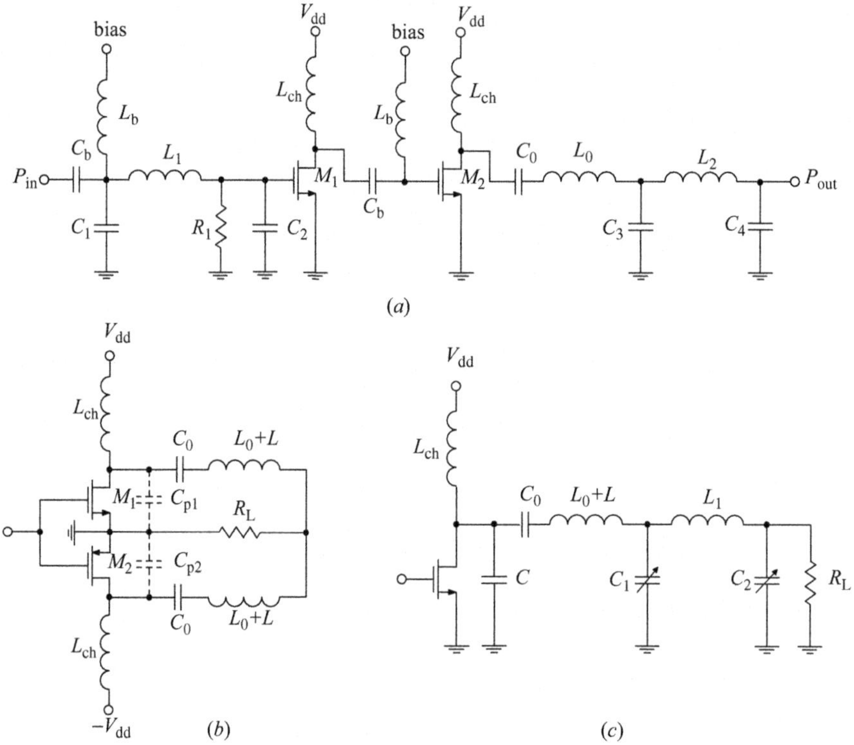

**FIGURE 5.36**

Schematics of CMOS Class-E power amplifiers.

inductors can be used instead of low-$Q$ on-chip inductors, which especially makes sense under high-current conditions. A square driving waveform can be realized by using a high-pass interstage-matching section, employing a high-frequency filter to minimize the second harmonic of the driving signal, or by overdriving the final stage with a large driving signal, provided there is sufficient power gain.

While most of the Class-E power amplifiers use a single-ended topology due to practical implementation issues and available power devices, a differential counterpart, which exhibits smaller harmonic distortion since even harmonics cancel each other in a purely symmetrical structure, can be easily implemented using a CMOS technology with nMOS and pMOS devices. The simplified topology of a Class-E complementary CMOS power amplifier is shown in Fig. 5.36(*b*) [70]. To minimize the saturation resistance of the devices, it is necessary to use transistors with large gate width, resulting in large parasitic output capacitances. In some cases, the device output parasitic capacitances $C_{p1}$ and $C_{p2}$ can fully represent the shunt capacitances required for a nominal Class-E operation mode. The identical series resonant circuits are each composed of a high-$Q_L$ series $L_0C_0$ filter

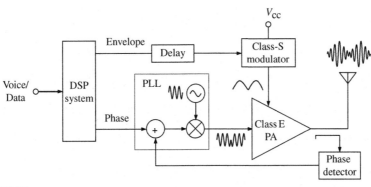

**FIGURE 5.37**

Polar transmitter architecture.

and an optimum Class-E inductance $L$. As a result, the ideal Class-E switching conditions are provided for MOS devices, while the current flowing into the load $R_L$ is free from the even harmonics.

Modern wireless communication systems require the power amplifier to amplify linearly an input signal whose RF amplitude varies with time. In this case, there is a tradeoff between power-amplifier efficiency and linearity, with improvement in one coming at the expense of the other. In addition, the power amplifiers in communications systems, such as GSM/EDGE, WCDMA/LTE, or CDMA2000, are required to linearly cover a dynamic range of the transmitter output power up to 80 dB. As a result, being designed for the highest power level with maximum available efficiency, the power amplifier tends to operate less efficiently at lower power levels, which leads to shortening the battery life and talk time reduction. In this case, to maximize efficiency of the power-amplifier operation, the load resistance for different output power levels should be different in order to provide a collector voltage amplitude close to the value of the supply voltage. Despite the fact that the shunt capacitance stays constant under different load conditions, resulting in a non-optimum Class-E operation, the efficiency at a lower output-power level can be significantly increased. Figure 5.36($c$) shows the architecture of the power-controllable Class-E power amplifier whose matching network consists of two electrically tunable shunt capacitors and a series inductor [71]. This $\pi$-type matching circuit can transform the standard load of 50 $\Omega$ into different loads whose resistance values are greater than the optimum Class-E load.

Another possibility to improve efficiency of the power amplifier for RF signals with non-constant envelope is to use the traditional analog envelope elimination and restoration (EER) approach combined with a digital signal processing (DSP) system, whose polar architecture is shown in Fig. 5.37. Here, the two separate signals, one with amplitude (envelope) information and the other with phase information only, are generated by DSP. The phase signal, in which the amplitude variations are cancelled, modulates the phase of the phase-locked loop (PLL).

Then, the output signal of the PLL is amplified by the highly efficient Class-E power amplifier. The amplitude signal is fed to the Class-S modulator to effectively vary the supply voltage according to the envelope amplitude. A Class-S modulator is a high-efficiency low-frequency power amplifier based on pulse-width modulation, and its output is a baseband envelope signal, including a dc component. To minimize misalignment between phase and amplitude, a delay line is required. Adding the output-phase feedback circuit allows the intermodulation distortion to be reduced. The measured results from a prototype test chip operating at 835 MHz indicate that phase correcting feedback reduces the 30° phase distortion of the Class-E power amplifier down to 4° when it delivers an output power of 443 mW to the load at a supply voltage of 2.4 V, and the total power efficiency of the linearized Class-E power amplifier reaches 65% [69].

Generally, the dependence of the output voltage on the dc supply voltage is linear enough that amplitude modulation can be achieved by linearly varying the supply voltage, both for bipolar and MESFET devices [72,73]. Variations from perfect linearity can be caused by variations in the transistor saturation voltage being a function of the supply voltage, input-to-output feedthrough of the driving signal via the feedback capacitance, and voltage-variable output capacitance of the transistor. Since efficiency degrades at low dc supply voltage, it can be improved with simultaneous extension of dynamic range by reducing the on-duty ratio by adaptive gate bias voltage optimization [74]. In this case, the maximum operating frequency $f_{max}$ of the Class-E power amplifier derived under nominal zero-voltage and zero voltage-derivative conditions for fixed values of the shunt capacitance, supply voltage, and output power is approximately ten times higher at 25% duty ratio than that of at 50% duty ratio [75].

In addition, the size of a Class-S modulator can be too large for MMIC implementation, and the modulator can be inefficient at high frequencies. Note that, for such a polar architecture, a high degree of amplitude and phase tracking is required to achieve high depth of modulation. For instance, a combination of 0.2-dB amplitude and 3° of phase tracking error will result in a maximum modulation dynamic range of only 20 dB. There should be a tradeoff between the switching frequency in a Class-S modulator and the order of a low-pass filter to minimize the intermodulation distortion. An experimental improved 3.9-MHz EER system with a peak envelope power (PEP) of 26 W and level of the third-order intermodulation distortion lower than −40 dB achieved the overall drain efficiency of 90% (95% in the Class-E power amplifier and 95% in the 1-MHz dc−dc converter that supplied the time-varying collector supply voltage to the RF power amplifier) [76]. All harmonics in the RF output were −60 dBc or smaller.

## REFERENCES

1. Wood JW. "High Efficiency Class C Amplifier," U.S. Patent 3,430,157, February 1969.
2. Khmelnitsky EP. *Operation of Vacuum-Tube Generator on Detuned Resonant Circuit* (in Russian). Moskva: Svyazizdat; 1962.

3. Ewing GD. *High-Efficiency Radio-Frequency Power Amplifiers*, Ph.D. Dissertation, Oregon State University; June 1964.

4. Lohrmann DR. Amplifiers has 85% efficiency while providing up to 10 watts power over a wide frequency band. *Electronic Design*. March 1966;14:38−43.

5. Artym AD. Switching mode of high frequency power amplifiers (in Russian). *Radiotekhnika*. June 1969;24:58−64.

6. Gruzdev VV. Calculation of circuit parameters of single-ended switching-mode power amplifier (in Russian). *Trudy MEIS*. 1969;2:124−128.

7. Grebennikov A. Class E high-efficiency power amplifiers: historical aspect and future prospect. *Applied Microwave & Wireless*. July 2002;14:64−71, August 2002; 64−72.

8. Kazimierczuk M. Class E tuned power amplifier with nonsinusoidal output voltage. *IEEE J. Solid-State Circuits*. August 1986;SC-21:575−581.

9. Kozyrev VB. Single-ended switching-mode power amplifier with filtering circuit (in Russian). *Poluprovodnikovye Pribory v Tekhnike Svyazi*. 1971;8:152−166.

10. Sokal NO, Sokal AD. Class E − a new class of high-efficiency tuned single-ended switching power amplifiers. *IEEE J. Solid-State Circuits*. June 1975;SC-10:168−176.

11. Sokal NO, Sokal AD. "High-Efficiency Tuned Switching Power Amplifier," U.S. Patent 3,919,656, November 1975.

12. Sokal NO. Class E high−efficiency power amplifiers, from HF to microwave. *1998 IEEE MTT-S Int. Microwave Sym. Dig.* 2:1109−1112.

13. Raab FH. Idealized operation of the Class E tuned power amplifier. *IEEE Trans. Circuits and Systems*. December 1977;CAS-24:725−735.

14. Molnar B. Basic limitations on waveforms achievable in single-ended switching-mode tuned (Class E) power amplifiers. *IEEE J. Solid-State Circuits*. February 1984;SC-19:144−146.

15. Mader TB, Bryerton EW, Marcovic M, Forman M, Popovic Z. Switched-mode high-efficiency microwave power amplifiers in a free-space power-combiner array. *IEEE Trans. Microwave Theory Tech*. October 1998;MTT-46:1391−1398.

16. Kazimierczuk M, Puczko K. Exact analysis of Class E tuned power amplifier at any $Q$ and switch duty cycle. *IEEE Trans. Circuits and Systems*. February 1987;CAS-34:149−158.

17. Sokal NO. Class-E High-efficiency RF/Microwave power amplifiers: principles of operation, design procedure, and experimental verification. In: Huijsing JH, Steyaert M, van Roermund A, eds. *Analog Circuit Design: Scalable Analog Circuit Design, High Speed D/A Converters, RF Power Amplifiers*. The Netherlands: Kluwer Academic Publishers; 2002:269−301.

18. Popov IA. Switching mode of single-ended transistor power amplifier (in Russian). *Poluprovodnikovye Pribory v Tekhnike Svyazi*. 1970;5:15−35.

19. Grebennikov AV. A simplified CAD approach to analyzing lumped elements. *RF Design*. June 2000;23:22−38.

20. Grebennikov AV. *RF and Microwave Power Amplifier Design*. New York: McGraw-Hill; 2004.

21. Negra R, Bachtold W. Lumped-element load-network design for Class-E power amplifiers. *IEEE Trans. Microwave Theory Tech*. June 2006;MTT-54:2684−2690.

22. Zomorrodian V, Pei Y, Mishra UK, York RA. High-efficiency Class E MMIC power amplifiers at 4.0 GHz using AlGaN/GaN HEMT technology. *2010 IEEE MTT-S Int. Microwave Symp. Dig.* 513−516.

23. Kalim D, Erguvan D, Negra R. Study on CMOS Class-E power amplifiers for LTE applications. *Proc. 5<sup>th</sup> German Microwave Conf.* 2010;186−189.
24. Raab FH, Sokal NO. Transistor power losses in the Class E tuned power amplifier. *IEEE J. Solid-State Circuits*. December 1978;SC-13:912−914.
25. Popov IA, ed. *Transistor Generators of Harmonic Oscillations in Switching Mode* (in Russian). Moskva: Radio i Svyaz; 1985.
26. Kessler DJ, Kazimierczuk MK. Power losses and efficiency of Class-E power amplifier at any duty ratio. *IEEE Trans. Circuits and Systems − I: Regular Papers*. September. 2004;CAS-I-51:1675−1689.
27. Kazimierczuk M. Effect of the collector current fall time on the Class E tuned power amplifier. *IEEE J. Solid-State Circuits*. April 1983;SC-18:181−193.
28. Blanchard JA, Yuan JS. Effect of collector current exponential decay on power efficiency for Class E tuned power amplifier. *IEEE Trans. Circuits and Systems − I: Fundamental Theory Appl.* January 1994;CAS-I-41:69−72.
29. Tu SH-L, Toumazou C. Effect of the loaded quality factor on power efficiency for CMOS Class-E RF tuned power amplifiers. *IEEE Trans. Circuits and Systems − I: Fundamental Theory Appl.* May 1999;CAS-I-46:628−634.
30. Islam SS, Anwar AFM. High frequency GaN/AlGaN HEMT Class-E power amplifier. *Solid-State Electronics*. October 2002;46:1621−1625.
31. Chudobiak MJ. The use of parasitic nonlinear capacitors in Class E amplifiers. *IEEE Trans. Circuits and Systems − I: Fundamental Theories Appl.* December 1994;CAS-I-41:941−944.
32. Alinikula P, Choi K, Long SI. Design of Class E power amplifier with nonlinear parasitic output capacitance. *IEEE Trans. Circuits and Systems − II: Analog and Digital Signal Process.* February 1999;CAS-II-46:114−119.
33. Suetsugu T, Kazimierczuk MK. Analysis and design of Class E amplifier with shunt capacitance composed of nonlinear and linear capacitances. *IEEE Trans. Circuits and Systems − I: Regular Papers*. July 2004;CAS-I-51:1261−1268.
34. Suetsugu T, Kazimierczuk MK. Comparison of Class-E amplifier with nonlinear and linear shunt capacitance. *IEEE Trans. Circuits and Systems − I: Fundamental Theory Appl.* August 2003;CAS-I-50:1089−1097.
35. Mediano A, Molina-Gaudo P, Bernal C. Design of Class E amplifier with nonlinear and linear capacitances for any duty cycle. *IEEE Trans. Microwave Theory Tech.* March 2007;MTT-55:484−492.
36. Suetsugu T, Kazimierczuk MK. Output characteristics of Class E amplifier with nonlinear shunt capacitance versus supply voltage. *Proc. 2007 IEEE Int. Symp. Circuits and Systems* 541−544.
37. Suetsugu T, Kazimierczuk MK. Power efficiency calculation of Class E amplifier with nonlinear shunt capacitance. *Proc. 2010 IEEE Int. Symp. Circuits and Systems* 2714−2717.
38. Bruevich AN. About optimum parameters of switching-mode tuned power amplifier with filtering resonant circuit (in Russian). *Poluprovodnikovaya Elektronika v Tekhnike Svyazi*. 1977;18:43−48.
39. Kozyrev VB, Shkvarin VV. Optimum operation mode of single-ended switching-mode power amplifier with forming resonant circuit (in Russian). *Radiotekhnika*. October 1982;37:90−93.

40. Raab FH. Suboptimum operation of Class-E power amplifiers. *Proc. RF Technology Expo '89*, Santa Clara, CA; February 1989:85−98.

41. Raab FH. Effects of circuit variations on the Class E tuned power amplifier. *IEEE J. Solid-State Circuits*. April 1978;SC-13:239−246.

42. Suetsugu T, Kazimierczuk MK. Design procedure of Class-E amplifier for off-nominal operation at 50% duty cycle. *IEEE Trans. Circuits and Systems − I: Regular Papers*. July 2006;CAS-I-53:1468−1476.

43. Suetsugu T, Kazimierczuk MK. Off-nominal operation of Class-E amplifier at any duty cycle. *IEEE Trans. Circuits and Systems − I: Regular Papers*. June 2007;CAS-I-54:1389−1397.

44. Raab FH. Effects of VSWR upon the Class-E RF-power amplifier. *Proc. RF Expo East '88*, Philadelphia, PA; October 1988:99−309.

45. Cripe DW. "Resonant Push−Pull Switching Power Amplifier," U.S. Patent 5,327,337, July 1994.

46. Ma S-W, Wong H, Yam Y-O. Optimal design of high output power Class E amplifier. *Proc. 4$^{th}$ IEEE Int. Caracas Conf. Devices Circuits Syst.* 2002;012-1−012-3.

47. Wong S-C, Tse CK. Design of symmetrical Class E power amplifiers for very low harmonic-content applications. *IEEE Trans. Circuits and Systems − I: Regular Papers*. August 2005;CAS-I-52:1684−1690.

48. Tayrani R, Meyers CW. "Efficiency broadband switching-mode amplifier," U.S. Patent 6,949,978, September 2005.

49. Mader TB, Popovic ZB. The transmission-line high-efficiency Class-E amplifier. *IEEE Microwave and Guided Wave Lett.* September 1995;5:290−292.

50. Raab FH. Class-E, Class-C, and Class F power amplifiers based upon a finite number of harmonics. *IEEE Trans. Microwave Theory Tech.* August 2001;MTT-49:1462−1468.

51. Ortega-Gonzalez FJ, Jimenez-Martin JL, Asensio-Lopez A, Torregrosa-Penalva G. High-efficiency load−pull harmonic controlled Class-E power amplifier. *IEEE Microwave and Guided Wave Lett.* Oct. 1998;8:348−350.

52. Aflaki P, Bae HG, Negra R, Ghannouchi FM. Novel compact transmission-line output network topology for Class-E power amplifiers. *Proc. 38$^{th}$ Europ. Microwave Conf.* 2008;238−241.

53. Choi Y-B, Cheng K-KM. Generalised frequency-domain analysis of microwave Class-E power amplifiers. *IEE Proc. Microwave Antennas Propag.* December 2001;148:403−409.

54. Watson P, Neidhard R, Kehias L, et al. Ultra-high efficiency operation based on an alternative Class-E mode. *2000 IEEE GaAs IC Symp. Dig.* 53−56.

55. Quach TK, Watson PM, Okamura W, et al. Ultrahigh-efficiency power amplifier for space radar applications. *IEEE J. Solid-State Circuits*. September 2002;SC-37:1126−1134.

56. Wilkinson AJ, Everard JKA. Transmission-line load-network topology for Class-E power amplifiers. *IEEE Trans. Microwave Theory Tech.* June 2001;MTT-49:1202−1210.

57. Negro R, Ghannouchi FM, Bachtold W. Study and design optimization of multiharmonic transmission-line load-networks for Class-E and Class-F $K$-Band MMIC power amplifiers. *IEEE Trans. Microwave Theory Tech.* June 2007;MTT-55:1390−1397.

58. Lee J, Kim S, Nam J, Kim J, Kim I, Kim B. Highly efficient LDMOS power amplifier based on Class-E topology. *Microwave and Optical Technology Lett.* April 2006;48: 789–791.
59. Lee Y-S, Lee M-W, Jeong Y-H. A 1-GHz GaN HEMT based Class-E power amplifier with 80% efficiency. *Microwave and Optical Technology Lett.* November 2008;50: 2989–2992.
60. Lee Y-S, Jeong Y-H. "A high-efficiency Class-E GaN HEMT power amplifier for WCDMA applications,". *IEEE Microwave and Wireless Components Lett.* August 2007;17:622–624.
61. Lee Y-S, Jeong Y-H. Applications of GaN HEMTs and SiC MESFETs in high efficiency Class-E power amplifier design for WCDMA applications. *2007 IEEE MTT-S Int. Microwave Symp. Dig.* 1099–1102.
62. Grebennikov A. A high-efficiency transmission-line GaN HEMT Class E power amplifier. *High Frequency Electronics.* December 2009;8:16–24.
63. Davis JF, Rutledge V. A low-cost Class-E power amplifier with sine-wave drive. *1998 IEEE MTT-S Int. Microwave Symp. Dig.* 2:1113–1116.
64. Frey R. 500 W, Class E 27.12 MHz amplifier using a single plastic MOSFET. *1999 IEEE MTT-S Int. Microwave Symp. Dig.* 1:359–362.
65. Zirath H, Rutledge D. An LDMOS VHF Class E power amplifier using a high Q novel variable inductor *1999 IEEE MTT-S Int. Microwave Symp. Dig.* 1:367–370.
66. Franco M, Katz A. Class-E silicon carbide VHF power amplifier. *2007 IEEE MTT-S Int. Microwave Symp. Dig.* 19–22.
67. You F, He S, Tang X, Cao T. Performance study of a Class-E power amplifiers with tuned series-parallel resonance network. *IEEE Trans. Microwave Theory Tech.* October 2008;MTT-56:2190–2200.
68. Tayrani R. A monolithic X-Band Class-E power amplifier for space based radar systems. *RF Design.* November 2003;26:D14–D19.
69. Tu SH-L. Class E RF tuned power amplifiers in CMOS technologies: theory and circuit design considerations. *IEEE Communications Mag.* September 2004;42:S6–S11.
70. Tu SH-L, Toumazou C. Low-distortion CMOS complementary Class E RF tuned power amplifiers. *IEEE Trans. Circuits and Systems – I: Fundamental Theory Appl.* May 2000;CAS-I-47:774–779.
71. Tu SH-L, Toumazou C. Design of highly-efficient power-controllable CMOS Class E RF power amplifier. *1999 IEEE Int. Circuits and Systems Symp. Dig.* 2:602–605.
72. Sowlati T, Andre C, Salama T, Sitch J, Rabjohn G, Smith D. Low voltage, high efficiency GaAs Class E power amplifiers for wireless transmitters. *IEEE J. Solid-State Circuits.* October 1995;SC-30:1074–1080.
73. Kazimierczuk M. Collector amplitude modulation of the Class E tuned power amplifier. *IEEE Trans. Circuits and Systems.* June 1984;CAS-31:543–549.
74. You F, He S, Tang X. Efficiency enhancement of Class-E power amplifiers at low drain voltage. *IEEE Trans. Microwave Theory Tech.* April 2010;MTT-58:788–794.
75. Suetsugu T, Kazimierczuk MK. Maximum operating frequency of Class-E amplifier at any duty ratio. *IEEE Trans. Circuits and Systems – II: Express Briefs.* August 2008; CAS-II-55:768–770.
76. Sokal NO, Sokal AD. "Amplifying and Processing Apparatus for Modulated Carrier Signal," U.S. Patent 3,900,823, August 1975.

# Class-E with Finite DC-Feed Inductance

## 6

## INTRODUCTION

In this chapter, the switchmode second-order Class-E configurations with one capacitor and one inductor and generalized load network including the finite dc-feed inductance, shunt capacitance, and series reactance are discussed and analyzed. The results of the Fourier analysis and derivation of the equations governing the operation in an idealized operation mode are presented. Based on these equations, the required voltage and current waveforms and load network parameters are determined for both general case and particular circuits, corresponding to the subharmonic Class-E, parallel-circuit Class-E, and even-harmonic Class-E modes. The effect of the device output bondwire inductance on the optimum circuit parameters is demonstrated. The possibilities for realizing a parallel-circuit Class-E approximation with transmission lines are shown and discussed. The operating power gain achieved with a parallel-circuit Class-E power amplifier is evaluated and compared with the operating power gain of a conventional Class-B power amplifier. The circuit design examples and practical implementations of the Class-E power amplifiers with finite DC-Feed inductance using a CMOS technology are given.

## 6.1 Class-E with one capacitor and one inductor

The Class-E load network with a shunt capacitor and a series inductor represents the simplest load network used for a switchmode operation because it can be analytically described by a first-order differential equation resulting in exact values

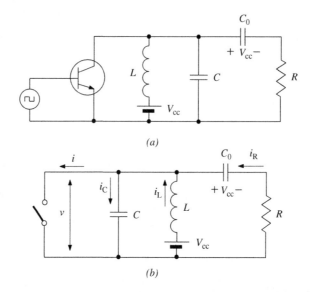

**FIGURE 6.1**

Basic circuits of a Class-E power amplifier with one capacitor and one inductor.

for the load-network parameters. However, a switchmode tuned Class-E power amplifier can also be realized with only one inductor and one capacitor connected in parallel in the load network [1−4]. In this case, the load-network topology is the same as in the classical current-source Class-C power amplifier, but the operation of these two circuits is entirely different. In a Class-C power amplifier, the tank circuit provides sinusoidal voltage on the device collector, being tuned to the fundamental frequency. In a Class-E power amplifier, this circuit is tuned between the fundamental frequency and second harmonic component providing a nonsinusoidal collector voltage waveform to minimize overlapping between the collector current and voltage waveforms, thus reducing the power losses on the device and increasing the collector efficiency.

The basic circuit of a switchmode Class-E power amplifier with one parallel $LC$ circuit is shown in Fig. 6.1($a$). In this case, the load network consists of a parallel inductor $L$, a parallel capacitor $C$, a dc-blocking capacitor $C_0$, and a load $R$. In a common case, the parallel capacitance $C$ can represent the intrinsic device output capacitance and external circuit capacitance added to the load network. The active device is considered an ideal switch that is driven to provide the device switching between its on-state and off-state operation modes. As a result, the collector voltage waveform is determined by the transient response of the load network when the switch is turned off.

To simplify an analysis of a Class-E power amplifier with one capacitor and one inductor, which can be represented by an equivalent circuit shown in Fig. 6.1($b$), consider the same idealized assumptions as for the Class-E power amplifier with shunt capacitance:

- The transistor has zero saturation voltage, zero saturation resistance, infinite off-state resistance, and its switching action is instantaneous and lossless.
- The total parallel capacitance is independent of the collector-emitter voltage and is assumed linear.
- There are no losses in the circuit except into the load $R$.
- For an optimum operation mode, a 50% duty cycle is used (other values can also be used if desired, as part of a trade-off among several performance parameters).

For an idealized theoretical analysis, an active device is replaced by the ideal switch, as shown in Fig. 6.1($b$). The moments of switch-on is $\omega t = 0$ and switch-off is $\omega t = \pi$ with a period of repeatability of $\omega T = 2\pi$ determined by the drive input to the power amplifier. For lossless operation mode, it is necessary to provide the following idealized optimum or nominal conditions for voltage across the switch just prior to the start of switch-on at the moment $\omega t = 2\pi$, when the transistor is voltage-saturated:

$$v(\omega t)|_{\omega t=2\pi} = 0 \tag{6.1}$$

$$\frac{dv(\omega t)}{d\omega t}\bigg|_{\omega t=2\pi} = 0 \tag{6.2}$$

where $v(\omega t)$ is the voltage across the switch.

When the switch is turned on for $0 \leq \omega t < \pi$, the voltage $v(\omega t)$ across the switch and the current $i_C(\omega t)$ flowing through the capacitance $C$ are equal to zero,

$$v(\omega t) = V_{cc} - v_L(\omega t) = 0 \tag{6.3}$$

$$i_C(\omega t) = \omega C \frac{dv(\omega t)}{d\omega t} = 0. \tag{6.4}$$

The current $i_R$ flowing into the load $R$ can generally be determined by

$$i_R(\omega t) = \frac{V_{cc}}{R} \exp\left(-\frac{\omega t}{\omega\tau}\right) \tag{6.5}$$

where $\tau = RC_0$. However, for $\tau \to \infty$ when the series capacitor $C_0$ serves as a dc-blocking capacitor with a large value of $C \to \infty$, the current $i_R$ can be simply rewritten as $i_R(\omega t) = V_{cc}/R$. Then, to describe the circuit operation in the time domain, the following system of equations can be obtained:

$$V_{cc} = \omega L \frac{di_L(\omega t)}{d\omega t} \tag{6.6}$$

$$i = i_L + i_R. \tag{6.7}$$

Taking into account that $i(0) = 0$ and

$$i_L(\omega t) = \frac{1}{\omega L} \int_0^{\omega t} v_L(\omega t)\, d\omega t + i_L(0) \qquad (6.8)$$

where $v_L = V_{cc}$ and $i_L(0) = -V_{cc}/R$, the current $i(\omega t)$ flowing through the switch can be defined as

$$i(\omega t) = \frac{V_{cc}}{\omega L}\omega t. \qquad (6.9)$$

When the switch is turned off for $\pi \le \omega t < 2\pi$, the following system of equations can be written:

$$V_{cc} = \omega L \frac{di_L(\omega t)}{d\omega t} \qquad (6.10)$$

$$i_C = i_L + i_R \qquad (6.11)$$

where the current $i_C(\omega t) = i_L(\omega t) + i_R(\omega t)$ flowing through the capacitance $C$ can be rewritten as

$$\omega C \frac{dv(\omega t)}{d\omega t} = \frac{1}{\omega L} \int_\pi^{\omega t} [V_{cc} - v(\omega t)]\, d(\omega t) + i_L(\pi) + \frac{V_{cc} - v(\omega t)}{R} \qquad (6.12)$$

under the initial off-state conditions $v(\pi) = 0$ and

$$\omega C \frac{dv}{d\omega t}\bigg|_{\omega t = \pi} = i(\pi) = \frac{V_{cc}\pi}{\omega L}.$$

The solution of the integro-differential equation given by Eq. (6.12) can be found by applying the Laplace transform method [2]. However, another approach is to apply the methods of solving the differential equations directly. As a result, Eq. (6.12) can be represented in the form of the linear nonhomogeneous second-order differential equation given by

$$\omega^2 LC \frac{d^2 v(\omega t)}{d(\omega t)^2} + \frac{\omega L}{R}\frac{dv(\omega t)}{d\omega t} + v(\omega t) = V_{cc} \qquad (6.13)$$

the general solution of which can be obtained in the form

$$v(\omega t) = \exp(a\omega t)[C_1 \cos(b\omega t) + C_2 \sin(b\omega t)] + V_{cc} \qquad (6.14)$$

where

$$a = -\frac{1}{2Q}$$

$$b = \frac{1}{2Q}\sqrt{(2Qq)^2 - 1}$$

$$q = \frac{1}{\omega\sqrt{LC}} \quad Q = \omega RC.$$

The coefficients $C_1$ and $C_2$ are determined from the initial off-state conditions by

$$C_1 = -V_{cc}\exp(-a\pi)\left[\cos b\pi + \frac{\pi q^2 + a}{b}\sin b\pi\right] \qquad (6.15)$$

$$C_2 = -V_{cc}\exp(-a\pi)\left[\sin b\pi - \frac{\pi q^2 + a}{b}\cos b\pi\right]. \qquad (6.16)$$

To solve Eq. (6.14) with regard to two unknown parameters $q$ and $Q$, it is necessary to apply two idealized optimum conditions given by Eqs (6.1) and (6.2). Let us consider the oscillatory case $2Qq > 1$ for a second-order differential equation. As a result, the numerical solution gives the following values:

$$q = 1.542 \qquad (6.17)$$

$$Q = 1.025. \qquad (6.18)$$

Thus, the idealized optimum values of a parallel inductance and a parallel capacitance corresponding to the switchmode Class-E-load network with only one inductor and one capacitor can be calculated from

$$L = 0.41\frac{R}{\omega} \qquad (6.19)$$

$$C = \frac{1.025}{\omega R} \qquad (6.20)$$

which are similar to those obtained in [2,3].

The optimum load $R$ and initial phase shift $\varphi$ for the fundamental-frequency component can be determined using the load-network impedance at the fundamental seen by the switch, whose phase angle $\phi$ can be represented as a function of the load-network parameters by

$$\phi = \tan^{-1}\left(\frac{R}{\omega L} - \omega RC\right) = \tan^{-1}\left[Q(q^2 - 1)\right] = 54.729°. \qquad (6.21)$$

However, the fundamental-frequency current $i_1(\omega t)$ flowing through the switch consists of two quadrature components, active $i_{R1}$ and reactive $i_{X1}$, whose amplitudes can be found using Fourier formulas and Eq. (6.9) by

$$I_{R1} = \frac{1}{\pi}\int_0^{2\pi} i(\omega t)\sin(\omega t + \varphi)\,d(\omega t) = \frac{V_{cc}}{\pi\omega L}(\pi\cos\varphi - 2\sin\varphi) \qquad (6.22)$$

$$I_{X1} = -\frac{1}{\pi}\int_0^{2\pi} i(\omega t)\cos(\omega t + \varphi)\,d(\omega t) = \frac{V_{cc}}{\pi\omega L}(\pi\sin\varphi + 2\cos\varphi) \qquad (6.23)$$

where $\varphi$ is the initial phase shift of the fundamental-frequency current.

Hence,

$$\tan\phi = \frac{I_{X1}}{I_{R1}} = \frac{\pi \sin\varphi + 2\cos\varphi}{\pi\cos\varphi - 2\sin\varphi} \tag{6.24}$$

and the phase shift $\varphi$ can be finally calculated as

$$\varphi = \tan^{-1}\left(\frac{\pi\tan\phi - 2}{2\tan\phi + \pi}\right) = 22.25°. \tag{6.25}$$

The dc supply current $I_0$ can be obtained using Fourier formula from Eq. (6.9) as

$$I_0 = \frac{1}{2\pi}\int_0^{2\pi} i(\omega t)\,d\omega t = \frac{\pi V_{cc}}{4\omega L}. \tag{6.26}$$

By using Eq. (6.22), the optimum load resistance $R$ for the specified values of supply voltage $V_{cc}$ and output power $P_1$ at the fundamental frequency can be written as

$$R = \frac{2P_1}{I_{R1}^2} = 2\left(\frac{\pi\omega L}{V_{cc}}\right)^2 \frac{P_1}{(\pi\cos\varphi - 2\sin\varphi)^2}. \tag{6.27}$$

Then, by taking into account Eqs (6.19) and (6.25), Eq. (6.27) can be simplified to

$$R = 1.394\frac{V_{cc}^2}{P_1}. \tag{6.28}$$

Figure 6.2 shows the normalized (a) fundamental-frequency load current (not including the harmonic components), (b) collector voltage, and (c) collector current waveforms for an idealized optimum Class-E mode with one capacitor and one inductor. From the collector voltage and current waveforms it follows that, when the transistor is turned on, there is no voltage across the switch, and the current $i(\omega t)$, consisting of the total load current and inductor current, flows through the switch. However, when the transistor is turned off, this current flows through the shunt capacitor. As a result, there is no nonzero voltage and current simultaneously, which means a lack of power losses and gives an idealized collector efficiency of 100%. The normalized currents flowing through the (a) shunt capacitor $C$, (b) parallel inductor $L$, and (c) load resistor $R$ for an idealized optimum Class-E operation mode are given in Fig. 6.3.

The peak collector current $I_{max}$ can be directly determined from Eq. (6.9) using Eq. (6.26), while the peak collector voltage $V_{max}$ is calculated numerically from Eq. (6.14), resulting in

$$I_{max} = 4.000\ I_0 \tag{6.29}$$

$$V_{max} = 3.849V_{cc}. \tag{6.30}$$

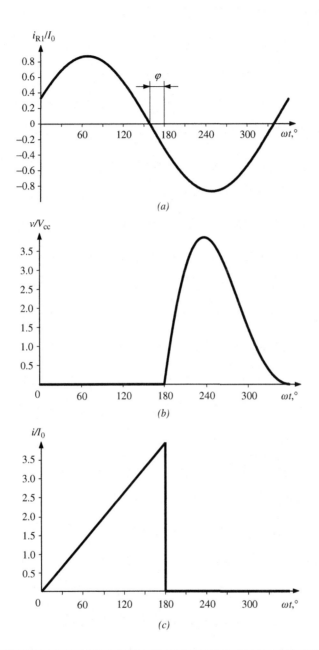

**FIGURE 6.2**

Normalized (*a*) load current and collector (*b*) voltage and (*c*) current waveforms for idealized optimum Class-E with one capacitor and one inductor.

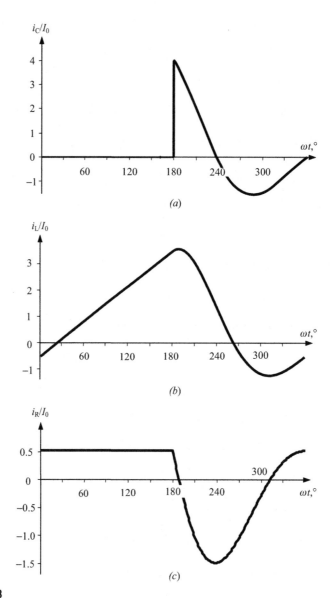

**FIGURE 6.3**

Normalized currents flowing through (*a*) shunt capacitor, (*b*) parallel inductor and (*c*) load resistor.

The numerical calculation shows that the harmonic level of the output signal spectrum is significant. The ratio of the fundamental-frequency output power $P_1$, second-harmonic output power $P_2$ and third-harmonic output power $P_3$ is

$$P_1/P_2/P_3 = 1/0.3156/0.0405 \qquad (6.31)$$

which shows that the suppressions of the second and third harmonics relative to the fundamental-frequency component in the output power spectrum are approximately $-5$ dBc and $-14$ dBc, respectively [2]. Note that the harmonic level in the output power spectrum in the Class-E power amplifier with shunt capacitor and series inductor of Fig. 5.2 is lower than that in the Class-E power amplifier with one capacitor and one inductor of Fig. 6.1, being less than 10% of the fundamental-frequency power [5,6]. Therefore, this version of the Class-E load network is suitable for applications in which the power delivered to the load $R$ is allowed to have a fairly large harmonic content and phase-modulation noise because of the absence of any additional filtering. However, this circuit can be effectively used as a test bench for extracting the parameters of a simple active device output-port model such as saturation resistance and output capacitance [7]. This approach can provide the designers with a quick way to find out if a transistor will be useful in a Class-E operation and with what maximum operating frequency.

## 6.2 Generalized Class-E load network with finite DC-Feed inductance

In practice, it is impossible to realize RF choke with infinite impedance at the fundamental frequency and other harmonic components. Moreover, using a finite DC-Feed inductance has the advantage of minimizing size, cost, and complexity of the overall circuit. The detailed approach to analyzing the effect of a finite DC-Feed inductance on the idealized Class-E mode with shunt capacitance and series filter was first described in [8]. An analysis was based on the Laplace-transform technique to solve a second-order differential equation describing the behavior of a Class-E load network with finite DC-Feed inductance. Later this approach was extended to the load network with finite $Q_L$-factor of the series filter and finite device saturation resistance [9,10]. The obtained results can be summarized as follows [10]:

- For a smaller DC-Feed inductance, smaller detuning of a series-tuned circuit is necessary, which is very attractive for radio-transmitter applications where the amplitudes of the two sidebands should not be substantially different.
- Using a small DC-Feed inductance is attractive when both the dc voltage and load resistance are specified and higher output power is needed.
- $Q_L$-factor of a series filter affects the power-amplifier performance less when the DC-Feed inductance is small.

- Maximum output power increases as a DC-Feed inductance decreases for large values of $Q_L$-factor and behaves oppositely for small values of $Q_L$-factors.
- Device saturation resistance $r_{sat}$ should be taken into account, especially when it is larger than $0.1R$, where $R$ is an optimum Class-E load resistance.

However, since the results of excessive analytical and numerical calculations are given only for a few particular cases, it is difficult to define the basic behavior of the load-network elements and derive simple equations for their parameters. Generally, based on the composing of the circuit equations in the form of a system of the first-order differential equations for currents and voltages and setting the design specifications, the optimum Class-E load-network parameters can be numerically calculated taking into account the finite DC-Feed inductance, drain current fall time, finite $Q_L$-factor, nonzero device saturation resistance, and nonlinear operation of any passive element simultaneously [11]. More physical insight into Class-E load network specifics was given when Class-E load network with finite DC-Feed inductance was analyzed for a duty ratio of 50% based on the idealized optimum Class-E conditions [12]. It was analytically shown that the series excessive reactance can be either inductive or capacitive depending on the values of the DC-Feed inductance and shunt capacitance. In a similar detailed analytical derivation procedure with particular examples, the finite device saturation resistance is taken into account to predict the power losses on the active device [13]. Based on a certain number of cases, a Lagrange polynomial interpolation was used to obtain explicit directly usable design equations for an idealized Class-E mode with small DC-Feed inductance and series inductive reactance [14].

Now let us consider a general case of the Class-E load network with finite dc-feed inductance. The main goal is to demonstrate an analytical design procedure in a clear mathematical form and to represent the design equations in a simple explicit analytical or graphical form with special highlighting of the specific particular cases. The generalized second-order load network of a switchmode Class-E power amplifier with finite DC-Feed inductance is shown in Fig. 6.4(*a*) [15]. The load network consists of a shunt capacitor $C$, a series inductor $L_b$, a parallel inductor $L$, a series reactance $X$, a series resonant $L_0C_0$ circuit tuned to the fundamental frequency, and a load resistor $R$. In a common case, a shunt capacitance $C$ can represent the intrinsic device output capacitance and external circuit capacitance added by the load network, a series inductance $L_b$ can be considered as the bondwire and lead inductance, a parallel inductance $L$ represents the finite DC-Feed inductance, and a series reactance $X$ can be positive (inductance), or negative (capacitance), or zero, depending on the particular Class-E mode. The active device is considered an ideal switch that is driven to provide the device switching between its on-state and off-state operation conditions. As a result, the collector voltage waveform is determined by the transient response of the load network when the switch is off.

To simplify an analysis of the general-circuit Class-E power amplifier, a simplified equivalent circuit of which is shown in Fig. 6.4(*b*), it makes sense to

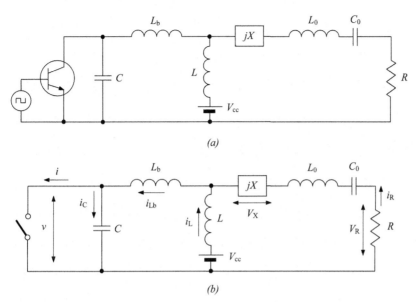

**FIGURE 6.4**

Equivalent circuits of the Class-E power amplifiers with generalized load network.

introduce the preliminary assumptions similar to those for the Class-E power amplifier with shunt capacitance. The moment of switch-on is $\omega t = 0$ and switch-off is $\omega t = \pi$ with a period of repeatability of the input driving signal $\omega T = 2\pi$ determined by the input drive to the power amplifier. Assume the losses in the reactive circuit elements are negligible and the loaded quality factor of the series $L_0 C_0$ circuit is sufficiently high. For lossless operation, it is necessary to provide the optimum zero-voltage and zero voltage-derivative conditions for voltage $v(\omega t)$ across the switch just prior to the start of switch-on at the moment $\omega t = 2\pi$, when the transistor is saturated, given by Eqs (6.1) and (6.2).

The output current flowing through the load is written as sinusoidal by

$$i_R(\omega t) = I_R \sin(\omega t + \varphi) \tag{6.32}$$

where $I_R$ is the load current amplitude and $\varphi$ is the initial phase shift.

When the switch is turned on for $0 \leq \omega t < \pi$, the voltage on the switch is

$$v(\omega t) = V_{cc} - v_{L_b}(\omega t) - v_L(\omega t) = 0 \tag{6.33}$$

where

$$v_{L_b}(\omega t) = \omega L_b \frac{di(\omega t)}{d\omega t} \tag{6.34}$$

$$v_L(\omega t) = \omega L \frac{di_L(\omega t)}{d\omega t}. \tag{6.35}$$

Since the current flowing through the capacitance $C$ is

$$i_C(\omega t) = \omega C \frac{dv(\omega t)}{d\omega t} = 0 \tag{6.36}$$

the current flowing through the switch can be written using Eqs (6.32), (6.33), and (6.35) as

$$
\begin{aligned}
i(\omega t) &= i_L(\omega t) + i_R(\omega t) \\
&= \frac{1}{\omega L} \int_0^{\omega t} [V_{cc} - v_{L_b}(\omega t)] d\omega t + i_L(0) + I_R \sin(\omega t + \varphi).
\end{aligned} \tag{6.37}
$$

Substituting Eq. (6.34) into Eq. (6.37) results in

$$i(\omega t) = \frac{V_{cc}}{\omega(L + L_b)} \omega t + \frac{\omega L}{\omega(L + L_b)} [i_L(0) + I_R \sin(\omega t + \varphi)]. \tag{6.38}$$

Since $i(0) = 0$, the initial value for the current $i_L(\omega t)$ flowing through the dc-feed inductance $L$ at $\omega t = 0$ can be found using Eq. (6.32) by

$$i_L(0) = i(0) - i_R(0) = -I_R \sin \varphi. \tag{6.39}$$

As a result,

$$i(\omega t) = \frac{V_{cc}}{\omega(L + L_b)} \omega t + \frac{\omega L I_R}{\omega(L + L_b)} [\sin(\omega t + \varphi) - \sin \varphi] \tag{6.40}$$

When the switch is turned off for $\pi \le \omega t < 2\pi$, the current $i(\omega t) = 0$, and the current $i_C(\omega t) = i_L(\omega t) + i_R(\omega t)$ flowing through the capacitance $C$ can be rewritten as

$$
\begin{aligned}
i_C(\omega t) &= \omega C \frac{dv(\omega t)}{d\omega t} = \omega C \frac{d[V_{cc} - v(\omega t) - v_{L_b}(\omega t)]}{d\omega t} \\
&= \frac{1}{\omega L} \int_\pi^{\omega t} [V_{cc} - v(\omega t) - v_{L_b}(\omega t)] \, d\omega t + i_L(\pi) + I_R \sin(\omega t + \varphi)
\end{aligned} \tag{6.41}
$$

under the initial off-state conditions $v(\pi) = 0$ and

$$i_L(\pi) = i(\pi) - i_R(\pi) = \frac{V_{cc}\pi - \omega L I_R \sin \varphi}{\omega(L + L_b)}.$$

Since the voltage $v_{L_b}(\omega t)$ across the inductor $L_b$ can be obtained from

$$v_{L_b}(\omega t) = \omega L_b \frac{di_C(\omega t)}{d\omega t} = \omega^2 L_b C \frac{d^2 v(\omega t)}{d(\omega t)^2} \tag{6.42}$$

Eq. (6.41) can be rewritten in the form

$$
\begin{aligned}
\omega C \frac{dv(\omega t)}{d\omega t} &= \frac{1}{\omega L} \int_\pi^{\omega t} \left[ V_{cc} - v(\omega t) - \omega L_b C \frac{d^2 v(\omega t)}{d(\omega t)^2} \right] d\omega t \\
&\quad + i_L(\pi) + I_R \sin(\omega t + \varphi).
\end{aligned} \tag{6.43}
$$

As a result, by differentiating of both sides of Eq. (6.43) and combining similar terms, it can be represented in the form of the linear nonhomogeneous second-order differential equation

$$\omega^2 (L + L_b) C \frac{d^2 v(\omega t)}{d(\omega t)^2} + v(\omega t) - V_{cc} - \omega L I_R \cos(\omega t + \varphi) = 0 \qquad (6.44)$$

the general solution of which can be obtained in the normalized form

$$\frac{v(\omega t)}{V_{cc}} = C_1 \cos(q\omega t) + C_2 \sin(q\omega t) + 1 - \frac{q^2 p}{1 - q^2} \cos(\omega t + \varphi) \qquad (6.45)$$

where

$$q = \frac{1}{\omega \sqrt{C(L + L_b)}} \qquad (6.46)$$

$$p = \frac{\omega L I_R}{V_{cc}} \qquad (6.47)$$

and the coefficients $C_1$ and $C_2$ are determined from the initial off-state conditions by

$$C_1 = -(\cos q\pi + q\pi \sin q\pi)$$

$$-\frac{qp}{1 - q^2} \left[ q \cos \varphi \cos q\pi - (1 - 2q^2) \sin \varphi \sin q\pi \right] \qquad (6.48)$$

$$C_2 = (q\pi \cos q\pi - \sin q\pi)$$

$$-\frac{qp}{1 - q^2} \left[ q \cos \varphi \sin q\pi + (1 - 2q^2) \sin \varphi \cos q\pi \right]. \qquad (6.49)$$

The dc supply current $I_0$ can be found using Fourier formula and Eq. (6.40) by

$$I_0 = \frac{1}{2\pi} \int\limits_0^{2\pi} i(\omega t) \, d\omega t = \frac{I_R}{2\pi} \left( \frac{\pi^2}{2p} + 2 \cos \varphi - \pi \sin \varphi \right) \Big/ \left( 1 + \frac{L_b}{L} \right). \qquad (6.50)$$

In an idealized Class-E operation mode, there is no nonzero voltage and current simultaneously that means a lack of power losses and gives an idealized collector efficiency of 100%. This implies that the dc power and the fundamental output power are equal,

$$I_0 V_{cc} = \frac{V_R^2}{2R} \qquad (6.51)$$

where $V_R = I_R R$ is the fundamental voltage amplitude across the load resistance $R$.

As a result, by using Eqs (6.50) and (6.51) and taking into account that $R = V_R^2/2P_{out}$, the optimum load resistance $R$ for the specified values of a supply voltage $V_{cc}$ and a fundamental output power $P_{out}$ can be obtained by

$$R = \frac{1}{2}\left(\frac{V_R}{V_{cc}}\right)^2 \frac{V_{cc}^2}{P_{out}} \tag{6.52}$$

where

$$\frac{V_R}{V_{cc}} = \frac{1}{\pi}\left(\frac{\pi^2}{2p} + 2\cos\varphi - \pi\sin\varphi\right)\bigg/\left(1 + \frac{L_b}{L}\right). \tag{6.53}$$

The normalized load-network inductance $L$ and capacitance $C$ as a function of the ratio $L_b/L$ can be appropriately defined using Eqs (6.46), (6.47), and (6.50) by

$$\frac{\omega L}{R} = p\left(1 + \frac{L_b}{L}\right)\bigg/\left(\frac{\pi}{2p} + \frac{2}{\pi}\cos\varphi - \sin\varphi\right) \tag{6.54}$$

$$\omega C R = \frac{1}{q^2\left(1 + \dfrac{L_b}{L}\right)\dfrac{\omega L}{R}}. \tag{6.55}$$

The series reactance $X$, which may have an inductive, capacitive, or zero reactance in special particular cases depending on the load-network parameters, can be generally calculated using two quadrature fundamental-frequency voltage Fourier components

$$V_R = -\frac{1}{\pi}\int_0^{2\pi}[v(\omega t) + v_{L_b}(\omega t)]\sin(\omega t + \varphi)\,d\omega t \tag{6.56}$$

$$V_X = -\frac{1}{\pi}\int_0^{2\pi}[v(\omega t) + v_{L_b}(\omega t)]\cos(\omega t + \varphi)\,d\omega t. \tag{6.57}$$

Generally, Eq. (6.45) for normalized collector voltage contains three unknown parameters $q$, $p$ and $\varphi$, which must be determined. In a common case, the parameter $q$ can be considered a variable, and the other two parameters $p$ and $\varphi$ are determined from a system of two equations resulting from applying two optimum zero-voltage and zero voltage-derivative conditions given by Eq. (6.1) and (6.2) to Eq. (6.45). Figure 6.5 shows the dependencies of the optimum parameters $p$ and $\varphi$ versus $q$ for Class-E with finite DC-Feed inductance when $L_b = 0$.

However, there is a specific point of $q = 1$ in Eq. (6.45) when the denominator of its last term becomes zero. Therefore, it is necessary to define the optimum parameters corresponding to this special point separately. Such a case when $q = 1$ means that the parallel circuit comprising a DC-Feed inductor $L$ and a shunt capacitor $C$ is tuned to the fundamental frequency. This implies in turn that the

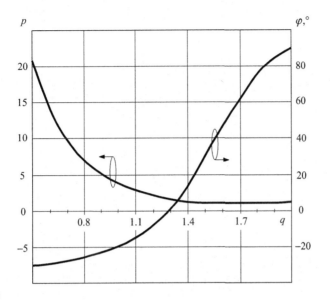

**FIGURE 6.5**

Optimum Class-E parameters $p$ and $\varphi$ versus $q$.

imaginary fundamental-frequency collector current component is equal to zero. Generally, the fundamental-frequency current flowing through the switch consists of the two quadrature components, the amplitudes of which can be found using Fourier formulas and Eq. (6.40) by

$$I_R = \frac{1}{\pi} \int_0^{2\pi} i(\omega t) \sin(\omega t + \varphi)\, d\omega t$$

$$= \frac{I_R}{\pi} \left[ \frac{\pi \cos\varphi - 2 \sin\varphi}{p} + \frac{\pi}{2} - \sin 2\varphi \right] \tag{6.58}$$

$$I_X = -\frac{1}{\pi} \int_0^{2\pi} i(\omega t) \cos(\omega t + \varphi)\, d\omega t$$

$$= \frac{I_R}{\pi} \left[ \frac{\pi \sin\varphi + 2 \cos\varphi}{p} - 2 \sin^2\varphi \right] = 0. \tag{6.59}$$

Hence, the optimum parameters $p$ and $\varphi$ can be calculated as

$$p = -\frac{4}{\pi \sin\varphi} = 3.84 \tag{6.60}$$

$$\varphi = \tan^{-1}\left(-\frac{2\pi}{\pi^2 + 8}\right) = -19.4°. \tag{6.61}$$

Based on the calculated optimum parameters $p$ and $\varphi$ as functions of $q$, the optimum load-network parameters of the Class-E load network with finite DC-Feed inductance can be determined from Eqs (6.52) to (6.55). The series reactance $X$ can be calculated through the ratio of two quadrature fundamental-frequency voltage Fourier components given in Eqs (6.56) and (6.57) as

$$\frac{X}{R} = \frac{V_X}{V_R}. \tag{6.62}$$

The dependences of the normalized optimum DC-Feed inductance $\omega L/R$ and series reactance $X/R$ are shown in Fig. 6.6(*a*), while the dependences of the normalized optimum shunt capacitance $\omega CR$ and load resistance $RP_{out}/V_{cc}^2$ are plotted in Fig. 6.6(*b*). Here, we can see that the subharmonic case of $q = 0.5$ is very close to Class-E mode with shunt capacitance since the value of the normalized inductance $\omega L/R$ is sufficiently high and the variations of normalized values of $\omega CR$ and $RP_{out}/V_{cc}^2$ are insignificant. The value of the series reactance $X$ changes its sign from positive to negative, which means that the inductive reactance is followed by the capacitive reactance. As a result, there is a special case of the load network with a parallel circuit and a load resistor only when $X = 0$ at $q = 1.412$. In this case, the maximum value of the optimum load resistance $R$ can be provided for the same supply voltage and output power, thus simplifying the matching with the standard load of 50 $\Omega$. In addition, the values of a DC-Feed inductance $L$ become sufficiently small, making Class-E very attractive for monolithic applications. The maximum operation frequency $f_{max}$ is realized at $q = 1.468$ where the normalized optimum shunt capacitance $\omega CR$ reaches its maximum.

The graphical solutions for the optimum load-network parameters can be replaced by the analytical design equations represented in terms of the simple second-order and third-order polynomial functions given by Tables 6.1, 6.2, and 6.3 for different ranges of the parameter $q$ from 0.6 to 1.9 [16]. The maximum difference between the polynomial approximations and exact numerical solutions given in the graphic form is about 2%.

Now let us consider the particular cases of the Class-E operation mode such as a subharmonic Class-E with $q = 0.5$, a parallel-circuit Class-E with $X = 0$, and an even-harmonic Class-E with $q = 2n$ assuming $L_b = 0$, which can be easily and explicitly described analytically. Then, the effect of a bondwire inductor $L_b$ on the parameters of a Class-E load network will be described based on the example of a parallel-circuit Class-E.

## 6.3 Subharmonic Class-E

The basic circuit of a switchmode subharmonic Class-E power amplifier is shown in Fig. 6.7. The load network consists of a parallel inductor $L$, a shunt capacitor $C$,

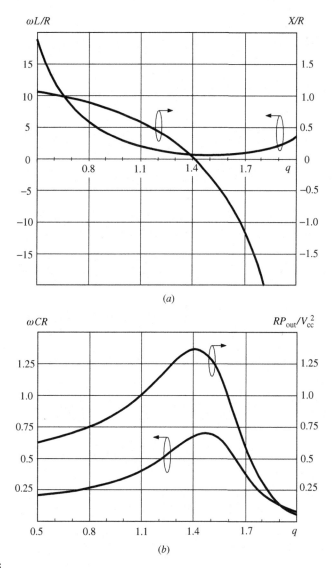

**FIGURE 6.6**

Normalized optimum Class-E load-network parameters.

a series $L_0C_0$ resonant circuit tuned to the fundamental frequency, and a load $R$. The condition of a subharmonic tuning implies that the inductance $L$ resonates with the capacitance $C$ at half the fundamental frequency $f_0$. In this case, to provide an inductive reactance at the fundamental frequency seen by the device collector, it is necessary to include a series inductance $L_X$. Hence, the first unknown parameter $q$ can be set to

$$q = 0.5. \qquad (6.63)$$

**Table 6.1** Load-Network Parameters for $0.6 < q < 1.0$

| Parameter | Design Equation |
|---|---|
| $\dfrac{\omega L}{R}$ | $44.93q^2 - 94.32q + 52.46$ |
| $\omega CR$ | $0.426q^2 - 0.379q + 0.3$ |
| $\dfrac{P_{out}R}{V_{cc}^2}$ | $-0.73q^2 + 0.411q + 1.03$ |
| $\dfrac{X}{R}$ | $0.74q^2 - 0.6q + 0.76$ |

**Table 6.2** Load-Network Parameters for $1.0 < q < 1.65$

| Parameter | Design Equation |
|---|---|
| $\dfrac{\omega L}{R}$ | $8.085q^2 - 24.53q + 19.23$ |
| $\omega CR$ | $-6.97q^3 + 25.93q^2 - 31.071q + 12.48$ |
| $\dfrac{P_{out}R}{V_{cc}^2}$ | $-2.9q^3 + 8.8q^2 - 10.2q + 5.02$ |
| $\dfrac{X}{R}$ | $-11.9q^3 + 42.753q^2 - 49.63q + 19.7$ |

**Table 6.3** Load-Network Parameters for $1.65 < q < 1.9$

| Parameter | Design Equation |
|---|---|
| $\dfrac{\omega L}{R}$ | $16.17q^2 - 52.26q + 42.94$ |
| $\omega CR$ | $2.55q^2 - 10.53q + 10.92$ |
| $\dfrac{P_{out}R}{V_{cc}^2}$ | $6.25q^2 - 24.73q + 24.56$ |
| $\dfrac{X}{R}$ | $-16.84q^2 + 51.38q - 39.83$ |

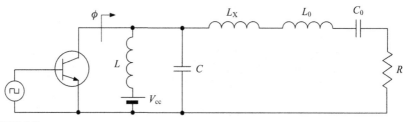

**FIGURE 6.7**

Equivalent circuit of a subharmonic Class-E power amplifier.

The coefficients $C_1$ and $C_2$ are determined from

$$C_1 = -\frac{\pi}{2} + \frac{p}{3} \sin \varphi \qquad (6.64)$$

$$C_2 = -1 - \frac{p}{3} \cos \varphi. \qquad (6.65)$$

Then, the other two parameters $\varphi$ and $p$ can be found by applying the optimum Class-E condition given by Eqs (6.1) and (6.2) to Eq. (6.45) and using Eqs (6.64) and (6.65) as

$$\varphi = \tan^{-1}\left(-\frac{\pi + 4}{2\pi + 6}\right) = -30.2°. \qquad (6.66)$$

$$p = \frac{3\pi + 9}{\cos \varphi} = 21.3. \qquad (6.67)$$

Figure 6.8 shows the normalized collector (a) voltage and (b) current waveforms for idealized optimum subharmonic Class-E mode. From collector voltage and current waveforms it follows that, when the transistor is turned on, there is no voltage across the switch, and current $i(\omega t)$ consisting of the load sinusoidal and inductive current flows through the device. The collector voltage and current waveforms are very similar to those of a Class-E with shunt capacitance.

As a result, the optimum load resistance $R$, finite DC-Feed inductance $L$, and shunt capacitance $C$ for a subharmonic Class-E mode can be obtained using Eqs (6.52) to (6.55) by

$$R = 0.635 \frac{V_{cc}^2}{P_{out}} \qquad (6.68)$$

$$L = 18.9 \frac{R}{\omega} \qquad (6.69)$$

$$C = \frac{0.212}{\omega R}. \qquad (6.70)$$

To define the phase-shifting series inductance $L_X$, it is necessary to solve Eq. (6.57) for an imaginary component of the fundamental-frequency voltage amplitude $V_X$, which can be simplified to

$$\frac{V_X}{V_{cc}} = \frac{4}{3\pi}[(C_1 - C_2)\sin \varphi + 0.5(C_1 + C_2)\cos \varphi] - \frac{2 \sin \varphi}{\pi} + \frac{p}{6}. \qquad (6.71)$$

Thus, using Eqs (6.53), (6.64), and (6.65) yields

$$L_X = \frac{V_X}{V_R} \frac{R}{\omega} = 1.058 \frac{R}{\omega}. \qquad (6.72)$$

The peak collector current $I_{max}$ and peak collector voltage $V_{max}$ can be determined from Eqs (6.40) and (6.45) using Eq. (6.50) as

$$I_{max} = 2.843 \ I_0 \qquad (6.73)$$

$$V_{max} = 3.571 \ V_{cc}. \qquad (6.74)$$

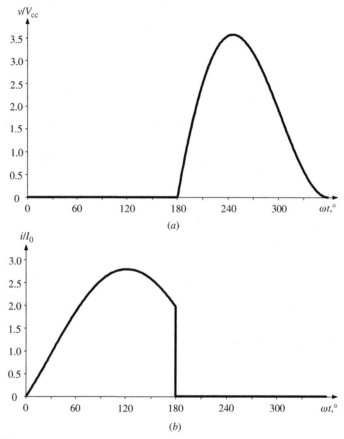

**FIGURE 6.8**

Normalized collector (*a*) voltage and (*b*) current waveforms for idealized optimum subharmonic Class-E.

From Eqs (6.68), (6.70), and (6.72) it follows that reducing a value of the dc-feed inductance to the level when it resonates with the shunt capacitance at half the fundamental frequency contributes to small variations of the optimum load-network parameters $C$, $L_X$, and $R$ derived for a Class-E with shunt capacitance, within approximately 10% only.

## 6.4 Parallel-circuit Class-E

The theoretical analysis of a switchmode parallel-circuit Class-E power amplifier using a series filter with calculation of the voltage and current waveforms and some graphical results was first published by Kozyrev [17,18]. Let us analyze the

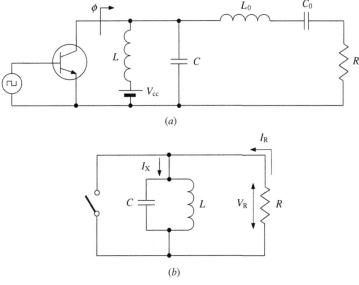

**FIGURE 6.9**

Equivalent circuits of a parallel-circuit Class-E power amplifier.

parallel-circuit Class-E mode in more detail. The basic circuit of a switchmode parallel-circuit Class-E power amplifier is shown in Fig. 6.9(a). The load network consists of a finite DC-Feed inductor $L$, a shunt capacitor $C$, a series $L_0C_0$ resonant circuit tuned to the fundamental frequency and a load resistor $R$. In this case, the switch sees a parallel connection of the load resistor $R$ and parallel $LC$ circuit at the fundamental frequency shown in Fig. 6.9(b), where also the real and imaginary collector fundamental-frequency current components $I_X$ and $I_R$ and the real collector fundamental-frequency voltage component $V_R$ are indicated.

For a subharmonic Class-E mode, to compensate for the resulting phase shift provided by the finite DC-Feed inductance and shunt capacitance tuned to half the fundamental frequency, it is required to include an additional series phase-shifting inductance $L_X$. In the case of a parallel-circuit Class-E load network without series phase-shifting reactance, since the parameter $q$ is unknown, generally it is necessary to solve a system of three equations to define the three unknown parameters $q$, $p$, and $\varphi$. In this case, first two equations are the result of applying the two optimum zero-voltage and zero voltage-derivative Class-E conditions given by Eq. (6.1) and (6.2) to Eq. (6.45). Since the fundamental-frequency collector voltage is fully applied to the load, this means that its reactive part must have zero value resulting in an additional equation

$$V_X = -\frac{1}{\pi} \int_0^{2\pi} v(\omega t)\cos(\omega t + \varphi)\, d\omega t = 0. \qquad (6.75)$$

Solving the system of three equations with three unknown parameters numerically gives the following values [19,20]:

$$q = 1.412 \tag{6.76}$$

$$p = 1.210 \tag{6.77}$$

$$\varphi = 15.155°. \tag{6.78}$$

Figure 6.10 shows the normalized (a) load current, (b) collector voltage, and (c) collector current waveforms for an idealized optimum parallel-circuit Class-E operation. From collector voltage and current waveforms it follows that, similar to other Class-E subclasses, there is no nonzero voltage and current simultaneously. When this happens, no power loss occurs and an idealized collector efficiency of 100% is achieved. The normalized currents flowing through the load-network (a) shunt capacitance $C$ and (b) finite DC-Feed inductance $L$ for an idealized optimum parallel-circuit Class-E operation mode are given in Fig. 6.11.

By using Eqs (6.52) to (6.55), the idealized optimum (or nominal) load resistance $R$, parallel inductance $L$, and shunt capacitance $C$ can be appropriately obtained by

$$R = 1.365 \frac{V_{cc}^2}{P_{out}} \tag{6.79}$$

$$L = 0.732 \frac{R}{\omega} \tag{6.80}$$

$$C = \frac{0.685}{\omega R}. \tag{6.81}$$

The dc supply current $I_0$ can be calculated from Eq. (6.50) as

$$I_0 = 0.826 I_R. \tag{6.82}$$

The phase angle $\phi$ seen from the device collector at the fundamental frequency can be represented either through the two quadrature fundamental-frequency current Fourier components $I_X$ and $I_R$ or as a function of the load-network elements by

$$\phi = \tan^{-1}\left(\frac{R}{\omega L} - \omega R C\right) = 34.244°. \tag{6.83}$$

If the calculated value of the optimum Class-E resistance $R$ is too small or differs significantly from the standard load impedance $R_L$ (usually equal to 50 $\Omega$), it is necessary to use an additional matching circuit to deliver maximum output power to the load. It should be noted that, among a family of the Class-E load networks, a parallel-circuit Class-E load network offers the largest value of $R$, thus simplifying the final matching design procedure. In this case, the first series element of such matching circuits should be the inductor to provide high impedance conditions for harmonics, as shown in Fig. 6.12.

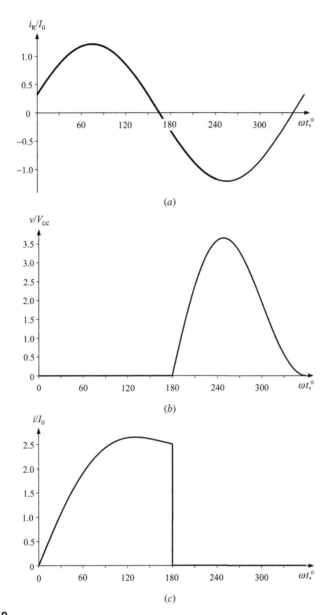

**FIGURE 6.10**

Normalized (*a*) load current and collector (*b*) voltage and (*c*) current waveforms for idealized optimum parallel-circuit Class-E.

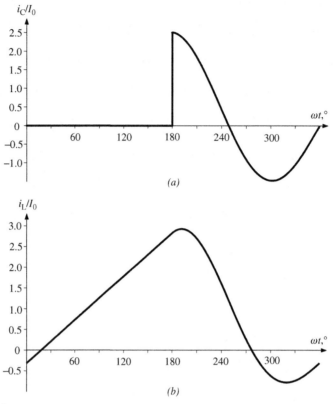

**FIGURE 6.11**

Normalized currents flowing through load network parallel (a) capacitance and (b) inductance for idealized optimum parallel-circuit Class-E.

The peak collector current $I_{max}$ and peak collector voltage $V_{max}$ can be determined from Eqs (6.40) and (6.45) using Eq. (6.82) as

$$I_{max} = 2.647I_0 \tag{6.84}$$

$$V_{max} = 3.647V_{cc}. \tag{6.85}$$

When realizing the optimum Class-E operation mode, it is important to know the maximum operation frequency $f_{max}$, which such an efficient operation mode can achieve. In this case, it needs to establish a relationship between the maximum frequency $f_{max}$, device output capacitance $C_{out}$ and supply voltage $V_{cc}$. The device output capacitance $C_{out}$ gives the main limitation of the maximum operation frequency, as it is an intrinsic device parameter and cannot be reduced for a

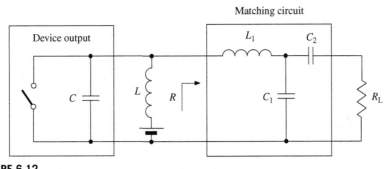

**FIGURE 6.12**

Parallel-circuit Class-E power amplifier with lumped matching circuit.

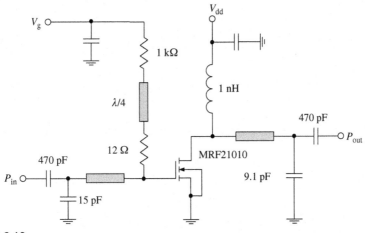

**FIGURE 6.13**

Schematic of Class-E power amplifier with transmission-line matching.

given active device. As a result, using Eqs (6.79) and (6.81) when $C = C_{out}$ gives the value of the maximum operation frequency of

$$f_{max} = 0.0798 \frac{P_{out}}{C_{out} V_{cc}^2} \qquad (6.86)$$

which is 1.4 times higher than the maximum operational frequency for an optimum Class-E power amplifier with shunt capacitance [21]. As a result, the parallel-circuit Class-E power amplifiers characterized by higher load resistance, higher maximum operating frequency and lower DC-Feed inductance, are better suited to the EER transmitters; with the same transistors providing much wider modulation bandwidth, better modulation linearity, and ease in practical implementation [22].

Figure 6.13 shows the circuit schematic of a 1-GHz LDMOSFET parallel-circuit Class-E power amplifier achieving a drain efficiency of 70.4% and an output power of more than 38 dBm at a supply voltage of 12 V [23]. In this case, the series $LC$ resonant circuit is replaced by a low-pass $L$-type output matching circuit with a series transmission line to match the low optimum Class-E resistance of 3.4 $\Omega$ to a 50-$\Omega$ load, having almost zero series excessive reactance $X$. The quarter-wavelength transmission line in the gate bias circuit provides RF isolation from the dc-voltage supply, and the 12-$\Omega$ gate resistor is required for stability.

## 6.5 Even-harmonic Class-E

The well-defined analytic solution based on an assumption of the even-harmonic resonant conditions when the finite DC-Feed inductance and parallel capacitance are tuned to any even-harmonic component was given in [24]. The load network of an even-harmonic Class-E is shown in Fig. 6.14, where the series capacitor $C_X$ is needed to compensate for the excessive inductive reactance caused by the preliminary choice of particular load-network parameters. The value of this capacitance can be found from the consideration of two fundamental-frequency voltage quadrature components across the switch given by Eqs (6.56) and (6.57).

Since, for an even-harmonic Class-E operation mode, the DC-Feed inductance is restricted to values that satisfy an even-harmonic resonance condition and it is assumed the fundamental-frequency voltage across the switch and output voltage across the load have a phase difference of $\pi/2$, the two unknown parameters can be set in this particular case as

$$q = 2n \tag{6.87}$$

$$\varphi = 90 \tag{6.88}$$

where $n = 1, 2, 3, \ldots$.

The third parameter $p$ can be found using an idealized optimum zero voltage-derivative condition given by Eq. (6.2) as

$$p = \frac{4n^2 - 1}{8n^2}\pi. \tag{6.89}$$

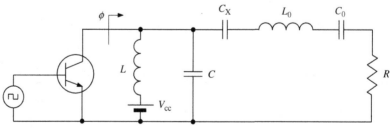

**FIGURE 6.14**

Equivalent circuit of the even-harmonic Class-E power amplifier.

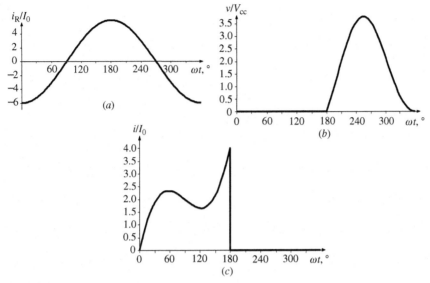

**FIGURE 6.15**

Normalized (a) load current and collector (b) voltage and (c) current waveforms for idealized optimum even-harmonic Class-E.

The dc supply current $I_0$ can be found from Eq. (6.50) when $L_b = 0$ by

$$I_0 = \frac{1}{2\pi} \int_0^{2\pi} i(\omega t) \, d\omega t = \frac{1}{2(4n^2 - 1)} I_R. \qquad (6.90)$$

As a result, the normalized steady-state collector voltage waveform for $\pi \le \omega t < 2\pi$ and current waveform for period of $0 \le \omega t < \pi$ are

$$\frac{v(\omega t)}{V_{cc}} = 1 - \frac{\pi}{2} \sin \omega t + \frac{\pi}{4n} \sin(2n\omega t) - \cos(2n\omega t) \qquad (6.91)$$

$$\frac{i(\omega t)}{I_0} = 2 \left[ \frac{8n^2}{\pi} \omega t - 4n^2 + 1 + (4n^2 - 1)\cos \omega t \right]. \qquad (6.92)$$

Figure 6.15 shows the normalized (a) load current, (b) collector voltage, and (c) collector current waveforms for an idealized optimum even-harmonic Class-E mode. If the collector voltage waveform corresponding to even-harmonic Class-E is very similar to the collector voltage waveform corresponding to Class-E with shunt capacitance, then the behavior of the collector current waveform is substantially different. So, for even-harmonic Class-E configuration, the collector current reaches its peak value, which is four times as high as the dc current, at the end of the conduction interval. Consequently, in the case of a sinusoidal driving signal it is impossible to provide close to the maximum collector current when the input base current is smoothly reducing to zero.

The optimum load-network parameters for the most practical case when $n = 1$ can be calculated from

$$R = \frac{1}{18} \frac{V_{cc}^2}{P_{out}} = 0.056 \frac{V_{cc}^2}{P_{out}} \tag{6.93}$$

$$L = \frac{9\pi}{8} \frac{R}{\omega} = 3.534 \frac{R}{\omega} \tag{6.94}$$

$$C = \frac{2}{9\pi} \frac{1}{\omega R} = 0.071 \frac{1}{\omega R} \tag{6.95}$$

$$C_X = \frac{4\pi}{32 + 3\pi^2} \frac{1}{\omega R} = 0.204 \frac{1}{\omega R}. \tag{6.96}$$

The main problem of an even-harmonic Class-E operation mode is a substantially small value of the load resistance $R$, which is over an order of magnitude smaller than for a Class-E with shunt capacitance and much smaller than for a parallel-circuit Class-E.

The phase angle $\phi$ between the fundamental-frequency voltage and current components seen by switch is equal to

$$\phi = \frac{3}{4} \frac{R}{\omega L} \frac{1 + (\omega C_X R)^2}{(\omega C_X R)^2} - \frac{1}{\omega C_X R} = 22.302° \tag{6.97}$$

while the maximum frequency $f_{max}$, up to which an idealized optimum even-harmonic Class-E mode can be realized, is calculated from

$$f_{max} = \frac{2}{\pi^2} \frac{P_{out}}{C_{out} V_{cc}^2} = 0.203 \frac{P_{out}}{C_{out} V_{cc}^2} \tag{6.98}$$

where $C_{out}$ is the device output capacitance.

## 6.6 Effect of bondwire inductance

At higher frequencies when the packaged power transistors are used in hybrid power amplifier integrated circuits, it is necessary to take into account the device output bondwire and lead inductance. Its influence may be significant, especially in particular cases of the high output power level and low supply voltage. For instance, the effect of a bondwire inductance for even-harmonic Class-E configuration may result in unrealistically small values for the optimum load resistance and DC-Feed inductance when typical values for the bondwire inductance of about 1 nH constitute most, if not all, of the required DC-Feed inductance, even at ultra-high frequencies [25].

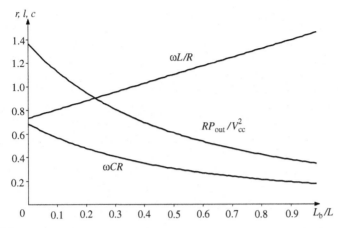

**FIGURE 6.16**

Normalized optimum load-network parameters versus normalized bondwire inductance $L_b/L$ for parallel-circuit Class-E.

The exact values for the idealized optimum load-network parameters can be easily obtained from Eqs (6.52) to (6.55), which were derived analytically for a generalized Class-E load network. Figure 6.16 shows the dependences of the normalized parallel inductance $l = \omega L/R$, shunt capacitance $c = \omega CR$, and load resistance $r = RP_{out}/V_{cc}^2$ for a parallel-circuit Class-E mode as functions of the normalized bondwire inductance $L_b/L$. From Fig. 6.16 it follows that an increasing effect of the bondwire inductance $L_b$ leads to the significantly reduced Class-E optimum values for the load resistance $R$ and shunt capacitance $C$ and increased optimum value for the finite DC-Feed inductance $L$.

## 6.7 Load network with transmission lines

At ultra-high and microwave frequencies, generally all inductances in the load-network circuits of the power amplifiers are normally replaced by the transmission lines to minimize power losses and effects of the parasitic capacitances. The load-network circuit can be composed of any type of transmission line including open-circuit or short-circuit stubs to provide the required matching and harmonic-suppression conditions. In some cases, for example, for compact small-size power-amplifier modules designed for handset wireless transmitters, the series microstrip lines and shunt chip capacitors are usually used in the external output matching circuits. In this case, a lumped-distributed Class-E load-network structure with short-length transmission lines can be very effective and helpful to increase the efficiency of a microwave power amplifier [26]. Generally, the transmission-line modeling (TLM) technique in the time domain can be used to simulate the Class-E power amplifier when the load-network elements can be modeled by the open-circuit and short-circuit stubs [27]. However, the TLM technique cannot be used to evaluate the required optimum parameter values directly to obtain the desired performance and the optimum conditions cannot be ensured. Idealized calculation of the collector voltage and current waveforms in the frequency domain, required for an optimum Class-E mode, which implies the availability of an infinite number of the harmonics with their optimum amplitudes and phases in the output spectrum.

First, consider the contribution of any harmonic component to the collector waveforms corresponding to a parallel-circuit Class-E mode similar to that for a Class-E with shunt capacitance. The Fourier-series expansion of the collector voltage $v(\omega t)$ is defined as

$$v(\omega t) = V_{cc} + \sum_{k=1}^{\infty} [V_{Rk} \sin k(\omega t + \varphi) + V_{Xk} \cos k(\omega t + \varphi)] \qquad (6.99)$$

where the dc supply voltage $V_{cc}$ and the real and imaginary harmonic components $V_{Rk}$ and $V_{Xk}$ are respectively obtained from

$$V_{cc} = \frac{1}{2\pi} \int_0^{2\pi} v(\omega t) \, d\omega t \qquad (6.100)$$

$$V_{Rk} = \frac{1}{\pi} \int_0^{2\pi} v(\omega t) \sin k(\omega t + \varphi) \, d\omega t \qquad (6.101)$$

$$V_{Xk} = \frac{1}{\pi} \int_0^{2\pi} v(\omega t) \cos k(\omega t + \varphi) \, d\omega t. \qquad (6.102)$$

The same Fourier analysis can be applied to the collector current $i(\omega t)$. Figure 6.17 shows the normalized ideal collector (a) voltage and (b) current waveforms plotted by solid lines, where $I_0$ is the dc current. In this case, as clearly seen from Fig. 6.17, a good approximation to parallel-circuit Class-E mode can be obtained with only the fundamental-frequency and second-harmonic components fully determining the collector voltage and current waveforms, which are indicated by dashed lines. The same conclusion, which is very important for the practical design of microwave Class-E power amplifiers, was obtained for a Class-E mode with shunt capacitance [28]. Consequently, the high-efficiency parallel-circuit Class-E power-amplifier circuit can also be effectively used in monolithic microwave integrated circuit (MMIC) design provided the optimum Class-E conditions at the fundamental frequency and second harmonic are fulfilled.

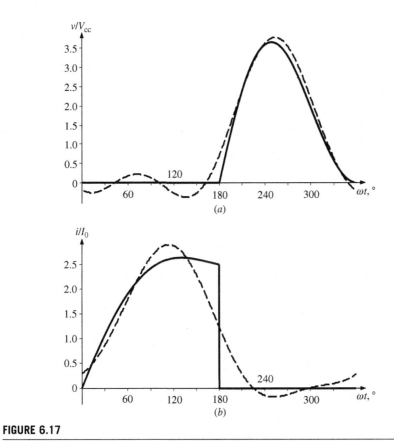

**FIGURE 6.17**

Two-harmonic approximation to parallel-circuit Class-E mode.

As a first step, the parallel inductance $L$ at microwaves should be replaced by a short-length short-circuited transmission line $TL$ shown in Fig. 6.18($a$) according to

$$Z_0 \tan \theta = \omega L \qquad (6.103)$$

where $Z_0$ and $\theta$ are the characteristic impedance and electrical length of the transmission line $TL$, respectively [29]. To approximate the idealized optimum parallel-circuit Class-E operation conditions for a microwave power amplifier, it is necessary to design the load network satisfying the required idealized optimum input impedance at the fundamental frequency

$$Z_{\text{net}}(\omega_0) = \frac{R}{1 - j \tan 34.244^\circ} \qquad (6.104)$$

which can be obtained from Eqs (6.80), (6.81), and (6.83).

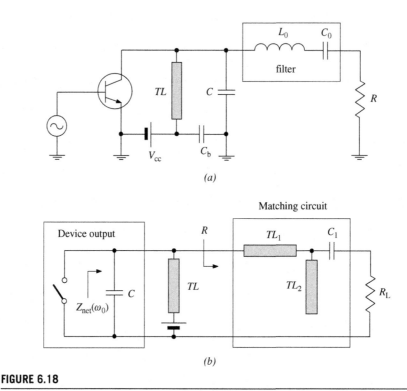

**FIGURE 6.18**

Equivalent circuits of a transmission-line parallel-circuit Class-E power amplifier.

By using Eq. (6.80), defining the idealized optimum (or nominal) parallel inductance $L$ for a parallel-circuit Class-E mode, Eq. (6.103) can be rewritten as

$$\tan \theta = 0.732 \frac{R}{Z_0}. \tag{6.105}$$

In practical circuits, when the impedance transformation between the optimum Class-E load resistance $R$ and standard load resistance of 50 $\Omega$ is required, the series $L_0 C_0$ filter should be replaced by the output matching circuit, the input impedance of which needs to be sufficiently high at the second- and higher-order harmonics. For example, the series $L_0 C_0$ filter can be replaced by a lumped $T$-transformer containing a series inductor and two (shunt and series) capacitors. For a transmission-line realization, the output matching circuit can be composed with any types of the transmission lines including open-circuit or short-circuit stubs to provide the required matching and harmonic-suppression conditions. However, to maintain the optimum-switching conditions at the fundamental frequency, such an output matching circuit should contain the series transmission line as the first element, as shown in Fig. 6.18(b).

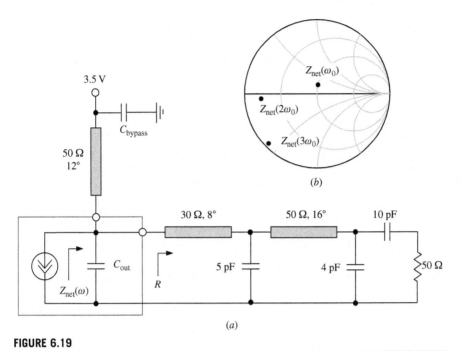

**FIGURE 6.19**

Transmission-line load network of a parallel-circuit Class-E power amplifier for handset application.

Figure 6.19(a) shows an example of the transmission-line Class-E load network of a two-stage 1.75-GHz GaAs HBT power amplifier with an output power of 33 dBm, which was designed for a cellular handset transmitter power amplifier and includes the series microstrip line with two shunt chip capacitors [19,30]. However, because of the finite electrical lengths of the transmission lines, it is impossible to simultaneously realize the required inductive impedance at the fundamental frequency with the purely capacitive reactances at higher-order harmonics. For example, at the second harmonic, the real part of the load network impedance $Z_{net}(2\omega_0)$ is sufficiently high, as shown in Fig. 6.19(b). Nevertheless, even such an approximation provides a good proximity to the parallel-circuit Class-E operation mode, resulting in a high operating efficiency of the power amplifier. In this case, there is no need to use an additional RF choke for dc supply current, since its function can be performed by the same short-length parallel microstrip line which is necessary to provide optimum inductive impedance at the fundamental frequency.

Comparison of the different Class-E load network topologies in terms of their parameters shows a clear preference of the parallel-circuit Class-E mode, especially for low supply voltage application with high level of the circuit integration. For such a load network, a DC-Feed inductance $L$ is sufficiently low, which can be provided by a short-length transmission line, and an optimum load resistance *R* (for the same output power and supply voltage) is greater by approximately 2.4 times and 24 times compared to the Class-E with shunt capacitance and even-harmonic Class-E, respectively. In addition, the parallel-circuit Class-E configuration can be easily implemented in a high-efficiency broadband high-power or low-voltage power amplifier design. In this case, it is only necessary to satisfy the required phase angle seen by the device collector at the fundamental frequency and to choose the proper low-$Q_L$ factor of the series resonant $L_0C_0$ circuit [31,32].

Figure 6.20(*a*) shows the circuit schematic and module structure of a 500-MHz single-stage parallel-circuit Class-E LDMOSFET power amplifier with a supply voltage of 28 V, an output power of 42 dBm, a linear power gain of 15 dB, and a *PAE* of 67% [19]. The input and output matching circuits represent *T*-type transformers with series 50-$\Omega$ microstrip lines fabricated by using a laminate substrate with a dielectric permittivity of 4.7 and a thickness of 0.4 mm. The required fundamental-frequency load-network phase angle is provided with the device output capacitance and parallel 50-$\Omega$ microstrip line with an electrical length of 25°. The overall size of a hybrid module is $45 \times 20$ mm$^2$ including an eutectic-attached 1.25-$\mu$m LDMOSFET die with a total gate width of $28 \times 1.44$ mm which is characterized by a small-signal transconductance $g_m = 0.6$ A/V and a transition frequency $f_T = 4.5$ GHz. The die is connected to the input microstrip line by four bondwires and to the output microstrip line by five bondwires, each having a length of 1.5 mm.

The circuit schematic of a two-stage InGaP/GaAs HBT power amplifier intended to operate in the WCDMA handset transmitters is shown in Fig. 6.20(*b*) [31]. The MMIC part of this power amplifier contains the transistors with emitter areas of the first and second stage as large as 540 $\mu$m$^2$ and 3600 $\mu$m$^2$, input matching circuit, interstage matching circuit, and bias circuits on a die with dimensions of less than 1 mm$^2$. Without any tuning of the output matching circuit, a saturated output power greater than 30 dBm and a *PAE* greater than 50% were obtained. Using high-$Q$ capacitors in output matching circuit can improve the power-added efficiency by about 8%. At the same time, this power amplifier without any additional tuning can provide the high-linearity performance for WCDMA band (1920−1980 MHz) at a 3.5-dB backoff output power of 27 dBm with a power gain of 22.6 dB and a sufficiently high efficiency. The measured *PAE* reached value of 38.3% at center bandwidth frequency of 1.95 GHz with an adjacent channel leakage power ratio (*ACLR*) of −37 dBc at a 5-MHz offset.

As an example of a close proximity to the idealized optimum parallel-circuit Class-E mode, the simulated collector (*a*) voltage and (*b*) current waveforms shown in Fig. 6.21 and collector efficiency of about 90% were achieved for a 1.75-GHz InGaP/GaAs HBT power amplifier at a supply voltage of 5 V [20]. During off-state operation mode, only the current flowing through the device collector capacitance defines the total collector current. The collector voltage

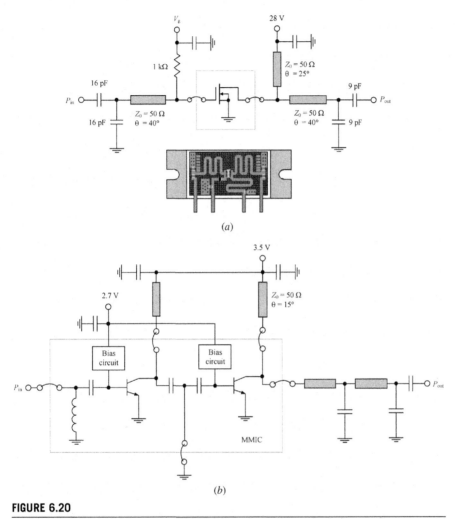

**FIGURE 6.20**

Circuit schematics and module design of a parallel-circuit Class-E high-voltage LDMOSFET and low-voltage GaAs HBT power amplifiers.

waveform is very similar to the ideal one, with a peak factor of about three. The main reason for the efficiency degradation from an ideal 100% is a high value of the saturation voltage of about 0.5-0.8 V, which can result in 15-20% collector efficiency reduction for a supply voltage of 3.5 V. Generally, the further decrease in collector efficiency in a real circuit can be explained by a violation of the required optimum impedance conditions; due to the transmission-line effect at the second- and higher-order harmonics, device finite switching time and parasitics, and the power losses in the load network.

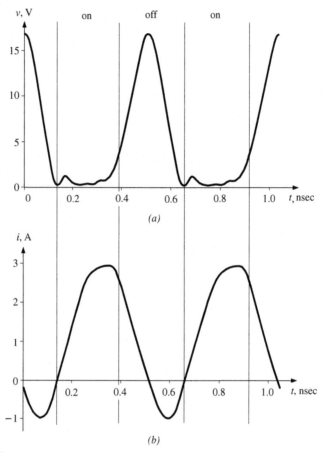

**FIGURE 6.21**

Simulated collector (*a*) voltage and (*b*) current waveforms of a transmission-line parallel-circuit Class-E low-voltage InGaP/GaAs HBT power amplifier.

## 6.8 Operation beyond maximum Class-E frequency

To obtain an optimum Class-E mode beyond its maximum operation frequency when the shunt capacitance becomes excessive, the value of the parallel DC-Feed inductor has to be chosen so as to compensate for the negative effect of this excessive capacitance not only at the fundamental frequency, but also at the higher-order harmonic components. Figure 6.22(*a*) shows the circuit schematic of a parallel-circuit Class-E power amplifier, where $C$ is the nominal Class-E capacitance, $C_{ex}$ is the excess capacitance, and $L_{new}$ is the required frequency-varying

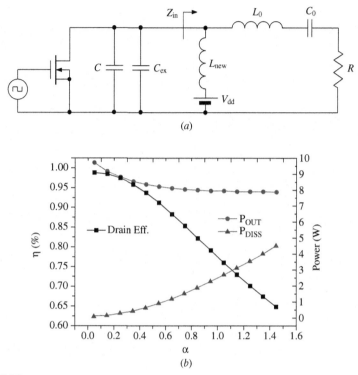

**FIGURE 6.22**

Parallel-circuit Class-E power amplifier with excess capacitance.

DC-Feed inductance. Adding the susceptances of $C_{ex}$ and $L_{new}$ should result in the susceptance of an optimum Class-E DC-Feed inductance $L$ according to

$$\frac{1}{j\omega L} = j\omega C_{ex} + \frac{1}{j\omega L_{new}} \qquad (6.106)$$

from which it follows that

$$L_{new} = \frac{L}{1 + \left(\dfrac{\omega}{\omega_{ex}}\right)^2} \qquad (6.107)$$

where $\omega_{ex} = 1/\sqrt{LC_{ex}}$. In this case, an inductor $L_{new}$ becomes a frequency-dependent element, whose value is different for corresponding harmonic component and which can be realized by means of an appropriate arrangement of lumped components. The number of lumped components depends on the accuracy of the approximation to the optimum parallel-circuit in Class-E mode.

Taking into account that the output series resonant circuit tuned to the fundamental frequency $\omega_0$ is characterized by a high-$Q_L$ factor, the input impedance $Z_{in}(\omega_0)$ can be written as

$$Z_{in}(\omega_0) = \frac{j\omega_0 L_{new} R}{R + j\omega_0 L_{new}} = \frac{j\omega_0 R \frac{L}{1 + (\omega_0/\omega_{ex})^2}}{R + j\omega_0 \frac{L}{1 + (\omega_0/\omega_{ex})^2}}. \tag{6.108}$$

Introducing the excess factor $\alpha = C_{ex}/C$ and using Eqs (6.80) and (6.81), we can write

$$\omega_{ex} = \omega_0 \sqrt{\frac{2}{\alpha}}. \tag{6.109}$$

As a result, Eq. (6.108) as a function of the excess factor $\alpha$ can be simplified using Eq. (6.80) for an optimum Class-E inductance $L$ at the fundamental frequency to

$$Z_{in}(\omega_0) = R \frac{1 + j1.366(1 + 0.5\alpha)}{1 + j1.866(1 + 0.5\alpha)^2} \tag{6.110}$$

and at the second- and higher-order harmonics $(n \geq 2)$ to

$$Z_{in}(n\omega_0) = j \frac{n\omega_0 L}{1 + 0.5\alpha n^2}. \tag{6.111}$$

In order to predict the effect of an excess capacitance $C_{ex}$ on the parallel-circuit Class-E power-amplifier performance, the drain efficiency $\eta$, output power $P_{out}$, and power dissipation $P_{diss}$ shown in Fig. 6.22(b) have been obtained by varying the excess factor $\alpha$, while keeping the rest of the circuit components constant [33].

Figure 6.23(a) shows the single-resonator topology of a parallel-circuit Class-E power amplifier, which can compensate for the excess capacitance at the second harmonic. The same network was used to approximate the three-harmonic impedance conditions of a Class-F power amplifier [34]. The equations relating the values of the load-network parameters to the excess factor $\alpha$ can be obtained by equating the impedance of the $L_1 L_2 C_2$ network to the impedance in Eq. (6.111) for $n = 1$ and $n = 2$, respectively, as

$$\omega_0 L_1 + \frac{\omega_0 L_2}{1 - \omega_0^2 L_2 C_2} = \frac{\omega_0 L}{1 + 0.5\alpha} \tag{6.112}$$

$$2\omega_0 L_1 + \frac{2\omega_0 L_2}{1 - 4\omega_0^2 L_2 C_2} = \frac{2\omega_0 L}{1 + 2\alpha} \tag{6.113}$$

where the resonant frequency $\omega_2$ of the $L_2 C_2$ resonator is located within $\omega_0 < \omega_2 < 2\omega_0$.

By introducing a new parameter $\gamma$ defining the ratio between $\omega_0$ and $\omega_2$,

$$\gamma = \left(\frac{\omega_0}{\omega_2}\right)^2 \tag{6.114}$$

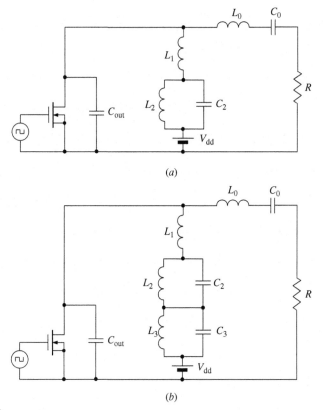

**FIGURE 6.23**

Single- and double-resonator topologies of a parallel-circuit Class-E power amplifier.

Equations (6.112) and (6.113) can now be rewritten as

$$\omega_0 L_1 + \frac{\omega_0 L_2}{1 - \gamma} = \frac{\omega_0 L}{1 + 0.5\alpha} \tag{6.115}$$

$$2\omega_0 L_1 + \frac{2\omega_0 L_2}{1 - 4\gamma} = \frac{2\omega_0 L}{1 + 2\alpha} \tag{6.116}$$

resulting finally in separate equations for the load-network parameters given by

$$L_1 = \frac{L(2\gamma + \alpha)}{\gamma(2\alpha^2 + 5\alpha + 2)} \tag{6.117}$$

$$L_2 = -\frac{\alpha L(4\gamma^2 - 5\gamma + 1)}{\gamma(2\alpha^2 + 5\alpha + 2)} \tag{6.118}$$

$$C_2 = \frac{\gamma}{w_0^2 L_2}. \tag{6.119}$$

Figure 6.23(*b*) shows the double-resonator topology of a parallel-circuit Class-E power amplifier, which can compensate for the excess capacitance at the second and third harmonics. In this case, the lumped load network comprises a series inductor and two parallel *LC* resonators. The resonance frequency $w_2$ of the $L_2 C_2$ resonator is located between $w_0$ and $2w_0$, whereas the resonant frequency $w_3$ of the $L_3 C_3$ resonator is located between $2w_0$ and $3w_0$ [33]. Fixing the value of these resonances and selecting appropriate values for $L_1$, $L_2$, and $L_3$ allows the right value of harmonic reactances to be obtained.

The equations for the parameters of the $L_1 L_2 C_2 L_3 C_3$ network can be written as

$$L_1 = \frac{L[\alpha^2 + 2\alpha(\gamma_2 + \gamma_3) + 4\gamma_2\gamma_3]}{\gamma_2\gamma_3(18\alpha^3 + 49\alpha^2 + 28\alpha + 4)} \tag{6.120}$$

$$L_2 = \frac{\alpha L(\alpha + 2\gamma_3)(-36\gamma_2^3 + 49\gamma_2^2 - 14\gamma_2 + 1)}{\gamma_2(\gamma_2 - \gamma_3)(18\alpha^3 + 49\alpha^2 + 28\alpha + 4)} \tag{6.121}$$

$$L_3 = \frac{\alpha L(\alpha + 2\gamma_2)(36\gamma_3^3 - 49\gamma_3^2 + 14\gamma_3 - 1)}{\gamma_3(\gamma_2 - \gamma_3)(18\alpha^3 + 49\alpha^2 + 28\alpha + 4)} \tag{6.122}$$

$$C_2 = \frac{\gamma_2}{w_0^2 L_2} \tag{6.123}$$

$$C_3 = \frac{\gamma_3}{w_0^2 L_3} \tag{6.124}$$

where *L* represents an optimum inductance in the classical parallel-circuit Class-E load network obtained by Eq. (6.80) and

$$\gamma_2 = \left(\frac{w_0}{w_2}\right)^2 \quad \text{where} \quad \frac{1}{4} < \gamma_2 < 1 \tag{6.125}$$

$$\gamma_3 = \left(\frac{w_0}{w_3}\right)^2 \quad \text{where} \quad \frac{1}{9} < \gamma_3 < \frac{1}{4}. \tag{6.126}$$

The simulated voltage and current drain waveforms obtained for $\alpha = 1.174$ with an ideal switch demonstrate that, if the switch current exhibits a small negative peak during turn on for a single-resonator topology, as shown in Fig. 6.24(*a*), this peak disappears for a double-resonator topology, as shown in Fig. 6.24(*b*), thus clearly approximating the ideal optimum Class-E switching conditions. The component values of the series $L_0 C_0$ resonant circuit were obtained by selecting $Q_L = 8$ and by setting the resonance frequency equal to 434 MHz.

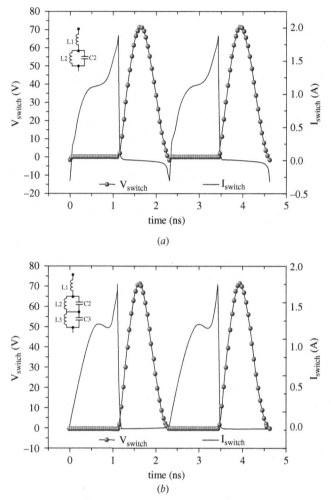

**FIGURE 6.24**

Simulated waveforms using lumped Class-E load networks with excess capacitance.

## 6.9 **Power gain**

To realize the idealized switching conditions, the load network corresponding to any type of a Class-E mode must be tuned to provide inductive or capacitive impedance at the fundamental frequency, thus violating the conjugate matching conditions required for conventional Class-B operation to provide maximum power delivery to the load. This means that generally the output voltage and current waves consist of both incident and reflected components. Besides, the power gain in a switchmode operation is normally lower than in a conventional mode because it requires higher driving voltage to realize a device voltage-saturation mode. In this case, the ratio of the power gain in a Class-E power amplifier to that in a Class-B power amplifier for the same output power is inversely proportional to a squared ratio of their voltage peak factors [35]. For a Class-E with one inductor and one capacitor, a maximum operating power gain $G_{P(E)max}$ of a single-stage bipolar power amplifier can be estimated by

$$G_{P(E)max} \approx \frac{\pi f_T}{2 f_0} \frac{V_{cc}}{V_{th}} \tag{6.127}$$

where $f_0$ is the operating frequency, $V_{th}$ is the threshold voltage, and it is assumed that the effect of a nonzero fall time is negligible [36].

However, it is very important to qualitatively compare the power gains of the Class-B and Class-E power amplifiers as functions of the device and load-network parameters. In this case, a parallel-circuit Class-E mode looks very attractive since it provides the highest value of load resistance compared with others Class-E alternatives. The operating power gain $G_P$, expressed through the active-device $Y$-parameters and load admittance $Y_L$, can be obtained by

$$G_P = \frac{|Y_{21}|^2}{Re Y_{in}} \frac{G}{|Y_{22} + Y_L|^2} \tag{6.128}$$

where $Y_{21}$ and $Y_{22}$ are the device transfer and driving-point output admittances, $Y_{in}$ is the input admittance of the loaded device, and $Y_L = G + jB$ [37]. Since the power amplifier is operated in a nonlinear mode, the admittance $Y$-parameters of the active device are considered linearized at the fundamental frequency. For example, for a power amplifier operating at the same conduction angle over various bias and drive conditions, these $Y$-parameters remain constant, and operating power gain becomes the function of load admittance only.

Equation (6.128) can be rewritten as

$$G_P = \frac{|Y_{21}|^2}{Re Y_{in}} \frac{R}{\left(1 + \dfrac{G_{22}}{G}\right)^2 + \left(\dfrac{B_{22} + B}{G}\right)^2} \tag{6.129}$$

where $G_{22} = Re Y_{22}$, $B_{22} = Im Y_{22}$, and $R = 1/G$ is the load resistance. Without significant loss of accuracy, the output conjugate-matching condition between imaginary parts of the device output admittance and the load admittance can be replaced by a simple condition of $B_{22} + B = 0$, which means a resonance tuning of the load network including the device output susceptance $B_{22}$. Besides, normally the device output conductance $G_{22}$ is significantly smaller than the load conductance $G$ for both MOSFET and bipolar transistors, especially at frequencies well below the device transition frequency $f_T$.

Consequently, the simplified ratio between the operating power gain $G_{P(E)}$ of a parallel-circuit Class-E power amplifier and the operating power gain $G_{P(B)}$ of a conventional Class-B power amplifier with conjugate-matched load can be written as

$$\frac{G_{P(E)}}{G_{P(B)}} = \frac{1}{1 + (B_{22} + B)^2 R_{(E)}^2} \frac{R_{(E)}}{R_{(B)}} \qquad (6.130)$$

where $R_{(E)}$ is the load resistance of a Class-E power amplifier and $R_{(B)}$ is the load resistance of a Class-B power amplifier.

For a nominal parallel-circuit Class-E operation mode with a 100% collector efficiency, from Eq. (6.53) for $L_b = 0$ it follows that

$$V_{R(E)} = 1.652 V_{cc}. \qquad (6.131)$$

For the same output power $P_{out}$ and taking into account that $V_{R(B)} = V_{cc}$ in a conventional Class-B mode with zero saturation voltage, we can write

$$\frac{R_{(E)}}{R_{(B)}} = \frac{V_{R(E)}^2}{V_{R(B)}^2} = 2.729 \qquad (6.132)$$

which shows the significantly higher value for the load resistance in a nominal parallel-circuit Class-E operation mode.

As a result, the power gain ratio given by Eq. (6.130) can be rewritten in the form

$$\frac{G_{P(E)}}{G_{P(B)}} = \frac{2.729}{1 + \tan^2 \phi} = 1.865 \qquad (6.133)$$

where $\phi = 34.244°$ is the phase angle between the fundamental-frequency voltage and current components at the device output required for the nominal parallel-circuit Class-E mode.

The result given by Eq. (6.133) means that ideally the operating power gain for an idealized optimum parallel-circuit Class-E mode compared to a conventional Class-B mode is almost the same and even slightly greater, despite the detuning of the load network. This can be explained by the larger value of the load resistance required for the nominal parallel-circuit Class-E load network. For example, for a Class-E mode with shunt capacitance, the operating power gain is smaller compared to a Class-B power amplifier because its nominal Class-E load resistance is about 2.4 times smaller than that of a parallel-circuit Class-E mode [38]. The idealized conditions for a switchmode operation can be achieved with instant on/off active device switching, which requires the rectangular input driving signal compared with a sinusoidal driving signal for a conventional Class-B mode. However, the power losses due to the nonzero switching times are sufficiently small and, for example, for switching time of $\tau_s = 0.35$ or $20°$ they are only about 1% [20]. Consequently, a slight overdrive of the active device is needed when the input power should be increased by $1 \div 2$ dB to minimize the switching time and maximize the collector efficiency of a switchmode parallel-circuit Class-E power amplifier. As a result, its resulting operating power gain becomes approximately equal to the operating power gain of a conventional Class-B power amplifier.

## 6.10 CMOS Class-E power amplifiers

Recent progress in CMOS technology has shown their promising future for RF power application. Much progress has been achieved at the research level, and the obvious possibility to minimize the cost and size of the integrated circuits for RF handset transmitters, especially power-amplifier MMICs, makes CMOS technology very feasible and brings considerable economical benefits. However, realizing high-efficiency operation of power amplifiers is limited by some technology issues, such as high value of the device saturation resistance, low value of the breakdown voltage, and lossy silicon substrate. Therefore, it is vital to apply high-efficiency techniques in the design of CMOS power amplifiers.

For example, a 900-MHz cascode power amplifier based on a 0.25-$\mu$m CMOS technology with active die area of $2 \times 2 \, \text{mm}^2$ can provide an output power of 0.9 W and a *PAE* of 41% using a Class-E load network with finite DC-Feed inductance, the circuit schematic of which is shown in Fig. 6.25 [39]. Minimizing the value of the DC-Feed inductance is necessary to minimize the die size, resulting also in higher optimum values of the load resistance and shunt capacitance required for a Class-E with finite DC-Feed inductance. For the same value of the saturation resistance, this contributes to lower power loss on the active device and can absorb a larger value of the device output capacitance. Cascode configuration

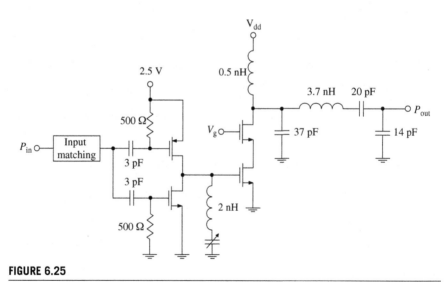

**FIGURE 6.25**

Cascode Class-E power amplifier with finite DC-Feed inductance.

and thick-oxide transistors are used to eliminate the effects of oxide breakdown voltage and hot carrier degradation effect, allowing the supply voltage $V_{dd}$ to be as high as 1.8 V. Since a cascode switch has higher on-resistance per unit channel width than a single common-source switch during the on-state mode, wider devices of 15-mm gate widths are used. The interstage bondwire inductor of 2 nH and external variable capacitor are used to resonate with the gate capacitance of the cascode device. Since the quality factor of on-chip spiral inductors provided by a typical CMOS technology is low because of a large loss in the silicon substrate and metal layers, bonding wires can be successfully used instead of spiral inductors. This provides less than 5% of inductance variation and less than 6% of $Q$-factor variation as a result of the wire-bonding process. Therefore, the complete power-amplifier load network consists of two aluminum bondwire inductors and one on-chip (37 pF) and two off-chip (20 pF and 14 pF) capacitors. The implemented power amplifier is differential, and baluns were used at both input and output to combine the two single-ended paths. By using an injection-locked oscillator technique and differential circuit topology implemented in a 0.35-μm CMOS technology, the cascode Class-E power amplifier was capable to provide a *PAE* of 41% and an output power of 1 W at the operating frequency of 1.98 GHz [40].

Figure 6.26 shows the circuit schematic of a parallel-circuit Class-E power amplifier with a 1:$n$ output transformer implemented in a 0.18-μm CMOS technology, where $n$ is the turns ratio, $M$ is the mutual inductance between the primary and secondary windings, and $Z_T$ is the impedance looking into the transformer [41]. In this case, the real and imaginary parts of the impedance $Z_T$ must be equal to $R$ and $\omega L_0$, respectively, where $R$ is the optimum Class-E load resistance and $L_0$ is the series inductance tuned to the fundamental frequency together with a series capacitance $C_0$. The optimum Class-E shunt capacitance $C$ is fully

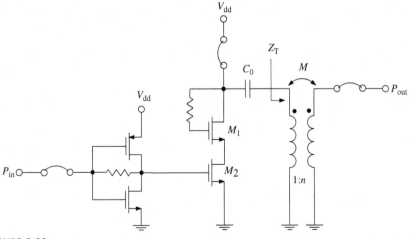

**FIGURE 6.26**

Cascode Class-E power amplifier with output transformer.

represented by the output CMOS capacitance, and the bondwire inductance connected to the dc supply represents the optimum Class-E DC-Feed inductance. A complete three-stage fully integrated CMOS power amplifier consists of the two push–pull parallel-circuit Class-E power amplifiers in a final stage with a $2 \times 1{:}2$ on-chip output transformer, where a two-turn secondary winding was selected to provide the necessary impedance transformation ratio to achieve 2 W of output power. As a result, a compressed power gain of around 19.5 dB and a *PAE* of 31% were achieved in a single-ended 50-$\Omega$ input and output conditions at the operating frequency of 1.8 GHz, with overall die size of $1.8 \times 1.65 \text{ mm}^2$.

Optimizing the cascode topology requires a proper setting of the bias voltage $V_g$ of the common-gate transistor shown in Fig. 6.27(a) to minimize the voltage

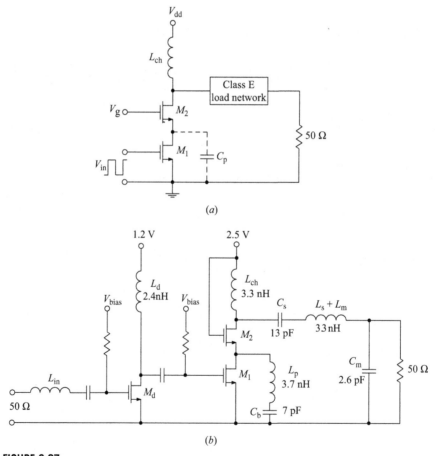

**FIGURE 6.27**

Cascode Class-E power amplifier with a compensating inductor.

drop across the oxide of each transistor $M_1$ and $M_2$ when these voltage drops become equal, allowing the use of approximately twice the supply voltage [42]. However, there is an additional power-loss mechanism as a specific property of a cascode configuration in a switching Class-E mode when the common-source device $M_1$ is turned off, that is associated with charging and discharging processes of the shunt parasitic capacitor $C_p$ representing the drain-bulk capacitance of the device $M_1$ and gate-source and source-bulk capacitances of the device $M_2$. This results in a nonzero switching time of the common-gate device $M_2$ when it cannot be instantly switched from the saturation mode to the pinch-off mode and operates in the active region, when output current and output voltage are simultaneously positive, causing power dissipation within the device. The parasitic capacitance $C_p$ can be three to four times larger than the drain-bulk capacitance of device $M_2$, resulting in a power loss as large as 20% of the output power.

A simple and effective way to minimize this power-loss contribution is to use a parallel inductor $L_p$ resonating the parasitic capacitor $C_p$ at the operating frequency, as shown in Fig. 6.27(b), where $C_b$ is the dc-blocking capacitor. The series-resonant circuit required to provide a sinusoidal current flowing to the load is replaced by the series inductor $L_m$ and shunt capacitor $C_m$ forming an L-type lumped impedance transformer to match the nominal or optimized Class-E load resistance with a standard 50-$\Omega$ load. As a result, a two-stage cascode Class-E power amplifier with a compensating inductor implemented in a 0.13-$\mu$m CMOS technology achieved a drain efficiency of 71%, a *PAE* of 67%, and an output power of 23 dBm at an operating frequency of 1.7 GHz with a dc-supply voltage of 2.5 V. The driving stage with a supply voltage of 1.2 V is biased in Class C. The value of the DC-Feed inductor $L_d$ is chosen to compensate for the gate-source capacitance of the device $M_1$. The measured power-added efficiency was higher than 60% over the frequency bandwidth of $1.4 \div 2.0$ GHz.

In a cascode configuration, the voltage across each transistor reduces twice with the same current, thus resulting in a higher overall saturation resistance and power loss. To overcome this problem, a common-gate Class-E power amplifier, the equivalent circuit of which is shown in Fig. 6.28, can be used [43]. Here, the input signal is directly applied to the device source, and power losses can be reduced for the same voltage across the device by choosing an inductor with a high-$Q$ factor. Since CMOS monolithic inductors are known for their low-quality factors due to high substrate losses and high parasitics, a bondwire inductor can be used to maximize the inductor quality factor. However, these bondwire inductors must be modeled very accurately for a particular design since their geometry is difficult to predetermine. For a two-stage power amplifier, it is very important to minimize the power consumption of the driving stage to minimize its effect of reducing the overall efficiency. In this case, the driver stage can include a positive feedback to operate as an injection-locked oscillator to maximize a power gain. When using a differential structure of a CMOS Class-E power amplifier, a driver stage can be designed by utilizing a cross-coupled differential pair to form a positive feedback. As a result, such a two-stage differential 2.45-GHz Class-E power

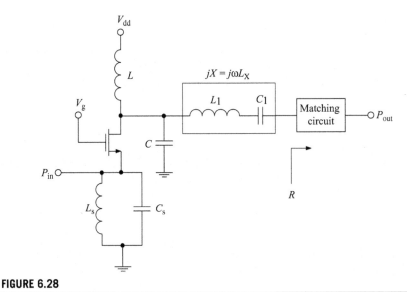

**FIGURE 6.28**

Common-gate Class-E power amplifier.

amplifier with finite DC-Feed inductance, implemented in a 0.25-μm standard CMOS process with chip area of $0.8 \times 0.9$ mm$^2$, achieved a *PAE* of 34.5% and an output power of 18 dBm (63 mW) at a supply voltage of 1 V. The power-added efficiency was at least 33% over the frequency range of 2.4-2.48 GHz.

For a given geometry of a CMOS device, there is an optimum output power which provides the maximum power-added efficiency. Maximizing the output power by lowering an optimum load resistance will result in a higher current flowing through the saturation resistance, increasing the power dissipation. However, lowering the output power by increasing the optimum load resistance will decrease the power-added efficiency because of a reduced power gain. Besides, there is an optimum device size that provides the maximum power-added efficiency. A larger transistor size will lower the saturation resistance, but will require more power from the driver stage and consumes more silicon area. In this case, it is more difficult to provide the input matching due to increasing gate-source capacitance and lower gate resistance with a sufficiently small DC-Feed inductance, thus resulting in implementation difficulties [44]. For example, for a simulated 1-GHz CMOS Class-E power amplifier fabricated using a 0.35-μm standard CMOS technology, a maximum *PAE* of about 62% is achieved for device gate widths of 6000−8000 μm with an output power of about 900 mW, a DC-Feed inductance of 2 nH, and a load resistance of 2 Ω [45].

Figure 6.29 shows the circuit schematic of a two-stage fully differential Class-E power amplifier designed in a 0.18-μm standard CMOS technology [46]. The driver stage is used to convert the sinusoidal input signal into its square-wave

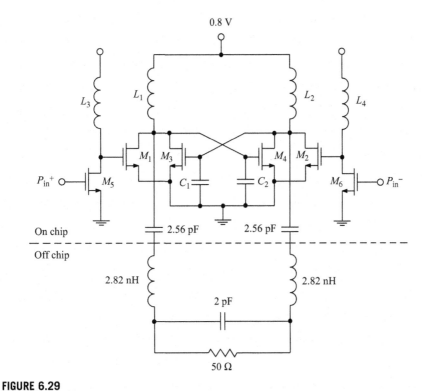

**FIGURE 6.29**

Two-stage injection-locked Class-E power amplifier.

approximation to drive the output stage. In some cases, the interstage circuit can contain parallel-tank circuits tuned to the fundamental-frequency and third-harmonic components, realizing a Class-F mode of the driving stage and making the driving voltage waveform closer to a square wave, thus minimizing the power loss due to nonzero switching time [47,48]. To obtain high operating efficiency, bondwires are used to implement the inductors $L_1$ and $L_2$, while the on-chip spiral inductors $L_3$ and $L_4$ are used in the driver stage for a better matching. The nMOS transistors $M_3$ and $M_4$ in the output stage are cross-coupled to assist the operation of the nMOS transistors $M_1$ and $M_2$, respectively, thus resulting in a significant reduction of the input-driving requirement due to injection-locking operating mode. The sizes of the nMOS devices were optimized to compromise the efficiency and output power. The shunt capacitors $C_1$ and $C_2$ can be used if the values of the gate-source capacitances are not enough to get the self-oscillations at the operating frequency to realize an injection-locking mode. As a result of an accurate simulation and optimization of design procedure, such a two-stage Class-E power amplifier was able to deliver an output power of 84.8 mW with a drain efficiency of 69.1% at an operating frequency of 2.4 GHz. The results of practical implementation of a two-stage injection-locked differential CMOS power amplifier in a 0.25-μm standard CMOS technology demonstrated the possibility of achieving an output power of 26 dBm with a *PAE* of 62% at a carrier frequency of 1 GHz, using a dc power supply of 1 V [49].

As it is very important to reduce the voltage peak factor of a switchmode Class-E power amplifier in general, and for CMOS power amplifiers with limited breakdown voltage in particular, the innovative design technique where the parallel *LC* circuit tuned to the second harmonic, and connected in series to the load can be used [50]. In this case, by obtaining a flat-top transistor-voltage waveform, its peak value reduces to 81% of a conventional Class-E with shunt capacitance.

## REFERENCES

1. Sokal NO. Class E high-efficiency switching-mode tuned power amplifier with only one inductor and one capacitor in load network — approximate analysis. *IEEE J. Solid-State Circuits*. August 1981;SC-16:380−384.
2. Kazimierczuk M. Exact analysis of Class E tuned power amplifier with only one inductor and one capacitor in load network. *IEEE J. Solid-State Circuits*. April 1983; SC-18:214−221.
3. Popov IA, ed. *Transistor Generators of Harmonic Oscillations in Switching Mode* (in Russian). Moskva: Radio i Svyaz; 1985.
4. Degtev VI. Switching-mode power amplifier with forming circuit (in Russian). *Radiotekhnika*. August 1985;40:77−80.
5. Gruzdev VV. Calculation of circuit parameters of single-ended switching-mode power amplifier (in Russian). *Trudy MEIS*. 1969;2:124−128.
6. Kazimierczuk M. Class E tuned power amplifier with nonsinusoidal output voltage. *IEEE J. Solid-State Circuits*. August 1986;SC-21:575−581.
7. Gaudio PM, Bernal C, Mediano A. Exact analysis of a simple Class E circuit version for device characterization. *2003 IEEE MTT-S Int. Microwave Sym. Dig.* 3:1737−1740.
8. Zulinski RE, Steadman JW. Class E power amplifiers and frequency multipliers with finite DC-Feed inductance. *IEEE Trans. Circuits and Systems*. September 1987;CAS-34:1074−1087.
9. Avratoglou CP, Voulgaris NC, Ioannidou FI. Analysis and design of a generalized Class E tuned power amplifier. *IEEE Trans. Circuits and Systems*. August 1989;CAS-36:1068−1079.
10. Smith GH, Zulinski RE. An exact analysis of Class E power amplifiers with finite DC-Feed inductance at any output *Q*. *IEEE Trans. Circuits and Systems*. April 1990; CAS-37:530−534.
11. Sekiya H, Sasase I, Mori S. Computation of design values for Class E amplifiers without using waveform equations. *IEEE Trans. Circuits and Systems − I: Fundamental Theory Appl.* July 2002;CAS-I-49:966−978.
12. Li C-H, Yam Y-O. Maximum frequency and optimum performance of Class E power amplifiers. *IEE Proc. Circuits Devices Syst.* June 1994;141:174−184.
13. Ho CK, Wong H, Ma SW. Approximation of non-zero transistor ON resistance in Class-E amplifiers. *Proc. 5th IEEE Int. Caracas Conf. Devices Circuits Syst.* 2004:90−93.

14. Milosevic D, van der Tang J, van Roermund A. Explicit design equations for Class-E power amplifiers with small DC-Feed inductance. *Proc. 2005 Europ. Conf. Circuit Theory and Design* 3:101−104.

15. Grebennikov A. Load network design techniques for Class E RF and microwave amplifiers. *High Frequency Electronics.* July 2004;3:18−32.

16. Acar M, Annema AJ, Nauta B. Analytical design equations for Class-E power amplifiers. *IEEE Trans. Circuits and Systems − I: Regular Papers.* December 2007;CAS-I-54:2706−2717.

17. Kozyrev VB. Single-ended switching-mode power amplifier with filtering circuit (in Russian). *Poluprovodnikovye Pribory v Tekhnike Svyazi.* 1971;8:152−166.

18. Grebennikov A. Class E high-efficiency power amplifiers: historical aspect and future prospect. *Applied Microwave & Wireless.* July 2002;14:64−71:pp. 64−72, August 2002.

19. Grebennikov AV, Jaeger H. Class E with parallel circuit − A new challenge for high-efficiency RF and microwave power amplifiers. *2002 IEEE MTT-S Int. Microwave Sym. Dig.* 3:1627−1630.

20. Grebennikov A. Switched-mode RF and microwave parallel-circuit Class E power amplifiers. *Int. J. RF and Microwave Computer-Aided Eng.* January/February 2004;14:21−35.

21. Kazimierczuk MK, Tabisz WA. Class C-E high-efficiency tuned power amplifier. *IEEE Trans. Circuits and Systems.* March 1989;CAS-36:421−428.

22. Heiskanen A, Rahkonen T. Comparison of two Class-E amplifiers for EER transmitter. *2005 IEEE Int. Circuits and Systems Symp. Dig.* 704−707.

23. Xu Y, Zhu X, You C. Analysis and design of Class-E power amplifier using equivalent LDMOS model with drift region effect. *Microwave and Optical Technology Lett.* August 2010;52:1836−1842.

24. Iwadare M, Mori S, Ikeda K. Even harmonic resonant Class E tuned power amplifier without RF choke. *Electronics and Communications in Japan.* January 1996;79:23−30.

25. Choi DK, Long SI. Finite DC feed inductor in Class E power amplifiers − a simplified approach. *2002 IEEE MTT-S Int. Microwave Symp. Dig.* 3:1643−1646.

26. Zhang N, Yam Y-O, Gao B, Cheung C-W. A new type high frequency Class E power amplifier. *Proc. 1997 Asia Pacific Microwave Conf.* 3:1117−1120.

27. Yam Y-O, Cheung C-W, Li C-H. Transmission-line modelling of the Class E amplifier with an analytical method. *Int. J. Numerical Modelling: Electronic Networks, Devices and Fields.* November 1995;8:357−366.

28. Mader TB, Bryerton EW, Marcovic M, Forman M, Popovic Z. Switched-mode high-efficiency microwave power amplifiers in a free-space power-combiner array. *IEEE Trans. Microwave Theory Tech.* October 1998;MTT-46:1391−1398.

29. Grebennikov AV, Jaeger H. High efficiency transmission line tuned power amplifier. U.S. Patent 6,552,610, April 2003.

30. Grebennikov AV. *RF and Microwave Power Amplifier Design.* New York: McGraw-Hill; 2004.

31. Jaeger H, Grebennikov AV, Heaney EP, Weigel R. Broadband high-efficiency monolithic InGaP/GaAs HBT power amplifiers for 3G handset applications. *2002 IEEE MTT-S Int. Microwave Symp. Dig.* 2:1035−1038.

32. Jaeger H, Grebennikov A, Heaney E, Weigel R. Broadband high-efficiency monolithic InGaP/GaAs HBT power amplifiers for wireless applications. *Int. J. RF and Microwave Computer-Aided Eng.* November/December 2003;13:496−510.

33. Cumana J, Grebennikov A, Sun G, Kumar N, Jansen RH. An extended topology of parallel-circuit Class-E power amplifier to account for larger output capacitances. *IEEE Trans. Microwave Theory Tech.* December 2011;MTT-59:3174−3183.

34. Grebennikov AV. Circuit design technique for high efficiency Class F amplifiers. *2000 IEEE MTT-S Int. Microwave Symp. Dig.* 2:771−774.

35. Popov IA. Switching mode of single-ended transistor power amplifier (in Russian). *Poluprovodnikovye Pribory v Tekhnike Svyazi.* 1970;5:15−35.

36. Kazimierczuk M. Charge-control analysis of Class E tuned power amplifier with only one inductor and one capacitor in load network. *IEEE Trans. Electron Devices.* March 1984;ED-31:366−373.

37. Krauss HL, Bostian CW, Raab FH. *Solid State Radio Engineering.* New York: John Wiley & Sons; 1980.

38. Modzelewski J. Power gain of Class E and Class B VHF tuned power amplifiers. *Proc. 15th Int. Conf. Microwaves Radar Wireless Commun.* 2004;1:41−44.

39. Yoo C, Huang Q. A common-gate switched 0.9-W Class-E power amplifier with 41% PAE in 0.25-μm CMOS. *IEEE J. Solid-State Circuits.* May 2001;SC-36:823−830.

40. Tsai K-C, Gray PR. A 1.9-GHz, 1-W CMOS Class-E power amplifier for wireless communications. *IEEE J. Solid-State Circuits.* July 1999;SC-34:962−970.

41. Lee O, An KH, Kim H, et al. Analysis and design of fully integrated high-power parallel-circuit Class-E CMOS power amplifiers. *IEEE Trans. Circuits and Systems − I: Regular Papers.* March 2010;CAS-I-57:725−734.

42. Mazzanti A, Larcher L, Brama R, Svelto F. Analysis of reliability and power efficiency in cascode Class-E PAs. *IEEE J. Solid-State Circuits.* May 2006;SC-41:1222−1229.

43. Ho K-W, Luong HC. A 1-V CMOS power amplifier for Bluetooth applications. *IEEE Trans. Circuits and Systems − II: Analog and Digital Signal Process.* August 2003;CAS-II-50:445−449.

44. Wang C, Larson LE, Asbeck PM. Improved design technique of a microwave Class-E power amplifier with finite switching-on resistance. *Proc. 2002 IEEE Radio and Wireless Conf.* 241−244.

45. Reynaert P, Mertens KLR, Steyaert MSJ. A state-space behavioral model for CMOS Class E power amplifiers. *IEEE Trans. Computer-Aided Design Integrated Circuits Syst.* February 2003;CAD-22:132−138.

46. Xu Z, El-Masry EI. Design and optimization of CMOS Class-E power amplifier. *2003 IEEE Int. Circuits and Systems Symp. Dig.* 1:325−328.

47. Sowlati T, Andre C, Salama T, Sitch J, Rabjohn G, Smith D. Low voltage, high efficiency GaAs Class E power amplifiers for wireless transmitters. *IEEE Trans. Solid-State Circuits.* October 1995;SC-30:1074−1080.

48. Sowlati T, Greshishchev Y, Andre C, Salama T. 1.8 GHz Class E power amplifier for wireless communications. *Electronics Lett.* September 1996;32:1846−1848.

49. Hung TT, El-Gamal MN. Class-E CMOS power amplifiers for RF applications. *2003 IEEE Int. Circuits and Systems Symp. Dig.* 1:449−452.

50. Mediano A, Sokal NO. Class-E RF power amplifier with a flat-top transistor-voltage waveform. *2012 IEEE MTT-S Int. Microwave Symp. Dig.*

# Class-E with Quarterwave Transmission Line

This chapter presents the results of exact time-domain analysis of switchmode Class-E power amplifiers with a quarterwave transmission line. The load network parameters are derived analytically. The ideal collector voltage and current waveforms demonstrate a possibility of 100% efficiency without overlapping each other. The load network implementation, including output matching circuit at RF and microwave frequencies using lumped and transmission-line elements, is considered with the accurate derivation of the matching circuit parameters. The switchmode Class-E power amplifiers with a quarterwave transmission line offer a new possibility for RF and microwave power amplification to provide the high-efficiency and harmonic-suppression operation conditions simultaneously.

## 7.1 Load network with parallel quarterwave line

The ideal Class-F load network with a quarterwave transmission line and a series $L_0C_0$ filter tuned to the fundamental frequency can provide a collector efficiency of 100% when the open-circuit conditions for odd harmonic components and short-circuit conditions for even harmonic components are realized. However, in practice, the idealized collector rectangular voltage and half-sinusoidal current waveforms corresponding to a Class-F operation mode provided by using a quarterwave transmission line in the load network can be realized at sufficiently low frequencies when an effect of the device output capacitance shown in Fig. 7.1 is negligibly small. Generally, the effect of the device output capacitance contributes

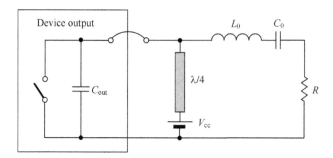

**FIGURE 7.1**

Class-F load network with parasitic shunt capacitance and bondwire inductance.

to a finite switching time resulting in time periods when the collector voltage and collector current exist at the same time. As a result, such a load network with a quarterwave transmission line and a shunt capacitor cannot provide the switch-mode operation with an instantaneous transition from the device pinch-off mode to saturation mode. Hence, during a finite time interval, the active device operates in the active region as a current source with a reverse-biased collector-base junction, and the collector current is provided by this current source. In this case, the required optimum conditions can be provided for only the fundamental frequency and several higher-order harmonic components. Moreover, it is necessary to take into account, at higher frequencies, the effects of the bondwire inductor connecting physically the device and with the off-chip board. In a packaged device, it also necessary to consider the effect of the external lead inductance. Therefore, a special case is a load network with a quarterwave transmission line when the loss-less transformation of a dc power to a fundamental-frequency output power can be provided. Fortunately, as it will be further demonstrated, the collector efficiency can be increased and the effect of the collector capacitance can be compensated with an inclusion of a series inductors between the shunt capacitor and the quarterwave transmission line realizing the switching Class-E operation conditions. The obvious advantage of such a load network is a combination of high operating efficiency corresponding to a Class-E mode and an even-harmonic suppression due to a quarterwave transmission line used in a Class-F mode.

The possibility to include a quarterwave transmission line into the Class-E load network with a shunt capacitance instead of an RF choke was first considered in 1975 [1]. However, such a location for a quarterwave transmission line with a straight connection to the device collector violates the required capacitive reactance conditions at even harmonics by short-circuiting them. As a result, an optimum Class-E operation mode cannot be realized when the shapes of the collector current and voltage waveforms provide a condition at which the high current and high voltage do not overlap simultaneously. Moreover, the

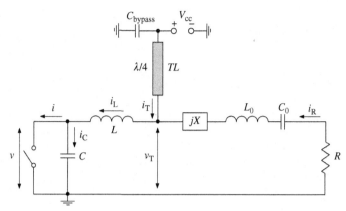

**FIGURE 7.2**

Equivalent circuit of a Class-E power amplifier with quarterwave transmission line.

larger the value of the shunt capacitance, the smaller the collector efficiency that can be achieved.

Figure 7.2 shows the Class-E load network consisting of a shunt capacitor $C$, a series inductor $L$, a quarterwave transmission line $TL$, a series reactance $X$, a series resonant $L_0C_0$ circuit tuned to the fundamental frequency, and a load resistor $R$. The bottom end of the quarterwave transmission line is connected between the series inductor $L$ and the series reactance $X$, while its top end is connected to the dc power supply which is RF grounded through the bypass capacitor. In a common case, a shunt capacitance $C$ can represent the intrinsic device output capacitance and external circuit capacitance added by the load network. The series reactance $X$ generally can be positive (inductance), negative (capacitance), or zero depending on the values of the shunt capacitance $C$ and series inductance $L$. The active device is considered an ideal switch that is driven so as to provide the device switching between its on-state and off-state operating conditions. As a result, the collector voltage waveform is determined by the switch when it is turned on and by the transient response of the load network when the switch is turned off.

To simplify an analysis of a Class-E power amplifier with a quarterwave transmission line, the following assumptions are introduced:

- The transistor has zero saturation voltage, zero saturation resistance, infinite off-resistance, and its switching action is instantaneous and lossless.
- The total shunt capacitance is independent of the collector and is assumed linear.
- The loaded quality factor $Q_L = \omega L_0/R = 1/\omega C_0 R$ of the series-resonant $L_0C_0$ circuit tuned to the fundamental frequency is high enough that the output current is sinusoidal at the switching frequency.

- There are no losses in the circuit except in the load $R$.
- For an optimum switching operation mode, a 50% duty cycle is used.

Let the output current flowing into the load be sinusoidal,

$$i_R(\omega t) = I_R \sin(\omega t + \varphi) \tag{7.1}$$

where $\varphi$ is the initial phase shift due to the shunt capacitance and series inductance.

For a lossless operation, it is necessary to provide the following optimum conditions for voltage across the switch at the turn-on instant of $\omega t = 2\pi$, when the transistor is voltage saturated:

$$v(\omega t)|_{\omega t = 2\pi} = 0 \tag{7.2}$$

$$\frac{dv(\omega t)}{d\omega t}\bigg|_{\omega t = 2\pi} = 0 \tag{7.3}$$

where $v$ is the voltage across the switch.

When the switch is turned on for $0 \le \omega t \le \pi$, the current flowing through the shunt capacitance $i_C(\omega t) = 0$ and, consequently,

$$i(\omega t) = i_L(\omega t) = i_T(\omega t) + i_R(\omega t). \tag{7.4}$$

When the switch is turned off for $\pi \le \omega t \le 2\pi$, there is no current flowing through the switch when $i(\omega t + \pi) = 0$, and the current flowing through the shunt capacitance $C$ is

$$i_C(\omega t + \pi) = i_L(\omega t + \pi) = i_T(\omega t + \pi) + i_R(\omega t + \pi). \tag{7.5}$$

To link both Eqs (7.4) and (7.5), each corresponding to one-half period, it is necessary to use a basic equation for the current flowing into the quarterwave transmission line given by Eq. (3.70) as

$$i_T(\omega t) = i_T(\omega t + \pi) \tag{7.6}$$

which means that the period of a signal flowing into the quarterwave transmission line is equal to $\pi$ because it contains only even harmonics.

Then,

$$i_L(\omega t) = i_T(\omega t + \pi) + i_R(\omega t) = i_L(\omega t + \pi) + i_R(\omega t) - i_R(\omega t + \pi) \tag{7.7}$$

resulting in

$$i_L(\omega t) - i_L(\omega t + \pi) = 2i_R(\omega t). \tag{7.8}$$

The current $i_L(\omega t + \pi) = i_C(\omega t + \pi)$ can be expressed through the voltages $v_T(\omega t + \pi)$ and

$$v_L(\omega t + \pi) = \omega L \frac{di_L(\omega t + \pi)}{d\omega t} \tag{7.9}$$

as

$$i_L(\omega t + \pi) = \omega C \frac{dv(\omega t + \pi)}{d\omega t} = \omega C \frac{d}{d\omega t}\left[v_T(\omega t + \pi) - \omega L \frac{di_L(\omega t + \pi)}{d\omega t}\right]. \quad (7.10)$$

Now we can use the equation for a voltage at the input of the quarterwave transmission line corresponding to each of a period given by Eq. (3.69) as

$$v_T(\omega t) = 2V_{cc} - v_T(\omega t + \pi). \quad (7.11)$$

Hence, when the switch is turned on resulting in $v_T(\omega t) = v_L(\omega t)$,

$$v_T(\omega t + \pi) = 2V_{cc} - \omega L \frac{di_L(\omega t)}{d(\omega t)}. \quad (7.12)$$

Substituting Eq. (7.12) in Eq. (7.10) and using Eqs (7.1) and (7.8) yields a second-order nonhomogeneous differential equation, corresponding to half a period of $\pi \leq \omega t < 2\pi$ when $i_L(\omega t + \pi) = i_C(\omega t + \pi)$, in the form

$$\frac{d^2 i_C(\omega t + \pi)}{d(\omega t)^2} + \frac{q^2}{2} i_C(\omega t + \pi) - I_R \sin(\omega t + \varphi) = 0 \quad (7.13)$$

or

$$\frac{d^2 i_C(\omega t)}{d(\omega t)^2} + \frac{q^2}{2} i_C(\omega t) + I_R \sin(\omega t + \varphi) = 0 \quad (7.14)$$

where

$$q = \frac{1}{\omega\sqrt{LC}}. \quad (7.15)$$

The general solution of Eq. (7.14) in the normalized form can be written as

$$\frac{i_C(\omega t)}{I_R} = C_1 \cos\left(\frac{q\omega t}{\sqrt{2}}\right) + C_2 \sin\left(\frac{q\omega t}{\sqrt{2}}\right) + \frac{2}{2 - q^2} \sin(\omega t + \varphi) \quad (7.16)$$

where the coefficients $C_1$ and $C_2$ are determined from the initial off-state conditions.

The first initial condition is obtained from Eq. (7.8) as

$$i_C(\omega t)\big|_{\omega t = \pi} = 2i_R(\pi) \quad (7.17)$$

taking into account that $i_L(\pi) = i_C(\pi)$ and $i_L(2\pi) = i_C(2\pi) = 0$.

To obtain the second initial condition, let us substitute Eq. (7.9) in Eq. (7.12) and use Eq. (7.8). As a result,

$$\omega L \frac{di_L(\omega t + \pi)}{d\omega t} = 2V_{cc} - \omega L \frac{di_L(\omega t)}{d\omega t}$$

$$= 2V_{cc} - \omega L \frac{di_L(\omega t + \pi)}{d\omega t} - 2\omega L \frac{di_R(\omega t)}{d\omega t}. \quad (7.18)$$

Then, by taking into account that $i_L(\pi) = i_C(\pi)$, we can write

$$\frac{di_C(\omega t)}{d\omega t}\bigg|_{\omega t=\pi} = \frac{V_{cc}}{\omega L} - I_R \cos \varphi. \qquad (7.19)$$

As a result, applying the initial conditions given by Eqs (7.17) and (7.19) to Eq. (7.16) yields

$$C_1 = -\frac{\sqrt{2}}{qp}\sin\left(\frac{q\pi}{\sqrt{2}}\right) - \frac{q\sqrt{2}}{2-q^2}\sin\left(\frac{q\pi}{\sqrt{2}}\right)\cos\varphi - 2\frac{1-q^2}{2-q^2}\cos\left(\frac{q\pi}{\sqrt{2}}\right)\sin\varphi \quad (7.20)$$

$$C_2 = \frac{\sqrt{2}}{qp}\cos\left(\frac{q\pi}{\sqrt{2}}\right) + \frac{q\sqrt{2}}{2-q^2}\cos\left(\frac{q\pi}{\sqrt{2}}\right)\cos\varphi - 2\frac{1-q^2}{2-q^2}\sin\left(\frac{q\pi}{\sqrt{2}}\right)\sin\varphi \quad (7.21)$$

where

$$p = \frac{\omega L I_R}{V_{cc}}. \qquad (7.22)$$

The dc supply current $I_0$ can be written using the Fourier formula and Eqs (7.5) and (7.6) as

$$I_0 = \frac{1}{2\pi}\int_0^{2\pi} i_T(\omega t)d\omega t = \frac{1}{\pi}\int_0^{\pi} i_T(\omega t + \pi)d\omega t$$

$$= \frac{1}{\pi}\int_0^{\pi} [i_C(\omega t + \pi) - i_R(\omega t + \pi)]d\omega t. \qquad (7.23)$$

Then, substituting Eqs (7.1) and (7.16) in Eq. (7.23) results in

$$I_0 = \frac{I_R}{\pi}\left\{ C_1\frac{\sqrt{2}}{q}\left[\sin\left(q\pi\sqrt{2}\right) - \sin\left(\frac{q\pi}{\sqrt{2}}\right)\right] \right.$$

$$\left. - C_2\frac{\sqrt{2}}{q}\left[\cos\left(q\pi\sqrt{2}\right) - \cos\left(\frac{q\pi}{\sqrt{2}}\right)\right] - \frac{2q^2}{2-q^2}\cos\varphi \right\}. \qquad (7.24)$$

The voltage $v(\omega t)$ across the switch is produced by charging of the shunt capacitor $C$ by the current given by Eq. (7.16) according to

$$v(\omega t) = \frac{1}{\omega C}\int_\pi^{\omega t} i_C(\omega t)d\omega t = \frac{I_R\sqrt{2}}{q\omega C}\left\{ C_1\left[\sin\left(\frac{q\omega t}{\sqrt{2}}\right) - \sin\left(\frac{q\pi}{\sqrt{2}}\right)\right] \right.$$

$$\left. - C_2\left[\cos\left(\frac{q\omega t}{\sqrt{2}}\right) - \cos\left(\frac{q\pi}{\sqrt{2}}\right)\right] - \frac{q\sqrt{2}}{2-q^2}\left[\cos(\omega t + \varphi) + \cos\varphi\right] \right\}. \qquad (7.25)$$

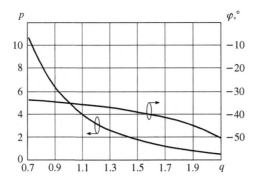

**FIGURE 7.3**

Optimum quarterwave-line Class-E parameters $p$ and $\varphi$ versus $q$.

Generally, Eq. (7.25) for the collector voltage contains the three unknown parameters $q$, $p$, and $\varphi$, which must be determined. In a common case, the parameter $q$ can be considered a variable, and the other two parameters $p$ and $\varphi$ are determined from a system of the two equations resulting from applying the two optimum zero-voltage and zero voltage-derivative conditions given by Eqs (7.2) and (7.3) to Eq. (7.25). Figure 7.3 shows the dependences of the optimum parameters $p$ and $\varphi$ versus $q$ for Class-E with a parallel quarterwave transmission line.

For the boundary cases when $q = 1/\sqrt{2}$ and $q = 3/\sqrt{2}$, the optimum parameters $p$ and $\varphi$ can easily be defined analytically, the exact values of which are

$$\varphi = \tan^{-1}\left(-\frac{2}{3}\right) = -33.69° \tag{7.26}$$

$$p = -\frac{6}{\sin\varphi} = 10.82 \tag{7.27}$$

for $q = 1/\sqrt{2} = 0.707$ and

$$\varphi = \tan^{-1}\left(-\frac{6}{5}\right) = -50.2° \tag{7.28}$$

$$p = -\frac{10}{27\sin\varphi} = 0.482 \tag{7.29}$$

for $q = 3/\sqrt{2} = 2.12$.

The current $i(\omega t) = i_L(\omega t)$ flowing through the inductor $L$ and the switch when the switch is turned on for $0 \le \omega t \le \pi$ can be written using Eqs (7.4) to (7.6) and (7.16) in a normalized form as

$$\frac{i(\omega t)}{I_R} = C_1 \cos\left[\frac{q}{\sqrt{2}}(\omega t + \pi)\right] + C_2 \sin\left[\frac{q}{\sqrt{2}}(\omega t + \pi)\right] + 2\frac{1 - q^2}{2 - q^2}\sin(\omega t + \varphi). \tag{7.30}$$

Then, the normalized voltage $v_T(\omega t)$ during this period can be obtained by

$$\frac{v_T(\omega t)}{V_{cc}} = \frac{\omega L I_R}{V_{cc}} \frac{di_L(\omega t)}{d\omega t} = p\left\{-C_1 \frac{q}{\sqrt{2}} \sin\left[\frac{q}{\sqrt{2}}(\omega t + \pi)\right]\right.$$

$$\left. + C_2 \frac{q}{\sqrt{2}} \cos\left[\frac{q}{\sqrt{2}}(\omega t + \pi)\right] + 2\frac{1 - q^2}{2 - q^2} \cos(\omega t + \varphi)\right\}. \tag{7.31}$$

However, when the switch is turned off for $\pi \leq \omega t \leq 2\pi$, it can be calculated from Eq. (7.11).

## 7.2 Optimum load-network parameters

To calculate the optimum load-network parameters for a Class-E power amplifier with a parallel quarterwave transmission line, first it is necessary to define the reactive quadrature fundamental-frequency Fourier component of the voltage $v_T(\omega t)$ according to

$$V_X = -\frac{1}{\pi}\int_0^\pi v_T(\omega t)\cos(\omega t + \varphi)d\omega t - \frac{1}{\pi}\int_\pi^{2\pi} [2V_{cc} - v_T(\omega t - \pi)]\cos(\omega t + \varphi)d\omega t.$$

$$\tag{7.32}$$

Then, the two real and imaginary quadrature fundamental-frequency Fourier components of the collector voltage $v(\omega t)$ are defined from

$$V_R = -\frac{1}{\pi}\int_0^{2\pi} v(\omega t)\sin(\omega t + \varphi)d\omega t \tag{7.33}$$

$$V_{L+X} = -\frac{1}{\pi}\int_0^{2\pi} v(\omega t)\cos(\omega t + \varphi)d\omega t. \tag{7.34}$$

Finally, as follows from Fig. 7.4, the optimum normalized series reactance $X$, series inductance $L$, and shunt capacitance $C$ can be calculated from

$$\frac{X}{R} = \frac{V_X}{V_R} \tag{7.35}$$

$$\frac{\omega L}{R} = \frac{V_{L+X}}{V_R} - \frac{X}{R} \tag{7.36}$$

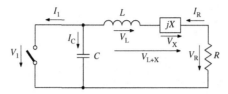

**FIGURE 7.4**

Equivalent quarterwave-line Class-E load network at fundamental frequency.

$$\omega CR = \frac{1}{q^2 \frac{\omega L}{R}}. \tag{7.37}$$

By taking into account that $R = V_R^2/2P_{out}$, the optimum load resistance $R$ for the specified values of a supply voltage $V_{cc}$ and an output power $P_{out}$ delivered to the load can be obtained by

$$R = \frac{1}{2}\left(\frac{V_R}{V_{cc}}\right)^2 \frac{V_{cc}^2}{P_{out}} \tag{7.38}$$

or, using Eq. (7.22), by

$$R = \frac{1}{2}\left(\frac{p}{l}\right)^2 \frac{V_{cc}^2}{P_{out}} \tag{7.39}$$

where $l = \omega L/R$.

The dependences of the normalized optimum series inductance $\omega L/R$ and series reactance $X/R$ are shown in Fig. 7.5(a), while the dependences of the normalized optimum shunt capacitance $\omega CR$ and load resistance $RP_{out}/V_{cc}^2$ are plotted in Fig. 7.5(b). Here, it is clearly seen that greater value of the normalized series inductance $\omega L/R$ corresponds to lower value of $q$. To compensate for such an increased inductive value, the reactance $X$ should have a negative capacitive value. Generally, the value of the series reactance $X$ changes its sign from negative to positive, which means that the capacitive reactance is followed by the inductive reactance, and it is required to add an additional inductance at higher values of $q$. As a result, there is a special case of a load network with $X = 0$ when there is no need for additional phase compensation. The variations of the normalized values of $\omega CR$ and $RP_{out}/V_{cc}^2$ versus $q$ are not so significant.

Figure 7.6 shows the circuit schematics of the Class-E power amplifiers with a parallel quarterwave transmission line corresponding to different values of the optimum parameter $q$. The inclusion of a series capacitance $C_x$ is necessary to compensate for the excess inductive reactance at the fundamental frequency when the series reactance $X$ is negative, as shown in Fig. 7.6(a). However, when the

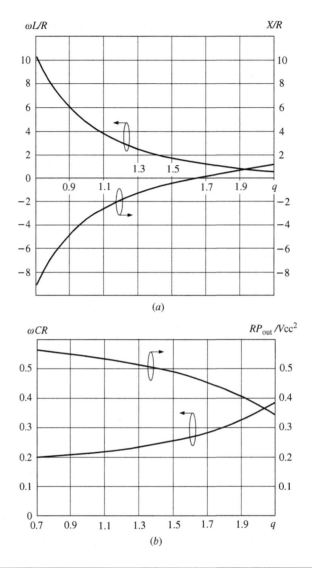

**FIGURE 7.5**

Normalized optimum quarterwave-line Class-E load network parameters.

series reactance $X$ is positive, an additional series inductance $L_X$ shown in Fig. 7.6(c) is necessary to increase the total series inductance at the fundamental frequency. The special case is a zero series reactance $X$ corresponding to the circuit schematic shown in Fig. 7.6(b), which we will consider later in more detail.

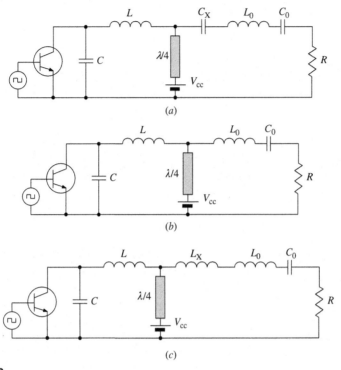

**FIGURE 7.6**

Schematic of Class-E power amplifiers with a quarterwave line.

## 7.3 **Load network with zero series reactance**

In this case, the Class-E load network consists of a shunt capacitor $C$, a series inductor $L$, a parallel quarterwave transmission line connected to the dc voltage supply, a series $L_0 C_0$ resonant circuit tuned to the fundamental frequency, and a load resistor $R$. Because the parameter $q$ corresponding to a zero reactance $X$ is unknown *a priori*, generally it is necessary to solve a system of three equations to define the three unknown parameters $q$, $p$, and $\varphi$. Two equations are the result of applying the two optimum zero-voltage and zero voltage-derivative conditions given by Eqs (7.2) and (7.3) to Eq. (7.25). Since the fundamental component of the voltage $v_T(\omega t)$ is fully applied to the load, this means that its reactive part must have zero value, resulting in an additional equation

$$V_X = -\frac{1}{\pi}\int_0^\pi v_T(\omega t)\cos(\omega t + \varphi)d\omega t - \frac{1}{\pi}\int_\pi^{2\pi}[2V_{cc} - v_T(\omega t - \pi)]\cos(\omega t + \varphi)d\omega t = 0.$$

$$(7.40)$$

As a result, the following exact values can be obtained numerically for the unknown parameters [2,3]:

$$q = 1.649 \tag{7.41}$$

$$p = 1.302 \tag{7.42}$$

$$\varphi = -40.8°. \tag{7.43}$$

Figure 7.7 shows the normalized (*a*) load current, (*b*) collector voltage, and (*c*) current waveforms for idealized optimum (or nominal) Class-E mode with a parallel quarterwave transmission line. From the collector voltage and current waveforms it follows that, when the transistor is turned on, there is no voltage across the switch and the collector current consisting of the load sinusoidal current shown in Fig. 7.7(*a*) and transmission-line current shown in Fig. 7.8(*a*) flows through the switch. However, when the transistor is turned off, this current with the waveform shown in Fig. 7.8(*b*) then flows through the shunt capacitor *C*, the charging process of which produces the collector voltage.

In an idealized Class-E operation mode, there is no nonzero voltage and current simultaneously that means a lack of power losses and gives an idealized collector efficiency of 100%. This implies that the dc power and the fundamental output power are equal,

$$I_0 V_{cc} = \frac{I_R V_R}{2} \tag{7.44}$$

where $V_R = I_R R$ is the fundamental voltage amplitude across the load resistor *R*.

By using Eqs (7.22) and (7.44), the normalized inductance can be defined as

$$\frac{\omega L}{R} = \frac{p}{2} \left( \frac{I_0}{I_R} \right)^{-1}. \tag{7.45}$$

As a result, the exact values of the optimum series inductance *L*, shunt capacitance *C*, and load resistance *R* can be calculated by using Eqs (7.24), (7.37), and (7.39) from

$$L = 1.349 \frac{R}{\omega} \tag{7.46}$$

$$C = \frac{0.2725}{\omega R} \tag{7.47}$$

$$R = 0.465 \frac{V_{cc}^2}{P_{out}}. \tag{7.48}$$

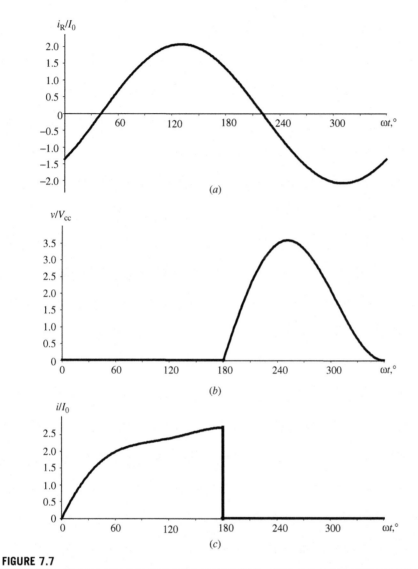

**FIGURE 7.7**

Voltage and current waveforms of a quarterwave-line Class-E power amplifier.

The peak collector current $I_{max}$ and peak collector voltage $V_{max}$ can be determined directly from Eqs (7.25) and (7.30) using Eq. (7.24) from numerical calculations that gives

$$I_{max} = 2.714I_0 \tag{7.49}$$

$$V_{max} = 3.589V_{cc}. \tag{7.50}$$

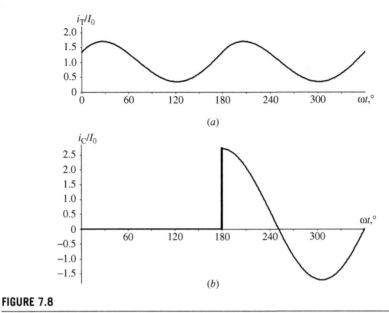

**FIGURE 7.8**

Current waveforms of a quarterwave-line Class-E power amplifier.

Using Eqs (7.47) and (7.48) when $C = C_{out}$, where $C_{out}$ is the device output capacitance, gives the value of a maximum operation frequency $f_{max}$ determined by

$$f_{max} = 0.093 \frac{P_{out}}{C_{out} V_{cc}^2} \tag{7.51}$$

which is 1.63 times as high as the maximum operation frequency for an optimum Class-E mode with shunt capacitance [4].

In Table 7.1, the optimum impedances seen by the device collector at the fundamental-frequency and higher-order harmonic components are illustrated by the appropriate circuit configurations. It can be seen that a Class-E mode with a quarterwave transmission line shows different impedance properties at even and odd harmonics. At odd harmonics, the optimum impedances can be established by the shunt capacitance that is required for all harmonic components in a Class-E with shunt capacitance. At even harmonics, the optimum impedances are realized by using a parallel $LC$ circuit that is required for all harmonic components in a Class-E with finite dc-feed inductance. Thus, the frequency properties of a grounded parallel quarterwave transmission line with its open-circuit conditions at odd harmonics and short-circuit conditions at even harmonics enable a Class-E with quarterwave transmission line to combine simultaneously

**Table 7.1** Optimum Impedances at Fundamental and Harmonics for Different Class-E Load Networks

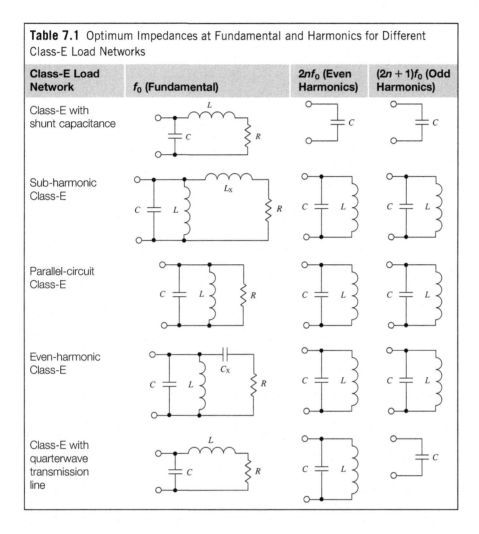

| Class-E Load Network | $f_0$ (Fundamental) | $2nf_0$ (Even Harmonics) | $(2n + 1)f_0$ (Odd Harmonics) |
|---|---|---|---|
| Class-E with shunt capacitance | | | |
| Sub-harmonic Class-E | | | |
| Parallel-circuit Class-E | | | |
| Even-harmonic Class-E | | | |
| Class-E with quarterwave transmission line | | | |

the harmonic impedance conditions typical for both Class-E with shunt capacitance and Class-E with finite dc-feed inductance.

## 7.4 Matching circuit with lumped elements

The theoretical results obtained for the Class-E power amplifier with a quarterwave transmission line show that it is enough to use a very simple load network to realize the optimum impedance conditions even for four harmonics. As follows from Fig. 7.8($a$), the current flowing into the quarterwave transmission line is very close to the sinusoidal second-harmonic current, which means that the level

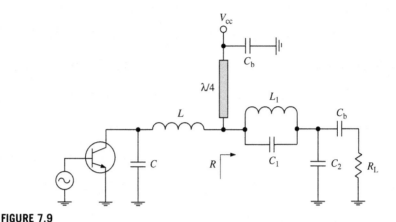

**FIGURE 7.9**

Schematic of a quarterwave-line Class-E power amplifier with lumped matching circuit.

of fourth- and higher-order harmonic components is negligible because of the significant shunting effect of the capacitance $C$. In this case, as the shunt capacitor $C$ and series inductor $L$ provide optimum inductive impedance at the fundamental frequency and the quarterwave transmission line realizes the shorting of even harmonics, it is only required to provide an open-circuit condition at the third harmonic component. Consequently, when the ideal series $L_0C_0$ circuit is replaced by the output matching circuit, the optimum impedance conditions for a Class-E load network with a quarterwave transmission line can be practically fully realized by simply providing an open-circuit condition at the third harmonic component.

Figure 7.9 shows the circuit schematic of a lumped Class-E power amplifier with a parallel quarterwave transmission line, where the parallel resonant $L_1C_1$ circuit tuned to the third harmonic component is used and $C_b$ represents the blocking or bypass capacitor. Since the reactance of the parallel third-harmonic tank circuit is inductive at the fundamental, it is enough to use the shunt capacitance $C_2$ composing the $L$-type low-pass matching circuit to provide the required impedance matching of the nominal Class-E load resistance $R$ with the standard load impedance of $R_L = 50\ \Omega$. In this case, it is assumed that $R < R_L$, which is normally the case for high-power or low-voltage power amplifiers.

To calculate the parameters of the matching-circuit elements, consider the loaded quality factor $Q_L = \omega C_2 R_L$, which also can be expressed through the resistances $R$ and $R_L$ as [3,5]

$$Q_L = \sqrt{\frac{R_L}{R} - 1}. \tag{7.52}$$

As a result, the matching-circuit parameters can be calculated at the fundamental frequency $f_0 = \omega_0/2\pi$ from

$$C_2 = \frac{Q_L}{\omega_0 R_L} \qquad (7.53)$$

$$L_1 = \frac{8}{9} \frac{Q_L R}{\omega_0} \qquad (7.54)$$

$$C_1 = \frac{1}{9\omega_0^2 L_1}. \qquad (7.55)$$

## 7.5 **Matching circuit with transmission lines**

At ultra-high frequencies and microwaves, the output lumped matching circuit is replaced by the transmission-line load network where the transmission lines are used instead of lumped inductors. In this case, the optimum Class-E inductance $L$ can be composed of the internal package bondwire inductance, external inductance of the device package lead, and inductance of the short-length transmission line required to properly connecting the transistor package to the transmission-line load-network fabricated on printed-circuit board (PCB). At the same time, the shunt capacitance $C$ represents a fully internal active-device output capacitance. Figure 7.10 shows the basic circuit schematic of a transmission-line Class-E power amplifier with a parallel quarterwave transmission line. The output matching circuit represents an $L$-type low-pass matching circuit consisting of a series transmission lines with electrical length $\theta_X$, which is the required for optimum Class-E mode to provide a positive reactance $X$ (can be replaced by a series capacitor if negative reactance $X$ is required), a series transmission line with electrical length $\theta_1$, which provides a required inductive reactance, and a shunt open-circuit stub with an electrical length $\theta_2$ of less than 90°, which provides a capacitive reactance.

Usually, the characteristic impedance $Z_1$ of a transmission line (often equal to 50 Ω) is much higher than the required optimum Class-E load network resistance $R$ for high-power or low-voltage applications. Consequently, the input impedance $Z_{in}$ of the loaded series transmission line with the characteristic impedance $Z_0$ and electrical length $\theta_X$ under the condition

$$\frac{R \tan \theta_X}{Z_0} \ll 1 \qquad (7.56)$$

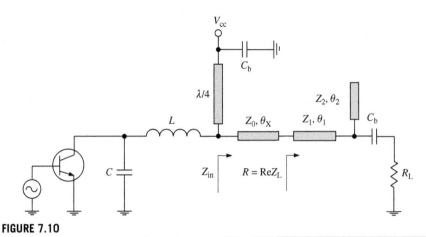

**FIGURE 7.10**

Schematic of a transmission-line Class-E power amplifier.

when the electrical length of a sufficiently short transmission line is much less than 45°, is determined by

$$Z_{in} = Z_0 \frac{R + jZ_0 \tan \theta_X}{Z_0 + jR \tan \theta_X} = Z_0 \frac{\frac{R}{Z_0} + j \tan \theta_X}{1 + j\frac{R}{Z_0} \tan \theta_X} \approx R + jZ_0 \tan \theta_X. \qquad (7.57)$$

As a result, the required optimum value of the electrical length $\theta_X$ for nominal Class-E mode with a quarterwave transmission line and series reactance using Eq. (7.57) can be obtained from $X = Z_0 \tan\theta_X$ as

$$\theta_X = \tan^{-1}\left(\frac{X}{Z_0}\right) \qquad (7.58)$$

where an electrical length $\theta_X$ can be positive or negative depending on the inductive or capacitive reactance $X$, respectively.

The output matching circuit is necessary to match the required nominal Class-E resistance $R$ calculated in accordance with Eq. (7.39) in a general case or Eq. (7.48) in a particular case with zero series reactance to the standard load resistance of 50 Ω. In addition, it is required to provide an open-circuit condition at the third-harmonic component. This can be easily done using the output matching topology in the form of an $L$-type low-pass transformer with the series transmission line and open-circuit stub [3,6]. In this case, the electrical length of the total series transmission line with combined electrical length $\theta_X + \theta_1$ should be chosen to provide an open circuit at its input at the third harmonic, and the electrical length of the open-circuit stub with electrical length $\theta_2$ should be chosen to be 30°. If both transmission lines have the same characteristic impedance, their total electrical length $\theta_X + \theta_1$ should be equal to 30° at the fundamental.

However, the stepped-impedance transmission line can change the position of poles of its input impedance which depends on the characteristic impedance ratio and position of their sections [7]. The load impedance $Z_L$ can be written as

$$Z_L = Z_1 \frac{R_L(Z_2 - Z_1 \tan \theta_1 \tan \theta_2) + jZ_1Z_2 \tan \theta_1}{Z_1Z_2 + jR_L(Z_1 \tan \theta_2 + Z_2 \tan \theta_1)} \tag{7.59}$$

where $Z_1$ and $\theta_1$ are the characteristic impedance and electrical length of the series transmission line, and $Z_2$ and $\theta_2$ are the characteristic impedance and electrical length of the open-circuit stub.

Hence, the complex-conjugate matching with the load at the fundamental can be provided by proper choice of the characteristic impedances $Z_1$ and $Z_2$. Separating Eq. (7.59) into real and imaginary parts and taking into account that $\mathrm{Re}Z_L = R$ and $\mathrm{Im}Z_L = 0$, the following system of two equations with two unknown parameters can be obtained as

$$R[R_L(1 + \tan^2\theta_1) - R] - Z_1^2 \tan^2 \theta_1 = 0 \tag{7.60}$$

$$(Z_1^2 - R^2)Z_2 \tan \theta_1 - (R^2 + Z_1^2 \tan^2 \theta_1)Z_1 \tan \theta_2 = 0 \tag{7.61}$$

which enables the characteristic impedances $Z_1$ and $Z_2$ to be properly calculated.

This system of two equations can be explicitly solved as a function of the parameter $r = R_L/R$ and electrical length $\theta_1$ when $\theta_2 = 30°$, resulting in

$$\frac{Z_1}{R_L} = \frac{\sqrt{r(1 + \tan^2 \theta_1) - 1}}{r \tan \theta_1} \tag{7.62}$$

$$\frac{Z_1}{Z_2} = \frac{\sqrt{3}}{\tan \theta_1} \frac{r - 1}{r} \tag{7.63}$$

where the electrical length $\theta_1$ generally affects the electrical length $\theta_X$. In a simple case of zero series reactance $X$ when $\theta_X = 0$ and $\theta_1 = 30°$, Eqs (7.62) and (7.63) can be rewritten as

$$\frac{Z_1}{R_L} = \frac{\sqrt{4r - 3}}{r} \tag{7.64}$$

$$\frac{Z_1}{Z_2} = 3\left(\frac{r - 1}{r}\right). \tag{7.65}$$

Consequently, for the specified value of the parameter $r$ with the required Class-E optimum load resistance $R$ and standard load $R_L = 50\ \Omega$, the characteristic impedance $Z_1$ is calculated from Eq. (7.64), and then the characteristic impedance $Z_2$ is calculated from Eq. (7.65).

Unlike the transmission-line Class-E load-network approximations with two-harmonic control [8] and with three-harmonic control [9], the Class-E load network with a quarterwave transmission line, which can provide the optimum impedance conditions for at least four harmonic components, is very simple in terms of circuit implementation and does not require an additional lumped RF choke element. In addition, there is no need to use the special computer simulation tools required to calculate the parameters of the existing Class-E transmission-line load-network topologies [9–11], since all parameters of the Class-E load network with a quarterwave transmission line and the output matching-circuit parameters are easily calculated explicitly from simple analytical equations. Besides, such a Class-E load network with a quarterwave transmission line is very useful in practical design providing simultaneously significant higher-order harmonic suppression.

## 7.6 Load network with series quarterwave line and shunt filter

Figure 7.11 shows an alternative configuration for the Class-E power amplifier with a parallel quarterwave transmission line, consisting of a shunt capacitor $C$, a series inductor $L$, a series quarterwave transmission line, a parallel-resonant $L_0C_0$ circuit tuned to the fundamental frequency, a shunt susceptance $B$, and a load resistor $R$. In a common case, the shunt capacitance $C$ can represent the intrinsic device output capacitance and external circuit capacitance added by the load network. The shunt susceptance $B$ can generally be positive (capacitance), negative (inductance), or zero depending on the value of the optimum load network parameter $q$. The RF choke is necessary to isolate the dc power supply from the RF circuit, while the blocking capacitor $C_b$ is necessary to separate the dc supply circuit from the load.

The operation principle of such a Class-E load network with a series quarter-wave line is similar to the Class-E load network with a parallel quarterwave line, assuming that the parallel $L_0C_0$ filter is ideal, having infinite impedance at the fundamental frequency and zero impedances at the second- and higher-order harmonic components. In this case, the short circuit on the load side of the quarter-wave transmission line produces a short circuit at its input for even harmonics

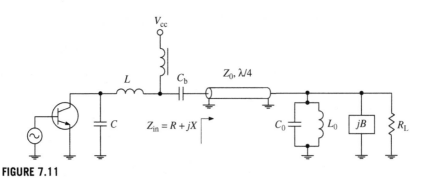

**FIGURE 7.11**

Class-E power amplifier with a series quarterwave transmission line and shunt filter.

and an open circuit at its input for odd harmonics, with resistive load at the fundamental frequency.

However, unlike the parallel quarterwave transmission line, a series quarterwave transmission line can also serve as an impedance transformer at the fundamental frequency. As a result, there is no need for an additional output matching circuit as required for the Class-E power amplifier with a parallel quarterwave transmission line. The impedance $Z_{in}$ at the input of the loaded quarterwave transmission line with the characteristic impedance $Z_0$ and electrical length $\theta$ at the fundamental frequency can be written as

$$Z_{in} = Z_0 \frac{Z_L + jZ_0 \tan \theta}{Z_0 + jZ_L \tan \theta}\bigg|_{\theta=90°} = \frac{Z_0^2}{Z_L} \qquad (7.66)$$

where

$$Z_L = \frac{R_L}{1 + jBR_L}. \qquad (7.67)$$

Substituting Eq. (7.67) to Eq. (7.66) yields

$$R + jX = Z_0^2 \frac{1 + jBR_L}{R_L}. \qquad (7.68)$$

Then, separating Eq. (7.68) into real and imaginary parts enables the standard load resistance $R_L$ and shunt susceptance $B$ to be expressed through the optimum load resistance $R$ and series reactance $X$ corresponding to the Class-E with a parallel quarterwave line as

$$R_L = \frac{Z_0^2}{R} \qquad (7.69)$$

$$B = \frac{X}{Z_0^2}. \qquad (7.70)$$

To calculate the optimum parameters of a Class-E load network with series quarterwave transmission line and shunt filter, it is necessary to use Eqs (7.35) to (7.39) for a Class-E load network with parallel quarterwave transmission line defining the optimum shunt capacitance $C$, series inductance $L$, series reactance $X$, and a load resistance $R$ in a common case or Eqs (7.46) to (7.48) in a particular case when $X = 0$.

## 7.7 Design example: 10-W, 2.14-GHz Class-E GaN HEMT power amplifier with parallel quarterwave transmission line

As an example, our objective is to design a high-efficiency single-stage hybrid Class-E GaN HEMT power amplifier with a parallel quarterwave transmission

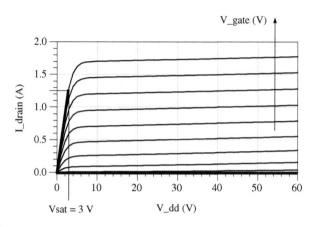

**FIGURE 7.12**

Device output current—voltage characteristics and load line.

line, operating at 2.14 GHz with an output fundamental-frequency power $P_{out} = 40$ dBm (10 W) at a supply voltage $V_{dd} = 28$ V. Using a GaN HEMT device can provide higher efficiency compared to other types of active devices and it is easier to apply a switchmode Class-E technique because of a sufficiently low output capacitance inherent to these devices. This kind of a power amplifier requires an operation in saturation mode resulting in a poor linearity, and therefore is not suitable to directly replace linear power amplifiers in conventional WCDMA/LTE or CDMA2000 communication transmitters with nonconstant envelope signal. However, to obtain both high efficiency and good linearity, a nonlinear high-efficiency power amplifier operating in a Class-E mode can be used in advanced transmitter architectures such as Doherty, LINC (linear amplification using nonlinear components), or ET (envelope tracking) with digital predistortion [12−14].

Assuming the drain efficiency of the Class-E power amplifier is 80%, the dc power is equal to $P_0 = 10$ W/0.8 = 12.5 W with the dc supply current $I_0 = P_0/V_{dd} = 12.5$ W/28 V ≈ 450 mA. The dc output current—voltage characteristics $I_{drain}(V_{dd})$ of a selected Cree CGH40010 GaN HEMT transistor for different gate bias voltage $V_{gate}$ varying from −3.0 to −0.5 V with a voltage step of 0.25 V are shown in Fig. 7.12 [15]. In this case, it is sufficient to evaluate the peak drain current for a Class-E mode with quarterwave transmission line and zero series reactance according to Eq. (7.49) with some margin, resulting in $I_{max} = 2.8I_0 \approx 1.25$ A. The idealized Class-E load line (with instant transition between pinch-off and saturation regions) is shown as a broken heavy line. It can be seen that the operating point moves along the horizontal $V_{dd}$ axis (pinch-off region) and then along the drain current saturated line (voltage saturation region) until $I_{drain} = I_{max} = 1.25$ A. At this final point, the saturation voltage can be found as equal to $V_{sat} \approx 3$ V, which means that the power loss due to the finite device saturation resistance can be calculated as $P_{sat} = I_0 V_{sat} = 1.35$ W with degradation in the drain efficiency of $P_{sat}/P_0 \approx 0.11$ or 11%.

In view of the device saturation voltage, the calculated parameters of the Class-E load network cannot generally be considered optimum unlike the ideal case of zero saturation voltage. However, they can be considered as a sufficiently accurate initial guess for final design and optimization when efficiency is sufficiently high. The drain-source capacitance $C_{ds} = 1.3$ pF is considered an output capacitance, while the series inductance is composed of the internal bondwire inductance and external package lead inductance, which can be estimated as approximately equal to 1 nH. To improve accuracy of simulation results, the device leads can be modeled as a strip line in air substrate with 45-mil width and 60-mil length. For $q = 2.1$, the optimum parameters of a Class-E load network with a quarterwave transmission line can be approximately estimated using corresponding plots in Figs. 7.5(a) and 7.5(b) in view of the saturation voltage $V_{sat}$ as

$$R = 0.38/\omega C = 0.34(V_{dd} - V_{sat})^2/P_{out} \approx 20 \ \Omega$$

$$X = 1.2R = 24 \ \Omega$$

$$L = 0.7R/\omega \approx 1 \text{ nH}.$$

In this case, the electrical length $\theta_X$ can be estimated for $Z_0 = 50 \ \Omega$ as

$$\theta_X = \tan^{-1}(X/Z_0) \approx 25°$$

which is sufficiently long compared with a total 30°. However, for $Z_0 = 100 \ \Omega$,

$$\theta_X = \tan^{-1}(X/Z_0) \approx 15°.$$

As a result, if the required optimum Class-E load resistance $R$ is equal to 20 $\Omega$ and standard load $R_L = 50 \ \Omega$ resulting in $r = R_L/R = 2.5$, the characteristic impedance of the series transmission line $Z_1$ calculated from Eq. (7.62) is equal to approximately 100 $\Omega$ and the characteristic impedance of the open-circuit stub $Z_2$ calculated from Eq. (7.63) is equal to 25 $\Omega$ for electrical length $\theta_1 = 15°$.

Figure 7.13 shows the simulated circuit schematic which approximates a 2.14-GHz transmission-line Class-E power amplifier based on a 28-V, 10-W Cree GaN HEMT power transistor CGH40010F and ideal transmission-line load network with the second- and third-harmonic tuning. The input matching circuit provides a complex-conjugate matching of the device input gate-source impedance $Z_{gs} \approx 5 + j2$ with the standard 50-$\Omega$ source. In this case, the input matching topology can simply be designed in the form of an $L$-type transformer with a series transmission line and an open-circuit stub [3,6]. The device output stripline lead is connected to a short-length wide metal pad of 50 mil × 120 mil on the substrate. It should be noted that approximated equivalent network of device output parasitics for a packaged 10-W GaN Cree device model includes not only the shunt capacitance $C_{ds}$ and series bondwire inductance but also two low-pass $LC$

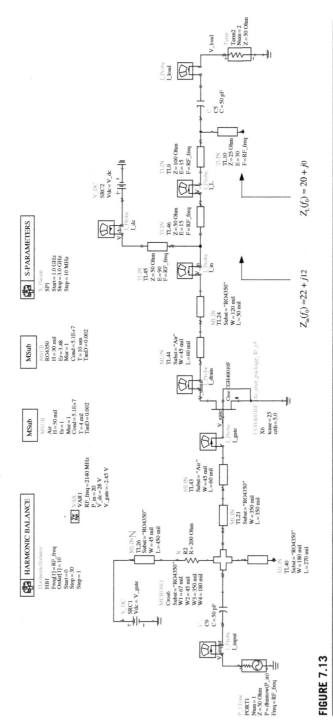

**FIGURE 7.13**

Simulation setup for a Class-E power amplifier with ideal transmission lines in load network.

sections corresponding to the internal package model [16]. Therefore, the parameters of the ideal transmission lines in the load network can be optimized to achieve maximum efficiency since the effect on the second- and third-harmonic impedance can be slightly different when considering a simplified model of the device output circuit with one shunt capacitance and one series inductance only.

Figure 7.14 shows the simulated results of a transmission-line Class-E GaN HEMT power amplifier with ideal transmission lines in the load network. The maximum output power of 40.5 dBm, power gain of 12.7 dB (linear gain over 18 dB), drain efficiency of 85.5%, and *PAE* of 81% are achieved at an operating frequency of 2.14 GHz with a supply voltage of 28 V and a quiescent current of 50 mA. The maximum drain efficiency was achieved at a gain compression point of about 5 dB.

Figure 7.15 shows the drain (*a*) voltage and (*b*) current waveforms corresponding to the transmission-line Class-E approximation. The drain voltage

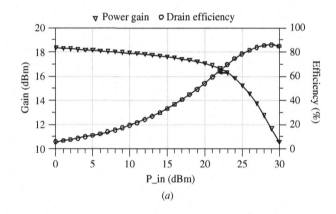

(*a*)

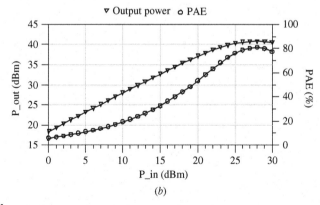

(*b*)

**FIGURE 7.14**

Simulated results of a Class-E power amplifier with ideal transmission lines.

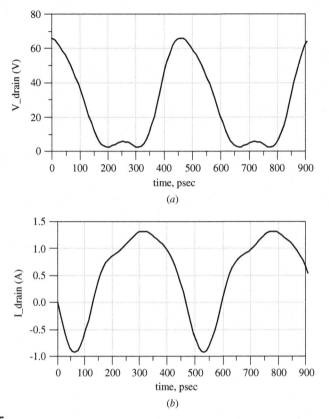

**FIGURE 7.15**

Drain voltage and current waveforms.

waveform is very similar to the idealized one shown in Fig. 7.7(*b*), with a peak factor of about 2.4 which is sufficiently small taking into account the device drain-source breakdown voltage of 120 V. The current flowing through the device output capacitance during off-state operation mode contributes to the total drain current waveform shown in Fig. 7.15(*b*), since it is measured at the output device terminal outside of the device package. In this case, the main contributor to efficiency degradation is the saturation resistance $r_{\text{sat}}$ reducing efficiency by 11%, whereas the finite fall time of the driving switching signal degrades efficiency by only a few percent. At the same time, to maximize drain efficiency, it is very important to choose the PCB board with a low-loss substrate such as RO4350.

Figure 7.16 shows the simulated circuit schematic of a 2.14-GHz transmission-line Class-E power amplifier, which is based on the nonlinear model for a 28-V, 10-W GaN HEMT power transistor CGH40010F and technical parameters for a 30-mil RO4350 substrate. The load network is slightly modified by

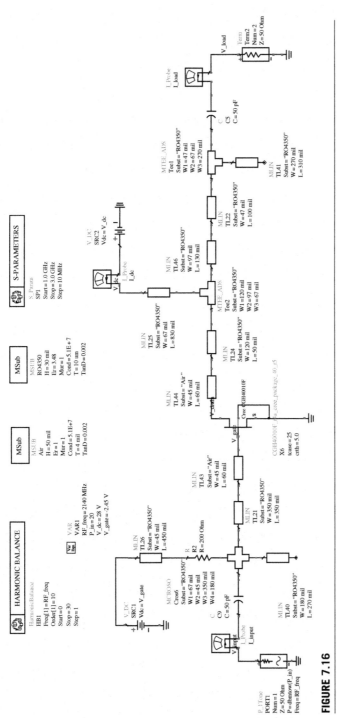

**FIGURE 7.16**

Simulation setup for a Class-E power amplifier with microstrip lines.

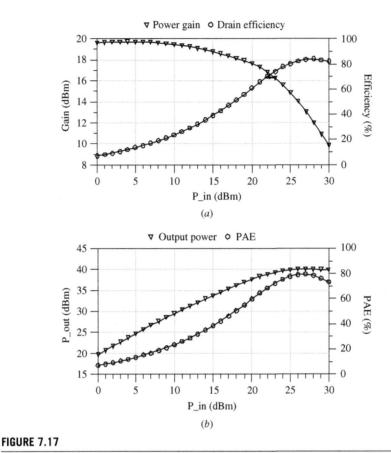

**FIGURE 7.17**

Simulated results of Class-E power amplifier with microstrip lines.

optimizing the parameters of the series microstrip lines and open-circuit stub because the series microstrip line with very high characteristic impedance results in excessive insertion losses. The physical parameters of the microstrip lines using a RO4350 substrate can easily be determined by using an ADS transmission-line calculator LineCalc, which can be found in Tools displayed by Project menu. The input matching circuit remains the same as for an idealized version of the Class-E power amplifier with a quarterwave transmission line shown in Fig. 7.13.

Figure 7.17 shows the simulated results of a transmission-line Class-E GaN HEMT power amplifier implemented on a 30-mil RO4350 substrate. In this case, the maximum output power of 40.0 dBm, power gain of 12.0 dB (linear gain close to 20 dB), drain efficiency of 83.8%, and *PAE* of 79.5% are achieved at an operating frequency of 2.14 GHz with a supply voltage of 28 V.

Due to the shorting effect for the second harmonic of the grounded quarterwave microstrip line and high-impedance condition provided by the open-circuit

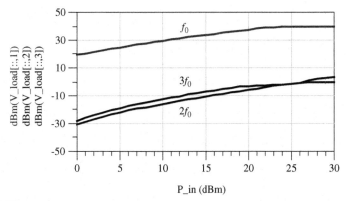

**FIGURE 7.18**

Simulated results of second- and third-harmonic suppression.

stub and series microstrip line with a quarterwave length for the third harmonic, the second- and third-harmonic suppression is achieved at the level of about 40 dB, as shown in Fig. 7.18.

# REFERENCES

1. Sokal NO, Sokal AD. High-efficiency tuned switching power amplifier. U.S. Patent 3,919,656, November 1975.
2. Grebennikov A. Load network design techniques for Class E RF and microwave amplifiers. *High Frequency Electronics*. July 2004;3:18−32.
3. Grebennikov AV. *RF and Microwave Power Amplifier Design*. New York: McGraw-Hill; 2004.
4. Kazimierczuk MK, Tabisz WA. Class C-E high-efficiency tuned power amplifier. *IEEE Trans. Circuits and Systems*. March 1989;CAS-36:421−428.
5. Grebennikov AV. A simplified CAD approach to analyzing lumped elements. *RF Design*. July 2000;23:22−38.
6. Grebennikov AV. Create transmission-line matching circuits for power amplifiers. *Microwaves & RF*. October 2000;39:113−122:172.
7. Grebennikov A. A simple stepped-impedance transmission line load network for inverse Class F power amplifiers. *Microwave and Optical Technology Lett*. May 2011;53:1157−1160.
8. Mader TB, Bryerton EW, Marcovic M, Forman M, Popovic Z. Switched-mode high-efficiency microwave power amplifiers in a free-space power-combiner array. *IEEE Trans. Microwave Theory Tech*. October 1998;MTT-46:1391−1398.
9. Ortega-Gonzalez FJ, Jimenez-Martin JL, Asensio-Lopez A, Torregrosa-Penalva G. High-efficiency load-pull harmonic controlled Class-E power amplifier. *IEEE Microwave and Guided Wave Lett*. October 1998;8:348−350.

10. Mader TB, Popovic ZB. The transmission-line high-efficiency Class-E amplifier. *IEEE Microwave and Guided Wave Lett.* September 1995;5:290–292.

11. Wilkinson AJ, Everard JKA. Transmission-line load-network topology for Class-E power amplifiers. *IEEE Trans. Microwave Theory Tech.* June 2001;MTT-49:1202–1210.

12. Lee Y-S, Lee M-W, Jeong Y-H. Highly efficient Doherty amplifier based on Class-E topology for WCDMA applications. *IEEE Microwave and Wireless Components Lett.* September 2008;18:608–610.

13. Chen C-T, Li C-J, Horng T-S, Jau J-K, Li J-Y. Design and linearization of Class-E power amplifier for nonconstant envelope modulation. *IEEE Trans. Microwave Theory Tech.* April 2009;MTT-57:957–964.

14. Ui N, Sano S. A 45% drain efficiency, −50 dBc ACLR GaN HEMT Class-E amplifier with DPD for W-CDMA base station. *2006 IEEE MTT-S Int. Microwave Symp. Dig.* 2:718–721.

15. Cree Inc. *CGH40010: 10 W RF Power GaN HEMT,* Rev. 3.0, May 2010.

16. Tasker PJ, Benedikt J. Waveform inspired models and the harmonic balance emulator. *IEEE Microwave Mag.* April 2011;12:38–54.

# Broadband Class-E

## INTRODUCTION

In this chapter, the reactance compensation technique applied to the circuits with lumped elements and transmission lines to provide a broadband operation of Class-E power amplifiers is introduced and analyzed. This technique can be directly used to design a parallel-circuit Class-E power amplifier because its load network configuration has exactly the same structure with shunt and series resonant circuits. Different circuit configurations and Class-E load-network techniques corresponding to broadband high-power RF power amplifiers operating in VHF and UHF bands, microwave monolithic integrated circuits of power amplifiers, and CMOS power amplifiers operating in a wide frequency range with high efficiency are given and described.

## 8.1 Reactance compensation technique

The conventional design of a high-efficiency switchmode Class-E power amplifier requires a high $Q_L$-factor to satisfy the necessary harmonic impedance conditions at the output device terminal. However, if a sufficiently small value of the loaded quality factor $Q_L$ is chosen, a high-efficiency broadband operation of the Class-E power amplifier can be realized. For example, a simple network consisting of a series resonant $LC$ circuit tuned to the fundamental frequency and a parallel inductor provides a constant load phase angle of 50° in a frequency range of about 50% [1]. From theoretical considerations it was found in the mid-1960s that the bandwidth response of a parametric amplifier can be improved using multiple-resonant bandpass filters for the signal and idling circuits rather than simple resonant circuits [2,3]. At the same time, it was analytically calculated that the

**387**

added resonant circuits should have an appropriate $Q_L$-factor to optimally reduce the rate of change of reactance of both the signal and idling circuits [4]. Adding additional resonators can increase the potential amplifier bandwidth even further, but the amount of improvement per additional resonator will decrease rapidly as the number of resonators is increased. Such a reactance compensation technique using a single-resonant circuit had also been applied to the varactor-tuned Gunn oscillator [5,6]. Moreover, it became possible to increase the tuning range of an oscillator by adding more stages of reactance compensation. For instance, for a resonant circuit having a 50-$\Omega$ load, an improvement of 4% in the tuning range can theoretically be achieved as a result of applying a double-resonant circuit reactance compensation, whereas an increase in the tuning range can reach 17% for a resonant circuit operating into 100-$\Omega$ load [7].

### 8.1.1 Load networks with lumped elements

To describe the reactance compensation circuit technique, let us consider the simplified equivalent load networks, one with a shunt resonant $L_pC_p$ circuit followed by a series resonant $L_sC_s$ circuit shown in Fig. 8.1(a) and the other with a series resonant $L_sC_s$ circuit followed by a shunt resonant $L_pC_p$ circuit shown in Fig. 8.1(b). In this case, all resonant circuits are tuned to the fundamental frequency and $R$ is the load resistance. The reactances of the series and shunt resonant circuits vary with frequency, increasing in the case of a series resonant circuit and reducing in the case

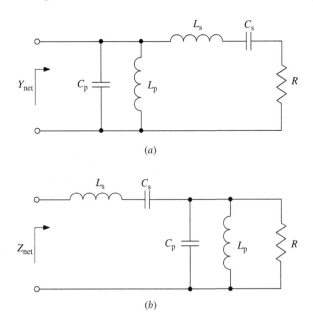

(a)

(b)

**FIGURE 8.1**

Single-reactance compensation circuits.

of a loaded parallel resonant circuit near the resonant frequency. As a result, near the resonant frequency of the series circuit with positive slope of its reactance, the slope of a shunt circuit reactance is negative. That reduces the overall reactance slope of the load network. By the correct choice of the components in the shunt circuit, the rate of change of reactance with frequency can be made exactly opposite to that of the series circuit, thus producing a zero total variation over a wide frequency bandwidth.

Consider the load-network admittance $Y_{net}$ corresponding to a single-reactance compensation circuit shown in Fig. 8.1($a$), which can be written as

$$Y_{net}(\omega) = \left( j\omega C_p + \frac{1}{j\omega L_p} + \frac{1}{R + j\omega' L_s} \right) \tag{8.1}$$

where

$$\omega' = \omega \left( 1 - \frac{\omega_0^2}{\omega^2} \right) \tag{8.2}$$

and $\omega_0 = 1/\sqrt{L_s C_s} = 1/\sqrt{L_p C_p}$ is the radian resonant frequency.

At the resonant frequency when $\omega' = 0$, the load-network admittance $Y_{net}(\omega)$ reduces to

$$Y_{net}(\omega) = \left( j\omega C_p + \frac{1}{j\omega L_p} + G \right) \tag{8.3}$$

where $G = 1/R$ is the load conductance.

The frequency bandwidth with zero susceptance will be maximized if, at a resonant frequency $\omega_0$,

$$\frac{dB_{net}(\omega)}{d\omega} \bigg|_{\omega=\omega_0} = 0 \tag{8.4}$$

where

$$B_{net}(\omega) = \mathrm{Im}\, Y_{in}(\omega) = \omega C_p - \frac{1}{\omega L_p} - \frac{\omega' L_s}{R^2 + (\omega' L_s)^2} \tag{8.5}$$

is the load-network susceptance.

As a result, an additional equation can be written as

$$C_p + \frac{1}{\omega_0^2 L_p} - \frac{2L_s}{R^2} = 0 \tag{8.6}$$

based on which the values of the series components $L_s$ and $C_s$ can, respectively, be obtained through the values of the shunt components $L_p$ and $C_p$ by

$$L_s = C_p R^2 \tag{8.7}$$

$$C_s = \frac{L_p}{R^2}. \tag{8.8}$$

In a similar manner, it may be shown that, for the load network with a series resonant $L_sC_s$ circuit followed by a shunt resonant $L_pC_p$ circuit shown in Fig. 8.1(b), maximum bandwidth with zero reactance can be achieved if

$$\left.\frac{dX_{net}(\omega)}{d\omega}\right|_{\omega=\omega_0} = 0 \qquad (8.9)$$

where

$$X_{net}(\omega) = \text{Im}Z_{net}(\omega) = \omega L_s - \frac{1}{\omega C_s} - \frac{\omega' C_p}{G^2 + (\omega' C_p)^2} \qquad (8.10)$$

is the load-network reactance, resulting in Eqs (8.7) and (8.8). From Eq. (8.7), it follows that the loaded quality factor of the shunt circuit $Q_L = \omega C_p R$ is equal to the loaded quality factor of the series compensating circuit $Q_L = \omega L_s/R$.

Figure 8.2(a) shows an example of a susceptance compensation load network, whose conductance $\text{Re}Y_{net}$ is almost constant across the frequency range of 40%, from 4 to 6 MHz, as shown in Fig. 8.2(b). The susceptance $\text{Im}Y_{net}$ of a shunt circuit varies with frequency, as shown in Fig. 8.2(c) by curve 1, with the gradient at $\omega_0$ being equal to $2C_p$. The addition of a series circuit with the same resonant frequency of 5 MHz between the shunt circuit and the load of the shunt circuit gives an additional susceptance term with a negative slope, as shown in Fig. 8.2(c) by curve 2. Selection of the series component enables the magnitude of the two slopes to be made identical, so that the total susceptance slope around the resonance is zero in an octave frequency range from 3.5 to 7 MHz, as shown in Fig. 8.2(c) by curve 3.

The load network which provides reactance compensation is shown in Fig. 8.3(a), where the shunt resonant circuit is connected between the series resonant circuit and the load. In this case, the resistance and reactance curves, the frequency behavior of which is similar to that for the conductance and susceptance curves characterizing the behavior of a susceptance compensation load network, are shown in Fig. 8.3(b) and 8.3(c), respectively. Here, the reactance of a series resonant circuit with a positive slope is shown by curve 1, the reactance of a shunt resonant circuit with a negative slope is shown by curve 2, and the total reactance slope shown by curve 3 is zero from 3.5 to 7 MHz.

Wider frequency bandwidth can be achieved using a double-susceptance compensation circuit shown in Fig. 8.4(a), where $L_sC_s$ and $L_1C_1$ are the series and parallel compensating circuits, respectively. In this case, a system of two additional equations to maximize the frequency bandwidth can be used, where the first and the third derivatives are set to zero according to

$$\left.\frac{dB_{net}(\omega)}{d\omega}\right|_{\omega=\omega_0} = \left.\frac{d^3 B_{net}(\omega)}{d\omega^3}\right|_{\omega=\omega_0} = 0 \qquad (8.11)$$

as the second derivative cannot provide an appropriate analytical expression.

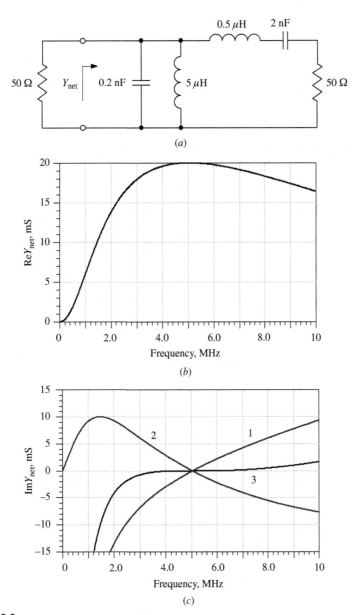

**FIGURE 8.2**

Single-susceptance compensation circuit and admittances.

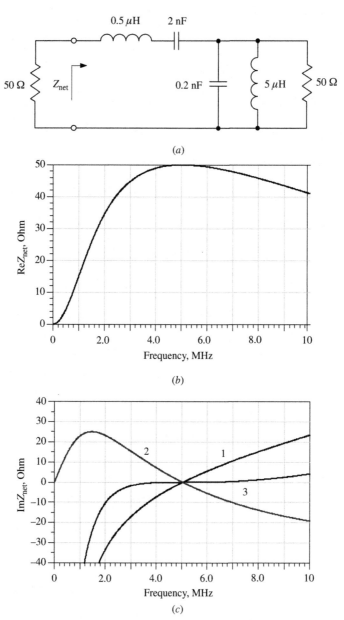

**FIGURE 8.3**

Single-reactance compensation circuit and impedances.

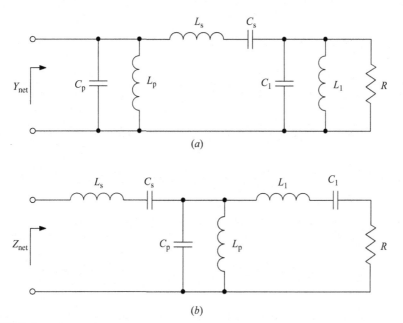

**FIGURE 8.4**

Double-reactance compensation circuits.

To determine the load-network parameters for a double-susceptance compensation circuit with the load-network susceptance

$$B_{net}(\omega) = \omega C_p - \frac{1}{\omega L_p} + \omega' \frac{C_1 R^2 [1 - (\omega')^2 L_s C_1] - L_s}{R^2 [1 - (\omega')^2 L_s C_1]^2 + (\omega' L_s)^2} \qquad (8.12)$$

where $B_{net} = \mathrm{Im} Y_{net}$, it is necessary to solve simultaneously the two following equations at the resonant frequency $\omega_0$:

$$C_{p1} + \frac{1}{\omega_0^2 L_p} - 2 \frac{C_1 R^2 - L_s}{R^2} = 0 \qquad (8.13)$$

$$\frac{1}{\omega_0^2 L_p} + \frac{C_1 R^2 - L_s}{R^2} - 8\omega_0^2 L_s \left[ C_1^2 + \frac{(C_1 R^2 - L_s)(L_s - 2C_1 R^2)}{R^4} \right] = 0. \qquad (8.14)$$

As a result, the parameters of the series and shunt compensating resonant circuits with the corresponding loaded quality factors $Q_s = \omega_0 L_s/R$ and $Q_1 = \omega_0 C_1 R$, which are close to unity and greater, can be calculated as a starting point for circuit optimization from

$$L_s = \frac{R}{\omega_0} \frac{2}{\sqrt{5} - 1} \qquad C_s = \frac{1}{\omega_0^2 L_s} \qquad (8.15)$$

$$C_1 = \frac{L_s}{R^2} \frac{3 - \sqrt{5}}{2} \qquad L_1 = \frac{1}{\omega_0^2 C_1}. \tag{8.16}$$

Similarly, the elements for the double-reactance compensation load network shown in Fig. 8.4(b) can be calculated from

$$L_1 = L_s \frac{\sqrt{5} - 1}{2} \qquad C_1 = C_s \frac{2}{\sqrt{5} - 1} \tag{8.17}$$

$$L_p = C_s \frac{2R^2}{\sqrt{5} + 1} \qquad C_p = L_s \frac{\sqrt{5} + 1}{2R^2} \tag{8.18}$$

where an inductance $L_s$ and a capacitance $C_s$ are known in advance [7]. An example of the load network which provides double-reactance compensation is shown in Fig. 8.5(a), whose resistance $\mathrm{Re}Z_{net}$ shown in Fig. 8.5(a) by curve 1 provides less deviation from 50 $\Omega$ in a slightly wider frequency bandwidth compared to the single-resonance load network (curve 2) with $L_s = 0.5 \ \mu\mathrm{H}$, $C_s = 2 \ \mathrm{nH}$, $L_p = 5 \ \mu\mathrm{H}$, and $C_p = 0.2 \ \mathrm{nF}$. The reactance $\mathrm{Im}Z_{net}$ of a double-reactance compensation circuit shown in Fig. 8.5(c) by curve 1 is close to zero, near resonance across the frequency range from 3 to 8 MHz, which is wider than that for a single-resonance compensation circuit (curve 2).

## 8.1.2 Load networks with transmission lines

The reactance compensation circuit technique can also be used for bandwidth improvement of microwave transistor amplifiers because the input and output transistor impedances can generally be represented by series or shunt *RLC* circuits. For compensating the reactive part and transforming the real part of the equivalent output transistor impedance to the conventional load impedance at the fundamental frequency, the quarter- and half-wavelength transmission lines can be used. For the first time, a quarter-wavelength transmission-line transformer was used for active reactance compensation when, by connecting two identical active devices together with a quarter-wavelength transformer, the inverted impedance of one device compensates the impedance of the other by reducing the total circuit reactance [8].

Let us consider the characteristics of the transmission line as an element of a susceptance compensation circuit shown in Fig. 8.6. For a parallel equivalent circuit, which represents the device output, the load-network input susceptance $B_{net} = \mathrm{Im}Y_{net}$ can be defined as

$$B_{net}(\omega) = \omega L_p C_p \left(1 - \frac{\omega_0^2}{\omega^2}\right) + \frac{\tan \theta}{Z_0} \frac{R_L^2 - Z_0^2}{R_L^2 + Z_0^2 \tan^2 \theta} \tag{8.19}$$

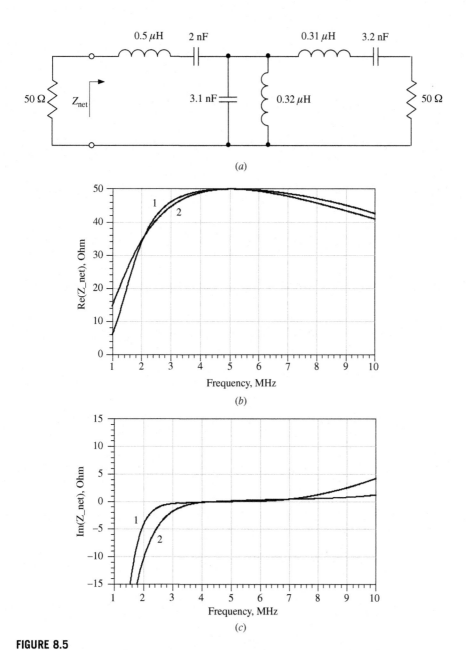

**FIGURE 8.5**

Double-reactance compensation circuit and impedances.

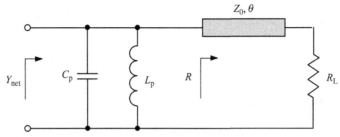

**FIGURE 8.6**

Transmission-line susceptance compensation circuit.

where

$$\theta = \frac{\pi}{2}\frac{f}{f_0}k \tag{8.20}$$

is the transmission-line electrical length, $Z_0$ is the transmission-line characteristic impedance, $f_0 = \omega_0/2\pi$ is the transmission-line resonant frequency, $k = 1, 2, \ldots, \infty$.

Applying the zero susceptance-derivative condition given by Eq. (8.4) to Eq. (8.19) allows us to obtain the susceptance-compensation circuit parameters for different electrical lengths of a transmission line in accordance with

$$2C_p + \frac{\pi}{2Z_0\omega_0}\frac{R_L^2 - Z_0^2}{\cos^2\theta}\frac{R_L^2 - Z_0^2\tan^2\theta}{(R_L^2 + Z_0^2\tan^2\theta)^2} = 0. \tag{8.21}$$

For a quarter-wavelength transmission line when $k = 1$ and $\theta = \pi/2$, the susceptance compensation will be performed under the condition $Z_0 < R_L$ with the characteristic impedance $Z_0$ defined from a quadratic equation

$$Z_0^2 + 4\frac{QR_L}{\pi}Z_0 - R_L^2 = 0 \tag{8.22}$$

where $Q = \omega_0 C_p R$ and $R = Z_0^2/R_L$.

As a result, the required value of the characteristic impedance $Z_0$ is obtained by

$$Z_0 = R_L\left(-\frac{2Q}{\pi} + \sqrt{\left(\frac{2Q}{\pi}\right)^2 + 1}\right) \tag{8.23}$$

or

$$Z_0 = R\bigg/\left(-\frac{2Q}{\pi} + \sqrt{\left(\frac{2Q}{\pi}\right)^2 + 1}\right). \tag{8.24}$$

Using the quarter- and half-wavelength transformers, the reactance- or susceptance-compensation load network generally can be realized differently for shunt and series equivalent output transistor circuits, as shown in Table 8.1 along with respective design equations [9,10]. The two most important device parameters in the

**Table 8.1** Transmission-Line Reactance Compensation Circuits and Design Equations

| Output Circuit Type | Matching Network | Design Equation |
|---|---|---|
| $Q = \omega_0 RC$ | | $Z_1 = \sqrt{RR_L}$ <br> $T = \left(\dfrac{2Q}{\pi} - \dfrac{1}{2}\right)\left(\dfrac{R_L}{Z_1} - \dfrac{Z_1}{R_L}\right)$ <br> $Z_2 = -\dfrac{TR_L}{2} + \sqrt{\left(\dfrac{TR_L}{2}\right)^2 + R_L^2}$ |
|  | | $Z_1 = \sqrt{RR_L}$ <br> $Z_2 = \dfrac{\pi Z_1 R_L^2}{\pi Z_1^2 - \pi R_L^2 - 4QZ_1 R_L}$ |
| $Q = 1/\omega_0 RC$ | | $Z_1 = AZ_2$ <br> $Z_2 = BR_L$ <br> $A = \sqrt{R/R_L}$ <br> $B = \dfrac{A}{1+A}\left[\dfrac{2Q}{\pi} + \sqrt{\left(\dfrac{2Q}{\pi}\right)^2 + \dfrac{(1+A)^2}{A}}\right]$ |
|  | | $Z_2 = \sqrt{RR_L}$ <br> $Z_1 = \dfrac{\left(\dfrac{TZ_2}{R_L}\right)^2}{\left(\dfrac{R_L}{Z_2} - \dfrac{Z_2}{R_L}\right) + \dfrac{4Q}{\pi}}$ <br> $T = \dfrac{2}{\pi\cos\theta_2}(\theta_1 + \theta_2)$ <br> $\theta_1 + \theta_2 = \dfrac{n\pi}{2} \rightarrow n = 1$ or $3$ |

equations are the loaded quality factor $Q$ and the real part $R$ of the equivalent device output impedance or admittance. Depending on the values of the transmission-line characteristic impedances $Z_1$ and $Z_2$, each circuit provides either positive or negative parallel-resonant slope-reactance compensation.

Figure 8.7(a) shows an example of a single-susceptance compensation load network with a series quarterwave transmission line having a characteristic impedance of 61.2 $\Omega$ to match a 50-$\Omega$ real part of the device equivalent output admittance to a 75-$\Omega$ load and an electrical length of $90°$ at 50 MHz. The combination of the resistances of a shunt $LC$ circuit (curve 1) and a series quarterwave transmission line (curve 2) provides minimum variations of the total resistance $\mathrm{Re}Z_{\mathrm{net}}$ shown in Fig. 8.7(b) by curve 3; around 50 $\Omega$ in a very wide frequency range. The susceptance $\mathrm{Im}Y_{\mathrm{net}}$ of a shunt circuit having a resonant frequency of 50 MHz varies with frequency with a positive slope, as shown in Fig. 8.7(c) by curve 1. The addition of a series quarter-wavelength transmission-line transformer between the shunt circuit and the load results in a negative slope providing an additional susceptance, as shown in Fig. 8.7(c) by curve 2. Selection of the proper characteristic impedance of the series quarterwave transmission line and the load resistance enables the magnitude of two slopes to be made identical, so that the total susceptance slope around resonance is zero, in a frequency range from 45 to 65 MHz, as shown in Fig. 8.7(c) by curve 3.

From Eq. (8.23) it follows that the maximum value of the characteristic impedance $Z_0$ is limited by the load resistance $R_{\mathrm{L}}$, and its value in some cases, especially for high value of $Q$, can be substantially smaller than 50 $\Omega$; which causes a problem in the practical implementation of a transmission line. In this case, it is best to apply a single-frequency equivalence technique when a quarterwave transmission line can be replaced by a symmetrical $\pi$-type low-pass transmission-line section with two equal shunt capacitances at a frequency $\omega_0$, as shown in Fig. 8.8.

The *ABCD*-matrix for a quarterwave transmission line can be written as

$$ABCD_{90°} = \begin{bmatrix} \cos 90° & jZ_0 \sin 90° \\ j\dfrac{\sin 90°}{Z_0} & \cos 90° \end{bmatrix} = \begin{bmatrix} 0 & jZ_0 \\ j\dfrac{1}{Z_0} & 0 \end{bmatrix} \qquad (8.25)$$

whereas, for a $\pi$-type low-pass transmission-line section, we can write

$$
\begin{aligned}
ABCD_\pi &= \begin{bmatrix} 1 & 0 \\ j\omega C_{\mathrm{T}} & 1 \end{bmatrix} \begin{bmatrix} \cos\theta_{\mathrm{T}} & jZ_{\mathrm{T}} \sin\theta_{\mathrm{T}} \\ j\dfrac{\sin\theta_{\mathrm{T}}}{Z_{\mathrm{T}}} & \cos\theta_{\mathrm{T}} \end{bmatrix} \begin{bmatrix} 1 & 0 \\ j\omega C_{\mathrm{T}} & 1 \end{bmatrix} \\[2mm]
&= \begin{bmatrix} \cos\theta_{\mathrm{T}} - \omega C_{\mathrm{T}} Z_{\mathrm{T}} \sin\theta_{\mathrm{T}} & jZ_{\mathrm{T}} \sin\theta_{\mathrm{T}} \\ \dfrac{j}{Z_{\mathrm{T}}}(2Z_{\mathrm{T}}\omega C_{\mathrm{T}}\cos\theta_{\mathrm{T}} + \sin\theta_{\mathrm{T}} - Z_{\mathrm{T}}^2\omega^2 C_{\mathrm{T}}^2\sin\theta_{\mathrm{T}}) & \cos\theta_{\mathrm{T}} - \omega C_{\mathrm{T}} Z_{\mathrm{T}}\sin\theta_{\mathrm{T}} \end{bmatrix}.
\end{aligned}
$$

$$(8.26)$$

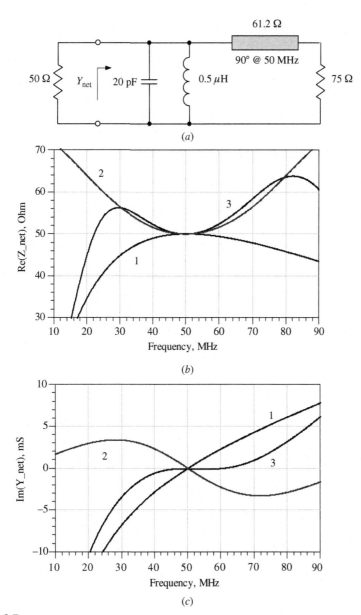

**FIGURE 8.7**

Susceptance compensation circuit with a quarter-wavelength transmission line.

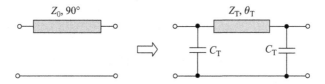

**FIGURE 8.8**

Transmission-line single frequency equivalence technique.

Hence, equating $A$ and $B$ elements from each matrix yields

$$Z_T = \frac{Z_0}{\sin \theta_T} \tag{8.27}$$

$$C_T = \frac{\cos \theta_T}{\omega Z_0}. \tag{8.28}$$

As a result, the electrical length of the transmission line can be reduced significantly with the increase of its characteristic impedance. Also, such a transformation is very important when the value of the device output capacitance exceeds the required optimum value for the optimum Class-E operation. In this case, the excess capacitance can be used as a part or entire shunt capacitance in the $\pi$-type low-pass section, and the optimum switching Class-E conditions will be completely satisfied at the fundamental frequency.

## 8.2 Broadband Class-E with shunt capacitance

In the basic circuit of a Class-E amplifier with a shunt capacitor shown in Fig. 8.9(a), the harmonic impedance of the series fundamentally tuned $L_0C_0$ circuit is assumed to be high due to its high loaded quality factor. The value of the shunt capacitor $C$ must also be correct to produce the correct voltage when the switch is turned off and to satisfy the steady-state switching conditions. In this case, the load phase angle of the series tuned circuit composed of the total inductor $(L + L_0)$ and capacitor $C_0$ determining the optimum angle for producing the correct voltage waveform can be obtained according to Eq. (5.31) at the resonant radian frequency $\omega_0 = 1/\sqrt{L_0 C_0}$ as

$$\theta = \tan^{-1}\left(\frac{\omega_0 L}{R}\right) = \tan^{-1} 1.1525 = 49.052°. \tag{8.29}$$

If the load network is designed without incorporating a shunt capacitor, a simple broadband network with an optimum load angle $\theta = 49.052°$ given in Eq. (8.29) can be designed. Then, this phase angle reduces to the required angle $\phi = 35.945°$ given by Eq. (5.34) when a shunt capacitor is added. The circuit schematic of a simple load network capable of presenting a constant load angle over a very large bandwidth is shown in Fig. 8.9(b) [1]. The load network consists

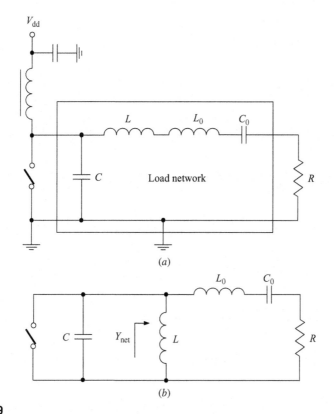

**FIGURE 8.9**

Load networks of Class-E with shunt capacitance.

of a low-$Q$ series $L_0C_0$ circuit connected in parallel with an inductor that allows a constant susceptance to be maintained over a wide bandwidth. The frequency behavior of the conductance $\mathrm{Re}\,Y_{net}$ and susceptance $\mathrm{Im}\,Y_{net}$ of this load network with parameters $L = 42$ nH, $L_0 = 30$ nH, and $C_0 = 35$ pF are shown in Fig. 8.10(a) and 10(b), respectively, where combining of the susceptance of the series resonant circuit with a negative slope (curve 1) and the susceptance of the shunt inductor with a positive slope (curve 2) provides a constant total susceptance over a very wide frequency range (curve 3).

In order to maintain the load angle constant in a wide frequency range, the slope of the susceptance provided by the inductance $L$ should be cancelled by the slope provided by the resonant $L_0C_0$ circuit. The load-network admittance of Fig. 8.9(b) can be written as

$$Y_{net} = -\frac{j}{\omega L} + \frac{1}{R + j\left(\omega L_0 - \dfrac{1}{\omega C_0}\right)} \qquad (8.30)$$

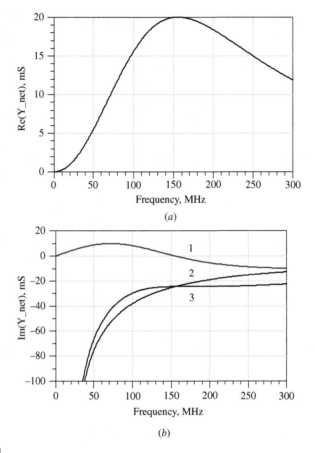

**FIGURE 8.10**

Conductance and susceptance of a broadband circuit.

which reduces at the resonant frequency to

$$Y_{net} = -\frac{j}{\omega L} + \frac{1}{R}. \qquad (8.31)$$

For slope cancellation, it is necessary to apply a zero-derivative condition given by Eq. (8.4) to Eq. (8.30) for the load-network susceptance $B_{net} = \text{Im} Y_{net}$ at the resonant frequency $\omega_0$. As a result,

$$\frac{1}{\omega_0^2 L} = \frac{2}{\omega_0^2 C_0 R^2}. \qquad (8.32)$$

Thus, the design equations to calculate the parameters of a broadband Class-E load network providing maximum flatness can be calculated from

$$L = \frac{R}{\omega_0 \tan \theta} \qquad (8.33)$$

$$C_0 = \frac{2L}{R^2} \qquad (8.34)$$

$$L_0 = \frac{1}{\omega_0^2 C_0}. \qquad (8.35)$$

To reduce the output power at the harmonics, such a simple load network can be combined with a broadband matching network and a bandpass filter. As an example, a complete circuit based on a low-pass L-type matching section and a third-order Chebyshev bandpass filter, as shown in Fig. 8.11(a), was designed to deliver 12 W into a 50-$\Omega$ load across the frequency bandwidth from 130 to 180 MHz using a 12-V power supply [1]. From Fig. 8.11(b), it follows that this load network presents a constant magnitude of input impedance of 12 $\Omega$ (curve 1) and a load phase angle of around 36° (curve 2) over the required wide frequency range. As a result, the broadband MOSFET Class-E power amplifier was capable of providing a fairly constant efficiency at approximately 60% with suppression of the second, third, and fourth harmonics better than 45 dB below fundamental. The drain efficiency of a GaN HEMT power amplifier with a Butterworth bandpass filter in the load network can be increased to more than 80% in a frequency bandwidth from 600 to 800 MHz with an output power of more than 45 dBm [11,12]. To provide the frequency bandwidth of 30% around the center bandwidth frequency of 1 GHz, the load network can be composed of a series transmission line and a shunt open-circuit stub [13].

Figure 8.12(a) shows the example of a reactance-compensation load network for Class-E power amplifier with shunt capacitance including a series transmission line and a parallel resonant circuit. In this case, the reactance of a Class-E load network with shunt capacitance and series inductance varies, similar to the series resonant with positive slope, whereas the required negative slope is provided by the parallel resonant circuit. Selection of the proper characteristic impedance and electrical length of the series transmission line enables the magnitude of the two slopes to be made identical, so as to achieve a constant total reactance and phase of the load network impedance $Z_{net}$ over a wide frequency range. The simulation results at the fundamental frequency show that the resistance Re$Z_{net}$ varies from 35 $\Omega$ at 30 MHz to 68 $\Omega$ at 70 MHz, as shown in Fig. 8.12(b) by curve 1, whereas the load-network phase varies between 27 and 40° in more than octave bandwidth from 33 to 80 MHz (curve 2).

Generally, the design of a practical multisection LC filter is based on some approximate equivalence between lumped and distributed elements, which can be established by applying a Richards's transformation [14]. This implies that the

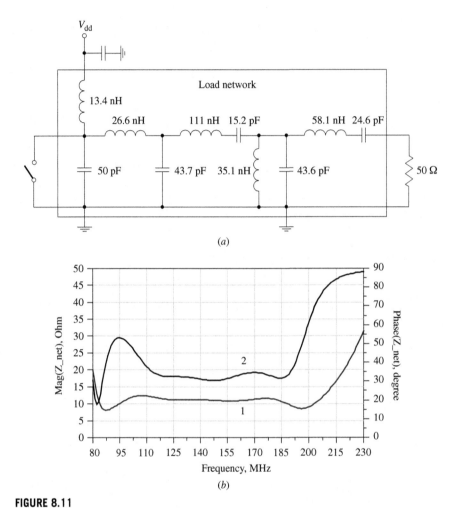

**FIGURE 8.11**

Broadband Class-E load network with bandpass filter and impedance.

distributed circuits composed of equal-length open- and short-circuited transmission lines can be treated as lumped elements under the transformation

$$s = j \tan \frac{\pi \omega}{2\omega_0} \qquad (8.36)$$

where $s = j\omega/\omega_c$ is the conventional normalized complex frequency variable, $\omega_c$ is the cutoff radian frequency, and $\omega_0$ is the radian frequency, for which the transmission lines are a quarter wavelength [15].

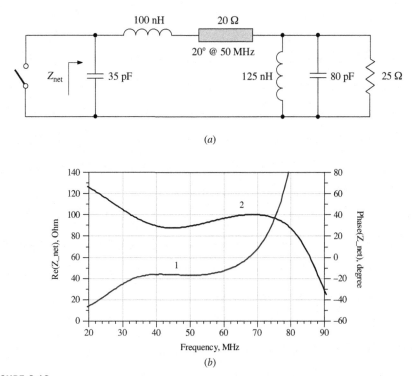

**FIGURE 8.12**

Class-E reactance compensation circuit with lumped elements and a transmission line.

As a result, for a unity characteristic impedance and cutoff frequency, the one-port impedance of a short-circuited transmission line corresponds to the reactive impedance of a lumped inductor $Z_L$ as

$$Z_L = sL = jL \tan \frac{\pi \omega}{2\omega_0}. \tag{8.37}$$

Similarly, the one-port admittance of an open-circuited transmission line corresponds to the reactive admittance of a lumped capacitor $Y_C$ as

$$Y_C = sC = jC \tan \frac{\pi \omega}{2\omega_0}. \tag{8.38}$$

The results obtained by Eqs (8.37) and (8.38) show that an inductor $L$ can be replaced with a short-circuit stub of electrical length $\theta = \pi \omega / 2\omega_0$ and characteristic impedance $Z_0 = L$, while a capacitor $C$ can be replaced with an open-circuit stub of electrical length $\theta = \pi \omega / 2\omega_0$ and characteristic impedance $Z_0 = 1/C$.

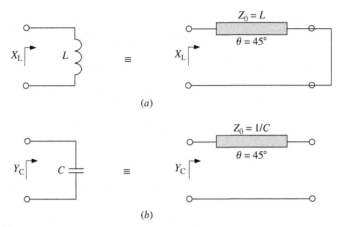

**FIGURE 8.13**

Equivalence between lumped elements and transmission lines.

From Eq. (8.36), it follows that the cutoff occurs when $\omega = \omega_c$, resulting in

$$\tan\frac{\pi\omega_c}{2\omega_0} = 1 \tag{8.39}$$

which gives a stub length $\theta = 45°$ (or $\pi/4$) with $\omega_c = \omega_0/2$. Hence, the inductors and capacitors of a lumped-element filter can be replaced with short-circuit and open-circuit stubs, as shown in Fig. 8.13. Since the lengths of all stubs are the same and equal to $\lambda/8$ at the cutoff frequency $\omega_c$, these lines are called the commensurate lines. At the frequency $\omega = \omega_0$, the transmission lines will be a quarter-wavelength long, resulting in an attenuation pole. However, at any frequency away from $\omega_c$, the impedance of each stub will no longer match the original lumped-element impedances, and the filter response will differ from the desired filter prototype response. Note that the response will be periodic in frequency, repeating every $4\omega_c$.

Figure 8.14(a) shows the idealized simulation setup of a 10-W, 28-V broadband Class-E power amplifier circuit designed to operate over a frequency bandwidth from 1.7 to 2.7 GHz and based on a GaN HEMT CGH40010 device, where both the input matching circuit and load network are composed of ideal transmission lines. To provide an input broadband matching, it is possible to use a multi-section matching transformer consisting of the stepped transmission-line sections with different characteristic impedances and electrical lengths [16,17]. Such an input matching structure is convenient in practical implementation since there is no need to use any tuning capacitors. The nominal Class-E load resistance can be calculated for $P_{\text{out}} = 15$ W, $V_{\text{dd}} = 28$ V, and $V_{\text{sat}} = 2.5$ V according to Eq. (5.33) as

$$R = 0.5768\frac{(V_{\text{dd}} - V_{\text{sat}})^2}{P_{\text{out}}} = 25\ \Omega \tag{8.40}$$

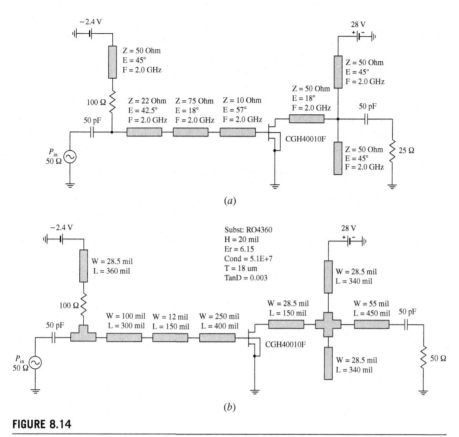

**FIGURE 8.14**

Idealized circuit schematic of a broadband GaN HEMT Class-E power amplifier.

where $P_{out}$ is the output power at the fundamental frequency, $V_{dd}$ is the drain supply voltage, and $V_{sat}$ is the saturation voltage defined from the device output current-voltage characteristics. In this case, the parallel resonant circuit in the broadband Class-E load network connected in parallel to a 25-$\Omega$ load is represented by the open- and short-circuit stubs, each having a characteristic impedance of 50 $\Omega$ and electrical length of 45° at 2 GHz. Simulation results show that drain efficiencies of 75% and greater can be achieved over whole required frequency bandwidth with a power gain of about 11 dB and an output power more than 42 dBm.

Figure 8.14(b) shows the implementation of an idealized circuit of a broadband GaN HEMT Class-E power amplifier shown in Fig. 8.14(a) into a RO4360 substrate, where an additional series transmission line with low characteristic impedance is used to match an idealized 25-$\Omega$ load with a standard 50-$\Omega$ load. As a result, an output power around 42 dBm with a power gain of more than 10 dB was simulated for an input power of 31 dBm, as shown in Fig. 8.15(a). In this case, the drain efficiency of over 72% was achieved across the required frequency range from 1.7 to 2.7 GHz, as shown in Fig. 8.15(*b*). Previously, a *PAE* above 60% was achieved between 1.87 and 2.11 GHz with an output power varying from 20 to 23 dBm for a medium-power broadband pHEMT Class-E power amplifier using a transmission-line parallel resonant circuit with short- and open-circuit stubs [18]. For a harmonically tuned GaN HEMT broadband power amplifier using open-circuit stubs in the load network incorporating a three-section bandpass filter, an output power of around 100 W with a drain efficiency of more than 65% in a frequency bandwidth from 1.55−2.25 GHz was achieved [19]. By using a six-order low-pass output matching circuit in a GaN HEMT Class-E power amplifier where the series inductors are replaced by the short-length transmission lines and the shunt capacitors are replaced by the open-circuit stubs, a drain efficiency of 63−89% with an output power of 10−20 W and a power gain of 10−13 dB was measured in a frequency bandwidth from 0.9 to 2.2 GHz at a supply voltage of 26 V [20].

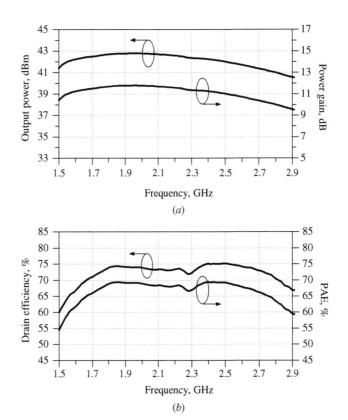

**FIGURE 8.15**

Output power, power gain, and efficiency versus frequency.

## 8.3 Broadband parallel-circuit Class-E

The susceptance compensation technique can be directly applied to the switch-mode parallel-circuit Class-E power amplifier because its load network configuration has exactly the same structure as shunt and series resonant circuits, as shown in Fig. 8.16(a) [21,22]. In this case, the nominal load resistance $R$ and phase angle $\phi$ of the parallel-circuit Class-E load network can be obtained from Eqs (6.79) and (6.83) in Chapter 6, respectively. The parallel inductance $L$ and shunt capacitance $C$ required for an optimum switchmode parallel-circuit Class-E operation are calculated as functions of the load resistance $R$ at the operating radian frequency $\omega$ from Eqs (6.80) and (6.81) in Chapter 6, respectively. The parameters

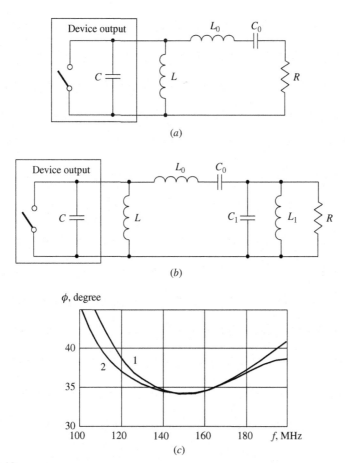

(a)

(b)

(c)

**FIGURE 8.16**

Single- and double-susceptance compensation circuits.

of the series resonant $L_0C_0$ circuit must be chosen to provide a constant phase angle of the load network over a required wide frequency bandwidth.

As a result, by substituting Eqs (6.80) and (6.81) into Eq. (8.6), the series capacitance $C_0$ and inductance $L_0$ can be calculated at the center bandwidth frequency $\omega_0$ by

$$L_0 = 1.026 \frac{R}{\omega_0} \tag{8.41}$$

$$C_0 = \frac{1}{\omega_0^2 L_0}. \tag{8.42}$$

Wider frequency bandwidth with high-efficiency performance can be achieved using a double-susceptance compensation circuit shown in Fig. 8.16(b), where $L_0C_0$ and $L_1C_1$ are the series and parallel resonant circuits, respectively [23]. In this case, similarly to the broadband design in a Class-E mode with shunt capacitance using a double-susceptance compensation, the parameters of the series and shunt resonant circuits for the broadband design in a parallel-circuit Class-E mode with the corresponding loaded quality factors $Q_0 = \omega_0 L_0/R$ and $Q_1 = \omega_0 C_1 R$, which are close to unity and greater, can approximately be calculated from Eqs (8.15) and (8.16), where the load angle $\theta = \tan^{-1}(R/\omega_0 L)$ using Eq. (6.80) is taken into account. Such a load network can be considered as a broadband matching-forming circuit which simultaneously provides the Class-E switching conditions and matching with a standard 50-$\Omega$ load over wide frequency bandwidth [24].

The circuit simulations for these two types of susceptance compensation load networks were performed at a center bandwidth frequency $f_0 = 150$ MHz for a standard load resistance $R = 50$ $\Omega$. Figure 8.16(c) shows the frequency dependencies of the load-network phase angle $\phi$ for the single-susceptance (curve 1) and double-susceptance (curve 2) compensation circuits, demonstrating their broadband operation capability. Using just a single-susceptance load network yields a significant widening of the operating frequency bandwidth with a minimum deviation of the magnitude and phase of the load-network impedance. A double-susceptance compensation load network obtains a maximum deviation from the optimum value of about $34°$ by only $3°$ in a frequency range from 120 to 180 MHz.

To achieve the high-efficiency broadband operation mode with a high-power gain in VHF frequency band, it is best to design the power amplifier based on silicon LDMOSFET devices. It is easy to provide a broadband input matching using lossy-matching circuit, especially at operating frequencies about 10 times lower than the device transition frequency $f_T$. Figure 8.17 shows the circuit schematic of an LDMOSFET power amplifier designed for operation in a 2:1 frequency bandwidth from 100 to 200 MHz using a double-susceptance compensation load network with broadband matching properties at the fundamental frequency [17]. The input lossy-matching circuit includes a simple $L$-transformer connected in parallel

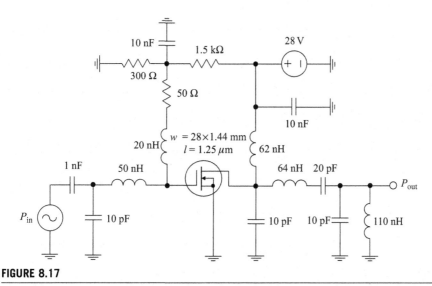

**FIGURE 8.17**

Simulated broadband Class-E LDMOSFET power amplifier.

with a series circuit consisting of an inductance of 20 nH and a resistance of
50 Ω. This provides a minimum input return loss at 200 MHz of about 15 dB and
an input *VSWR* less than 1.4 over the entire frequency bandwidth from 100 to
200 MHz.

From Fig. 8.18(*a*) it follows that, for such an octave-band VHF Class-E power
amplifier with an input power of 1 W using a 1.25-μm LDMOSFET device with a
total gate width of 28 × 1.44 mm, a power gain of 10 dB with deviation of only
±0.5 dB (curve 2) can be achieved with a drain efficiency of about 70% and
higher (curve 1). An analysis of the simulated drain voltage and current waveforms at the center bandwidth frequency of 150 MHz shown in Fig. 8.18(*b*)
demonstrates that the broadband operating mode is very close to a nominal parallel-circuit Class-E operation mode, although the impedance conditions at higher
harmonics are not controlled properly. As seen from the plots when the transistor
is turned on, high values of drain current (up to 1.3 A) are achieved with small
saturation voltages of 0−4 V. However, when the transistor is turned off, the
drain current continues to flow, but now through the device gate-drain capacitance $C_{gd}$ and drain-source capacitance $C_{ds}$ but not through the active channel.
A drain efficiency of 74% with an output power of 8 W across the frequency
range from 136 to 174 MHz with a power flatness of 0.7 dB were measured for a
parallel-circuit Class-E LDMOSFET power amplifier with a low supply voltage
of 7.2 V [25]. A power-added efficiency can be increased to 80% and more in a
frequency range from 140 to 180 MHz with an output power of 34.4 ± 1.5 dBm
using a GaN HEMT device [26].

Similarly, the transmission-line susceptance compensation technique can also
be applied to a parallel-circuit Class-E power amplifier where the series

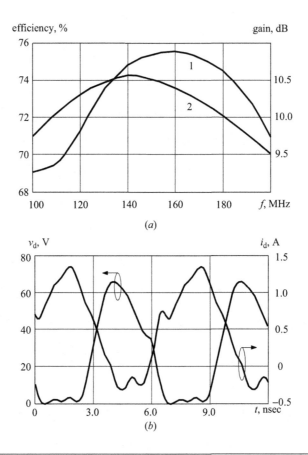

**FIGURE 8.18**

Broadband performance of a Class-E LDMOSFET power amplifier.

transmission line of a quarter wavelength at the center bandwidth frequency can be used instead of a series $L_0 C_0$ resonant circuit, as shown in Fig. 8.19(a). In some practical cases, the series quarterwave line can be replaced by an equivalent low-pass $\pi$-type circuit consisting of a series transmission line with higher characteristic impedance and electrical length much less than 90° and two shunt capacitors when the capacitance adjacent to the device output can be counted within the total shunt capacitance required for a nominal parallel-circuit Class-E mode. If it is necessary to additionally provide an output matching between the nominal Class-E resistance and 50-$\Omega$ load, a series quarterwave line can be replaced by a low-pass $L$-type matching circuit with a series transmission line and a shunt capacitor, as shown in Fig. 8.19(b).

Figure 8.20(a) shows an example of the transmission-line broadband Class-E load network where the parallel inductor is replaced by a short-length short-circuited transmission line which can be easily implemented on a printed-circuit

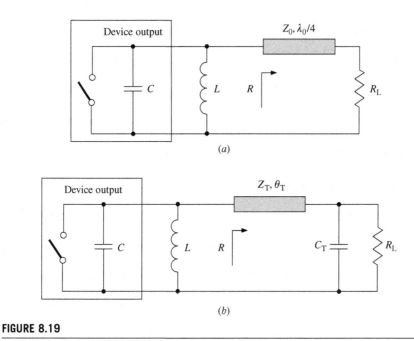

**FIGURE 8.19**

Transmission-line susceptance compensation circuit.

board providing minimum insertion loss. The electrical lengths of the transmission lines are given at the center bandwidth frequency of 300 MHz. In this case, the input load-network resistance varies from 17 $\Omega$ at 225 MHz to 47.5 $\Omega$ at 400 MHz, as shown in Fig. 8.20(*b*) by curve 1, with much less variation from 18.5 $\Omega$ to 27 $\Omega$ in a frequency range from 250 to 350 $\Omega$. The phase stays almost constant around 33° in a frequency range from 250 to 350 $\Omega$ and varies from 22.5° at 225 MHz to 39.5° at 400 MHz (curve 2).

Figure 8.21(*a*) shows the circuit schematic of a broadband high-efficiency microstrip LDMOSFET power amplifier with an output power of around 20 W and a power gain of more than 12 dB in a frequency range from 225 to 400 MHz at a dc-supply voltage of 28 V. Here, to approximate the parallel-circuit Class-E mode in a wide frequency range, the load network was designed to realize a single-susceptance compensation technique using a parallel short-length transmission line in conjunction with a single *L*-type transmission-line transformer, since a ratio between the device equivalent output resistance required for an optimum Class-E operation and the standard load of 50 $\Omega$ is not significant. The input matching circuit includes two low-pass *L*-type matching sections to compensate for the device input capacitance over the entire frequency range. A lossy parallel resistance of 75 $\Omega$ is necessary to simplify the matching procedure and improve the input return loss. As a first step, each matching network structure is calculated at the center-band frequency based on the technical requirements and device

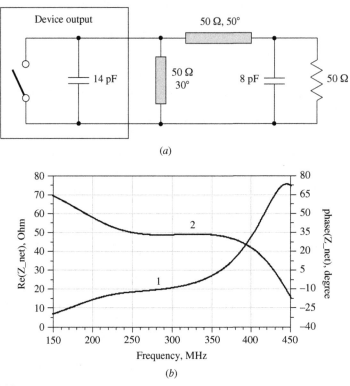

**FIGURE 8.20**

Transmission-line Class-E load network with a susceptance compensation.

equivalent circuit parameters. Then, to optimize the power amplifier performance over the entire frequency band, the simplest and fastest way is to apply an optimization procedure using computer simulators to satisfy certain criteria. For such a broadband power amplifier, the minimum output power ripple and input return loss with maximum power gain and efficiency can be chosen as the criteria. Generally, applying a nonlinear broadband optimization technique and setting the ranges of electrical length of the transmission lines between 0 and 90° and parallel capacitances from 0 to 100 pF, we can obtain the parameters of the input matching and output load network.

However, to speed up this procedure, it is best to optimize circuit parameters separately for the input and output circuits. In this case, the input matching circuit is loaded by the device equivalent input series *RC* circuit, consisting of its gate resistance and gate-source capacitance. The load network must include at its input the device equivalent output shunt *RC* circuit consisting of an optimum Class-E load resistance required for a specified output power and supply voltage and drain-source capacitance. In this case, it is sufficient to use a fast linear

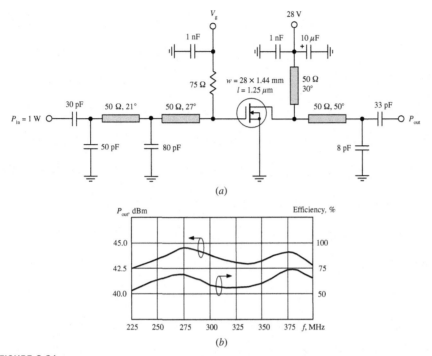

**FIGURE 8.21**

Broadband high-efficiency microstrip LDMOSFET power amplifier.

optimization process, which will take only a few minutes to complete the circuit design procedure. Finally, the resulting optimized values are incorporated into the overall power amplifier circuit for each element and final optimization is performed using a nonlinear active device model. The optimization process is finalized by choosing the nominal level of input power with optimizing elements in narrower ranges of their values of about 10–20% for most critical elements. For practical convenience, it is advisable to choose the characteristic impedances of all transmission lines of 50 Ω. Figure 8.21(b) shows the simulated broadband high-efficiency power amplifier performance achieving an output power of 42.5–44.5 dBm, a power gain of 13.5 ± 1 dB, and a drain efficiency of 64 ± 10% in a frequency bandwidth from 225 to 400 MHz.

The circuit schematic of a broadband two-stage InGaP/GaAs HBT power amplifier intended to operate in the WCDMA handset transmitters is shown in Fig. 8.22 [21,22]. The MMIC part of this power amplifier contains the transistors with emitter areas of the first and second stage as large as 540 and 3600 $\mu m^2$, input matching circuit, interstage matching circuit, and bias circuits on a die with dimensions of less than 1 mm$^2$. The MMIC packaged in a $3 \times 3$ mm$^2$ package was mounted on a FR4 substrate which contaians the output matching circuit and

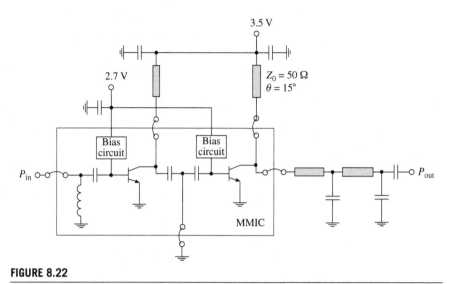

**FIGURE 8.22**

Circuit schematic of a parallel-circuit GaAs HBT Class-E MMIC power amplifier.

microstrip lines. Standard ceramic chip capacitors were used in the output matching circuit, and no further additional tuning was necessary. In this case, a very short microstrip line operating as a dc-feed inductance is required to approximate parallel-circuit Class-E switching conditions.

Figure 8.23(a) shows that the small-signal gain varies within $22.5 \pm 0.5$ dB and the input return loss is greater than 13 dB in a frequency range from 1.6 GHz to more than 2 GHz, thus confirming the broadband operation of the amplifier. Without any tuning of the output matching circuit, a saturated output power greater than 30 dBm and a *PAE* greater than 50% were obtained. These single-tone measurements shown in Fig. 8.23(b) were performed at the respective center-band frequencies 1.75 and 1.88 GHz. Using high-$Q$ capacitors in output matching circuit can improve the power-added efficiency by about 8%, resulting in a close to 60% result. At the same time, the power amplifier provides high-linearity performance in a handset WCDMA band (1920−1980 MHz) at a 3.5-dB backoff output power of 27 dBm with a power gain of 22.6 dB and a sufficiently high efficiency. The measured *PAE* reached a value of 38.3% at center bandwidth frequency of 1.95 GHz with an *ACLR* of −37 dBc at a 5-MHz offset and *ACLR* of −56 dBc at a 10-MHz offset.

## 8.4 High-power RF Class-E power amplifiers

The load network of a high-power broadband VHF Class-E power amplifier can be based on lumped low-pass *LC* matching sections to match the nominal Class-E

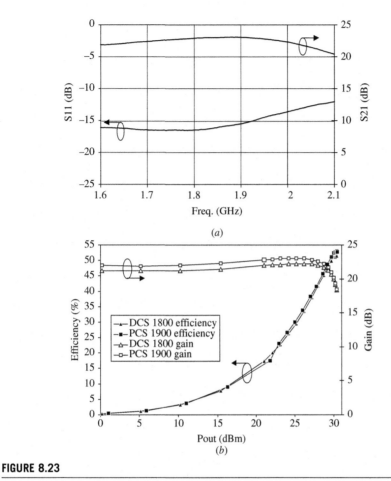

**FIGURE 8.23**

Input return loss, gain, and efficiency *versus* frequency.

load with the standard 50-Ω load. When the frequency bandwidth is not very wide, at about 50%, it is possible to use the simplest single-section low-pass π-transformer to provide a drain efficiency greater than 60% over a frequency bandwidth from 80 to 135 MHz with an average output power of 20 W using a 28-V, 60-W SiC MESFET device [27]. However, variation of both the output power (more than 3 dB) and drain efficiency (more than 15%) across this frequency bandwidth is significant, since such a simple load network cannot maintain the required real and imaginary part of the Class-E load network constant enough over the entire frequency range. In this case, the imaginary part of the load network increases from a low-band frequency to a high-band frequency, characterizing by the positive slope.

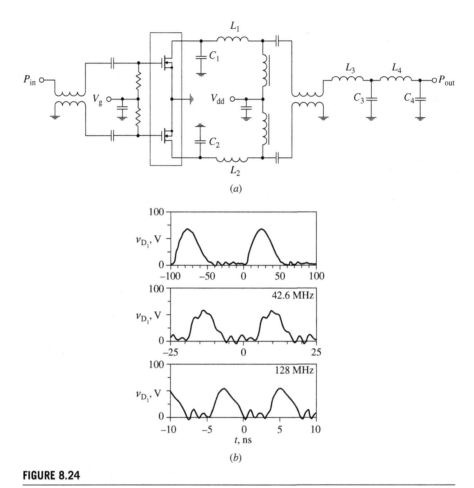

**FIGURE 8.24**

Push–pull MOSFET Class-E power amplifier and drain-voltage waveforms.

Figure 8.24(a) shows the circuit schematic of a push–pull Class-E power amplifier using a 50-V balanced MOSFET device, producing an output power up to 200 W and operating in a frequency range from 1.8 to 128 MHz [28]. The circuit incorporates a broadband input matching circuit so that the frequency changes require only retuning or switching of the output filter. In this case, the drain efficiency varies from 90% at 1.8 MHz to 70% at 128 MHz. The input signal is transformed and split by a broadband 4:1 Guanella transformer, whose outputs represent two 10-W, 6.25-$\Omega$ resistors shunting the MOSFET gates and providing the required resistive loads for the input transformer. Such an input circuit is operated in a broadband mode until the reactances of the MOSFET gates become less than 6.25 $\Omega$. Operation at higher frequencies requires the addition of inductors to cancel the MOSFET gate capacitances. The drain loads of 3.125 Ω in a Class-E mode with shunt capacitance allow about 100 W to be produced in each side of the MOSFET with a peak drain voltage below the breakdown voltage. The total drain capacitance is approximately correct for optimum Class-E operation at about 85 MHz. At lower frequencies, the drain-shunt capacitors are therefore added to achieve optimum Class-E operation. The broadband output transformer represents a balun that splits its 6.25-Ω load into two 3.125-Ω drain loads. The output impedance-transforming filter is a low-pass two-section ladder-type network where the shunt variable capacitors are responsible for load control and efficiency adjustment and the first series inductor provides sufficient reactance to ensure adequate suppression of the third harmonic. The drain-voltage waveforms for operation at 10 MHz, 42.6 MHz, and 128 MHz are shown in Fig. 8.24(b), where the ideal optimum Class-E waveform is satisfied at 10 MHz and close at 42.6 MHz. At 128 MHz, the power amplifier is operating in a suboptimum Class-E mode.

The third-order bandpass broadband Class-E load network can be optimized not only to provide required optimum inductive impedance at the fundamental frequency, but also purely capacitive optimized reactances at the second and third harmonics. Figure 8.25(a) shows such a third-order parallel-circuit Class-E load network, where the impedance-transforming network is used to transform the standard resistive 50-Ω load to the load required for proper termination of the Class-E load network [29]. As a result, a drain efficiency near 70% and better can be achieved over the frequency bandwidth from 95 to 135 MHz for a 12-W, 12-V MOSFET power amplifier with a 4:1 Ruthroff transmission-line transformer in an impedance-transforming network providing 12.5 Ω to the Class-E load network at a fundamental and several harmonics. Figure 8.25(b) shows the output circuit of a high-power broadband 28-V LDMOSFET power amplifier with a broadband low-loss impedance transformer located right at the output port of the power transistor [30]. The transformer is made of three rings of low impedance 15-Ω semi-rigid coaxial cable, where the cable outer jackets are connected in parallel to form the primary of the transformer and the inner conductors are connected series to form the secondary of the transformer. In order to reduce losses to a minimum, no magnetic core is used. A lumped network is located after the transformer to provide both proper loads at the harmonics and at the fundamental frequency for a nominal Class-E with shunt capacitance. In this case, high drain efficiencies up to 86% with output powers of greater than 120 W were measured in a frequency bandwidth of VHF broadcasting from 88 to 114 MHz.

## 8.5 Microwave monolithic Class-E power amplifiers

Generally, by providing an open-circuit termination for the second- and third-harmonic components, the collector efficiency of a microwave Class-E power

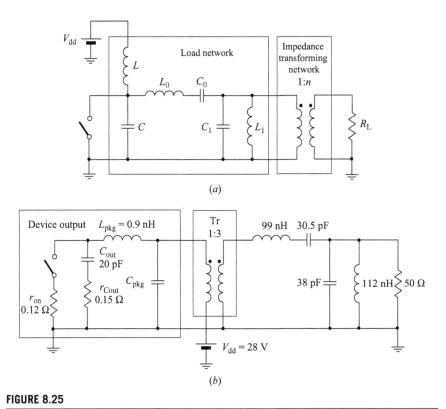

**FIGURE 8.25**

Output circuits of broadband VHF Class-E power amplifiers.

amplifier can be increased by 10% [31]. In this case, the second-harmonic termination has the most impact on the collector efficiency, while effect of an open-circuit termination for the fourth harmonic is negligible. Moreover, the variation of the second-harmonic load reflection coefficient by 10% in magnitude from 1 to 0.9 and $\pm 20°$ in phase angle results in an insignificant efficiency variation, within only 1%. Figure 8.26(a) shows the circuit schematic of a monolithic broadband Class-E power amplifier with a chip size of $2 \times 2.2$ m$^2$. The load network is a compromise solution between having a low insertion loss and meeting the necessary requirements for the optimum Class-E operation with nonzero voltage and voltage-derivative switching conditions [32]. This is accomplished by using two open-circuit stubs in conjunction with a shunt capacitor, where the first open-circuit stub in combination with the series transmission line presents broadband high-impedance terminations for the second harmonics within 18−22 GHz, while the combination of the second open-circuit stub and the shunt capacitor presents broadband low impedances at the third harmonics within 27−33 GHz and also transforms the optimum load impedances at the fundamental frequencies. The simulated loading conditions presented to the output of the device at the

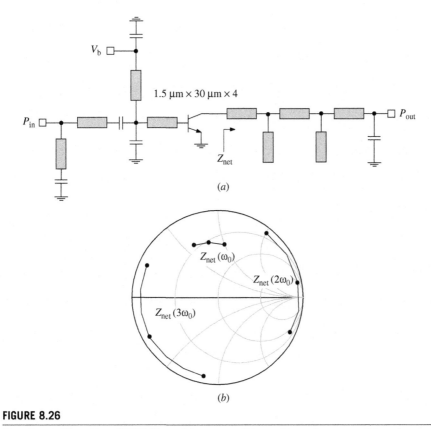

**FIGURE 8.26**

Broadband *X*-band in a DHBT Class-E power amplifier and impedance conditions.

fundamental (inductive impedance), second (high impedance), and third harmonic (low impedance) frequencies are shown in Fig. 8.26(*b*), which are proved to be adequate for broadband Class-E power amplifiers. As a result, using an indium phosphide (InP) double HBT (DHBT) technology, a *PAE* of 49−65% with an output power of 18−22 dBm was achieved over the frequency bandwidth from 9 to 11 GHz [31]. Based on an InP DHBT technology, a single-stage broadband *X*-band Class-E power amplifier can also achieve a *PAE* of 45−60% with an output power of 19−21.5 dBm and a power gain of 9−11.5 dB over a 34% bandwidth, from 8.2 to 11.6 GHz [33].

To increase the overall efficiency of a two-stage power amplifier, it may be assumed that it is worthwhile to optimize both amplifying stages to operate in a Class-E mode. For example, for a hybrid microwave GaAs MESFET Class-E power amplifier using the same devices in both stages, the maximum two-stage power-added efficiency achieved was as high as 52% (including connector loss)

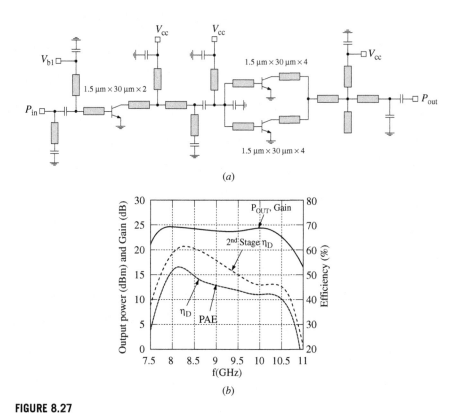

(a)

(b)

**FIGURE 8.27**

Broadband two-stage X-band in a DHBT Class-E power amplifier and its performance.

with a corresponding power gain of 16 dB and an output power of 20 dBm at a carrier frequency of 10 GHz and a supply voltage of 4.2 V [34]. However, due to Class-E operation mode of the driver stage, the overall power gain is sufficiently small, thus affecting the overall efficiency. Therefore, by using a Class-AB driver stage, similar efficiency can be achieved with substantially higher power gain. As a result, for a monolithic microwave two-stage high-efficiency InP DHBT power amplifier shown in Fig. 8.27(a) where the driver stage is operated in a Class-AB mode and the output stage is operated in a Class-E mode, a *PAE* of 52% with an output power of 24.6 dBm and a power gain of 24.6 dB was achieved at a carrier frequency of 8 GHz and a supply voltage of 4 V. The total emitter area of the driver-stage device was chosen to be 90 $\mu m^2$, providing a *PAE* of the driver stage above 40% and an adequate power to push the output stage deep into compression, as required for a switchmode Class-E operation. The output stage consists of two active devices with a total emitter area of 360 $\mu m^2$ combined in parallel reactively, taking care to provide odd-mode instability suppression resistors between the base and collector of each transistor. The power-added efficiency is

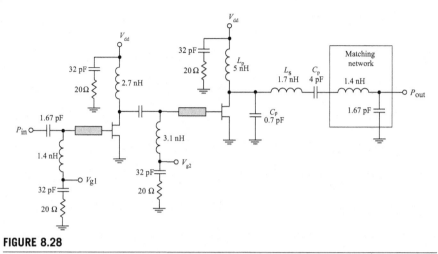

**FIGURE 8.28**

Circuit schematic of a broadband Class-E pHEMT power amplifier.

maintained at greater than 40% over a frequency bandwidth from 7.7 to 10.5 GHz, as shown in Fig. 8.27(*b*) [34].

Figure 8.28 shows the circuit schematic of a two-stage broadband Class-E power amplifier implemented in a 0.5-μm enhancement/depletion pHEMT process with a chip size of $2 \times 2$ mm$^2$, which is intended to operate in a frequency range from 1.5 to 3.8 GHz with a *PAE* of better than 62% and an output power of more than 27 dBm at $V_{dd} = 6$ V [35]. In this case, to provide high operation efficiency in a wide frequency range, the Class-E load network with reactance compensation technique followed by the low-pass matching network is used. The driver stage is designed to operate in a Class-AB mode with a small quiescent current for high gain and high efficiency when both input and interstage matching circuits are conjugately matched. For a 0.5-μm pHEMT two-stage broadband Class-E power amplifier with a chip size of $5.25 \times 2.8$ mm$^2$, a *PAE* of above 50% with an output power of more than 36 dBm at a drain supply voltage of 6 V was obtained in a frequency range from 3.0 to 3.75 GHz [36].

Figure 8.29 shows the circuit schematic of a compact single-stage broadband Class-E GaN HEMT power amplifier, where the load network is based on reactance compensation technique with a parallel circuit followed by the low-$Q$ series resonant circuit [37]. The use of a finite dc-feed inductance has advantages in terms of the output power and maximum frequency of operation, and results in a higher load resistance than in the classical Class-E configuration with infinite RF choke. To shape the gate voltage waveform, the second-harmonic signal is short-circuited at the gate by the series resonant circuit representing a second-harmonic trap. The high- and low-pass matching sections form a bandpass input matching circuit, where a 300-Ω parallel resistor with a 37-Ω bias resistor provide an unconditional stability of the power amplifier both at low and high frequencies.

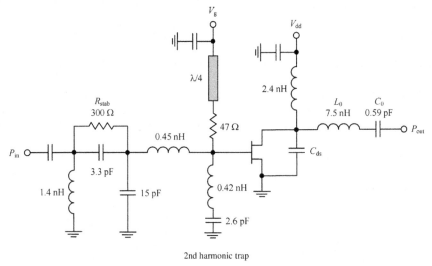

**FIGURE 8.29**

Circuit schematic of a broadband Class-E GaN HEMT power amplifier.

The actual size of a broadband Class-E power amplifier with the input matching circuit and load network implemented in a two-layer Rogers laminate with a dielectric permittivity of 3.5 is only $1.1 \times 1.6$ cm$^2$. As a result, a drain efficiency above 74% and an output power more than 7 W with an input power of 600 mW at $V_{dd} = 40$ V can be achieved across the bandwidth of 2.0–2.5 GHz. Wider frequency bandwidth of 2.1–2.7 GHz with a drain efficiency exceeding 63% and an output power above 9.3 W can be provided without input of the second-harmonic trap and retuning the finite dc-feed inductance. By using a two-section $LC$ ladder output matching circuit in a GaN HEMT Class-E power amplifier, a drain efficiency over 68% with an output power of 42 to 65 W in a frequency bandwidth from 1.7 to 2.3 GHz at a supply voltage of 35 V was achieved for a compact hybrid implementation of the power amplifier with effective area of $2 \times 2$ cm$^2$ where the bondwire inductors and MIM capacitors are used [38].

## 8.6 CMOS Class-E power amplifiers

To realize a broadband high-efficiency operation of the fully integrated CMOS Class-E power amplifier, a broadband and low-loss 1:4 Ruthroff-type transmission-line transformer based on the broadside-coupled transmission lines can provide an impedance transformation from $12.2 \pm 01$ $\Omega$ to 50 $\Omega$ [39]. In a six-layer 0.18-μm CMOS process, the thickest top metal 6 is used as the primary winding, the identical thick metal stacked from metal 1 to metal 4 is used as the secondary winding to improve insertion loss, and both windings are wound in loops keeping

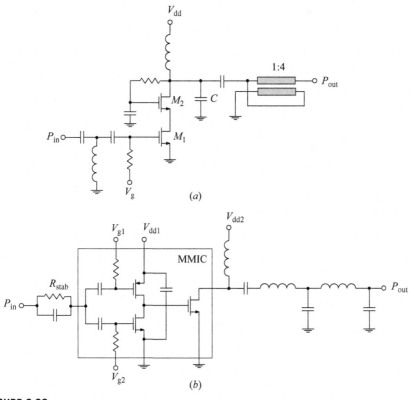

**FIGURE 8.30**

Circuit schematics of broadband Class-E CMOS power amplifiers.

the 1:1 turns ratio to reduce the transformer size. Figure 8.30(*a*) shows the circuit schematic of a 0.18-μm CMOS Class-E power amplifier composed of the two nMOS transistors in a cascode configuration and one shunt capacitor in the load network required for optimum Class-E operation. Here, the series *LC* resonant circuit at the fundamental frequency of the Class-E power amplifier is replaced by the 1:4 transmission-line transformer operating as a broadband bandpass filter. To enhance the reliability of the transistors, the thick-oxide transistor $M_2$ is used for the common-gate stage, and the thin-oxide transistor $M_1$ is used for the common-source stage. The fully integrated CMOS Class-E power amplifier with a 1:4 transmission-line transformer exhibits a broadband and flat output power level of $24 \pm 0.2$ dBm from 2.4 to 3.5 GHz at a supply voltage of 3.6 V, with a maximum *PAE* of 33.2% at 2.6 GHz.

Figure 8.30(*b*) shows the circuit schematic of a two-stage broadband Class-E CMOS power amplifier, where the power output stage is formed by a high-voltage, extended-drain, thick-oxide nMOS device implemented in a standard 65-nm

CMOS technology [40]. The total gate width of the transistor is 3.84 mm and the channel length is 0.28 μm, realizing an on-resistance of 0.7 Ω, an off-resistance of 10 kΩ, and a drain-source capacitance of approximately 4.14 pF. To drive the output stage as a switch, a square-wave signal is generated by an inverter-based driver implemented using standard thick-oxide MOS devices with a gate length of 0.28 μm. To reduce the peak drain voltage and improve reliable operation, a suboptimum Class-E operation is applied. The broadband load network represents an off-chip two-section *LC* ladder circuit. As a result, a measured output power of $30.5 \pm 0.5$ dBm, a power gain of $16.5 \pm 0.5$ dB, a drain efficiency above 67%, and a *PAE* above 52% are achieved across the frequency bandwidth from 550 to 1050 MHz.

# REFERENCES

1. Everard JKA, King AJ. Broadband power efficient Class E amplifiers with a non-linear CAD model of the active MOS device. *J. IERE*. March 1987;57:52−58.
2. Matthaei GL. A study of the optimum design of wide-band parametric amplifiers and up-converters. *IRE Trans. Microwave Theory Tech*. January 1961;MTT-9:23−38.
3. DeJaeger JT. Maximum bandwidth performance of a nondegenerate parametric amplifier with single-tuned idler circuit. *IEEE Trans. Microwave Theory Tech*. July 1964; MTT-12:459−467.
4. Humphreys BL. Characteristics of broadband parametric amplifiers using filter networks. *Proc. IEE*. February 1964;111:264−274.
5. Aitchison CS, Gelsthorpe RV. A circuit technique for broadbanding the electronic tuning range of gunn oscillators. *IEEE J. Solid-State Circuits*. February 1977; SC-12:21−28.
6. Aitchison CS. Method of improving tuning range obtained from a varactor-tuned gunn oscillator. *Electronics Lett*. April 1974;10:94−95.
7. Gelsthorpe RV, Aitchison CS. Analytical evaluation of the components necessary for double reactance compensation of an oscillator. *Electronics Lett*. September 1976;12:485−486.
8. Chapman AG, Aitchison CS. Circuit technique for broadband impedance matching of passive loads. *IEE J. Microwaves Optics Acoustics*. March 1979;3:43−50.
9. Camargo E, Consoni D. Reactance compensation matches FET circuits. *Microwaves*. June 1985;24:93−95.
10. Soares R. *GaAs MESFET Circuit Design*. Boston: Artech House; 1988.
11. Al Tanany A, Sayed A, Boeck G. Broadband GaN switch mode Class E power amplifier for UHF applications. *2009 IEEE MTT-S Int. Microwave Symp. Dig*. 761−764.
12. Al Tanany A, Sayed A, Bengtsson O, Boeck G. Time domain analysis of broadband GaN switch mode Class-E power amplifier. *Proc. 5th German Microwave Conf*. 2010;254−257.
13. Rao Gudimetla VS, Kain AZ. Design and validation of the load networks for broadband Class E amplifiers using nonlinear device models. *1999 IEEE MTT-S Int. Microwave Symp. Dig*. 2:823−826.

14. Richards PI. Resistor-transmission-line circuits. *Proc. IRE.* February 1948;36:217−220.
15. Saal R, Ulbrich E. On the design of filters by synthesis. *IRE Trans Circuit Theory.* December 1958;CT-5:284−327.
16. Meschanov VP, Rasukova IA, Tupikin VD. Stepped transformers on TEM-transmission lines. *IEEE Trans. Microwave Theory Tech.* June 1996;MTT-44:793−798.
17. Grebennikov A. *RF and Microwave Power Amplifier Design.* New York: McGraw-Hill; 2004.
18. Qin Y, Gao S, Butterworth P, Korolkiewicz E, Sambell A. Improved design technique of a broadband Class-E power amplifier at 2 GHz. *Proc. 35th Europ. Microwave Conf.* 2005;1:1−4.
19. Al Tanany A, Gruner D, Boeck G. Harmonically tuned 100 W broadband GaN HEMT power amplifier with more than 60% PAE. *Proc. 41st Europ. Microwave Conf.* 2011;159−162.
20. Chen K, Peroulis D. Design of highly efficient broadband Class-E power amplifier using synthesized low-pass matching networks. *IEEE Trans. Microwave Theory Tech.* December 2011;MTT-59:3162−3173.
21. Jaeger H, Grebennikov AV, Heaney EP, Weigel R. Broadband high-efficiency monolithic InGaP/GaAs HBT power amplifiers for 3G handset applications. *2002 IEEE MTT-S Int. Microwave Symp. Dig.* 2:1035−1038.
22. Jaeger H, Grebennikov AV, Heaney EP, Weigel R. Broadband high-efficiency monolithic InGaP/GaAs HBT power amplifiers for wireless applications. *Int. J. RF and Microwave Computer-Aided Eng.* June 2003;13:496−519.
23. Grebennikov A. Simple design equations for broadband Class E power amplifiers with reactance compensation. *2001 IEEE MTT-S Int. Microwave Symp. Dig.* 3:2143−2146.
24. Degtev VI, Kozyrev VB. Transistor single-ended switching-mode power amplifier with forming circuit (in Russian). *Poluprovodnikovaya Elektronika v Tekhnike Svyazi.* 1986;26:178−188.
25. Kumar N, Prakash C, Grebennikov A, Mediano A. High-efficiency broadband parallel-circuit Class E power amplifier with reactance-compensation technique. *IEEE Trans. Microwave Theory Tech.* March 2008;MTT-56:604−612.
26. Khansalee E, Puangngernmak N, Chalermwisutkul S. Design of 140−170 MHz Class E power amplifier with parallel circuit on GaN HEMT. *Proc. Int. Electrical Eng./Electron. Computer Telecom. Inform. Technol. Conf.* 2010;570−574.
27. Chen W, Li X, Wang L, Feng Z, Xue X. A novel broadband VHF MESFET Class-E high power amplifier. *Microwave and Optical Technology Lett.* February 2010;52:272−276.
28. Raab FH. Broadband Class-E power amplifier for HF and VHF. *2006 IEEE MTT-S Int. Microwave Symp. Dig.* 902−905.
29. Ortega-Gonzalez FJ. Load-pull wideband Class-E amplifier. *IEEE Microwave and Wireless Components Lett.* March 2007;17:235−237.
30. Ortega-Gonzalez FJ. High power wideband Class-E power amplifier. *IEEE Microwave and Wireless Components Lett.* October 2010;20:569−571.
31. Quach TK, Watson PM, Okamura W, et al. Ultra-high efficiency power amplifier for space radar applications. *IEEE J. Solid-State Circuits.* September 2002; SC-37:1126−1134.

32. Watson P, Neidhard R, Kehias L, et al. Ultra-high efficiency operation based on an alternative Class-E mode. *2000 IEEE GaAs IC Symp. Dig.* 53−56.

33. Watson P, Quach T, Axtel H, et al. An indium phosphide X-Band Class-E power MMIC with 40% bandwidth. *2005 IEEE Compound Semiconductor Integrated Circuit Symp. Dig.* 220−223.

34. Pajic S, Wang N, Watson PM, Quach TK, Popovic Z. X-Band two-stage high-efficiency switched-mode power amplifiers. *IEEE Trans. Microwave Theory Tech.* September 2005;MTT-53:2899−2907.

35. Lin C-H, Chang H-Y. A high efficiency broadband Class-E power amplifier using a reactance compensation technique. *IEEE Microwave and Wireless Components Lett.* September 2010;20:507−509.

36. van Wanum M, van Dijk R, de Hek P, van Vliet FE. Broadband S-Band Class E HPA. *Proc. 4th Europ. Microwave Integrated Circuits Conf.* 2009;29−32.

37. van der Heijden MP, Acar M, Vromans JS. A Compact 12-W high-efficiency 2.1−2.7 GHz Class-E GaN HEMT power amplifier for base stations. *2009 IEEE MTT-S Int. Microwave Symp. Dig.* 657−660.

38. Shi K, Calvillo-Cortes DA, de Vreede LCN, van Rijs F. A compact 65 W 1.7−2.3 GHz Class-E GaN power amplifier for base stations. *Proc. 6th Europ. Microwave Integrated Circuits Conf.* 2011;542−545.

39. Liao H-Y, Pan M-W, Chiou H-K. Fully-integrated CMOS Class-E power amplifier using broadband and low-loss 1:4 transmission-line transformer. *Electronics Lett.* October 2010;46:1490−1491.

40. Zhang R, Acar M, van der Heijden MP, Apostolidou M, de Vreede LCN, Leenaerts MW. A 550−1050 MHz +30 dBm Class-E power amplifier in 65 nm CMOS. *2011 IEEE RFIC Symp. Dig*, pp. 289−292.

# Alternative and Mixed-Mode High-Efficiency Power Amplifiers

## INTRODUCTION

In this chapter, the different alternative and mixed-mode configurations of high-efficiency power amplifiers are presented. A Class-DE power amplifier is based on the combination of a voltage-switching Class-D mode with Class-E switching conditions, thus extending the switching Class-D operation to higher frequencies. Effects of the saturation resistance and nonlinear capacitance, driving waveforms, and some practical examples of Class-DE power amplifiers are discussed. The switchmode Class-FE or Class-E/F power amplifier can provide lower voltage peak factors when zero voltage and zero voltage-derivative conditions, corresponding to Class-E mode required to eliminate discharge loss of the shunt capacitance, are accompanied by harmonic tuning using the resonant circuits tuned to selected harmonic components realizing Class-F or inverse Class-F mode. Also, the biharmonic Class-$E_M$ mode is described, which can eliminate the efficiency degradation of a Class-E operation mode at higher frequencies due to the increased switching power losses with increasing values of the turn-off switching time. The requirements of both jumpless voltage and current waveforms and sinusoidal load waveform with nonzero output power delivered to the load can be provided by using nonlinear reactive elements in the load network to convert fundamental-frequency power to a desired harmonic frequency or by injecting the harmonic-frequency power into the load network from an external source. An inverse Class-E power amplifier represents an inverse version of a classical Class-E power amplifier with a shunt capacitance where the load-network inductor and capacitor replace each other. Generally, it is limited to low operating frequencies or low output powers since it is based on zero-current and zero current-derivative switching conditions. However, to compensate for the device output capacitor finite discharging process, it is necessary to provide zero voltage-derivative conditions at the same time, and collector efficiency will drop drastically if this capacitance is significant. Harmonic-control techniques for designing microwave power amplifiers are given with a description of a systematic procedure of multiharmonic load-pull simulation using the harmonic-balance method and active load-pull measurement system. Finally, outphasing modulation systems is considered where a variable envelope output is created by the sum of two constant-envelope signals with varying phases which, therefore, can be amplified by a highly efficient power amplifier operated in a switching Class-D, Class-E, or Class-F operation mode, or their combinations.

## 9.1  Class-DE power amplifier

The complementary voltage-switching Class-D power amplifier with a series filter, whose equivalent circuit is shown in Fig. 9.1(a), has the advantage of a lower voltage peak factor, compared with the current-switching Class-D power amplifiers. The maximum peak factor for the voltage-switching Class-D power amplifier, as a ratio of the maximum collector voltage, which is equal to the dc supply voltage, and the dc component of the collector voltage waveform does not exceed 2. This results in a higher output power which can be delivered to the load. Besides, the transitions from a pinch-off mode to a saturation mode and vice versa occur at zero collector currents, thus avoiding the device active modes and eliminating additional power losses. However, the frequency limitations of such a voltage-switching Class-D power amplifier are provided by the device parasitic collector shunt capacitances, as shown in Fig. 9.1(b), resulting in the increased switching transition times due to the capacitor charging and discharging processes.

A possible way to eliminate these power losses and to extend the voltage-switching class-D mode to higher frequencies is to introduce a dead time during the period when one device has already turned off but the other has not turned on yet, and the inductive load network is used to charge and discharge the shunt capacitances. This can be done by introducing the Class-E switching conditions when the switching loss during off-to-on transition is reduced to zero by the operating requirements of zero voltage at zero voltage slope at the end of the period. Since the shunt capacitor must be discharged at that exact time, an additional series inductor $L$ with optimum value should be included into the load network, as shown in Fig. 9.1(b). As a result, the switching current and voltage waveforms have the characteristics of both Class-D and Class-E operation modes. The series $L_0C_0$ filter is still necessary to suppress the harmonic components to allow only the sinusoidal signal to flow into the load $R$.

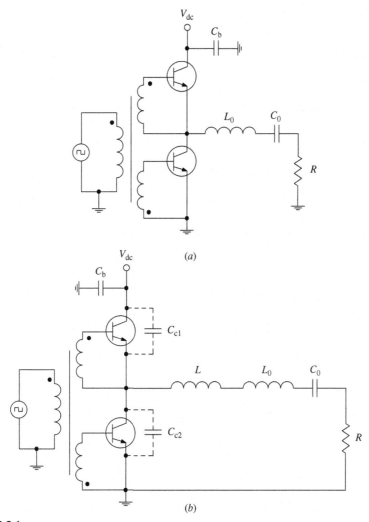

**FIGURE 9.1**

Complementary voltage-switching Class-D and Class-DE power amplifiers with a series filter.

Such a Class-DE power amplifier was first described by Zhukov and Kozyrev in [1] and has found some particular applications due to its high operation efficiency at higher operating frequencies [2,3]. The optimum parameters of a voltage-switching Class-DE power amplifier can be determined based on its steady-state collector voltage and current waveforms. Figure 9.2 shows the three different Class-DE switching circuits that occur during a switching cycle, the load network of which consists of the shunt collector capacitances $C_{c1}$ and $C_{c2}$, a series

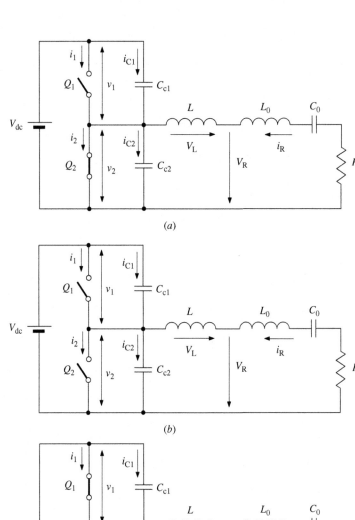

**FIGURE 9.2**

Equivalent circuits of a voltage-switching Class-DE power amplifier.

inductor $L$, a series fundamentally tuned $L_0C_0$ circuit, and a load resistor $R$. Both active devices $Q_1$ and $Q_2$ are considered an ideal switch that is driven so as to provide the device switching between its on-state and off-state operation conditions. The collector of the top device $Q_1$ is connected to the dc supply voltage and the emitter of the bottom device $Q_2$ is connected to the ground. To simplify the analysis of a Class-DE power amplifier, the following several assumptions are introduced:

- The transistors have zero saturation voltage, zero saturation resistance, infinite off-resistance, and their switching is instantaneous and lossless.
- The shunt capacitances are independent of the collector and are assumed linear.
- The loaded quality factor $Q_L = \omega L_0/R = 1/\omega C_0 R$ of the series resonant $L_0C_0$ circuit tuned to the fundamental frequency is high enough for the output current to be sinusoidal at the switching frequency.
- There are no losses in the circuit except into the load $R$.

The Class-E switching conditions for each device can be written as

$$v_1(\omega t)\Big|_{\omega t=\pi} = 0 \qquad \frac{dv_1(\omega t)}{d\omega t}\Big|_{\omega t=\pi} = 0 \tag{9.1}$$

$$v_2(\omega t)\Big|_{\omega t=2\pi} = 0 \qquad \frac{dv_2(\omega t)}{d\omega t}\Big|_{\omega t=2\pi} = 0 \tag{9.2}$$

where $v_1$ and $v_2$ are the voltages across switches $Q_1$ and $Q_2$, respectively.

The detailed theoretical analysis of a Class-DE power amplifier in a general form is given in [1], where the output current is assumed sinusoidal written as

$$i_R(\omega t) = I_R \sin(\omega t + \varphi) \tag{9.3}$$

where $I_R$ is the current amplitude and $\varphi$ is the initial phase shift.

The analysis is performed in the interval $0 \le \omega t \le 2\pi$ with dead time $\tau_d$ during which each of switches $Q_1$ and $Q_2$ are turned off. The basic Kirchhoff equations characterizing the electrical behavior of the equivalent circuits of a voltage-switching Class-DE power amplifier are

$$i_2(\omega t) + i_{C2}(\omega t) = i_1(\omega t) + i_{C1}(\omega t) + i_R(\omega t) \tag{9.4}$$

$$v_1(\omega t) = V_{dc} - v_2(\omega t) \tag{9.5}$$

$$i_{C1}(\omega t) = \omega C_{c1} \frac{dv_1(\omega t)}{d\omega t} \tag{9.6}$$

$$i_{C2}(\omega t) = \omega C_{c2} \frac{dv_2(\omega t)}{d\omega t}. \tag{9.7}$$

During the interval of $0 \le \omega t \le \pi - \tau_d$ before the switching instant, the switch $Q_1$ is turned off and the switch $Q_2$ is turned on, conducting the load current, as shown in Fig. 9.2(a), resulting in the current and voltage conditions

$$i_1(\omega t) = 0 \qquad (9.8)$$

$$i_2(\omega t) = i_R(\omega t) \qquad (9.9)$$

$$i_{C1}(\omega t) = i_{C2}(\omega t) = 0 \qquad (9.10)$$

$$v_1(\omega t) = V_{dc} \qquad (9.11)$$

$$v_2(\omega t) = 0 \qquad (9.12)$$

where $V_{dc} = 2V_{cc}$ is the dc supply voltage.

During the interval of $\pi - \tau_d \le \omega t \le \pi$ or dead time, the switch $Q_1$ is still turned off while the switch $Q_2$ is then turned off, as shown in Fig. 9.2(b), and currents continue to flow through the device shunt capacitors. The current discharges the capacitor of one device, while it charges the capacitor of the other device. If it is assumed that these capacitors are identical, then the load current is equally divided between them. Using Eqs (9.5) to (9.7) and an initial condition of $v_2(\pi - \tau_d) = 0$ yields

$$i_1(\omega t) = i_2(\omega t) = 0 \qquad (9.13)$$

$$i_{C2}(\omega t) - i_{C1}(\omega t) = i_R(\omega t) \qquad (9.14)$$

$$v_2(\omega t) = \frac{1}{\omega C} \int_{\pi - \tau_d}^{\omega t} i_{C2}(\omega t) d\omega t + v_2(\pi - \tau_d)$$

$$= -\frac{I_R}{\omega C} [\cos(\tau_d + \varphi) + \cos(\omega t + \varphi)] \qquad (9.15)$$

where $C = C_{c1} + C_{c2}$.

Equation (9.14) can also be written in a differential form using Eqs (9.3), (9.6), and (9.7) as

$$\frac{dv_2(\omega t)}{d\omega t} = \frac{I_R}{\omega C} \sin(\omega t + \varphi). \qquad (9.16)$$

Applying the switching conditions given in Eq. (9.2) to Eq. (9.16) results in

$$\sin(2\pi + \varphi) = 0 \qquad (9.17)$$

from which it follows that the initial phase $\varphi$ can be set to zero.

During the interval of $\pi \le \omega t \le 2\pi - \tau_d$ before the switching instant, the switch $Q_1$ is then turned on and the switch $Q_2$ is still turned off, as shown in Fig. 9.2(c), resulting in

$$i_1(\omega t) = -i_R(\omega t) \qquad (9.18)$$

$$i_2(\omega t) = 0 \qquad (9.19)$$

$$i_{C1}(\omega t) = i_{C2}(\omega t) = 0 \tag{9.20}$$

$$v_1(\omega t) = 0 \tag{9.21}$$

$$v_2(\omega t) = V_{dc}. \tag{9.22}$$

During the interval of $2\pi - \tau_d \le \omega t \le 2\pi$ or dead time, both switches $Q_1$ and $Q_2$ are turned off, as shown in Fig. 9.2($b$), and currents flow through the device shunt capacitors charging the upper one and discharging the lower one. Using Eqs (9.5) to (9.7) and an initial condition of $v_2(2\pi - \tau_d) = V_{ds}$ yields

$$i_1(\omega t) = i_2(\omega t) = 0 \tag{9.23}$$

$$i_{C2}(\omega t) - i_{C1}(\omega t) = i_R(\omega t) \tag{9.24}$$

$$v_2(\omega t) = \frac{1}{\omega C} \int_{2\pi - \tau_d}^{\omega t} i_{C2}(\omega t) d\omega t + v_2(2\pi - \tau_d)$$

$$= \frac{I_R}{\omega C}(\cos \tau_d - \cos \omega t) + V_{dc}. \tag{9.25}$$

Figure 9.3 shows the current and voltage waveforms of a voltage-switching Class-DE power amplifier in an optimum operation mode during the whole interval $0 \le \omega t \le 2\pi$ corresponding to its equivalent circuits shown in Fig. 9.2.

From the boundary condition $v_2(\pi) = V_{dc}$ or $v_2(2\pi) = 0$ by using the corresponding Eq. (9.15) or Eq. (9.25), one can write

$$\frac{\omega C V_{dc}}{I_R} = 1 - \cos \tau_d. \tag{9.26}$$

The fundamental-frequency component of the voltage $v_2(\omega t)$ across the switch $Q_2$ can be represented by the two quadrature components shown in Fig. 9.2, whose amplitudes can be found using Fourier formulas as

$$V_R = -\frac{1}{\pi} \int_{\pi - \tau_d}^{2\pi} v_2(\omega t) \sin \omega t d\omega t = \frac{1 + \cos \tau_d}{\pi} V_{dc} = I_R R \tag{9.27}$$

$$V_L = -\frac{1}{\pi} \int_{\pi - \tau_d}^{2\pi} v_2(\omega t) \cos \omega t d\omega t = \frac{\tau_d - \sin \tau_d \cos \tau_d}{\pi} \frac{I_R}{\omega C}. \tag{9.28}$$

As a result, the optimum normalized total shunt capacitance $C$ and series inductance $L$ as the functions of a dead time $\tau_d$ can be calculated using Eq. (9.26) from

$$\omega C R = \frac{\sin^2 \tau_d}{\pi} \tag{9.29}$$

$$\frac{\omega L}{R} = \frac{V_L}{V_R} = \frac{\tau_d - \sin \tau_d \cos \tau_d}{\sin^2 \tau_d}. \tag{9.30}$$

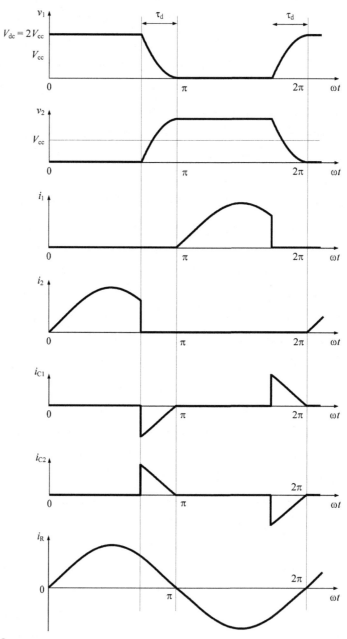

**FIGURE 9.3**

Waveforms for optimum Class-DE operation mode.

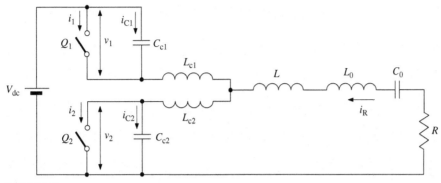

**FIGURE 9.4**

Equivalent circuit of a voltage-switching Class-DE power amplifier with parasitic inductors.

The optimum load resistance $R$ can be obtained using Eq. (9.27) for the supply voltage $V_{dc}$ and fundamental-frequency output power $P_{out}$ delivered to the load as

$$R = \frac{1}{2} \frac{V_R^2}{P_{out}} = \frac{(1 + \cos \tau_d)^2}{2\pi^2} \frac{V_{dc}^2}{P_{out}}. \tag{9.31}$$

The maximum fundamental-frequency output power $P_{out}$ delivered to the load as a function of the dead time $\tau_d$ and capacitance $C$ can be obtained using Eqs (9.29) and (9.31) for the supply voltage $V_{dc}$ as

$$P_{out} = \frac{1}{2\pi} \left( \frac{1 + \cos \tau_d}{\sin \tau_d} \right)^2 \omega C V_{dc}^2. \tag{9.32}$$

For practical design procedure, it is important to know the dependence

$$\tau_d = \cos^{-1} \frac{2\pi \dfrac{P_{out}}{\omega C V_{dc}^2} - 1}{2\pi \dfrac{P_{out}}{\omega C V_{dc}^2} + 1}. \tag{9.33}$$

It should be mentioned that the parasitic device lead inductors $L_{c1}$ and $L_{c2}$ shown in Fig. 9.4, which are located between the output shunt capacitance of each device ($C_{c1}$ and $C_{c2}$) and load network inductor $L$, do not cause the power losses if their quality factors are high enough. However, their presence results in an increased peak collector voltage, which will become more than $V_{dc}$, due to the parasitic oscillations with the oscillation frequency

$$\omega_1 = \frac{1}{\sqrt{(L_{c1} + L_{c2}) \dfrac{C_{c1} C_{c2}}{C_{c1} + C_{c2}}}} \tag{9.34}$$

when both switches $Q_1$ and $Q_2$ are turned off, and with the oscillation frequency

$$\omega_2' = \frac{1}{\sqrt{(L_{c1} + L_{c2})C_{c1}}} \tag{9.35}$$

or

$$\omega_2'' = \frac{1}{\sqrt{(L_{c1} + L_{c2})C_{c2}}} \tag{9.36}$$

when either switch $Q_1$ or switch $Q_2$ is turned on.

To minimize such an increase in the collector voltage peak factor to only 10%, their values must be sufficiently small at the switching frequency defined by

$$L_{c1} < \frac{0.01}{\omega^2 C_{c1}} \tag{9.37}$$

$$L_{c2} < \frac{0.01}{\omega^2 C_{c2}}. \tag{9.38}$$

Figure 9.5(a) shows the equivalent circuit of a voltage-switching Class-DE power amplifier using switching devices with nonzero saturation resistances $r_{sat1}$ and $r_{sat2}$. In this case, during the interval of $0 \leq \omega t \leq \pi - \tau_d$ when the device $Q_2$ is saturated, the collector voltage $v_2(\omega t)$ shown in Fig. 9.5(b) can be defined as $v_{2sat}(\omega t) = i_2(\omega t) r_{sat2} > 0$. Similar changes of the collector voltage shape occur during the interval of $\pi \leq \omega t \leq 2\pi - \tau_d$ when the device $Q_2$ is pinched off, which is a result of a nonzero saturation voltage across the device $Q_1$, equal to $v_{2sat}(\omega t) = i_2(\omega t) r_{sat2} > 0$. Hence, assuming identical devices when $r_{sat} = r_{sat1} = r_{sat2}$, the collector efficiency of the device $Q_2$ as function of a dead time $\tau_d$ can be written as

$$\eta = 0.5 + 0.5 \sqrt{1 - 16\xi^2(\tau_d) \frac{r_{sat} P_{out2}}{V_{dc}^2}} \tag{9.39}$$

where

$$\xi(\tau_d) = \frac{\pi}{\sqrt{(1 + \cos\tau_d)^2 + \left(\dfrac{\tau_d - \sin\tau_d \cos\tau_d}{1 - \cos\tau_d}\right)^2}} \tag{9.40}$$

is the ratio between the dc supply voltage $V_{dc}$ and the fundamental-frequency component of the collector voltage $v_2$ when $r_{sat} = 0$, and $P_{out2}$ is the output power from the device $Q_2$ [1].

The maximum operating frequency $f_{max}$ in an optimum mode is limited by the device collector capacitances $C_{c1}$ and $C_{c2}$. To correctly compare the collector efficiencies corresponding to the push−pull Class-DE and single-ended Class-E operation modes, it is better to determine $f_{max}$ for a Class-DE power amplifier through

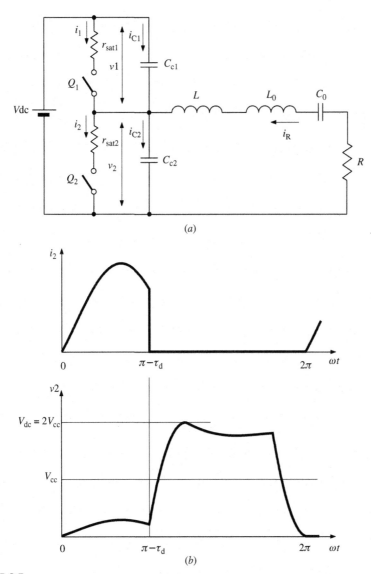

**FIGURE 9.5**

Equivalent circuit of a voltage-switching Class-DE power amplifier with saturation resistances.

the parameters of a single transistor. Then, by taking into account that $V_{cc} = V_{dc}/2$, $C_{c2} = C/2$, and $P_{out2} = P_{out}/2$, from Eq. (9.32) we can derive that

$$f_{max} = 0.25 \left( \frac{\sin \tau_d}{1 + \cos \tau_d} \right)^2 \frac{P_{out2}}{C_{c2} V_{cc}^2}.$$

(9.41)

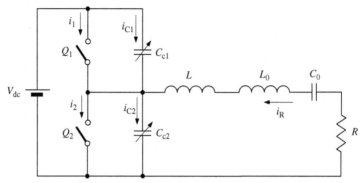

**FIGURE 9.6**

Equivalent circuit of a voltage-switching Class-DE power amplifier with nonlinear shunt capacitances.

The comparison of both Class-DE and Class-E modes as functions of the saturation time period $\tau_{sat} = \pi - \tau_d$ shows that a complementary voltage-switching Class-DE power amplifier provides a substantially lower maximum operating frequency in an optimum mode. For example, for $\tau_d = 60°$ and $\tau_{sat} = 120°$, the maximum operating frequency of a Class-E power amplifier is greater by a factor of more than three [1,4]. However, when using a Class-DE mode, the operating frequency range with collector (or drain) efficiencies of about 90% can be two to three times higher when compared with a Class-D mode [5].

Figure 9.6 shows the equivalent circuit of a voltage-switching Class-DE power amplifier with nonlinear shunt capacitances, since generally the output device capacitance is nonlinear. For example, the drain-source capacitance of a power MOSFET can be represented as a junction capacitance with linearly graded or abrupt junction, while the collector capacitance of a bipolar transistor is normally modeled as a junction capacitance with abrupt junction having a reverse square-root dependence on the voltage between the collector and the emitter [4,6]. The nonlinear behavior of the drain-source capacitance $C_{ds}$ of a MOSFET device can be described by

$$C_{ds} = \frac{C_{j0}}{\left(1 + \dfrac{v}{V_{bi}}\right)^m} \tag{9.42}$$

where $V_{bi}$ is the built-in potential, which typically ranges from 0.5 to 0.9 V, $v$ is the drain-source voltage, $C_{ds0}$ is the capacitance at $v = 0$, and $m$ is the grading coefficient [7]. The results of numerical calculations show that an abrupt junction nonlinearity of the output capacitances when $m = 2$ does not affect the value of a normalized inductance $\omega L/R$. However, the normalized total shunt capacitance $\omega CR$ is $\sqrt{2}$ times smaller compared to the case with constant capacitances [5]. It is shown analytically that the dc supply voltage and current are always

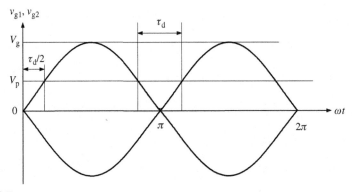

**FIGURE 9.7**

Sinusoidal 180° out-of-phase driving waveforms.

proportional to the amplitudes of the output voltage and current, and the grading coefficient $m$ is an important parameter to satisfy the Class-E switching conditions [8,9]. It should be mentioned that, for loaded quality factors of the series resonant $L_0C_0$ circuit $Q_L > 3$, the current flowing into the load is nearly sinusoidal. However, if $Q_L < 3$, the design and performance values of a Class-DE power amplifier are strongly affected $Q_L$, especially for values of $\tau_{\text{sat}}$ close to 0.5, resulting in the significant harmonic content as well [4].

Practically, it is difficult to drive a Class-DE power amplifier operating at high frequencies using a rectangular driving signal, since the dead time must be controlled with high accuracy (of the order of 1 ns) that also requires the very wideband input transformer. In this case, it is better and easier to apply a sinusoidal drive because it reduces the driving power requirements, and control of the dead time can be provided by the drive amplitude. Besides, the transition from a pinch-off mode to a saturation mode for the power switching MOSFETs occurs very fast once the device gate threshold (or pinch-off) voltage is crossed. Consider the 180° out-of-phase sinusoidal signals driving each device, as shown in Fig. 9.7 [2]. When the gate voltage of one device will cross its pinch-off voltage $V_p$ to turn that device off at angular time $\pi - 0.5\tau_d$, the gate voltage across the other device is still below threshold. Thus, during this crossover period from $\pi - 0.5\tau_d$ to $\pi + 0.5\tau_d$, the two devices are turned off simultaneously.

For a sinusoidal gate voltage across the gate-source capacitance,

$$v_g(\omega t) = V_g \sin \omega t \tag{9.43}$$

where $V_g$ is the voltage amplitude, this voltage is equal to the device pinch-off voltage $V_p$ at angular times $\tau_d$, $\pi - 0.5\tau_d$, and $\pi + 0.5\tau_d$, respectively, that gives

$$V_p = V_g \sin \frac{\tau_d}{2} = V_g \sin\left(\pi - \frac{\tau_d}{2}\right) = -V_g \sin\left(\pi + \frac{\tau_d}{2}\right). \tag{9.44}$$

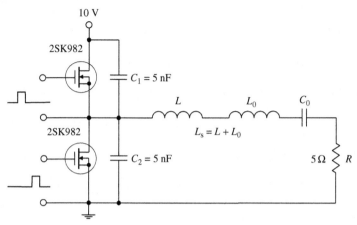

**FIGURE 9.8**

Circuit diagram of an experimental Class-DE power amplifier with rectangular drive.

Equation (9.44) can be rewritten as

$$V_g = \frac{V_p}{\sin\dfrac{\tau_d}{2}} \tag{9.45}$$

representing dependence of the required gate voltage amplitude $V_g$ for a given device dead time $\tau_d$.

However, it is important first to examine the range of dead times versus gate voltage amplitude of a sinusoidal drive. By rearranging Eq. (9.45), we can write

$$\tau_d = 2\sin^{-1}\left(\frac{V_p}{V_g}\right). \tag{9.46}$$

For example, if the peak gate voltage of the devices is limited to 20 V, then, assuming $V_p = 3.5$ V, the minimum dead time is equal to approximately 20°. The other limit occurs when the gate voltage becomes too low to turn on the device properly. Usually, this means that, in practice, the gate-voltage amplitude must not be lower than about 6.5 V. Hence, the maximum dead time that can be achieved is about 65°.

Figure 9.8 shows an experimental circuit of the Class-DE power amplifier operating at 1 MHz with rectangular drive [3]. To get the output power of 1 W at a supply voltage of 10 V, the two MOSFET devices 2SK982 were used as the switching active devices. Assuming that the device dead time $\tau_d = 90°$ and load network quality factor $Q = 10$, the load resistance $R$ can be found from Eq. (9.31) equal to 5.07 $\Omega$, the shunt capacitances $C_1$ and $C_2$ ($C_1 = C_2 = C/2$) can be determined from Eq. (9.29) as equal to 5 nF each, while the series inductance $L_s$

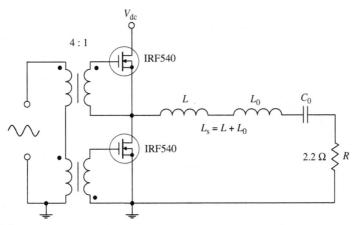

**FIGURE 9.9**

Circuit diagram of experimental Class-DE power amplifiers with sinusoidal drive.

combining the additional compensating inductance $L$ and series filter inductance $L_0$ can be calculated from $L_s = QR/\omega$ equal to 8.07 μH. In this case, the inductance $L_0$ can be found using Eq. (9.30) as

$$L_0 = L_s - L = \left(Q - \frac{\pi}{2}\right)\frac{R}{\omega} = 6.8 \ \mu\text{H}. \qquad (9.47)$$

Then, the series capacitance $C_0$ can be obtained by

$$C_0 = \frac{1}{\omega^2 L_0} = 1/\omega R \left(Q - \frac{\pi}{2}\right) = 3.73 \ \text{nF}. \qquad (9.48)$$

As a result, the measured drain efficiency was 96.2% with an output power of 0.51 W at an operating frequency of 1.04 MHz. The output power was reduced by increasing the operating frequency from 1 to 1.04 MHz. In this case, the smoothest switching voltage waveforms were observed. Generally, the output power could be changed from 30% to 136% by controlling the operating frequency within 10%.

Figure 9.9 shows the experimental circuit of a high-power Class-DE power amplifier operating at 1 MHz with sinusoidal drive [2]. The two gates of the two switching MOSFET devices IRF540 are driven from two transformers with serious-connected primary windings and secondary windings connected with opposite polarities to give the phase shift of 180° between the gate voltages. The two gates could be driven with a single transformer with one primary winding. However, using the separate primary windings has the advantage of eliminating a positive feedback mechanism due to the device feedback gate-source capacitances. The output power $P_{\text{out}} = 300$ W was achieved at an operating frequency of 13.56 MHz with a supply voltage $V_{dc} = 72.5$ V. The dc supply is bypassed by several 0.1-$\mu$F ceramic capacitors, which have a self-resonance at close to 13.56 MHz. Assuming the load current amplitude $I_R = 15.3$ A, the load resistance $R$ can be obtained as

$$R = \frac{2P_{out}}{I_R^2} = 2.56\,\Omega. \tag{9.49}$$

For a load network quality factor $Q = 10$, the series inductance $L_s$ can be calculated from

$$L_s = \frac{QR}{\omega} = 0.3\,\mu\text{H}. \tag{9.50}$$

The dead time $\tau_d$ can be found from Eq. (9.27) equal to approximately 45°. By using Eq. (9.30), the series inductance $L$ required to compensate for the device output capacitance of 500 pF (at $V_{ds} = 25$ V) is equal to 17.2 nH. Then, the series capacitance $C_0$ can be obtained as

$$C_0 = \frac{1}{\omega^2(L_s - L)} = 487\,\text{pF}. \tag{9.51}$$

As a result, the measured output power of 298.2 W with a drain efficiency of 94.2% was achieved at a supply voltage of 73.1 V. The total power dissipated in the gate circuit was only 9.6 W with a gate voltage amplitude of 7 V. By using two IRFP450 MOSFETs fed by supply voltages of +150 V and −150 V (300 V with the center point at ground, allowing a direct connection from the switching node to a grounded load), an output power of 1 kW can be achieved in a frequency range of 50 kHz to 5 MHz with efficiency of more than 91% [10].

## 9.2 Class-FE power amplifiers

In practice, the idealized collector voltage and current waveforms corresponding to Class-F mode can be realized at sufficiently low frequencies when the effect of the device collector capacitance is negligible. At higher frequencies, the effect of the collector capacitance contributes to a finite switching time, resulting in time periods when the collector voltage and collector current exist at the same time when simultaneously $v > 0$ and $i > 0$. As a result, such a load network with shunt capacitor cannot provide the switchmode operation with an instantaneous transition from the device pinch-off mode to saturation mode. Therefore, during a finite time interval the device is operated in the active region as a nonlinear current source with the reverse-biased collector-base junction, and the collector current is provided by this current source. To minimize power losses during transients, the optimum zero-voltage and zero voltage-derivative Class-E conditions can be applied to a Class-F circuit with an additionally connected shunt capacitor and series inductor, providing soft-switching operation mode, and that was proposed and called as Class Φ [11]. Experimental results demonstrate that high drain efficiency of more than 80% can be achieved with a Class-C biasing at an operating frequency of 13.56 MHz with maximum peak factor of about two, which is significantly lower than that in a switchmode Class-E mode. Therefore, it is important to expand the experimental results by providing an analytical insight to this mixed operation mode which can be simply called the Class-FE similar to Class-DE with the derivation of the optimum parameters for the load network and by deriving the additional Class-FE power-amplifier configurations [12].

It should be mentioned that the rectangular voltage and half-sine current collector waveform can be provided by a push–pull voltage-switching Class-D power amplifier with the fundamentally tuned series $L_0C_0$ filter, whose circuit schematic is shown in Fig. 9.10(a). In this case, the bipolar transistors must be driven so as to provide an alternate voltage switching between the on-state and off-state modes. However, in a single-ended Class-F power amplifier with a parallel quarterwave transmission line shown in Fig. 9.10(b), the symmetrizing action of the transmission-line behavior can easily provide a function of the voltage inverter, resulting in the same collector waveforms with voltage peak factor of two. The quarterwave line stores the voltage waveform in a traveling wave along its length, which returns delayed by one-half fundamental period and inverted, because of power-reflection conditions at the short-circuit termination. However, the presence of parasitic collector capacitances degrades the amplifying performances of both transistor stages due to finite time of the charging and discharging processes. As a result, the frequency limitation is provided by the device parasitic collector shunt capacitances, resulting in increased switching transition times due to the capacitor charging and discharging processes, and the fact that transistor switches charge and discharge the shunt capacitance, dissipating power in the charging and discharging processes.

A possible way to eliminate these power losses and increase efficiency by compensating for the device output capacitance is to introduce a dead time during the period when one device has already turned off but the other has not turned on yet, and the inductive load network is used to charge and discharge the shunt capacitances. This can be provided by introducing the Class-E switching conditions when the switching loss during off-to-on transition is reduced to zero by the operating requirements of zero voltage at the zero-voltage slope at the end of period. Since the shunt capacitor must be discharged at that instant, an additional series inductor $L$ with optimum value should be included into the load network, as shown in Fig. 9.11(a) for a Class-DE power amplifier. Similarly, a quarterwave transmission line can be used as a voltage inverter resulting in a single-ended Class-FE (or Class EF) power amplifier which combines a Class-F operation mode with Class-E switching conditions, as shown in Fig. 9.11(b).

The optimum parameters of a Class-FE power amplifier can be determined based on its steady-state collector voltage and current waveforms [12]. Figure 9.12 shows the two equivalent Class-FE switching circuits that occur

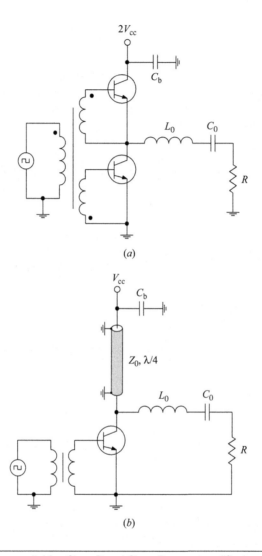

**FIGURE 9.10**

Schematics of voltage-switching Class-D and Class-F power amplifiers.

during a switching cycle, the load network of which consists of a grounded parallel quarterwave transmission line, a shunt capacitor $C$, a series inductor $L$, a series fundamentally tuned $L_0C_0$ circuit, and a load resistor $R$. The active device is considered an ideal switch that is driven so as to provide the device switching between its on-state and off-state operation conditions. To simplify the analysis of a Class-FE power amplifier, the same assumptions which were used to analyze a Class-DE power amplifier are introduced.

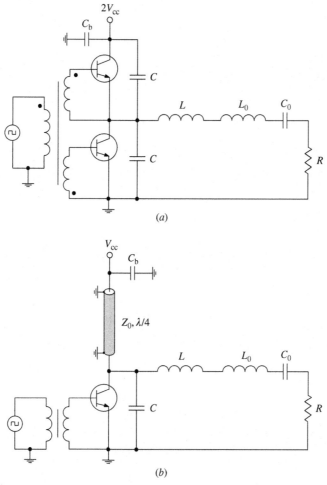

**FIGURE 9.11**

Schematics of voltage-switching Class-DE and Class-FE power amplifiers.

The Class-E switching conditions for the transistor switch can be written as

$$v(\omega t)|_{\omega t=2\pi} = 0 \tag{9.52}$$

$$\left.\frac{dv(\omega t)}{d\omega t}\right|_{\omega t=2\pi} = 0 \tag{9.53}$$

where $v$ is the voltage across the switch.

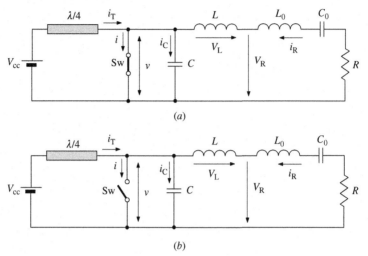

**FIGURE 9.12**

Equivalent circuits of a Class-FE power amplifier.

Let the output current flowing through the load is written as sinusoidal,

$$i_R(\omega t) = I_R \sin(\omega t + \varphi) \tag{9.54}$$

where $I_R$ is the current amplitude and $\varphi$ is the initial phase shift.

The basic Kirchhoff's equations characterizing the electrical behavior of the equivalent circuits of a Class-FE power amplifier are

$$i(\omega t) + i_C(\omega t) = i_T(\omega t) + i_R(\omega t) \tag{9.55}$$

$$i_C(\omega t) = \omega C \frac{dv(\omega t)}{d\omega t}. \tag{9.56}$$

During the interval of $0 \le \omega t \le \pi - \tau_d$ before the switching instant, where $\tau_d$ is the dead time, the switch is turned on conducting the load current, as shown in Fig. 9.12(a), resulting in the current and voltage conditions

$$i_C(\omega t) = 0 \tag{9.57}$$

$$i(\omega t) = i_R(\omega t) + i_T(\omega t) \tag{9.58}$$

$$v(\omega t) = 0. \tag{9.59}$$

During the interval of $\pi - \tau_d \le \omega t \le \pi$ or dead time, the switch is turned off, as shown in Fig. 9.12(b), and currents continue to flow through the shunt capacitance charging the capacitor. Using Eq. (9.56) and initial condition $v(\pi - \tau_d) = 0$ yields

$$i(\omega t) = i_T(\omega t) = 0 \tag{9.60}$$

$$i_C(\omega t) = i_R(\omega t) \tag{9.61}$$

$$
\begin{aligned}
v(\omega t) &= \frac{1}{\omega C} \int_{\pi - \tau_d}^{\omega t} i_C(\omega t) d\omega t + v(\pi - \tau_d) \\
&= -\frac{I_R}{\omega C} [\cos(\tau_d + \varphi) + \cos(\omega t + \varphi)].
\end{aligned}
\tag{9.62}
$$

During the interval of $\pi \le \omega t \le 2\pi - \tau_d$ before the switching instant, the switch is still turned off, resulting in

$$i_T(\omega t) = -i_R(\omega t) \tag{9.63}$$

$$i(\omega t) = i_C(\omega t) = 0 \tag{9.64}$$

$$v(\omega t) = 2V_{cc}. \tag{9.65}$$

During the interval of $2\pi - \tau_d \le \omega t \le 2\pi$ or dead time, the switch is turned off, and currents flow through the shunt capacitance discharging the capacitor. Using Eq. (9.56) and initial condition $v(2\pi - \tau_d) = 2V_{cc}$ yields

$$i(\omega t) = i_T(\omega t) = 0 \tag{9.66}$$

$$i_C(\omega t) = i_R(\omega t) \tag{9.67}$$

$$
\begin{aligned}
v(\omega t) &= \frac{1}{\omega C} \int_{2\pi - \tau_d}^{\omega t} i_C(\omega t) d\omega t + v(2\pi - \tau_d) \\
&= \frac{I_R}{\omega C} [\cos(\tau_d + \varphi) - \cos(\omega t + \varphi)] + 2V_{cc}.
\end{aligned}
\tag{9.68}
$$

Equation (9.67) can also be written in a differential form using Eqs (9.54) and (9.56) as

$$\frac{dv(\omega t)}{d\omega t} = \frac{I_R}{\omega C} \sin(\omega t + \varphi). \tag{9.69}$$

Applying the switching condition given in Eq. (9.53) to Eq. (9.69) results in

$$\sin(2\pi + \varphi) = 0 \tag{9.70}$$

from which it follows that the initial phase $\varphi$ can be set to zero.

Figure 9.13 shows the ideal current and voltage waveforms of a Class-FE power amplifier in an idealized optimum operation mode during the whole interval $0 \le \omega t \le 2\pi$ corresponding to its equivalent circuits shown in Fig. 9.12. The quarterwave transmission line has a half-wave symmetry, which means that the line attempts to do the same work in the first and second halves of the cycle. Therefore, the transmission-line half-repeating current must necessarily instantaneously fall to zero at $\pi - \tau_d$ and $2\pi - \tau_d$, being zero during both collector voltage edges. Note that, for the sinusoidal driving signals, the transistor must be

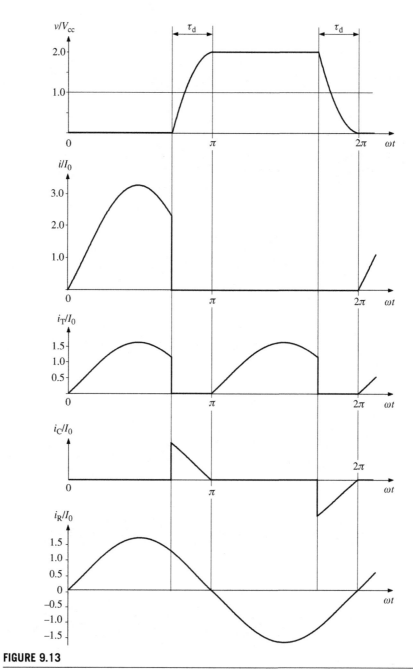

**FIGURE 9.13**

Waveforms for an optimum Class-FE operation mode.

biased in a Class-C mode to provide duty ratio $D$, less than 50% $(D<0.5)$, because the period of time when the device is turned off must exceed the period of time when the device is turned on.

From the boundary condition $v(\pi) = 2V_{cc}$ or $v(2\pi) = 0$ by using the corresponding Eq. (9.62) or Eq. (9.68), one can write

$$\frac{\omega C V_{cc}}{I_R} = \frac{1 - \cos \tau_d}{2}. \tag{9.71}$$

The fundamental-frequency component of the voltage $v(\omega t)$ across the switch can be represented by the two quadrature components shown in Fig. 9.12, whose amplitudes can be found using Fourier formulas by

$$V_R = -\frac{1}{\pi} \int_0^{2\pi} v(\omega t) \sin \omega t \, d\omega t = \frac{2(1 + \cos \tau_d)}{\pi} V_{cc} = I_R R \tag{9.72}$$

$$V_L = -\frac{1}{\pi} \int_0^{2\pi} v(\omega t) \cos \omega t \, d\omega t = \frac{\tau_d - \sin \tau_d \cos \tau_d}{\pi} \frac{I_R}{\omega C}. \tag{9.73}$$

In an idealized Class-FE operation mode, there is no nonzero voltage and current simultaneously that gives an idealized collector efficiency of 100% when the dc power $P_0$ and fundamental-frequency output power $P_{out}$ are equal. Consequently, the load current amplitude $I_R$ as a function of the dead time $\tau_d$ can be obtained from Eq. (9.72) as

$$I_R = \frac{\pi}{1 + \cos \tau_d} I_0. \tag{9.74}$$

As a result, the optimum normalized shunt capacitance $C$ and series inductance $L$ as functions of a dead time $\tau_d$ can be calculated using Eq. (9.71) from

$$\omega C R = \frac{\sin^2 \tau_d}{\pi} \tag{9.75}$$

$$\frac{\omega L}{R} = \frac{V_L}{V_R} = \frac{\tau_d - \sin \tau_d \cos \tau_d}{\sin^2 \tau_d}. \tag{9.76}$$

The optimum load resistance $R$ can be obtained using Eq. (9.72) for the dc supply voltage $V_{cc}$ and fundamental-frequency output power $P_{out}$ delivered to the load as

$$R = \frac{1}{2} \frac{V_R^2}{P_{out}} = \frac{2(1 + \cos \tau_d)^2}{\pi^2} \frac{V_{cc}^2}{P_{out}}. \tag{9.77}$$

The maximum fundamental-frequency output power $P_{out}$ delivered to the load as functions of the dead time $\tau_d$ and capacitance $C$ can be obtained from Eqs (9.75) and (9.77) for the supply voltage $V_{cc}$ as

$$P_{out} = \frac{2}{\pi} \left( \frac{1 + \cos \tau_d}{\sin \tau_d} \right)^2 \omega C V_{cc}^2. \tag{9.78}$$

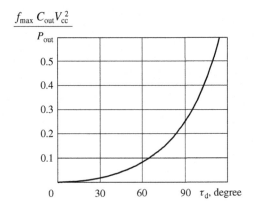

**FIGURE 9.14**

Normalized maximum operating frequency versus dead time.

The maximum operating frequency $f_{max}$ in an optimum Class-FE mode is limited to the device output (collector or drain-source) capacitance $C_{out}$. As a result, from Eq. (9.78) it follows that

$$f_{max} = 0.25 \left( \frac{\sin \tau_d}{1 + \cos \tau_d} \right)^2 \frac{P_{out}}{C_{out} V_{cc}^2}. \qquad (9.79)$$

Figure 9.14 shows the dependence of the normalized maximum operating frequency $(f_{max} C_{out} V_{cc}^2)/P_{out}$ on the dead time $\tau_d$. From Eq. (9.78) and Fig. 9.14, it follows that the maximum operating frequency $f_{max}$ increases with smaller dead times, and

$$f_{max} = 0.043 \frac{P_{out}}{C_{out} V_{cc}^2} \qquad (9.80)$$

for a $\tau_d = 45°$, which is the same as for a Class-DE and close to the maximum operating frequency of a Class-E with a 50% duty ratio. However, for a saturation period of $\pi - \tau_d = 180° - 45° = 135°$ when the transistor is turned on, the maximum operating frequency $f_{max}$ of a Class-E power amplifier is more than three times greater [4]. At the same time, for a saturation period of 135°, the collector peak voltage of a Class-E power amplifier is three times larger than the applied supply voltage $V_{cc}$ [13]. Note that the maximum collector voltage peak factor of a Class-FE power amplifier never exceeds a factor of two for any saturation periods.

In Table 9.1, the optimum impedances seen from the device collector at the fundamental and higher-order harmonic components are illustrated by the appropriate circuit configurations. It can be seen that Class-F mode with a quarter-wave transmission line shows purely resistive impedance at the fundamental,

**Table 9.1** Optimum Impedances at Fundamental and Harmonics

| High-Efficiency Mode | $f_0$ (Fundamental) | $2nf_0$ (Even Harmonics) | $(2n+1)f_0$ (Odd Harmonics) |
|---|---|---|---|
| Class-F with quarterwave line | | short | open |
| Class-E with shunt capacitance | | | |
| Class-FE with quarterwave line | | short | |

short-circuit conditions at even harmonics, and open-circuit conditions at odd harmonics. As opposed to Class-F mode, Class-E mode with shunt capacitance requires inductive impedance at the fundamental and capacitive reactance at any other harmonic component. As a result, Class-FE with a quarterwave transmission line simultaneously provides the harmonic impedance conditions typical for both Class-F with a quarterwave transmission line and Class-E with a shunt capacitance.

Figure 9.15 shows the simulation setup for an ideal Class-FE operation in frequency domain. Generally, using the frequency domain enables the overall simulation procedure to be much faster than that in the time domain and can take a few seconds. However, because the number of harmonic components is not infinite, the simulation waveforms and numerical results for the optimum load-network parameters are not so accurate. In this case, the input source represents a voltage source with Fourier-series expansion of a periodical rectangular wave with different pulsewidth characterized by a duty ratio used in a harmonic balance simulator. The harmonic order is chosen to 100.

The optimization procedure can also be applied with respect to the efficiency as an optimization parameter. Since the simulation time is very short, a number of iterations can be increased significantly to provide better calculation accuracy. Figure 9.16 shows the normalized switch (a) voltage and (b) current waveforms obtained for the idealized optimum or nominal parameters of the Class-FE load network given by Eqs (9.75) to (9.77) with a dead time $\tau_d = 45°$ or duty ratio (or cycle) $D = 0.375$ defined as the ratio of the switch-on time to one period of the switching frequency. The simulated waveforms of the currents flowing into the quarterwave transmission line and shunt capacitor are shown in Fig. 9.17(a) and Fig. 9.17(b), respectively.

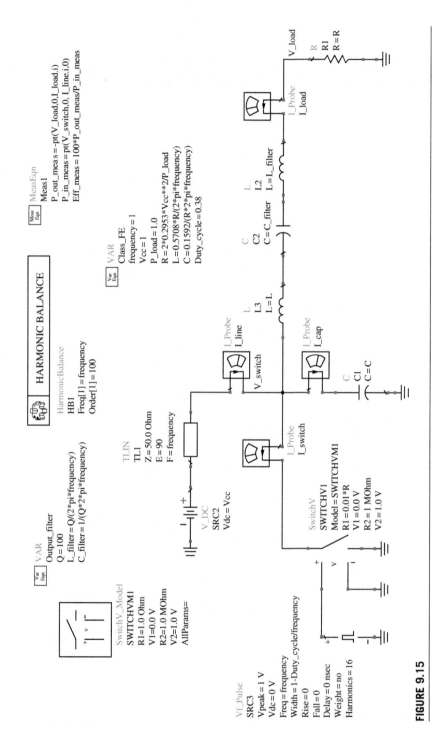

**FIGURE 9.15**

Simulation setup for an idealized Class-FE mode in frequency domain.

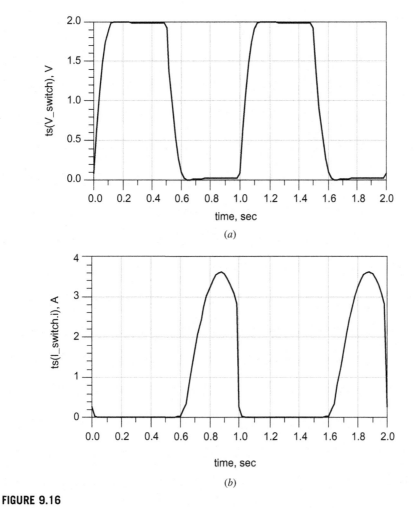

**FIGURE 9.16**

Simulated Class-FE switch voltage and current waveforms.

Unlike the time domain simulations, there are smoother transitions between the positions when the switch is turned on and the switch is turned off and vice versa. Nevertheless, for $r_{sat}/R = 0.01$ and lossless circuit elements except load resistance $R$, the efficiency is equal to 98.2%, and the simulated normalized switch and load-network voltage and current waveforms are similar to the same theoretical waveforms shown in Fig. 9.13.

Figure 9.18 shows the circuit schematic of a Class-FE power amplifier with a series quarterwave transmission line consisting of a shunt capacitor $C$, a series quarterwave transmission line loaded by a parallel resonant $L_0 C_0$ circuit tuned to

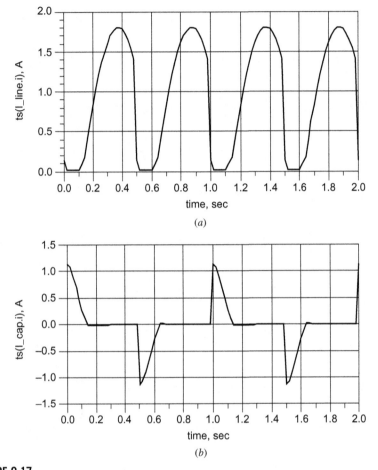

**FIGURE 9.17**

Simulated Class-FE load-network current waveforms.

the fundamental frequency, a shunt capacitor $C_L$, and a load resistor $R_L$. In a common case, a shunt capacitance $C$ can represent the intrinsic device output capacitance and external circuit capacitance added by the load network. The RF choke is necessary to isolate the dc power supply from the RF circuit, whereas the blocking capacitor $C_b$ is necessary to separate the dc supply circuit from the load network. The operation principle of such a Class-FE power amplifier with a series quarterwave line is similar to that with a parallel quarterwave line, assuming that the parallel $L_0C_0$ filter is ideal having infinite impedance at the fundamental frequency and zero impedances at the second- and higher-order harmonic components. In this case, the short circuit on the load side of the quarterwave

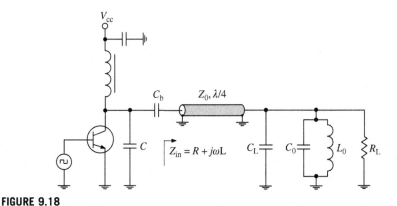

**FIGURE 9.18**

Schematic of a Class-FE power amplifier with series quarterwave transmission line.

transmission line produces a short circuit at its input for even harmonics and open circuit at its input for odd harmonics, with resistive load at the fundamental frequency.

Unlike the parallel quarterwave transmission line, a series quarterwave transmission line can serve as an impedance transformer at the fundamental frequency. In this case, there is no need for an additional output impedance matching circuit required for a Class-FE power amplifier with a parallel quarterwave transmission line. The impedance $Z_{in}$ at the input of the loaded quarterwave transmission line with the characteristic impedance $Z_0$ and electrical length $\theta$ at the fundamental frequency can be written as

$$Z_{in} = Z_0 \frac{Z_L + jZ_0 \tan\theta}{Z_0 + jZ_L \tan\theta}\bigg|_{\theta = 90°} = \frac{Z_0^2}{Z_L} \tag{9.81}$$

where

$$Z_L = \frac{R_L}{1 + j\omega C_L R_L}. \tag{9.82}$$

Separating Eq. (9.81) into real and imaginary parts enables the standard load resistance $R_L$ and shunt capacitance $C_L$ to be expressed through the optimum load resistance $R$ and series inductance $L$ corresponding to the Class-FE mode according to Eqs (9.76) and (9.77) for a particular value of the transmission-line characteristic impedance $Z_0$ as

$$R_L = \frac{Z_0^2}{R} \tag{9.83}$$

$$C_L = \frac{L}{Z_0^2}. \tag{9.84}$$

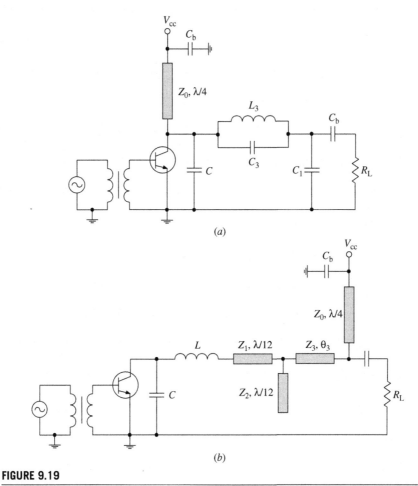

**FIGURE 9.19**

Schematics of Class-$E_3F$ and Class-$E_3F_2$ power amplifiers.

The total series inductances $L + L_0$ in the load network of a Class-FE power amplifier shown in Fig. 9.11(b) typically has a large value accompanied by high parasitic series resistance and low self-resonant frequency. Figure 9.19(a) shows the circuit schematic of a Class-$E_3F$ (or Class-$FE_3$) with a reduced synthesized inductance value where the tank $L_3C_3$ circuit is tuned to the third-harmonic component to provide an open-circuit condition [14]. Since the impedance of the tank circuit is inductive at the fundamental, this equivalent series inductor $L_1$ together with the shunt capacitor $C_1$ represents a simple $L$-type output matching circuit at the fundamental. In this case, close to the ideal Class-EF voltage waveform at the collector is provided by the quarterwave transmission line suppressing even harmonics and open-circuit tank for the third-harmonic open-circuit condition.

The values of an inductance $L_3$ and a capacitance $C_3$ can be calculated from

$$L_3 = \frac{8}{9}(L + L_1) \tag{9.85}$$

$$C_3 = \frac{1}{9\omega_0^2 L_3} \tag{9.86}$$

where an equivalent series inductance $L_1$ and a shunt capacitance $C_1$ match the optimum Class-FE load resistance $R$ defined by Eq. (9.77) and standard load resistance $R_L$ can be obtained by

$$L_1 = \frac{R}{\omega_0}\sqrt{\frac{R_L}{R} - 1} \tag{9.87}$$

$$C_1 = \frac{1}{\omega_0 R_L}\sqrt{\frac{R_L}{R} - 1}. \tag{9.88}$$

The transmission-line implementation of a Class-FE power amplifier is shown in Fig. 9.19(b) [14]. Since the transmission-line load network controls only the first three harmonics with an inductive impedance at the fundamental, a short-circuit termination for the second harmonic, and an open-circuit condition for the third harmonic, such a transmission-line Class-FE approximation is called the Class-$E_3F_2$ power amplifier. Unlike the schematics with a parallel quarterwave transmission line, this topology incorporates a series inductance $L$ so as to eliminate the effect that the parasitic device output bondwire and package lead inductance has on the amplifier performance at high frequencies. At the third-harmonic frequency, the second series transmission-line and open-circuit stub, both having an electrical length of 90° at $3f_0$, provide an open-circuit condition seen by the inductor $L$. At the second-harmonic frequency, the quarterwave transmission line creates a short-circuit condition for the series transmission line with an electrical length $\theta_3$, thus providing its input inductive reactance. Since the open-circuit stub has a capacitive reactance when its electrical length is 60°, both short-circuit and open-circuit transmission lines represent a second-harmonic tank. As a result, the first series open-circuited transmission line with an electrical length of 60° behaving as a capacitor provides with a series inductor $L$ a short circuit seen by the device collector. In this case, the transmission-line characteristic impedances $Z_1$ and $Z_2$ can be defined as

$$Z_1 = \sqrt{3}Z_3 \tan 2\theta_3 \tag{9.89}$$

$$Z_2 = 2\sqrt{3}\omega_0 L \tag{9.90}$$

where the inductance $L$ can be calculated from Eq. (9.76).

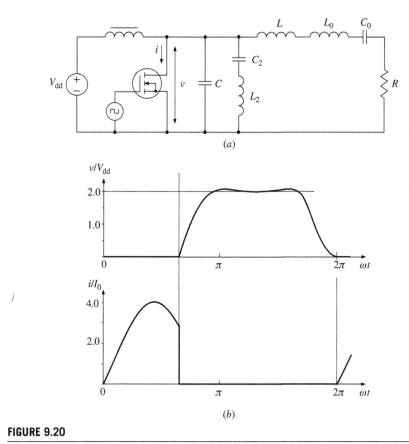

**FIGURE 9.20**

Class-EF$_2$ power amplifier and corresponding optimum drain waveforms.

In special applications when it is difficult to implement a quarterwave line, for example, in monolithic integrated circuits or at sufficiently low frequencies, it is possible to consider a Class-FE approximation based on the lumped resonant circuits. Figure 9.20 shows the circuit schematic of a Class-FE power amplifier with a second-harmonic control using an additional series resonant circuit $L_2C_2$ which provides a short circuit for the second harmonic at the device drain [15,16]. Such a load network represents a subclass of Class-FE and is called the Class EF$_2$ (or Class-F$_2$E) when, for a second-harmonic short-circuit termination, the idealized optimum Class-E zero-voltage and zero voltage-derivative conditions are satisfied at the end of period, as shown in Fig. 9.20(b) for the duty ratio $D = 0.35$. From comparison of the theoretical and simulated output voltage and current waveforms

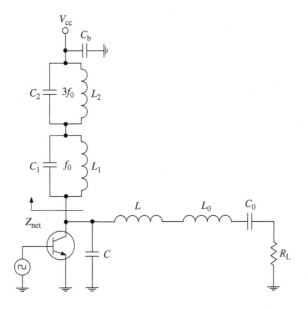

**FIGURE 9.21**

Class-FE approximation with lumped resonant circuits.

shown in Figs. 9.13 and 9.16 for a general Class-FE and in Fig. 9.20(*b*) for a second-harmonic Class EF$_2$, respectively, it follows that, if the voltage peak factors for both cases are practically the same, however the peak current value for the latter case is substantially higher. In addition, the higher duty ratio is used, the higher voltage peak factor of the Class-EF$_2$ mode has been achieved, approaching the factor of 4 for duty ratios close to $D = 0.5$ [15]. As a result, for a duty ratio $D = 0.3$, an output power of over 500 W with a drain efficiency above 92% was achieved at a switching frequency of 30 MHz using a 500-V vertical MOSFET device [16]. However, for duty ratios greater than 0.5, it is preferable to use a Class-E/F$_3$ mode where the series resonant circuit $L_2 C_2$ represents a third-harmonic short circuit, resulting in a third-harmonic approximation of a Class-E/F where the Class-E idealized optimum conditions are applied to an inverse Class-F. For example, when $D = 0.55$, the peak factor of the flattened drain voltage waveform is only slightly exceeds the factor of 3 [15].

Figure 9.21 shows the circuit schematic of a Class-FE power amplifier schematic with the two parallel resonant $LC$ circuits tuned to the fundamental and third harmonic components, respectively. In this case, to provide the short-circuit condition at the second-harmonic frequency, the parameters of these resonant circuits must be properly optimized.

The network impedance $Z_{net}$ at the input of the two cascaded tank circuits tuned to fundamental ($\omega = \omega_0$) and third harmonic ($\omega = 3\omega_0$) that provides the open-circuit conditions for these components, respectively, can be written as

$$Z_{net} = \frac{j\omega L_1}{1 - \omega^2 L_1 C_1} + \frac{j\omega L_2}{1 - \omega^2 L_2 C_2}. \tag{9.91}$$

To provide the short-circuit condition at the second harmonic ($\omega = 2\omega_0$), Eq. (9.91) can be rewritten in the form

$$Z_{net}(2\omega_0) = \frac{j2\omega_0 L_1}{1 - 4\omega_0^2 L_1 C_1} + \frac{j2\omega_0 L_2}{1 - 4\omega_0^2 L_2 C_2} = 0. \tag{9.92}$$

As a result, the ratios between the circuit elements for a certain value of $C_1$ are

$$L_1 = \frac{1}{\omega_0^2 C_1} \tag{9.93}$$

$$L_2 = \frac{5}{27} L_1 \tag{9.94}$$

$$C_2 = \frac{1}{9\omega_0^2 L_2}. \tag{9.95}$$

## 9.3 Class-E/F power amplifiers

When comparing Class-E to its two counterparts, conventional Class-F and inverse Class-F, both advantages and disadvantages can be found with respect to each class. For example, using a switchmode Class-E can result in high operation efficiency with a simple load network, whereas an efficiency improvement by using a Class-F is achieved at the expense of the circuit complexity. Even with tuning up to seventh-harmonic component, the collector (or drain) voltage and current waveforms exhibit significant overlapping, and the maximum theoretical collector efficiency cannot exceed 90%. The switching speed of the Class-E power amplifier is limited only by the active device, provided the optimum parameters of the load network are properly set. However, the switching speed of the Class-F power amplifier is defined by both the active device and a limited number of the controlled harmonic components composing the collector voltage and current waveforms. In addition, Class-E can incorporate the device output capacitance into the load-network topology, but a large capacitance becomes a limiting factor at high frequencies. In the presence of large output capacitance, it is possible to realize a simple Class-F topology with control of only a few harmonics. However, the collector-voltage peak factor in a Class-F mode cannot exceed a maximum value of 2, while the maximum collector-voltage peak factor in a Class-E mode can be greater than 3.5.

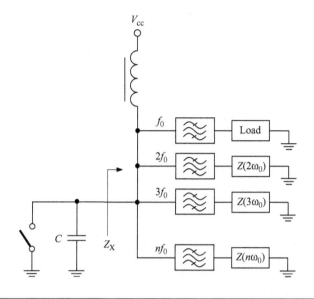

**FIGURE 9.22**

Generalized block schematic of a harmonic-tuned Class-E/F power amplifier.

Figure 9.22 shows the generalized block schematic of a switchmode Class-E/F power amplifier where zero-voltage and zero voltage-derivative conditions corresponding to Class-E mode can be used to eliminate discharge loss of the shunt capacitance, and harmonic tuning can be provided by using the resonant circuits tuned to selected harmonic components approximating Class-F mode with improved collector waveforms [17]. Each resonant circuit or filter presents a desired impedance at the corresponding harmonic component that it tunes, while leaving other harmonics unaffected. Consequently, the load network describes a family of switching power amplifiers for which the impedance seen by the switch is capacitive at all harmonics, except for a selected number of tuned harmonics.

To achieve Class-E optimum conditions, the choice of an inverse Class-F is natural to provide mixed Class-E/F mode since it provides a zero-voltage condition for a half-sinusoidal collector voltage waveform, unlike the conventional Class-F, which presents voltage discontinuity at the switching instant in the ideal case of infinite harmonic tuning. The Class-E/F name is chosen to indicate that this mixed mode is based on Class-E and inverse Class-F (or Class 1/F) rather than the conventional Class-F. Generally, some selection of even harmonics may be open-circuited, some selection of odd harmonics may be short-circuited, and the remaining harmonics are presented with a fixed capacitance. According to a Class-E mode, the load resistance $R$, shunt capacitance $C$, and series inductance $L$ shown in Table 9.2 are adjusted at the fundamental frequency to achieve the

**Table 9.2** Optimum Class-E/F Impedances at Fundamental and Harmonics

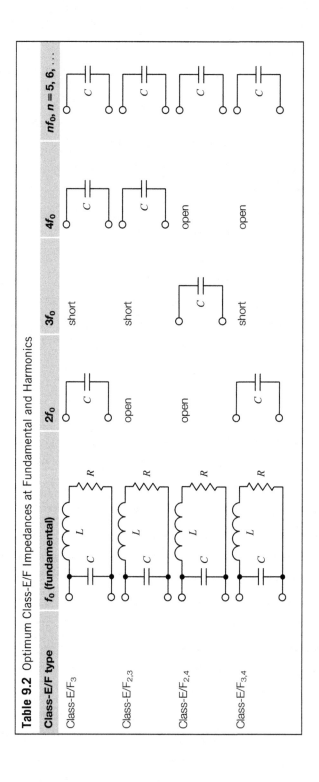

| Class-E/F type | $f_0$ (fundamental) | $2f_0$ | $3f_0$ | $4f_0$ | $mf_0, n = 5, 6, \ldots$ |
|---|---|---|---|---|---|
| Class-E/F$_3$ | $L$, $C$, $R$ | $C$ | short | $C$ | $C$ |
| Class-E/F$_{2,3}$ | $L$, $C$, $R$ | open | short | $C$ | $C$ |
| Class-E/F$_{2,4}$ | $L$, $C$, $R$ | open | $C$ | open | $C$ |
| Class-E/F$_{3,4}$ | $L$, $C$, $R$ | $C$ | short | open | $C$ |

required inductive impedance to obtain the overall zero-voltage and zero voltage-derivative conditions in an ideal case with different tuning at the harmonics. To differentiate these harmonic tunings between each other, different names can be used of the form Class-E/F$_{n1, n2, n3,...}$, where the numerical subscripts indicate the number of tuned harmonics in an inverse Class-F mode. Table 9.2 shows the load-network harmonic-impedance specifications of several Class-E/F power amplifiers. Even-harmonic tuning tends to primarily affect the current waveform, whereas odd-harmonic tuning tends to have the most effect on the voltage waveform. As a result, tuning of several harmonics can minimize the peak voltage amplitude with current waveform approximating a square wave. The minimum peak factor of 3.08 is realized for Class-E/F$_{2, 3, 4}$, which is lower than in an inverse Class-F of 3.12 and much lower than in a Class-E with shunt capacitance of 3.56.

### 9.3.1 Symmetrical push–pull configurations

In practice, it is difficult to eliminate the effects of the resonant circuits tuned to different harmonics, on each other, or fundamental-frequency impedance. For example, if a series resonator is tuned to create a short circuit for the third-harmonic component in a Class-E/F$_3$ mode, it will present a capacitance at the second harmonic, reducing the second-harmonic impedance and degrading voltage waveform. In this case, either a very high loaded quality factor of this resonator is required, or additional resonances must be introduced into the circuit. The easiest and most convenient way to separate the effects of even and odd harmonics is to use a symmetrical push–pull switching power-amplifier configuration. By using the effects of a virtual ground for odd harmonics and a virtual open for even harmonics, a push–pull Class-E/F$_3$ power amplifier can be constructed by placing the third-harmonic short circuit as a differential load. In this case, the shorting of the third harmonic can be provided by a low-$Q_L$ resonator without reducing the impedances at any even harmonics. A similar strategy may be used to selectively tune the impedances at even harmonics. Figure 9.23(a) shows the Class-E/F$_{odd}$ circuit implementation with short-circuiting of all odd harmonics by placing a differential short circuit between the two switches [17]. To avoid the short circuiting of a fundamental-frequency component, a bandstop filter such as a parallel $LC$-tank can be used to provide a differential short circuit at all odd harmonics other than the fundamental.

The bandstop filter between two switches forces the differential voltage $v_2 - v_1$ to be sinusoidal,

$$v_2(\omega t) - v_1(\omega t) = V\sin(\omega t + \varphi) \qquad (9.96)$$

where $V$ is the differential voltage amplitude and $\varphi$ is the arbitrary initial phase shift.

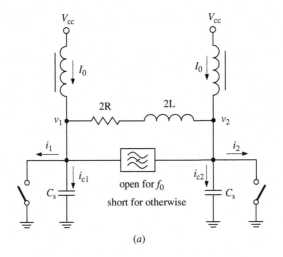

(a)

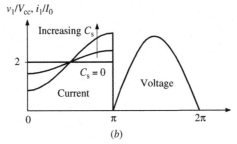

(b)

**FIGURE 9.23**

Class-E/F$_{odd}$ circuit implementation and collector waveforms.

Since one of the switches is always turned on during half a period and has zero voltage across it for an idealized case, the voltages $v_1$ and $v_2$ can be written, respectively, as

$$v_1(\omega t) = \begin{cases} 0 & 0 \leq \omega t \leq \pi \\ -V\sin(\omega t + \varphi) & \pi \leq \omega t \leq 2\pi \end{cases} \qquad (9.97)$$

and

$$v_2(\omega t) = \begin{cases} V\sin(\omega t + \varphi) & 0 \leq \omega t \leq \pi \\ 0 & \pi \leq \omega t \leq 2\pi \end{cases} \qquad (9.98)$$

where $V$ is the drain voltage amplitude and $V_{cc}$ is the supply voltage.

According to Class-E zero-voltage switching conditions, both voltages must be equal to zero at turn-on, resulting in the only trivial solution for $\varphi$ to be zero. This means that the properly tuned switch voltages $v_1$ and $v_2$ are purely half-sinusoidal. Since each switch is connected to the supply voltage through the RF choke, the dc voltage of each switch voltage waveform is the supply voltage $V_{cc}$. Then, the peak voltage for a half-sine waveform can be written as $V = \pi V_{cc}$.

As a result, the currents $i_{c1}$ and $i_{c2}$ flowing through linear shunt capacitors can be found, respectively, from

$$i_{c1}(\omega t) = C_s \frac{dv_1(\omega t)}{d\omega t} = \begin{cases} 0 & 0 \le \omega t \le \pi \\ -\pi V_{cc} C_s \cos \omega t & \pi \le \omega t \le 2\pi \end{cases} \tag{9.99}$$

and

$$i_{c2}(\omega t) = C_s \frac{dv_2(\omega t)}{d\omega t} = \begin{cases} \pi V_{cc} C_s \cos \omega t & 0 \le \omega t \le \pi \\ 0 & \pi \le \omega t \le 2\pi \end{cases} \tag{9.100}$$

During each one-half period, one of the transistors is open-circuited, while the other transistor is short-circuited. Hence, the switch currents $i_1$ and $i_2$ can be written, respectively, as

$$i_1(\omega t) = \begin{cases} 2I_0 - i_{c2}(\omega t) & 0 \le \omega t \le \pi \\ 0 & \pi \le \omega t \le 2\pi \end{cases} \tag{9.101}$$

and

$$i_2(\omega t) = \begin{cases} 0 & 0 \le \omega t \le \pi \\ 2I_0 - i_{c1}(\omega t) & \pi \le \omega t \le 2\pi \end{cases} \tag{9.102}$$

The switch voltage and current waveforms for various values of the shunt capacitance $C_s$ corresponding to a Class-E/F$_{odd}$ mode are shown in Fig. 9.23(b). If the shunt capacitance is zero, the switch current $i_1$ (or $i_2$) represents a rectangular waveform corresponding to a well-known current-switching Class-D mode. However, according to Eq. (9.99) and (9.100), the increasing shunt capacitance $C_s$ results in higher current peak values, also changing the shape of a current waveform.

The Class-E/F$_{odd}$ topology shown in Fig. 9.23(a) can be transformed to a similar circuit schematic with only one RF choke connected between the dc power supply and the center tap of the tank inductor $L_1$, where $C_b$ is the bypass capacitor, and the resonator composed of the tank inductor $L_1$ and capacitor $C$ is detuned to have the required inductance at the fundamental frequency to achieve Class-E switching conditions [18]. In some special cases, the RF choke can be replaced by an inductor $L_2$, as shown in Fig. 9.24(a), which can be tuned to resonate with the shunt capacitors at the second harmonic, providing an open-circuit to each switch at that frequency similar to a single-ended even-harmonic Class-E mode [19].

The voltage $v_{L2}$ across the inductor $L_2$ can be written as

$$v_{L2}(\omega t) = \frac{v_1(\omega t) + v_2(\omega t)}{2} = \begin{cases} \dfrac{\pi}{2} V_{cc} \sin \omega t & 0 \le \omega t \le \pi \\ -\dfrac{\pi}{2} V_{cc} \sin \omega t & \pi \le \omega t \le 2\pi \end{cases} \tag{9.103}$$

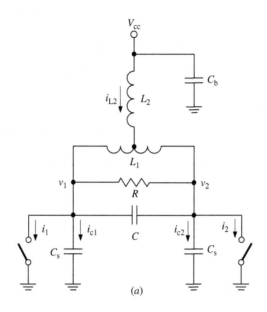

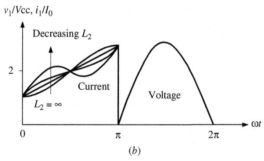

**FIGURE 9.24**

Class-E/F$_{2,odd}$ power amplifier and collector waveforms.

The current $i_{L2}$ flowing through the inductor $L_2$ during each one-half period can be obtained by

$$i_{L2}(\omega t) = \frac{1}{\omega L_2} \int [V_{cc} - v_{L2}(\omega t)] \, d\omega t$$

$$= \begin{cases} \dfrac{\pi}{2} \dfrac{V_{cc}}{\omega L_2} \left( \dfrac{2\omega t}{\pi} - 1 + \cos \omega t \right) & 0 \le \omega t \le \pi \\[3mm] \dfrac{\pi}{2} \dfrac{V_{cc}}{\omega L_2} \left( 2\dfrac{\omega t - \pi}{\pi} - 1 - \cos \omega t \right) & \pi \le \omega t \le 2\pi \end{cases} \qquad (9.104)$$

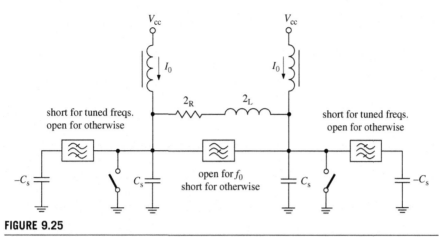

**FIGURE 9.25**

Generalized Class-E/F$_{x,odd}$ circuit implementation.

Then, the switch currents $i_1$ and $i_2$ can be found, respectively, from

$$i_1(\omega t) = \begin{cases} i_{L2}(\omega t) - i_{c2}(\omega t) & 0 \le \omega t \le \pi \\ 0 & \pi \le \omega t \le 2\pi \end{cases} \qquad (9.105)$$

and

$$i_2(\omega t) = \begin{cases} 0 & 0 \le \omega t \le \pi \\ i_{L2}(\omega t) - i_{c1}(\omega t) & \pi \le \omega t \le 2\pi \end{cases} \qquad (9.106)$$

where the capacitor currents $i_{c1}$ and $i_{c2}$ are defined by Eqs (9.99) and (9.100).

The switch voltage and current waveforms for various values of an inductance $L_2$ corresponding to a Class-E/F$_{2,odd}$ mode are shown in Fig. 9.24(b). For each inductance tuning, the voltage waveform is a half-sinusoidal. The current waveforms depend on the even harmonics tuned and the switch shunt capacitance. For high values of $L_2$ corresponding to a basic Class-E/F$_{odd}$ mode, the switch current represents a nearly trapezoidal waveform. However, for low values of $L_2$ with open-circuit even-harmonic tuning, the switch current approximates more closely the rectangular waveform with peak value at the turn-off switching instant, even for very large values of the shunt capacitance [17]. Classes E/F$_{odd}$ and E/F$_{2,odd}$ can tolerate twice the shunt capacitance of a Class-E for the same operating frequency, supply voltage, and output power, without having a negative switch current. As a result, the maximum operating frequency can be doubled, compared to a single-ended Class-E mode using the same transistor size.

The tuning of several even harmonics realizing a Class-E/F$_{x,odd}$ operation mode, where $x$ is a set of tuned even harmonics, can be accomplished with the generalized circuit implementation shown in Fig. 9.25, where the additional resonators are connected in parallel to each switch [17]. Each additional resonator can

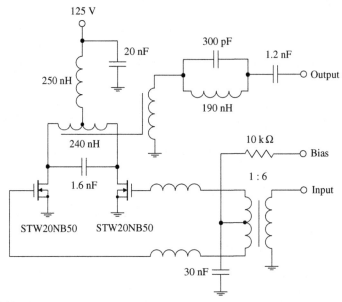

**FIGURE 9.26**

Schematic of Class-E/F$_{2,\text{odd}}$ power amplifier with third-harmonic trap.

be treated as a bandpass filter placed in series with negative capacitance $-C_s$. The filter connects this negative capacitance in parallel with the switch shunt capacitance at the tuned even harmonics, thus canceling the capacitance at these frequencies.

Figure 9.26 shows the schematic of a high-power Class-E/F$_{2,\text{odd}}$ MOSFET power amplifier designed to provide an output power of 1 kW at an operating frequency of 7 MHz [18]. The magnetizing inductance of an output air-core transformer was used as the tank inductance of 240 nH. To reduce a ringing transient superimposed on the drain voltage and current waveforms due to parasitic device package inductors resonating with the tank capacitor and drain-source capacitances, the loaded quality factor of the parallel $LC$ resonator was lowered to 3.6. The second-harmonic tuning helps to keep the drain current positive, providing an additional peaking in the early part of the current waveform where the ringing has the highest amplitude. To suppress the third-harmonic component to an acceptable level, a third-harmonic trap was added in series with the load. As a result, the second- and third-harmonic components in the output spectrum were 33 and 35 dB below the fundamental-frequency component, respectively, and all other higher-order harmonic components were at least 40 dB below the fundamental. The Class-E/F$_{2,\text{ odd}}$ power amplifier exhibited an output power over 1.1 kW with a drain efficiency of 85% and a power gain of 17 dB.

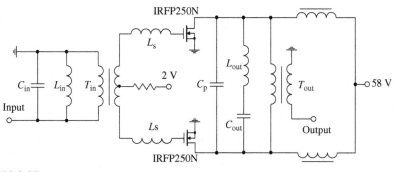

**FIGURE 9.27**

Schematic of dual-band Class-E/F$_{odd}$ power amplifier.

Figure 9.27 shows the high-power dual-band Class-E/F$_{odd}$ MOSFET power amplifier designed to operate in the 7- and 10-MHz frequency bands [20]. The input network consisting of the parallel capacitor $C_{in}$, parallel inductor $L_{in}$, balanced-to-unbalanced (balun) impedance transformer $T_{in}$, and two series inductors $L_s$ converts a source impedance of 50 Ω to balanced impedance conjugately matched to each transistor input. The load network consists of the parallel capacitor $C_p$, series tank with $L_{out}$ and $C_{out}$, output air-core transformer $T_{out}$ representing a 1:1 balun, the magnetizing inductance of which is used in the load-network tuning. Such a load network provides inductive, capacitive, and again inductive reactance as frequency is increased, with the inductive reactance at the fundamental required for zero-voltage switching operation at each frequency band. At all odd harmonics of each frequency of operation, the load network including the device drain-source capacitances provides a low capacitive reactance to ground compared to the load resistance. Due to the differential topology, each transistor is represented only by its own output capacitance at even harmonics. As a result, an output power of 250 W with a drain efficiency of 94% and a power gain of 16 dB was achieved at 7.15 MHz, whereas an output power of 225 W with a drain efficiency of 90% and a power gain of 15 dB was measured at 10.1 MHz. The bandwidth in both bands was sufficient to allow high-efficiency operation from 7.0- to 7.3-MHz and from 10.0- to 10.18-MHz bands.

### 9.3.2 Single-ended Class-E/F$_3$ mode

The optimum parameters of a single-ended Class-E/F$_n$ power amplifier can be determined based on an analytical derivation of its steady-state collector voltage and current waveforms [21]. Figure 9.28(a) shows the basic circuit configuration of a Class-E/F$_n$ power amplifier where the load network consists of a shunt capacitor $C$, a series resonant $L_nC_n$ circuit tuned to $n$th harmonic of the fundamental frequency and connected in parallel to the device output, a series inductor $L$,

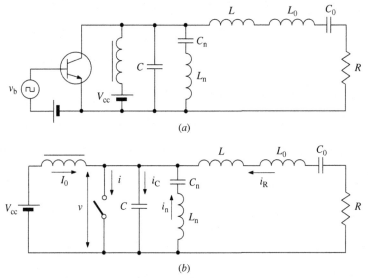

**FIGURE 9.28**

Basic circuits of single-ended Class-E/F$_n$ power amplifier.

a series fundamentally tuned $L_0C_0$ circuit, and a load resistor $R$. In a common case, a shunt capacitance $C$ can represent the intrinsic device output capacitance and external circuit capacitance added by the load network. The collector of the transistor is connected to the dc power supply by an RF choke with infinite reactance at the fundamental and any higher-order harmonic component. The active device is considered an ideal switch that is driven at the operating frequency so as to provide instantaneous switching between its on-state and off-state operation conditions.

To simplify the analysis of a Class-E/F$_n$ power amplifier, the same assumptions which were used to analyze a Class-DE or a Class-FE power amplifier are introduced. The nominal Class-E switching conditions for the transistor switch are defined by Eqs (9.52) and (9.53), where $v$ is the voltage across the switch. Note that these nominal Class-E switching conditions do not correspond to minimum dissipated power losses for the nonideal transistor switch with nonzero value of its saturation resistance.

Let the output current $i_R$ flowing through the load and $n$th-harmonic current $i_n$ be written as sinusoidal,

$$i_R(\omega t) = I_R \sin(\omega t + \varphi) \tag{9.107}$$

$$i_n(\omega t) = I_n \sin n\,\omega t \tag{9.108}$$

where $I_R$ is the fundamental-frequency current amplitude, $I_n$ is the $n$th-harmonic current amplitude, and $\varphi$ is the initial phase shift. It should be noted that generally a sum of sinusoidal harmonic components can be used in Eq. (9.108).

When the switch is turned on for of $0 \leq \omega t \leq \pi$, there is no current flowing through the capacitor $C$,

$$i_C(\omega t) = \omega C \frac{dv(\omega t)}{d\omega t} = 0 \tag{9.109}$$

and, consequently,

$$i(\omega t) = I_0 + I_n \sin n\,\omega t + I_R \sin(\omega t + \varphi) \tag{9.110}$$

where $I_0$ is the dc supply current.

For an initial on-state condition $i(0) = 0$, Eq. (9.110) reduces to

$$1 + \frac{I_R}{I_0} \sin \varphi = 0. \tag{9.111}$$

When the switch is turned off for $\pi \leq \omega t < 2\pi$, there is no current flowing through the switch, i.e. $i(\omega t) = 0$, and the current flowing through the capacitor $C$ can be written as

$$i_C(\omega t) = I_0 + I_n \sin n\,\omega t + I_R \sin(\omega t + \varphi) \tag{9.112}$$

producing the voltage across the switch by charging of this capacitor according to

$$v(\omega t) = \frac{1}{\omega C} \int_\pi^{\omega t} i_C(\omega t)d\omega t = \frac{I_0}{\omega C}\left\{ \omega t - \pi - \frac{I_n}{n}(1 + \cos n\,\omega t) - I_R[\cos(\omega t + \varphi) + \cos \varphi] \right\}. \tag{9.113}$$

Applying a zero-voltage switching condition given by Eq. (9.52) to Eq. (9.113) results in

$$\frac{\pi}{2} - \frac{I_n}{nI_0} - \frac{I_R}{I_0} \cos \varphi = 0 \tag{9.114}$$

where the relative amplitude of the $n$th-harmonic component $I_n/I_0$ can be obtained for any $n$ by

$$I_n = \frac{1}{\pi} \int_0^{2\pi} i(\omega t) \sin n\,\omega t d\omega t = \frac{2I_0}{n\pi}(1 - \cos n\pi) \tag{9.115}$$

which is reduced for odd $n$ to

$$\frac{I_n}{I_0} = \frac{4}{n\pi}. \tag{9.116}$$

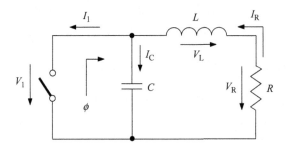

**FIGURE 9.29**

Equivalent Class-E/$F_n$ load network at fundamental frequency.

As a result, by substituting Eq. (9.111) and Eq. (9.116) into (9.114), the initial phase angle $\varphi$ can be explicitly obtained for odd $n$ by

$$\varphi = \tan^{-1}\left(\frac{2\pi n^2}{8 - \pi^2 n^2}\right). \tag{9.117}$$

The dc supply voltage $V_{cc}$ can be determined by applying a Fourier-series expansion to Eq. (9.113) as

$$V_{cc} = \frac{1}{2\pi}\int_0^{2\pi} v(\omega t)d\omega t = \frac{I_0}{2\pi\omega C}\left[\frac{\pi^2}{2} - \frac{\pi I_n}{n I_0} - \frac{I_R}{I_0}(2\sin\varphi + \pi\cos\varphi)\right]. \tag{9.118}$$

For an idealized collector efficiency of 100%, the dc power $P_0 = I_0 V_{cc}$ and fundamental-frequency output power $P_{out} = V_R^2/2R$ delivered to the load are equal, i.e.

$$I_0 V_{cc} = \frac{V_R^2}{2R} \tag{9.119}$$

where $V_R = I_R R$ is the voltage amplitude across the load resistor $R$.

The fundamental-frequency voltage $v_1(\omega t)$ across the switch consists of the two quadrature components, as shown in Fig. 9.29, whose amplitudes can be found using Fourier formulas and Eq. (9.113) by

$$V_R = -\frac{1}{\pi}\int_0^{2\pi} v(\omega t)\sin(\omega t + \varphi)\,d\omega t$$

$$= -\frac{2I_0}{\pi\omega C}\left[\sin\varphi + \cos\varphi\left(-\frac{\pi}{2} + \frac{1}{n}\frac{I_0}{I_n} + \frac{I_R}{I_0}\cos\varphi\right)\right] \tag{9.120}$$

$$V_L = -\frac{1}{\pi} \int_0^{2\pi} v(\omega t)\cos(\omega t + \varphi)\, d\omega t$$

$$= -\frac{I_0}{\pi \omega C}\left[\sin\varphi\left(\pi - \frac{2 I_n}{n I_0}\right) + 2\cos\varphi - \frac{I_R}{I_0}\left(\frac{\pi}{2} + \sin 2\varphi\right)\right]. \tag{9.121}$$

As a result, for a particular case of the third-harmonic tuning when $n = 3$,

$$\varphi = \tan^{-1}\left(\frac{18\pi}{8 - 9\pi^2}\right) = -34.98^\circ \tag{9.122}$$

$$\frac{I_R}{I_0} = -\frac{1}{\sin\varphi} = 1.7444 \tag{9.123}$$

$$\frac{I_n}{I_0} = \frac{4}{3\pi} = 0.4244. \tag{9.124}$$

Figure 9.30 shows the normalized collector voltage and current waveforms, the normalized waveform of the current flowing through the capacitor $C$, the third-harmonic and fundamental-frequency sinusoidal current waveforms for idealized optimum Class-E/F$_3$ mode during the whole interval $0 \le \omega t \le 2\pi$. From the collector voltage and current waveforms, it follows that, when the transistor is turned on, there is no voltage across the switch and the current consisting of the dc supply current, third-harmonic, and load sinusoidal currents flows through the switch. However, when the transistor is turned off, all this current flows through the capacitor $C$. In this case, there is no nonzero voltage and current simultaneously, which means a lack of the power losses that gives an idealized collector efficiency of 100% given by Eq. (9.119).

Hence, the optimum normalized series inductance $L$ and shunt capacitance $C$ can be calculated as

$$\frac{\omega L}{R} = \frac{V_L}{V_R} = 0.961 \tag{9.125}$$

$$\omega C R = \frac{\omega C V_R}{I_0} \Big/ \frac{I_R}{I_0} = 0.209. \tag{9.126}$$

The optimum load resistance $R$ can be obtained from Eqs (9.119) and (9.123) for the dc supply voltage $V_{cc}$ and fundamental-frequency output power $P_{out}$ delivered to the load as

$$R = \frac{1}{2}\frac{V_R^2}{P_{out}} = \frac{2 V_{cc}^2}{P_{out}} \Big/ \left(\frac{I_R}{I_0}\right)^2 = 0.657\frac{V_{cc}^2}{P_{out}}. \tag{9.127}$$

From Eqs (9.125) to (9.127) it follows that the load resistance $R$ is greater by 14% in nominal Class-E/F$_3$ mode compared with a conventional Class-E with shunt capacitance, whereas the series inductance $L$ and shunt capacitance $C$ are approximately the same for both classes for the same $V_{cc}$ and $P_{out}$.

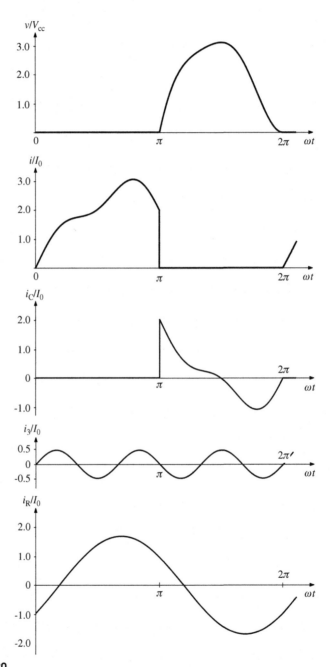

**FIGURE 9.30**

Ideal waveforms for optimum Class-E/F$_3$ operation mode.

**Table 9.3** Optimum Impedances at Fundamental and Harmonics

| High-Efficiency Mode | $f_0$ (Fundamental) | $2nf_0$ (Even Harmonics) | $(2n + 1)f_0$ (Odd Harmonics) |
|---|---|---|---|
| Inverse Class-F | R | open | short |
| Class-E with shunt capacitance | C L R | C | C |
| Class-E/F$_3$ | C L R | C | short |

The phase angle $\phi$ of the load network at fundamental seen by the switch, as shown in Fig. 9.29, and required for an idealized optimum (or nominal) Class-E/F$_3$ mode can be determined through the load-network parameters using Eqs (9.125) and (9.126) as

$$\phi = \tan^{-1}\left(\frac{\omega L}{R}\right) - \tan^{-1}\left(\frac{\omega CR}{1 - \frac{\omega L}{R}\omega CR}\right) = 29.2°. \qquad (9.128)$$

The peak collector voltage $V_{max}$ and current $I_{max}$ can be determined from Eqs (9.110) and (9.113) using Eqs (9.118) and (9.122) to (9.124) as

$$\frac{V_{max}}{V_{cc}} = 3.142 \qquad (9.129)$$

$$\frac{I_{max}}{I_0} = 3.056 \qquad (9.130)$$

that shows that the voltage peak factor is lower by 13.4% compared with a conventional Class-E with shunt capacitance.

The maximum operating frequency $f_{max}$ in nominal Class-E/F$_3$ mode is limited by the device output (collector or drain-source) capacitance $C_{out}$. As a result, from Eqs (9.126) and (9.127) it follows that

$$f_{max} = 0.0506 \frac{P_{out}}{C_{out} V_{cc}^2} \qquad (9.131)$$

which corresponds to the maximum operating frequency of a Class-E with shunt capacitance.

In Table 9.3, the optimum impedances seen from the device collector at the fundamental and higher-order harmonic components are illustrated by the appropriate circuit configurations. It can be seen that an inverse Class-F mode shows purely resistive impedance at the fundamental, open-circuit conditions at even harmonics, and short-circuit conditions at odd harmonics. As opposed to inverse Class-F mode, Class-E mode with shunt capacitance requires inductive impedance at the fundamental and capacitive reactance at any other harmonic component. As a result, an idealized mixed Class-E/F$_3$ mode provides simultaneously the harmonic impedance conditions typical for both Class-E with a shunt capacitance and inverse Class-F third-harmonic control.

Figure 9.31 shows the ADS simulation setup for a nominal Class-E/F$_3$ operation in the time domain, where an input source Vt_PulseDT represents a voltage source with pulse train defined at discrete time steps used in envelope and transient simulators. The active device is represented by a voltage-controlled switch with off-resistance of 1 MΩ and small finite on-resistance, the value of which can generally be varied. The use of a discrete time pulse source, as opposed to a standard pulse source, can guarantee that there is no timing jitter in the pulse edges due to the waveform being sampled asynchronously by a fixed time interval simulation. The simulation time is significantly faster than the period of a square wave.

To provide the circuit simulation in the time domain, the transient simulator is added to the simulation template. The stop time of 80 s is chosen for an operating frequency normalized to unity that is sufficient to reach a steady-state mode. The inductors and capacitors are lossless and the loaded quality factor $Q_L$ of the series resonant circuit is chosen to be as high as 10. The measure equations include the conditions when the switch voltage $V_{switch}$ and its voltage derivative must take zero values at the instant just before the switch is turned on. The efficiency is calculated in the 79th + 80th period since the products of instantaneous current and voltage are integrated over these two periods and divided by two. Since the RF choke inductance $L_{feed}$ should have a finite number in transient simulations, its value was chosen to be 10 times greater than that of a series inductance $L$. As a result, the value of a load resistance increases by 22% ($R_{factor} = 1.22$) compared to the nominal case to obtain the optimized switch voltage and current waveforms shown in Fig. 9.32 with an efficiency of 99.8%. Similar increase in optimum value of the load resistance for the reduced value of the RF choke inductance is observed in a Class-E mode with finite dc-feed inductance.

The simulation setup shown in Fig. 9.31 can also be used to obtain the nominal Class-E/F$_3$ operation with infinite RF choke in the frequency domain, where an input source Vt_Pulse represents a voltage source with Fourier series expansion of periodic rectangular wave with different pulse width characterized by a duty cycle used in a harmonic-balance simulator. Generally, using the frequency domain enables the overall simulation procedure to be much faster than that in the time domain and can take a few seconds. However, because the number of harmonic components is not infinite, the simulation waveforms and numerical results for the optimum load network parameters are not so accurate. The harmonic order is chosen to 100.

In this case, the optimization procedure can be applied with respect to efficiency as an optimization parameter. Since the simulation time is very short,

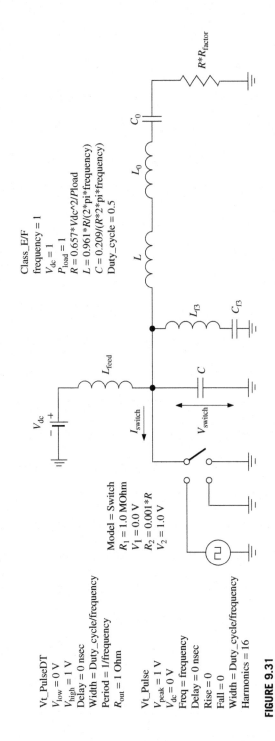

**FIGURE 9.31**

Simulation setup for idealized Class-E/F$_3$ mode in the time domain.

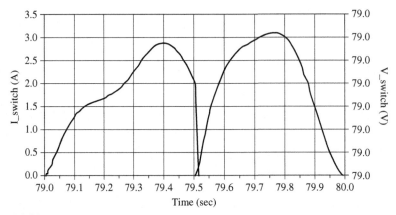

**FIGURE 9.32**

Simulated switch voltage and current waveforms in the time domain.

a number of iterations can significantly be increased for more accuracy. Figure 9.33 shows the normalized (a) switch voltage and current waveforms and (b) sinusoidal third-harmonic and fundamental-frequency waveforms obtained for the nominal parameters of the Class-E/F$_3$ load network given by Eqs (9.125) to (9.127) with 50% duty cycle and $R_{factor} = 1$.

Unlike the simulations in the time domain, the frequency-domain simulations are characterized by smoother transitions between the positions when the switch is turned on and the switch is turned off and *vice versa*. Nevertheless, for $r_{sat}/R = 0.001$, where $r_{sat}$ is the switch saturation resistance, and lossless circuit elements except load resistance $R$, the efficiency is equal to 98.2%, and the simulated normalized switch voltage and current waveforms shown in Fig. 9.33 are similar to the same theoretical waveforms shown in Fig. 9.30.

To design a lumped load network which corresponds to Class-E/F$_3$ mode, first it is necessary to choose the optimum parameters from Eqs (9.125) to (9.127) for the shunt capacitance, series inductance, and load resistance at the fundamental and then add an additional network to provide an open circuit at the second harmonic and a short circuit at the third harmonic. For example, it can be represented by a simple lumped network shown in Fig. 9.34, which consists of the two cascaded parallel resonant circuits ($L_1C_1$ and $L_2C_2$), tuned to the fundamental ($\omega = \omega_0$) and second harmonic ($\omega = 2\omega_0$) to provide the open-circuit conditions for these components, and a series inductor $L_3$.

The network impedance $Z_{net}$ at the input of this lumped network can be written as

$$Z_{net}(\omega) = \frac{j\omega L_1}{1 - \omega^2 L_1 C_1} + j\omega L_3 + \frac{j\omega L_2}{1 - \omega^2 L_2 C_2}. \qquad (9.132)$$

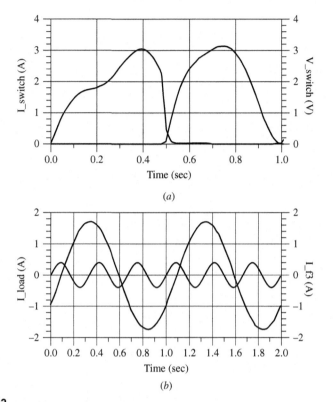

**FIGURE 9.33**

Simulated voltage and current waveforms in the frequency domain.

To provide a short-circuit condition at the third harmonic ($\omega = 3\omega_0$), Eq. (9.132) can be rewritten in the form

$$Z_{\text{net}}(3\omega_0) = \frac{j3\omega_0 L_1}{1 - 9\omega_0^2 L_1 C_1} + j3\omega_0 L_3 + \frac{j3\omega_0 L_2}{1 - 9\omega_0^2 L_2 C_2} = 0. \qquad (9.133)$$

As a result, the ratios between the network elements for certain values of $L_1$ and $L_2$ can be given by

$$C_1 = \frac{1}{\omega_0^2 L_1} \qquad (9.134)$$

$$C_2 = \frac{1}{4\omega_0^2 L_2} \qquad (9.135)$$

$$L_3 = 0.125 L_1 + 0.8 L_2. \qquad (9.136)$$

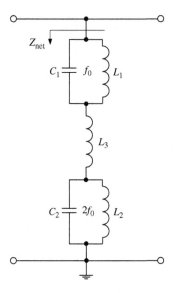

**FIGURE 9.34**

Lumped network for second- and third-harmonic tuning.

Figure 9.35 shows the simulated transfer characteristic (magnitude $S_{21}$ in decibels *versus* frequency) for the lumped network shown in Fig. 9.34, where its parameters were calculated according to Eqs (9.134) to (9.136) for $f_0 = 430$ MHz, $L_1 = 60$ nH, and $L_2 = 12.5$ nH as $L_3 = 17.5$ nH, $C_1 = 2.3$ pF, and $C_2 = 2.7$ pF. The inductor quality factors $Q$ for all three inductors were set to 50. The values of inductances $L_1$ and $L_2$ should be properly chosen to provide low impedance (first zero) enough far from both first (at fundamental) and second (at the second harmonic) poles, as shown in Fig. 9.35.

The lumped Class-E/F$_3$ power amplifier based on a 28-V 10-W Cree GaN HEMT power transistor CGH40010 was simulated at an operating frequency of 430 MHz according to the simulation setup shown in Fig. 9.36. In this case, the required optimum load resistance given by Eq. (9.127) is close to 50 $\Omega$, therefore it is enough to use a series high-$Q_L$ resonant circuit including an optimum series inductance given by Eq. (9.125). An additional external shunt capacitance of 0.6 pF is used because the device output capacitance of about 1.3 pF is slightly less than required by Eq. (9.126). Generally, it is necessary to take into account the device package parasitics, with the series bondwire and lead inductance as the most important, which can seriously affect the device performance at higher frequencies if the device drain-source capacitance is not optimum according to Eq. (9.126). As a result, for infinite inductor quality factors $Q$ in the load network, an output power of 40.1 dBm, a drain efficiency of 87%, and a *PAE* of 84.5% with a power gain of about 15 dB were simulated for an input power

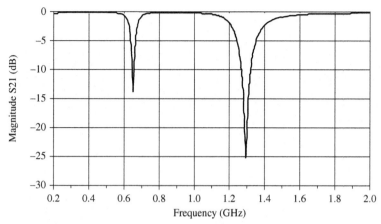

**FIGURE 9.35**

Frequency response of a lumped network for Class-E/F$_3$ mode.

$P_{in} = 25$ dBm, as shown in Fig. 9.37($a$). For more practical values of $Q = 50$, an output power of 39.3 dBm is achieved with a drain efficiency of 77.5% and a *PAE* of 74.9%. Figure 9.37($b$) shows the simulated drain waveform which is very close to the theoretical one shown in Fig. 9.30.

The circuit schematics of a high-efficiency transmission-line Class-E/F$_3$ power amplifier are shown in Fig. 9.38, where the optimum load resistance $R$ is matched to the standard load impedance $R_L = 50\ \Omega$ at the fundamental frequency. Here, both high impedance at the second harmonic and low impedance at the third harmonic are created at the device output by using a series transmission line $TL_1$ with electrical length of 45°, a series transmission line $TL_2$, the electrical length $\theta$ which depends on the value of the optimum series inductance $L$, a quarterwave short-circuit transmission line $TL_3$, and an open-circuit stub $TL_4$ with electrical length of 30°.

Figure 9.39($a$) shows the load network seen by the device multiharmonic current source at the fundamental. Here, the combined series transmission line $TL_1 + TL_2$ with the same characteristic impedance $Z_1$ together with an open-circuit capacitive stub $TL_4$ with an electrical length of 30° provides an impedance matching between the optimum load-network impedance seen from the device output $Z_{net} = R$ and the load resistance $R_L$ by proper choice of the transmission-line characteristic impedances $Z_1$ and $Z_2$, which can be explicitly calculated for a certain electrical length $\theta$.

The optimum load-network impedance $Z_{net}$ at the fundamental can be written as

$$Z_{net} = Z_1 \frac{R_L(Z_2 - Z_1 \tan 30° \tan \theta_1) + jZ_1Z_2 \tan \theta_1}{Z_1Z_2 + j(Z_1 \tan 30° + Z_2 \tan \theta_1)\ R_L} \qquad (9.137)$$

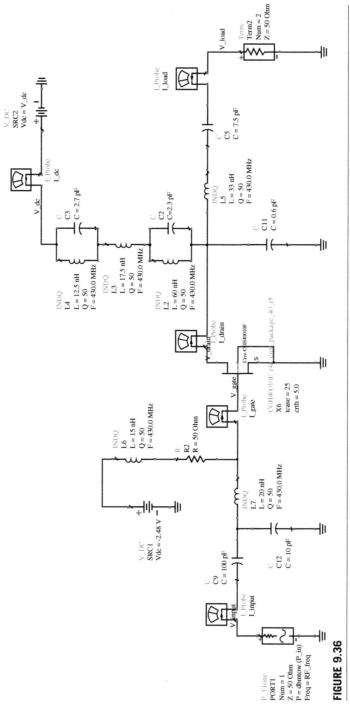

**FIGURE 9.36**

Simulation setup for a lumped Class-E/F$_3$ GaN HEMT power amplifier.

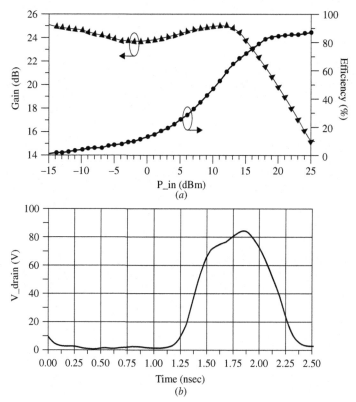

**FIGURE 9.37**

Simulated results of a lumped Class-E/F$_3$ GaN HEMT power amplifier.

where $\theta_1 = 45° + \theta$, $Z_1$ is the characteristic impedance of the series transmission line, and $Z_2$ is the characteristic impedance of the open-circuit stub with electrical length of 30°.

Separating Eq. (9.137) into real and imaginary parts, the following system of two equations with two unknown parameters is obtained:

$$(Z_1 \tan 30° + Z_2 \tan \theta_1)^2 R_L^2 R - (Z_1 Z_2)^2 [R_L(1 + \tan^2 \theta_1) - R] = 0 \qquad (9.138)$$

$$(Z_1 Z_2)^2 \tan \theta_1 + R_L^2 (Z_1 \tan 30° + Z_2 \tan \theta_1)(Z_2 - Z_1 \tan 30° \tan \theta_1) = 0 \quad (9.139)$$

which allows direct calculation of the characteristic impedances $Z_1$ and $Z_2$. This system of two equations can be explicitly solved as a function of the parameter $r = R_L/R$ resulting in

$$\frac{Z_1}{R_L} = \frac{\sqrt{r(1 + \tan^2 \theta_1) - 1}}{r \tan \theta_1} \qquad (9.140)$$

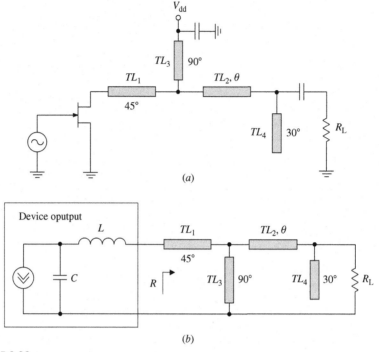

**FIGURE 9.38**

Schematics of a transmission-line Class-E/F$_3$ power amplifier.

$$\frac{Z_1}{Z_2} = \frac{r-1}{r\tan 30° \ \tan \theta_1} \tag{9.141}$$

which are reduced for $\theta = 15°$ and $\theta_1 = 45° + \theta = 60°$ to

$$\frac{Z_1}{R_L} = \frac{\sqrt{4r-1}}{\sqrt{3r}} \tag{9.142}$$

$$\frac{Z_1}{Z_2} = \frac{r-1}{r}. \tag{9.143}$$

Consequently, for specified value of the parameter $r$ with the required Class-E/F$_3$ optimum load resistance $R$ and standard load $R_L = 50\ \Omega$, the characteristic impedance $Z_1$ is calculated from Eq. (9.142) and then the characteristic impedance $Z_2$ is obtained by Eq. (9.143). Hence, if $R = 20\ \Omega$, resulting in $r = 2.5$, the characteristic impedance of the series transmission line is $Z_1 = 35\ \Omega$and the characteristic impedance of the open-circuit stub is $Z_2 = 58\ \Omega$.

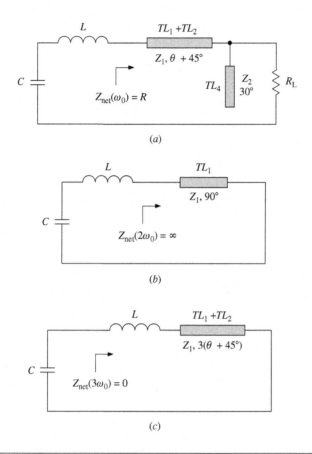

**FIGURE 9.39**

Load networks seen by a device output at harmonics.

The load network seen by the device output at the second harmonic is shown in Fig. 9.39(b), taking into account the shorting effect of the quarterwave short-circuit stub $TL_3$. Here, the transmission line $TL_1$ provides an open-circuit condition for the second harmonic at the device output. A similar load network at the third harmonic is shown in Fig. 9.39(c), due to the open-circuit effect of the short-circuited quarterwave line $TL_3$ and short-circuit effect of the open-circuited harmonic stub $TL_4$ at the third harmonic. In this case, the combined transmission line $TL_1 + TL_2$ together with the series inductance $L_{out}$ provides a short-circuit condition for the third harmonic at the device output being shorted at its right-hand side. Depending on the actual physical length of the device output electrode, on-board adjusting of the transmission lines $TL_1$ and $TL_2$ can easily provide the required open- and short-circuit conditions, as well as an impedance matching at the fundamental frequency, due to their series connection to the device output.

The required electrical length $\theta$ of the transmission line $TL_2$ can be defined from

$$3\omega_0 L + Z_1 \tan 3(\theta + 45°) = 0 \qquad (9.144)$$

where the total electrical length of the entire series transmission line $\theta_1 = 45° + \theta$ is equal to $\pi$ or $180°$ at the third harmonic component when $L = 0$. As a result,

$$\theta = \frac{\pi}{12} - \frac{1}{3} \tan^{-1} \frac{3\omega_0 L}{Z_1}. \qquad (9.145)$$

In this case, by combining Eqs (9.140) and (9.145), the resulting transcendental equation to implicitly determine the electrical length $\theta$ as one unknown parameter by iterative method can be obtained. Then, the characteristic impedance $Z_1$ is calculated from Eq. (9.140) that allow the characteristic impedance $Z_2$ to be defined from Eq. (9.141).

Figure 9.40 shows the basic simulated circuit schematic, which approximates a transmission-line Class-E/F$_3$ power amplifier based on a 28-V, 10-W Cree GaN HEMT power transistor CGH40010 and a transmission-line load network implemented on a 30-mil RO4350 substrate shown in Fig. 9.38($a$). The input matching circuit provides a complex-conjugate matching with the standard 50-$\Omega$ source. The load network was slightly modified by optimizing the parameters of the series transmission line since the device output capacitance $C_{out}$ and series inductance $L_{out}$ formed by drain bondwires and package lead do not match the required exact values of $C$ and $L$ for nominal Class-E/F$_3$ operation mode. Special stripline models for the device input and output package leads need to be added to the simulation setup to take into account their inductances, whose correct values depend on assembling procedure. The maximum output power of 40.8 dBm, power gain of 12.8 dB (linear gain of about 18 dB), drain efficiency of 82.5%, and *PAE* of 78.2% are achieved at an operating frequency of 2.14 GHz with a supply voltage of 28 V and a quiescent current of 40 mA.

## 9.4 Biharmonic Class-E$_M$ power amplifier

A basic limitation of a Class-E operation mode at higher frequencies is significant efficiency degradation due to the increased switching power losses with increasing values of the turn-off switching time. To minimize this undesirable effect, it is necessary to find a solution without an instant jump in an ideal collector current waveform at turn-off to allow efficient operation at frequencies; high enough that the switch turn-off transition would occupy a substantial fraction of the waveform period of 30% and more. However, the Class-E power amplifier can deliver nonzero output power only if at least one of the switch waveforms, either voltage or current, has a jump under the assumption that the circuit comprises an ideal

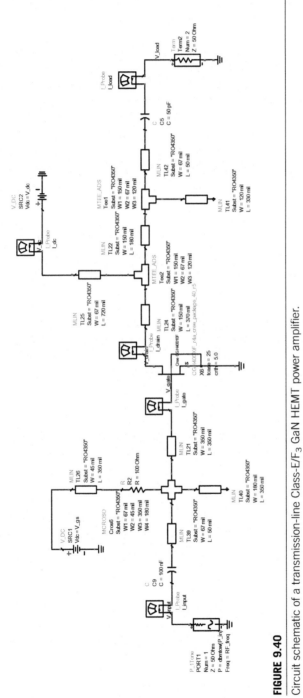

**FIGURE 9.40**

Circuit schematic of a transmission-line Class-E/F$_3$ GaN HEMT power amplifier.

switch and linear passive components [22]. To satisfy the requirements of both jumpless voltage and current waveforms and sinusoidal load waveform with non-zero output power delivered to the load, it is necessary to allow power flow in the system at two or more harmonically related frequencies. This can be done by using nonlinear reactive elements in the load network to convert the fundamental-frequency power to a desired harmonic frequency or by injecting the harmonic-frequency power into the load network from an external source.

The simplest low-order implementation approach having jumpless switch voltage and current waveforms, called the biharmonic Class-$E_M$ mode and described in [23], comprises the two-part output stage including:

- Main amplifier that consumes dc power equal to approximately 75% of the load power, and converts this power and the power generated by the auxiliary amplifier to power at the output frequency $f$.
- Smaller auxiliary amplifier (or varactor frequency multiplier), phased, locked to the main amplifier, which generates approximately 25% of the load power at the frequency $2f$.

The main amplifier has jumpless switch voltage and current waveforms, while the auxiliary amplifier can be a conventional Class-E power amplifier. If the frequency multiplier is fed from the output of the main amplifier, the load power is reduced by the amount of power converted by the frequency multiplier from frequency $f$ to frequency $2f$ to change the waveform shapes to continuous ones. The higher-order implementations can use harmonic components of orders higher than two, or multiple harmonics. For operation at higher frequencies, the biharmonic Class-$E_M$ power amplifier can be energetically superior to a conventional Class-E power amplifier using the same power device and supplying the same output power at the same operating frequency. That is because it can tolerate slow transistor turn-off with much less efficiency loss. In addition, less input drive is needed for the biharmonic Class-$E_M$ power amplifier because slower switching times are tolerable, taking into account that switching times are inversely proportional to the square root of the input driving power.

Figure 9.41 shows the circuit schematic of a biharmonic Class-$E_M$ MOSFET power amplifier designed to operate at a 3.5 MHz with second-harmonic power injection from an auxiliary amplifier operating at a 7 MHz. The derivation of the ideal drain waveforms of the main amplifier is based on the assumption that the resultant current of the active device, operating as a switch, and its shunt capacitor contains only dc, fundamental, and second-harmonic components written as

$$i(\omega t) = I_0 + I_{1A} \cos \omega t + I_{1B} \sin \omega t + I_{2A} \cos 2\omega t + I_{2B} \sin 2\omega t \qquad (9.146)$$

where $\omega$ is the angular fundamental frequency, $I_{1A}$ and $I_{1B}$ are the quadrature fundamental current components, and $I_{2A}$ and $I_{2B}$ are the quadrature second-harmonic current components, respectively. The shunt capacitors at the transistor drains can be composed of the device output capacitances and external capacitors.

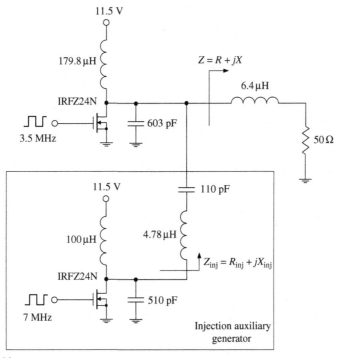

**FIGURE 9.41**

Biharmonic Class-E$_M$ power amplifier schematic.

For a 50% duty ratio when the switch is turned off during $0 < \omega t \le \pi$, the current through the switch $i(\omega t) = 0$, and the current $i_C(\omega t)$ flowing through the capacitor $C$ fully represents the current $i(\omega t)$ given in Eq. (9.146), reproducing the voltage across the switch by the charging of this capacitor according to

$$v(\omega t) = \frac{1}{\omega C} \int_0^{\omega t} i(\omega t)d\omega t. \tag{9.147}$$

The conditions for a biharmonic Class-E$_M$ optimum operation with jumpless voltage and current waveforms, $v(\omega t)$ and $i(\omega t)$, and unipolar switch current are

$$i(\omega t)|_{\omega t=0} = 0 \tag{9.148}$$

$$i(\omega t)|_{\omega t=\pi} = 0 \tag{9.149}$$

$$v(\omega t)|_{\omega t=\pi} = 0 \tag{9.150}$$

$$\left.\frac{v(\omega t)}{d\omega t}\right|_{\omega t=0} = 0. \tag{9.151}$$

Substituting Eq. (9.146) into Eq. (9.147) and applying the boundary conditions given by Eqs (9.148) to (9.151) yield

$$I_{1A} = 0 \tag{9.152}$$

$$I_{1B} = -\frac{\pi}{2}I_0 \tag{9.153}$$

$$I_{2A} = -I_0 \tag{9.154}$$

$$I_{2B} = \frac{\pi}{4}I_0. \tag{9.155}$$

As a result, the normalized steady-state ideal switch voltage waveform for a period of $0 \leq \omega t < \pi$ and current waveform for a period of $\pi \leq \omega t < 2\pi$ are

$$\frac{i(\omega t)}{I_0} = 1 - \frac{\pi}{2}\sin \omega t + \frac{\pi}{4}\sin 2\omega t - \cos 2\omega t \tag{9.156}$$

$$\frac{v(\omega t)}{V_{dd}} = \frac{2}{\pi}(8\omega t + 4\pi\cos \omega t - \pi\cos 2\omega t - 4\sin 2\omega t - 3\pi) \tag{9.157}$$

where $V_{dd}$ is the dc supply voltage.

Figure 9.42(b) shows the normalized switch voltage and current waveforms for an idealized optimum biharmonic Class-$E_M$ with a second-harmonic power injection. From switch voltage and current waveforms it follows that, when the transistor is turned on, there is no voltage across the switch and the current $i(\omega t)$ consisting of the dc, fundamental, and injected second-harmonic components flows through the device. However, when the transistor is turned off, this current flows through the shunt capacitance $C$. There is no jump in the switch current waveform at the instant of switching off compared to the switch current corresponding to a Class-E with shunt capacitance, the voltage and current waveforms of which are shown in Fig. 9.42(a). However, the voltage peak factor is higher in a biharmonic Class-$E_M$ mode exceeding a value of 4. It should be mentioned that injecting a higher-order harmonic component will generally increase the voltage peak factor even more. Also, there is no solution for a biharmonic Class-$E_M$ mode with third-harmonic injection and duty ratio of 50%. The voltage peak factor can exceed a value of 7 for a third-harmonic injection with a duty ratio of 33%. The voltage and current waveforms of the auxiliary amplifier are the usual waveforms corresponding to a switchmode Class-E with shunt capacitance.

In a biharmonic Class-$E_M$ mode, it is assumed that the dc power $P_0 = I_0 V_{dd}$ is equal to approximately 75% of the output load power $P_{out}$ delivered to the load that results in

$$I_0 V_{cc} = \frac{3}{4}\frac{P_{out}}{V_{dd}}. \tag{9.158}$$

The fundamental-frequency load-network impedance of the main amplifier and the second-harmonic injection-port impedance of the auxiliary amplifier can

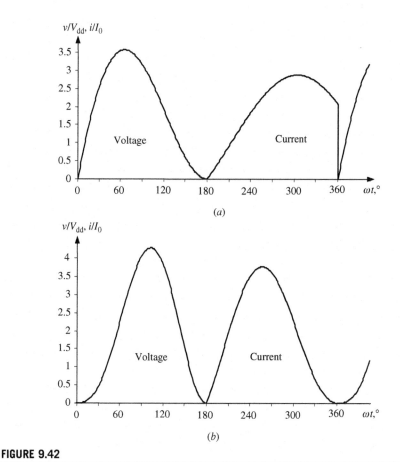

**FIGURE 9.42**

Normalized ideal switch waveforms of (a) Class-E with shunt capacitance and (b) biharmonic Class-E$_M$ with second-harmonic power injection.

be determined by a Fourier-series analysis of the voltage and current waveforms. As a result, the optimum shunt capacitance $C$ and load-network impedance $Z = R + jX$ for the main amplifier as a function of the dc supply voltage $V_{dd}$ and output power $P_{out}$ are written as

$$C = \frac{3\pi}{64} \frac{P_{out}}{\omega V_{dd}^2} \tag{9.159}$$

$$R = \frac{128}{9\pi^2} \frac{V_{dd}^2}{P_{out}} \tag{9.160}$$

$$X = \frac{32(3\pi^2 - 32)}{9\pi^3} \frac{V_{dd}^2}{P_{out}} \tag{9.161}$$

while the optimum injection-port impedance $Z_{inj} = R_{inj} + jX_{inj}$ for the auxiliary amplifier can be calculated from

$$R_{inj} = \frac{128}{9(\pi^2 + 16)} \frac{V_{dd}^2}{P_{out}}$$ (9.162)

$$X_{inj} = -\frac{16(3\pi^2 + 16)}{9\pi(\pi^2 + 16)} \frac{V_{dd}^2}{P_{out}}.$$ (9.163)

The measured output power of a second-harmonic Class-$E_M$ power amplifier was 13.2 W with overall *PAE* of 85.2% at an operating frequency of 3.5 MHz. The injected power at $2f$ necessary to achieve jumpless drain waveforms was measured as 29.8% of the total dc power of the main amplifier instead of the theoretical value of 25% due to the resistive power losses in reactive components, finite loaded quality factors of the series filters, and harmonic power conversions in the nonlinear device capacitances. To achieve simple and accurate design for the Class-$E_M$ power amplifier with higher-order circuits, the numerical design procedure can be applied [24].

Figure 9.43 shows the comparison between power-added efficiencies of the second-harmonic Class-$E_M$ and classical Class-E power amplifiers as functions of the normalized transistor switching time $\tau_s$. It is assumed that the switching time is inversely proportional to the input-drive power. The plots were simulated for power amplifiers delivering the output power 3.2 W at an operating frequency of 870 MHz using a pHEMT device with a gate periphery of 0.5 μm × 50 mm in the main amplifier. The peak value of a *PAE* for the biharmonic Class-$E_M$ power amplifier is 3.3% lower than that for the classical Class-E power amplifier. However, a *PAE* for the Class-$E_M$ power amplifier varies by just ±2% for all

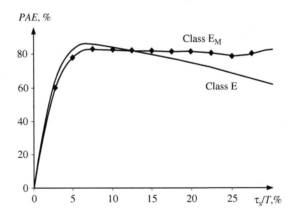

**FIGURE 9.43**

Efficiency *versus* switching time for Class-$E_M$ and Class-E power amplifiers.

switching times from 6% to 30% of the period, whereas a *PAE* for the Class-E power amplifier drops monotonically from its peak to 73.5% of its peak value for switching times of 30% of the period.

## 9.5 Inverse Class-E power amplifiers

Another approach to the design of the Class-E power amplifier with efficiency of 100% under idealized operation conditions is to use its configuration with shunt inductance [25]. Such a Class-E power amplifier represents an inverse version of a classical Class-E power amplifier with shunt capacitance, where the load-network inductor and capacitor replace each other. In this case, the storage element is a shunt inductor instead of a shunt capacitor, resulting in the inverse collector voltage and current waveforms. The basic circuit of a Class-E power amplifier with shunt inductance as a simplest version of an inverse Class-E mode is shown in Fig. 9.44(*a*), where the load network consists of an inductor $L$ shunting the transistor, a series capacitor $C$, a series fundamentally tuned $L_0C_0$ circuit, and a load resistor $R$. Figure 9.44(*b*) shows the equivalent circuit of such a switchmode power amplifier where the active device is considered an ideal switch that is driven so as to provide the device switching between its on-state and off-state operation conditions.

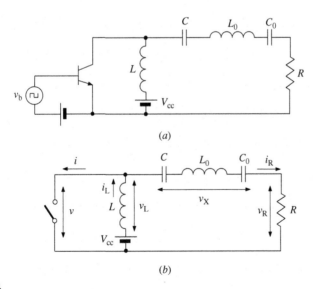

(*a*)

(*b*)

**FIGURE 9.44**

Basic circuits of a Class-E power amplifier with shunt inductance.

To simplify an analysis of a Class-E power amplifier with shunt inductance, the following several assumptions are introduced:

- The transistor has zero saturation voltage, zero saturation resistance, infinite off-resistance, and its switching action is instantaneous and lossless.
- The loaded quality factor $Q_L = \omega L_0/R = 1/\omega C_0 R$ of the series resonant $L_0 C_0$ circuit tuned to the fundamental frequency is high enough for the output current to be sinusoidal at the switching frequency.
- There are no losses in the circuit except in the load $R$.
- For an optimum operation mode, a 50% duty ratio is used.

In the class-E mode with shunt inductance similar to the other Class-E modes, it is possible to eliminate power losses during on-to-off transition by providing the following collector current conditions:

$$i(\omega t)|_{\omega t=2\pi} = 0 \tag{9.164}$$

$$\left.\frac{di(\omega t)}{d\omega t}\right|_{\omega t=2\pi} = 0 \tag{9.165}$$

where $i(\omega t)$ is the current flowing through the switch. These conditions are important to provide zero on-to-off switching time without any power dissipation and can be performed by proper choice of the load-network parameters.

For the simplified power amplifier circuit shown in Fig. 9.44(b), the relationships between the voltages and currents can be written as

$$i(\omega t) = i_L(\omega t) - i_R(\omega t) \tag{9.166}$$

$$v(\omega t) = V_{cc} - v_L(\omega t) \tag{9.167}$$

where the load current $i_R$ is assumed sinusoidal,

$$i_R(\omega t) = I_R \sin(\omega t + \varphi) \tag{9.168}$$

with the amplitude $I_R$ and initial phase $\varphi$.

When the switch is turned off during $0 \le \omega t \le \pi$,

$$i(\omega t) = 0 \tag{9.169}$$

then from Eqs (9.166) and (9.168) it follows that

$$i_L(\omega t) = i_R(\omega t) = I_R \sin(\omega t + \varphi). \tag{9.170}$$

At the same time, the voltage across the inductor $L$ can be written as

$$v_L(\omega t) = \omega L \frac{di_L(\omega t)}{d\omega t} = \omega L I_R \cos(\omega t + \varphi) \tag{9.171}$$

from which it follows that the voltage $v(\omega t)$ according to Eq. (9.167) is

$$v(\omega t) = V_{cc} - \omega L I_R \cos(\omega t + \varphi). \tag{9.172}$$

When the switch is turned on during $\pi \leq \omega t \leq 2\pi$,

$$v(\omega t) = 0 \tag{9.173}$$

then substituting Eq. (9.173) in Eq. (9.167) yields

$$v_L(\omega t) = V_{cc}. \tag{9.174}$$

By using Eqs (9.170), (9.171), and (9.174), the current flowing through the inductor $L$ can be obtained as

$$i_L(\omega t) = \frac{V_{cc}}{\omega L}(\omega t - \pi) - I_R \sin \varphi. \tag{9.175}$$

Hence, from Eqs (9.166), (9.168), and (9.175) it follows that

$$i(\omega t) = \frac{V_{cc}}{\omega L}(\omega t - \pi) - I_R[\sin(\omega t + \varphi) + \sin \varphi]. \tag{9.176}$$

Applying a zero-current condition given by Eq. (9.164) to Eq. (9.176) gives

$$I_R(\omega t) = \frac{\pi V_{cc}}{2\omega L \sin \varphi} \tag{9.177}$$

where $0 < \varphi < \pi$ because $I_R > 0$.

As a result, by substituting Eqs (9.177) into Eq. (9.176), the steady-state current waveform across the switch can be obtained as

$$i(\omega t) = \frac{V_{cc}}{\omega L}\left[\omega t - \frac{3\pi}{2} - \frac{\pi}{2\sin \varphi}\sin(\omega t + \varphi)\right] \tag{9.178}$$

Applying a zero current-derivative condition given by Eq. (9.165) to Eq. (9.178) gives

$$\tan \varphi = \frac{\pi}{2} \tag{9.179}$$

resulting in the phase angle equal to

$$\varphi = \tan^{-1}\left(\frac{\pi}{2}\right) = 57.518°. \tag{9.180}$$

From Eq. (9.179), it follows that

$$\sin \varphi = \frac{\pi}{\sqrt{\pi^2 + 4}} \tag{9.181}$$

$$\cos \varphi = \frac{2}{\sqrt{\pi^2 + 4}} \tag{9.182}$$

which allows us to rewrite Eq. (9.178) as

$$i(\omega t) = \frac{V_{cc}}{\omega L}\left(\omega t - \frac{3\pi}{2} - \frac{\pi}{2}\cos \omega t - \sin \omega t\right). \tag{9.183}$$

By using a Fourier-series expansion, the supply dc current is

$$I_0 = \frac{1}{2\pi} \int_0^{2\pi} i(\omega t)d\omega t = \frac{V_{cc}}{\pi \omega L} \tag{9.184}$$

whereas the fundamental-current amplitude can be obtained using Eqs (9.177), (9.181), and (9.184) as

$$I_R = \frac{\pi\sqrt{\pi^2 + 4}}{2}I_0 = 5.8499I_0. \tag{9.185}$$

As a result, the normalized steady-state collector current waveform for $\pi \le \omega t \le 2\pi$ and the normalized voltage waveform for $0 \le \omega t \le \pi$ are

$$\frac{i(\omega t)}{I_0} = \pi\left(\omega t - \frac{3\pi}{2} - \frac{\pi}{2}\cos\omega t - \sin\omega t\right) \tag{9.186}$$

$$\frac{v(\omega t)}{V_{cc}} = \frac{\pi}{2}\sin\omega t - \cos\omega t + 1. \tag{9.187}$$

From the voltage and current waveforms shown in Fig. 9.45, it follows that, when the switch is turned off, the switch current $i(\omega t)$ is zero, and the inductor current $i_L$ is determined by the sinusoidal load current $i_R$. However, when the switch is turned on, the switch voltage $v(\omega t)$ is zero, and the dc-supply voltage $V_{cc}$ produces the linearly increasing current $i_L$. The difference between the inductor current $i_L$ and the load current $i_R$ flows through the switch.

The peak collector current $I_{max}$ and voltage $V_{max}$ can be determined by differentiating the corresponding waveforms given by Eqs (9.186) and (9.187) and setting the results equal to zero, thus resulting in

$$I_{max} = \pi(\pi - 2\varphi)I_0 = 3.562I_0 \tag{9.188}$$

$$V_{max} = \left(\frac{\sqrt{\pi^2 + 4}}{2} + 1\right)V_{cc} = 2.8621V_{cc}. \tag{9.189}$$

Under the assumption of a sinusoidal load current, the load voltage is also sinusoidal,

$$v_R(\omega t) = V_R \sin(\omega t + \varphi) \tag{9.190}$$

where $V_R = I_R R$ is the voltage amplitude across the load resistor.

The voltage $v_X$ across the elements of the series resonant circuit is not sinusoidal. Consequently, its fundamental component across the capacitance $C$ is obtained due to an assumed high $Q_L$-factor as

$$v_C(\omega t) = -V_C \cos(\omega t + \varphi) \tag{9.191}$$

where $V_C = I_R/\omega C$ is the voltage amplitude on the capacitor.

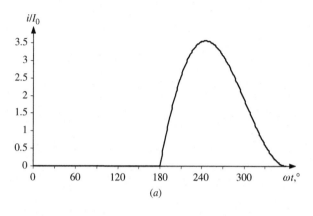

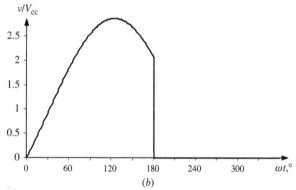

**FIGURE 9.45**

Normalized switch voltage and current waveforms for inverse Class-E mode.

As a result, the phase shift between the fundamental components of the capacitor and load-resistor voltages can be defined by

$$\tan \psi = -\frac{V_C}{V_R} = -\frac{1}{\omega RC}. \tag{9.192}$$

Using Eq. (9.187) for the idealized collector voltage waveform and applying a Fourier-series expansion yields

$$V_R = \frac{1}{\pi} \int_0^{2\pi} v(\omega t)\sin(\omega t + \varphi)d\omega t = \frac{4V_{cc}}{\pi\sqrt{\pi^2 + 4}} = 0.3419V_{cc} \tag{9.193}$$

$$V_C = -\frac{1}{\pi} \int_0^{2\pi} v(\omega t)\cos(\omega t + \varphi)d\omega t = \frac{(\pi^2 + 12)V_{cc}}{4\sqrt{\pi^2 + 4}} = 1.4681V_{cc}. \tag{9.194}$$

Substituting Eqs (9.184) and (9.185) into Eqs (9.193) and (9.194) results in

$$V_R = \frac{8}{\pi(\pi^2 + 4)}\omega L I_R \tag{9.195}$$

$$V_C = \frac{\pi^2 + 12}{2(\pi^2 + 4)}\omega L I_R \tag{9.196}$$

Hence, by using Eq. (9.192), the optimum series inductance $L$ and shunt capacitance $C$ can be calculated as

$$\frac{\omega L}{R} = \frac{\pi(\pi^2 + 4)}{8} = 5.4466 \tag{9.197}$$

$$\omega CR = \frac{16}{\pi(\pi^2 + 12)} = 0.2329 \tag{9.198}$$

with an optimum phase angle $\psi$ equal to

$$\psi = -\tan^{-1}\left[\frac{\pi(\pi^2 + 12)}{16}\right] = -76.891°. \tag{9.199}$$

The optimum phase angle $\phi$ of the entire load network with a shunt inductor $L$ and a series capacitor $C$ is calculated by using Eqs (9.197) and (9.198) as equal to

$$\phi = -35.945° \tag{9.200}$$

which indicates that the input impedance of such a load network is capacitive.

The optimum load resistance $R$ can be found using Eq. (9.193) by

$$R = \frac{8}{\pi^2(\pi^2 + 4)}\frac{V_{cc}^2}{P_{out}} = 0.05844\frac{V_{cc}^2}{P_{out}}. \tag{9.201}$$

The high-$Q_L$ assumption for the series resonant $L_0C_0$ circuit can lead to considerable errors if its value is substantially small in real circuits [26]. For example, for a 50% duty ratio, the values of the optimum shunt inductance $L$, collector-current peak factor, and second-harmonic component increase rapidly for $Q_L < 5$.

Figure 9.46 shows an alternative inverse Class-E load-network configuration with a series inductor $L$, a shunt capacitor $C$, and a parallel resonant $L_0C_0$ circuit tuned to the fundamental frequency [27]. Here, the RF choke is necessary to supply the dc current $I_0$, and the blocking capacitor $C_b$ is required to decouple the parallel resonant circuit from the dc path. Such an inverse Class-E with a series inductance and a shunt capacitance is dual to a classical Class-E with a shunt capacitance and a series inductance. In this case, the series resonant $L_0C_0$ circuit is replaced by its parallel-equivalent configuration. However, it is characterized by the same switch voltage and current waveforms, as shown in Fig. 9.45 for an inverse Class-E configuration with a shunt inductance and a series capacitance.

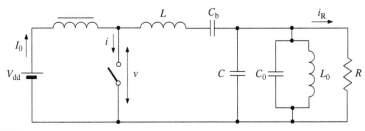

**FIGURE 9.46**

Equivalent circuit of an inverse Class-E power amplifier with a series inductance.

The optimum series inductance $L$, shunt capacitance $C$, and load resistance $R$ of the Class-E load network with a series inductance and a shunt capacitance can be calculated as

$$\frac{\omega L}{R} = \frac{8}{\pi(\pi^2 + 4)} = 1.1525 \tag{9.202}$$

$$\omega CR = \frac{\pi(\pi^2 - 4)}{16} = 0.1836 \tag{9.203}$$

$$R = \frac{\pi^2 + 4}{8}\frac{V_{dd}^2}{P_{out}} = 1.7337 \frac{V_{dd}^2}{P_{out}}. \tag{9.204}$$

where $V_{dd}$ is the drain supply voltage and $P_{out}$ is the fundamental-frequency output power delivered to the load [28].

The previous analysis was based on the assumption of a zero device output capacitance. However, at high frequencies and for high-power applications, this capacitance cannot be considered negligible and should be taken into account. In this case, the power amplifier could operate in a switching mode with both shunt and series capacitances when the required idealized operations conditions are realized with zero-current and zero current-derivative switching conditions. However, to compensate for the capacitor nonzero discharging process, it is necessary to provide a zero voltage-derivative condition at the same time. Therefore, the theoretical analysis illustrates the infeasibility of a zero-current switching Class-E mode to approach 100% collector efficiency with nonzero device output capacitance [29]. Moreover, the collector efficiency will drop drastically if the output device capacitance is significant. This means that, in the frequency domain, the harmonic impedance conditions should be different, being inductive at the fundamental frequency and capacitive for higher-order harmonics, which can be achieved only with zero-voltage switching conditions.

However, there is a class of the pHEMT or GaN HEMT devices, the output capacitance of which is sufficiently small and which can be used for a low-power high-frequency Class-E power amplifier applications. The typical value of the

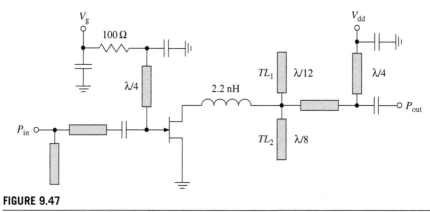

**FIGURE 9.47**

Circuit schematic of a transmission-line inverse Class-E power amplifier.

drain-source capacitance for a 0.15-μm or 0.5-μm pHEMT device with a total gate width of $8 \times 75$ μm is of about 0.25 pF, and, as a first approximation, it increases linearly with total gate width [28]. The simulation results at an operating frequency of 2.5 GHz show that, for the same device parameters when using output capacitance as a variable, the drain efficiency is just slightly affected when its value is less than 0.7 pF. However, the drain efficiency reduces significantly as the drain-source capacitance increases above 0.7 pF. Consequently, there is a limitation in the device size, above which it is not possible to achieve high drain efficiency. To estimate the maximum achievable output power, it is necessary to take into account the typical pHEMT dc current density of 300-mA/mm for a reliable operation. The measurements results of a hybrid single-stage power amplifier using a pHEMT device with the gate width of $6 \times 50$ μm and standard surface-mounted components in an inverse Class-E mode with series inductance exhibited an output power of 18 dBm with a power gain of 18 dB and a drain efficiency of 95% at an operating frequency of 870 MHz and a dc supply voltage of 2 V [27].

Figure 9.47 shows the circuit schematic of a transmission-line inverse Class-E power amplifier, where the shunt capacitor $C$ is replaced by two open-circuit shunt stubs to provide the second- and third-harmonic short-circuit terminations, thus dispensing with the need for a parallel-tuned $L_0 C_0$ circuit [30,31]. The electrical lengths of the open-circuit stubs are 45° and 30° at the fundamental frequency $f_0$, representing an overall capacitive reactance, and they are equivalent to 90° at $2f_0$ and $3f_0$, respectively. As a result, for a 0.3-μm GaAs MESFET with a gate width of 1200 μm, an output power of 22 dBm with a drain efficiency of 69% was obtained at an operating frequency of 2.3 GHz with a dc-supply voltage of 3 V.

## 9.6  **Harmonic tuning using load-pull techniques**

Generally, the transistor is operated as a nonlinear active device under large-signal conditions, resulting in a significant amount of harmonic components. Since the second harmonic is the largest harmonic component, it was found experimentally that just providing the proper source and load second-harmonic terminating impedances at the input and output ports of the transistor, respectively, can significantly improve the power-amplifier efficiency At microwave frequencies, the special network based on three coupled bars positioned between two ground planes can be used to realize a wide range of impedance matching at the fundamental frequency, while simultaneously presenting reactive impedance at the second-harmonic component, thus decoupling the second-harmonic signal from a 50-$\Omega$ load [32].

To maximize the power-amplifier efficiency, the optimum impedances at the harmonic components must be provided at the transistor output port. Those impedances must be measured or simulated very accurately at microwave frequencies, taking into account the device parasitic elements and difficulties in accurately defining the values of the device equivalent-circuit parameters under large-signal operation. Since it is necessary to realize zero or infinite harmonic impedance conditions seen by the internal device multiharmonic current source required for ideal conventional Class-F or inverse Class-F modes, the corresponding optimum reactances at the harmonics must be presented at the device external output port which can then be translated to those internal required for a switchmode operation. Once the optimum impedances at the device output for the second and third harmonics are established by accurate measurements or simulations, the next step is to properly design the load network using lumped or transmission-line elements to achieve those impedances.

Figure 9.48 shows the circuit configuration of a high-efficiency MESFET power amplifier using a harmonic-reaction technique based on the effect of the second-harmonic injection [33]. Unlike the balanced power amplifier, this configuration includes a second-harmonic path between the output ports of the devices. The fundamental-frequency output paths and second-harmonic path are designed to provide independent matching of the device output impedances, conjugate impedance matching with a 50-$\Omega$ load at the fundamental frequency, and optimum reactance matching at the second harmonic for maximum drain efficiency. Since the second-harmonic path has well-matched impedance characteristics, each MESFET device mutually injects without reflection a second-harmonic component into the other MESFET device through the second-harmonic path. However, it should be noted that a high-efficiency operation is possible only if both devices are driven with phase-coherent and equal-amplitude input signals. As a result, an output power of 5 W with a *PAE* of 70% and a drain efficiency of 75% was obtained at an operating frequency of 2 GHz and a supply voltage of 7 V.

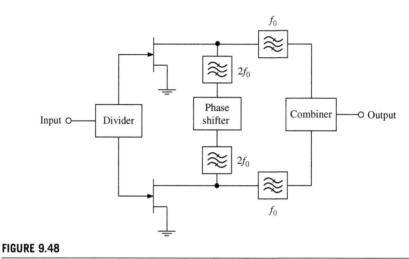

**FIGURE 9.48**

Circuit configuration of a harmonic-reaction amplifier.

Figure 9.49(*a*) shows the circuit schematic of a single-ended *X*-band MESFET power amplifier designed to provide simultaneously the optimum load impedance at the fundamental frequency and zero impedance at the second harmonic at the device output [34]. The load network includes the series parasitic drain lead inductor, whose value was optimized to provide the proper reactance to the device. An open-circuit stub with electrical length of 45° at the fundamental was used for the second-harmonic short circuit. Since it presents a capacitive reactance at the fundamental frequency, it becomes a part of the output matching circuit as a shunt capacitor. An output power of 450 mW and a maximum *PAE* of 61% with a drain efficiency of 76% were obtained at an operating frequency of 10 GHz, using a MESFET device with a gate geometry of 0.5 μm × 1200 μm. The same power amplifier without the second-harmonic tuning circuit had demonstrated a *PAE* of only 50%. Figure 9.49(*b*) shows the circuit schematic of an *X*-band 0.5-W MESFET power amplifier with the same device geometry having a load network with lumped elements tuned to optimally terminate the second and third harmonics [35]. As a result, the measured efficiency of the power amplifier with optimal harmonic tuning was 49.3%, while the maximum efficiency measured for a device tuned to only the fundamental was 44.5%.

Generally, the second-harmonic-tuned power-amplifier design strategy can be based on the control of the harmonic terminations at both input and output device ports, so as to properly shape the gate and drain voltage waveforms at 1-dB gain-compression point [36]. In this case, input matching circuit is responsible for assuring the fundamental conjugate-matching condition and shaping the device input waveform, thus controlling the generation of properly phase-related output drain-current harmonic components. The load network is responsible for shaping

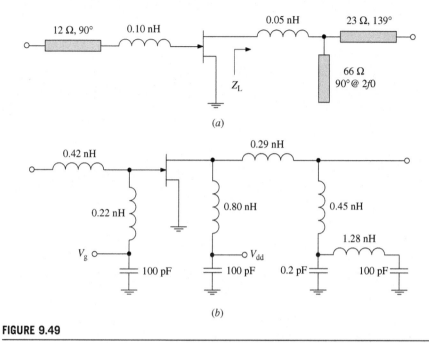

**FIGURE 9.49**

Circuit schematics of a Class-B power amplifier with harmonic tuning.

the device output voltage waveform and delivering maximum output power to the load at the fundamental frequency. For an integrated power amplifier structure, the input and output matching circuits can be designed using high-dielectric substrates to be implemented into a small-size package. In this case, the input matching circuit includes two quarterwave transmission lines to transform the standard 50-$\Omega$ source to the device input impedance and open-circuit stubs to provide a second-harmonic reflection phase of around 180° (short) at the gate. The output matching circuit comprises three transmission-line transformers with different characteristic impedances to transform the fundamental-frequency low optimum impedance at the device drain with the standard 50-$\Omega$ load and provide the second-harmonic impedance close to an open-circuit condition. As a result, a drain efficiency of 68% can be achieved for a 100-W GaN HEMT power amplifier (package cavity size is 14.3 × 15.2 mm) with a dc-supply voltage of 40 V at an operating frequency of 4 GHz [37].

Figure 9.50 shows the circuit schematic of a load network with distributed harmonic control, which was designed to operate in the frequency range from 1.8 to 2.3 GHz and provide a support of multi-band and multi-standard operation [38]. The load network is composed of two separate blocks: a fundamental matching circuit and a second-harmonic inductive termination circuit. In this case, the

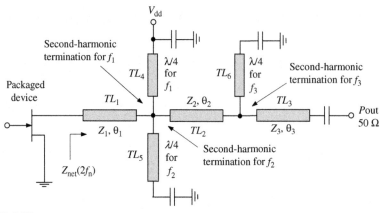

**FIGURE 9.50**

Circuit schematic of a load network with distributed second-harmonic control.

fundamental matching circuit consisting of three series transmission lines $TL_1$, $TL_2$, and $TL_3$ of different characteristic impedances and electrical lengths is relatively simple because the optimum impedance remains relatively constant over the required frequency range. Therefore, the fundamental matching impedance was optimized for a centre bandwidth frequency of 2.14 GHz. Since the optimum second-harmonic termination $Z_{net}$ varies significantly over the required frequency range, a distributed topology with multiple discrete harmonic terminations was designed and optimized for $f_1 = 1.85$ GHz, $f_2 = 1.96$ GHz, and $f_3 = 2.14$ GHz. As a result, drain efficiencies over 60% are achieved between 1.77 and 2.17 GHz for a 45-W GaN HEMT power amplifier. Note that generally high efficiency can be achieved over a range of second-harmonic termination phase, and, as the second harmonic termination is swept from capacitive to inductive, the optimal fundamental impedance sweeps from inductive to capacitive [39].

One popular and effective way to measure the fundamental optimum equivalent device output impedance is to use the passive or active load-pull measurement system [40,41]. The objective of the active load-pull measurement technique is to find the best output load in terms of the optimum value of the load reflection coefficient $\Gamma_L(f_0)$ presented to the transistor output at the fundamental frequency, such that the power gain and power-added efficiency are optimized. Although cumbersome and labor intensive, such an experimental technique can provide accurate information regarding the nonlinear operation of the active device. It should be mentioned that the transistor large-signal $S$-parameters measured at the certain bias conditions and output power provide insufficient information for the design of strongly driven power amplifiers with strongly nonlinear active device behavior. Therefore, when the transistor is operated in a nonlinear mode, measurement techniques are required that reproduce large-signal

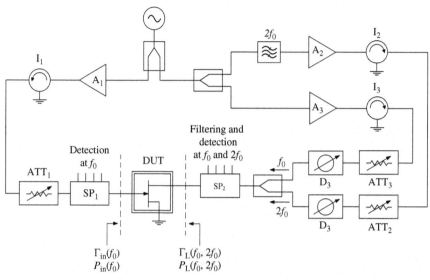

**FIGURE 9.51**

Harmonic active load-pull measurement system schematic.

operation in the form of the input and output impedances or a nonlinear active device model valid for all used bias conditions and frequency ranges.

Figure 9.51 shows the basic components of an active load-pull measurement system based on two six-port reflectometers [42]. At the input of the device under test (DUT), the six-port reflectometer $SP_1$ provides a measurement of the large-signal reflection coefficient $\Gamma_{in}(f_0)$ and the input power $P_{in}(f_0)$ absorbed by the DUT. At the output of the DUT, the measurements of the large-signal reflection coefficient $\Gamma_L(nf_0)$ and the output power $P_L(nf_0)$ for $n = 1, 2$ are achieved by the six-port reflectometer $SP_2$. The evaluation of $\Gamma_L(nf_0)$ and $P_L(nf_0)$ at a particular frequency requires that the power at that frequency be extracted from the output spectrum by means of splitting and filtering circuits inserted between all detection ports and the power sensors. Performing a fundamental load-pulling consists of varying the load $\Gamma_L(f_0)$ over the entire Smith chart with RF and dc measurements for each load, while the load $\Gamma_L(2f_0)$ is maintained constant and near 50 $\Omega$. However, the fundamental load-pull technique does not include the effects of harmonic loading. The second harmonic load-pull characterization consists in varying the load $\Gamma_L(2f_0)$, while maintaining the load $\Gamma_L(f_0)$ at a desired value. As a result, it was found that the best performance in terms of power gain and power-added efficiency is obtained for purely reactive second-harmonic loads, since this allows the elimination of the resistive power losses. For a specific case of the MESFET power amplifier designed to operate at 3.5 GHz, an improvement of power-added efficiency by about 8% was achieved by using of a second-harmonic active load-pull characterization technique. The active load-pull system based on

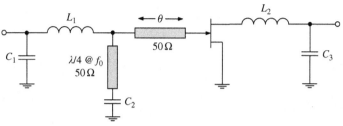

**FIGURE 9.52**

Circuit schematic of a Class-AB power amplifier with source-harmonic tuning.

a six-port reflectometer can also be used to optimize both linearity and output power or power-added efficiency for modulated signal [43].

In addition to harmonic loading at the device output, the harmonic tuning at the device input can be effective. For Class-AB power amplifiers, both the linearity and efficiency can be improved by the suppression of even-order harmonic components at the device input. Figure 9.52 shows an example of the circuit schematic of a microwave MESFET power amplifier with source second-harmonic tuning designed to operate in a Class-AB mode [44]. To terminate even harmonics at the device input, a quarter-wavelength shunt transmission line was short-circuited at its far end using a large-value capacitance $C_2$. The series transmission line with varying electrical length $\theta$ is necessary to provide a symmetrical gate-voltage waveform by compensating for the phase of terminating second-harmonic component, taking into account the device input gate-source capacitance and parasitic series gate inductance. The results of nonlinear circuit simulations show a linear (at 1-dB gain compression point) output power of 37.4 dBm at an operating frequency of 6 GHz with a *PAE* improved by 7% by using source-harmonic tuning. It should be noted that changes in the source second-harmonic impedance can vary *PAE* significantly, from 30% to 80% for a 5-GHz, 28.4-dBm MESFET power amplifier [45].

A systematic procedure of multiharmonic load-pull simulation using the harmonic-balance method includes two basic steps: finding the optimal loading at each harmonic component and checking the power levels of higher harmonics [46]. An impedance-sampling method can be used to find the optimal harmonic loading. Generally, only one impedance component $Z_i(k\omega_0)$ at $k$th harmonic and $i$th external port is sampled at the defined impedance plane, while the others are kept constant at each step. For a specific case of the MESFET power amplifier with input and output tuners, as shown in Fig. 9.53(*a*), the design procedure steps are as follows:

1. Start load-pull simulation by sampling the impedance at the fundamental frequency $Z_i(\omega_0)$ and find the optimal load $Z_{iopt}(\omega_0)$.
2. Fix $Z_i(\omega_0)$ at $Z_{iopt}(\omega_0)$ and check the output spectrum to observe the effects of higher harmonic loads on the power-amplifier response.

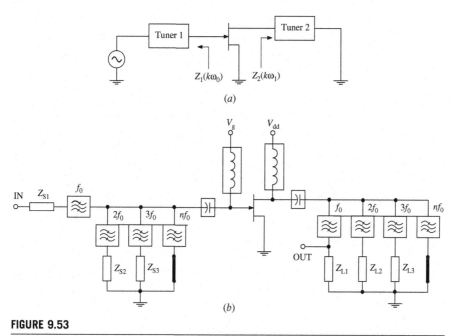

**FIGURE 9.53**

Circuit topologies for multiharmonic load-pull design procedure.

**3.** Perform load-pull simulation by sampling the $k$th harmonic impedance $Z_i(k\omega_0)$ and find the optimal load $Z_{iopt}(k\omega_0)$.

**4.** Fix $Z_i(k\omega_0)$ at $Z_{iopt}(k\omega_0)$ and check the output spectrum to observe the effects of higher harmonic loads on the power-amplifier response.

It should be noted that, if checking the output power spectrum shows that the power at higher harmonics is sufficiently small, then the design procedure is stopped.

For a 900-MHz power amplifier using a GaAs MESFET device with a gate periphery of 0.7 μm × 600 μm biased at 3 V with a quiescent current of approximately 10% of the total dc drain current, it was shown that, for a sinusoidal driving signal, the best power-added efficiencies are achieved at load phase angles close to ±180° for a second harmonic, corresponding to short-circuit condition, and at load phase angles close to zero at the third harmonic, corresponding to open-circuit condition that approximates a Class-F mode [47]. The simulation results demonstrated a strong dependence of efficiency with the second-harmonic termination, while the effect of the third harmonic is much weaker. The effect of the fourth- and fifth-harmonic terminations, stronger in the former case and weaker in the latter case, results in an efficiency variation of approximately 2.5%. It is necessary to minimize the resistive losses in harmonic terminations, intended to be purely reactive, that could be changed from reactive to resistive if the second- and higher-order harmonic terminations are excessively lossy.

The properly phased input and output voltage-harmonic components using a multiharmonic terminating scheme shown in Fig. 9.53(b) can be achieved by choosing the optimum impedances for each harmonic component [48,49]. For example, in a Class-F mode these output impedances can be purely resistive (low at even harmonics and high at odd harmonics), since the drain (collector) current second harmonic is always in-phase and the drain (collector) current third harmonic is always out-of-phase with respect to the fundamental component for conduction angles ranging from 180° (Class-B biasing) to 0° (Class-C biasing). Under slightly overdriven device operation when the drain current waveform represents a truncated half-sinusoid, the third harmonic becomes out-of-phase also for Class-AB biasing with conduction angles greater than 180°. The same result can be obtained by using a driver stage in a Class-F mode. However, in an inverse Class-F mode, the drain current second harmonic must be out-of-phase with respect to the fundamental-frequency component which can be realized by reactive input termination of the second harmonic providing the required phase shift. Fortunately, the main contribution to the harmonic-generating mechanism at the device input is given by its nonlinear input gate-source capacitor acting in a required reverse direction by generating the second- and third-harmonic components with proper phasing. As a result, optimum half-sinusoidal drain-voltage waveform corresponding to an inverse Class-F mode can be achieved by optimizing the input and output impedances at the second and third harmonics simultaneously.

Generally, it is very difficult to provide accurate source- and load-pull measurements at high microwave and millimeter-wave frequencies where de-embedding of waveguide transitions and parasitic discontinuities lead to significant measurement errors. It can be done accurately enough by using on-wafer measurements for an active device with limited output-power capability at the fundamental frequency. At the same time, the $S$-parameters of tuner structures can be accurately predicted using both linear and electromagnetic circuit simulators. The determination of the optimum impedances at the second harmonic can be done by computer source- and load-pull analysis. In this case, it is no problem to isolate the fundamental and second-harmonic paths to determine their optimum impedances. Figure 9.54 shows the schematic of a source- and load-pull simulation setup for a $V$-band InP HEMT power amplifier, where the ideal output duplexer separates the fundamental from the second-harmonic components [50]. By changing the ideal transformer turns ratio and the reactive component values, the output power, $PAE$, power gain, and stability circle contours can be simulated for various bias voltages and operating frequencies around center-band frequency of 60 GHz. The optimum second-harmonic termination resulted in a theoretical 6% improvement in $PAE$ at the 3-dB gain-compression point under Class-AB bias. However, in practice, a smaller improvement is expected because low-loss reactive harmonic loading at

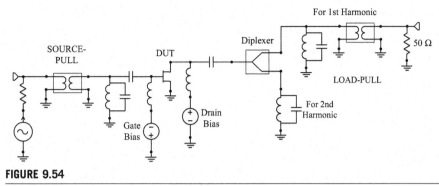

**FIGURE 9.54**

Schematic of a source- and load-pull simulation setup.

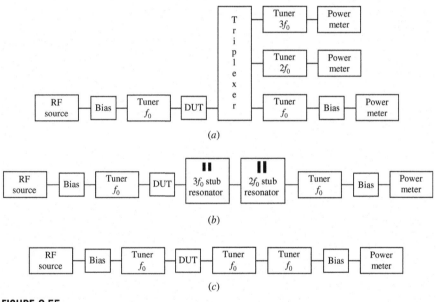

**FIGURE 9.55**

Block diagrams for multiharmonic load-pull design procedures.

120 GHz is difficult to physically implement without degrading the fundamental load match at 60 GHz.

Three basic methods of harmonic tuning have been offered commercially for load-pull systems with passive automated tuners [51,52]. The *triplexer tuning method* uses filters to separate the fundamental and harmonic signals so they can be tuned separately. The block diagram corresponding to this method is shown in Fig. 9.55(*a*), where a triplexer includes a low-pass filter for the fundamental frequency $f_0$, a bandpass filter for the second harmonic $2f_0$, and a bandpass or high-pass filter for the third harmonic $3f_0$. The *stub resonator method* is based on using open-circuit stubs with quarterwave lengths at the second- and third-harmonic components, connected to the center conductor with a sliding contact, according to the block diagram shown in Fig. 9.55($b$). The *cascaded tuner method* uses the two cascaded tuners shown in Fig. 9.55($c$) with 625 states each, producing nearly 400,000 available impedance states at the fundamental frequency with a variety of the impedances at the second harmonic. This allows the possibility to optimize the impedance at $2f_0$ with approximately constant impedance at $f_0$. Each method has its own advantages and disadvantages. For example, the triplexer method is the only approach with high tuning isolation, however with slightly more power loss at harmonic frequencies, especially at the third harmonic, compared to the stub-resonator method. At the same time, the stub-resonator method with dual stubs is operated over a very narrow bandwidth. The cascaded-tuner method has an advantage here, because no hardware needs to be changed, however its tuning isolation is very poor. Thus, the triplexer method looks the best because of the major advantages over the other methods, with typical return-loss isolation over 100 dB and insertion loss of 0.2 dB to 0.3 dB.

## 9.7 Chireix outphasing power amplifiers

The outphasing modulation technique was invented in the middle of 1930s in order to improve both the efficiency and linearity of AM-broadcast transmitters [53]. Substantially later in the 1970s, its application was extended up to microwave frequencies under the name LINC (*linear amplification using nonlinear components*) [54]. An outphasing transmitter is operated as a linear power amplifier system for amplitude-modulated signals having a linear transfer function over a wide range of the input signal levels by combining the outputs of two nonlinear power amplifiers that are driven with signals of constant amplitude but different time-varying phases corresponding to the envelope of the input signal.

A simple outphasing power amplifier system is shown in Fig. 9.56($a$) [55]. The signal component separator (SCS) generates from the input amplitude-modulated signal two sinewave signals of constant envelopes with different phases $+\phi(t)$ and $-\phi(t)$. These two signals are then separately amplified by the identical nonlinear power amplifiers and combined to produce the output amplitude-modulated signal. The peak output power is obtained with $\phi = 90°$ when currents from power amplifiers with equal amplitudes $I_L = I_1 = I_2$ are added in phase, similar to a push-pull operation. Zero output power corresponds to the signal with $\phi = 0°$ when equal currents from power amplifiers cancel each other resulting in $I_L = 0$. Intermediate values of phase in between $0° < \phi < 90°$ produce

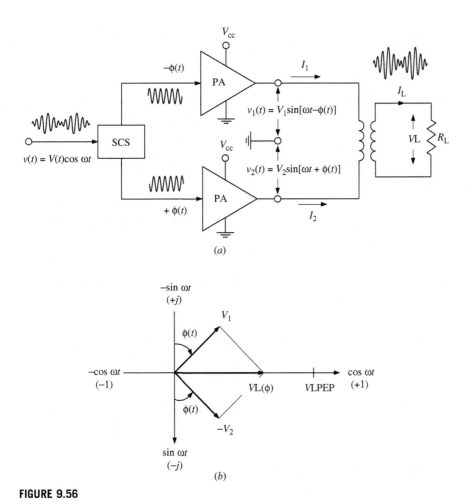

**FIGURE 9.56**

Simple outphasing power amplifier system.

intermediate values of the output voltage amplitude. As shown in Fig. 9.56(b), the time-varying phase $\phi$ can be written using the vector sum of the output voltages $V_1$ and $V_2$ by

$$\phi = \sin^{-1}\left(\frac{V_L}{V_{LPEP}}\right) \tag{9.205}$$

where $V_L = I_L R_L$ is the output voltage amplitude across the load resistance $R_L$ and $V_{LPEP}$ is the maximum output voltage amplitude at peak envelope power.

The instantaneous collector efficiency of a simple outphasing system with Class-B power amplifiers can be calculated from

$$\eta = \frac{\pi}{4} \frac{V_L}{V_{LPEP}} \tag{9.206}$$

having maximum value of 78.5% in saturation when $V_L = V_{LPEP}$ with $\phi = 90°$ and zero value when $V_L = 0$ with $\phi = 0°$. Thus, the efficiency of a simple power amplifier outphasing system is the same as that of an ideal Class-B power amplifier, reducing linearly with the output voltage amplitude. In this case, to accurately perform the required signal component separation, it is necessary to use the digital signal processing (DSP) technique [56].

The efficiency at lower output voltages can be significantly improved by using a lossless Chireix outphasing power amplifier system shown in Fig. 9.57(a),

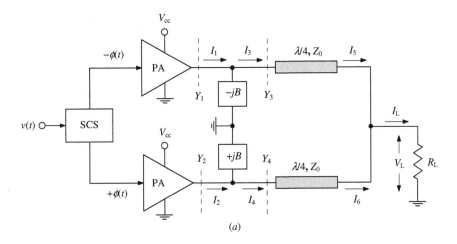

(a)

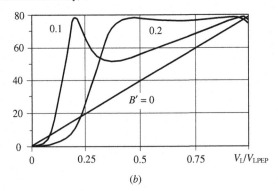

(b)

**FIGURE 9.57**

Chireix outphasing power amplifier system and instantaneous efficiencies.

which includes additional series quarterwave transmission lines and shunt reactances. Phasor analysis of the load network for the time-varying phase $\phi$ with an impedance-transforming quarterwave transmission line results in

$$Y_3 = \frac{2R_L}{Z_0^2} \frac{V_L}{V_{LPEP}} (\sin \phi + j\cos \phi) \qquad (9.207)$$

$$Y_4 = \frac{2R_L}{Z_0^2} \frac{V_L}{V_{LPEP}} (\sin \phi - j\cos \phi) \qquad (9.208)$$

where $Z_0$ is the characteristic impedance of the transmission lines [55]. From Eqs (9.207) and (9.208) it follows that the admittances $Y_3$ and $Y_4$ are purely resistive only for $\phi = 90°$ corresponding to the case of in-phase output currents. However, for most values of phase $\phi$, the power amplifiers have highly reactive loads that become completely reactive when $\phi = 0°$ with out-of-phase output currents. The effect of the reactive loads can be partially compensated by adding the corresponding shunt susceptances $-B$ and $+B$, respectively. In this case, the reactive parts of the admittances $Y_1 = Y_3 - jB$ and $Y_2 = Y_4 + jB$ can be zeroed at one specific output voltage amplitude by setting

$$B = \frac{2R_L}{Z_0^2} \frac{V_L}{V_{LPEP}} \sqrt{1 - \left(\frac{V_L}{V_{LPEP}}\right)^2} \qquad (9.209)$$

which can be obtained by substituting Eq. (9.205) into Eqs (9.207) and (9.208). As a result, for the case of a purely resistive load, the instantaneous collector efficiency of a Chireix outphasing system with ideal Class-B power amplifiers can reach the maximum value of

$$\eta = \frac{\pi}{4}. \qquad (9.210)$$

The instantaneous efficiencies of the Chireix outphasing system for different values of the normalized shunt susceptance $B' = BZ_0^2/2R_L$ are shown in Fig. 9.57(b), from which it follows that the selection of a proper value of $B$ increases efficiency at a specified medium level of the output voltage amplitude, however it is degraded at low and high amplitudes. Using a value $B' = 0.2$ can provide high efficiency over the upper 6 dB of the output voltage range. The case when $B' = 0$ corresponds to the collector efficiency variations of an ideal Class-B power amplifier. An improvement in the average efficiency calculated over a wide range of output voltages for various amplitude-modulated signals of up to a factor of two over that of an ideal Class-B power amplifier can be achieved by properly selecting the shunt susceptances in the outphasing power amplifier system. On the whole, to design such an outphasing system, it is necessary to simultaneously consider such factors as a complexity of the SCS circuit and sensitivity of the power amplifiers to the wide range of load impedances.

In a practical LINC transmitter, there are three main mechanisms that degrade the overall performance: power gain and phase imbalance between two RF amplifying paths and different nonlinear characteristics of both power amplifiers [57]. For example, if the gain imbalance between paths is about 1%, the output rejection may reduce to 45 dB and even further to only 28 dB, depending on the relative level of the input modulating signal amplitude. However, a phase error between amplifying paths as low as 2° diminishes the undesired response rejection to only 33 dB at best case. At the same time, the effect of the imbalance of the power amplifier nonlinearities is less meaningful. In terms of the Chireix power amplifier parameters, it is very important to minimize the combiner impedance mismatching, especially at microwave frequencies where the Chireix outphasing combiner should consist of two shunt stubs of equal and opposite reactances and two series quarterwave transmission lines [58]. To provide high amplitude and phase accuracy of the LINC system, a DSP-based architecture can be developed where the compensation of the amplitude and phase imbalances can be accomplished using calibration schemes [59]. For a Chireix outphasing system with a transmission-line combiner, an average efficiency of 30% with an *ACLR* of −45 dBc was achieved for a single-carrier WCDMA signal in a frequency range of 2.11−2.17 GHz, measured at a channel output power of 20 W [60].

Further efficiency improvement can potentially be achieved by using high-efficiency power amplifiers operating in different switching modes or their approximations. For example, a drain efficiency of a 800-MHz outphasing system based on the voltage-mode Class-D amplifiers implemented in a 0.18-μm SiGe BiCMOS process and transmission-line Chireix combiner was improved for a CDMA IS-95 signal with a *PAR* = 5.5 dB from 38.6% to 48% at an output power of 15.4 dBm [61]. Applying a single-carrier WCDMA signal, a drain efficiency of 51% with an output power of 21.6 dBm was measured for a 1.92-GHz outphasing system based on the voltage-mode Class-D amplifiers and integrated lumped Chireix combiner fully implemented in a 0.13-μm CMOS process [62].

Using a push−pull configuration with a rat-race balun for each saturated Class-B power amplifier based on 0.25-μm pHEMT devices in an outphasing system with a Chireix microstrip combiner had contributed to a system efficiency of 42.2% with a channel power of 31.2 dBm (−7 dB backoff) for a single-carrier 2.14-GHz WCDMA signal, which is more than two times better than the Wilkinson combiner system [63]. Note that the Chireix combiner when used with ideal sources leads to a linear LINC system. However, for 10-W GaN HEMT power amplifiers operating in a Class-F mode, more than 4-dB expansion in gain and 14° compression in phase are obtained in a Chireix outphasing system due to load-pulling effect [64]. Due to the concept of an outphasing system using an asymmetric Chireix transmission-line combiner with different electrical lengths and additional input phase adjustment where power amplifiers are based on 15-W LDMOSFET transistors operating in an inverse Class-F mode, a drain efficiency of 48% at 6-dB output-power backoff was achieved at an operating frequency of 2.14 GHz [65].

Figure 9.58($a$) shows the asymmetric architecture for the outphasing of Class-E power amplifiers, where the transmission lines $TL_1$ and $TL_2$ have electrical lengths of $\theta + \delta$ and $\theta - \delta$, respectively [66,67]. In this case, the transmission lines $TL_1$ and $TL_2$ rotate the impedance loci on the Smith chart to center them on the line at 65°, which corresponds to the maximum efficiency of the Class-E power amplifiers when $\theta = 147.5°$. As a result of centering the impedance loci, the amplitudes of the power-amplifier outputs ideally vary identically with difference in phase between the drive signals. The phases of the output signals are nearly the same over most of the amplitude range of $\Delta\phi$. The value of a differential line length $\delta$ can be chosen to alter the instantaneous-efficiency characteristics to optimize the average efficiency for a given signal. The circuit schematic of a lumped MOSFET prototype of the Chireix outphasing system with asymmetric combiner operating at 1.82 MHz is shown in Fig. 9.58($b$). Here, two identical Class-E power amplifiers using IRF510 MOSFETs and achieving a drain efficiency of 95% with an output power of 14 W each were implemented using lumped shunt capacitors and series lumped inductors. The asymmetric Chireix combiner represents a lumped low-pass $T$-network replacing the transmission lines $TL_1$ and $TL_2$, where the values of the $LC$ elements can be determined analytically using a single-frequency equivalence between the lumped and distributed circuits.

The transmission matrix $[ABCD]_{\mathrm{TL}}$ for a transmission line with electrical lengths of $\theta \pm \delta$ and the transmission matrix $[ABCD]_{\mathrm{LC}}$ for a low-pass $T$-type lumped circuit, consisting of a shunt capacitor and two series inductors, are written as

$$[ABCD]_{\mathrm{TL}} = \begin{bmatrix} cos(\theta \pm \delta) & jZ_0\sin(\theta \pm \delta) \\ j\dfrac{\sin(\theta \pm \delta)}{Z_0} & \cos(\theta \pm \delta) \end{bmatrix} \tag{9.211}$$

$$\begin{aligned} [ABCD]_{\mathrm{LC}} &= \begin{bmatrix} 1 & j\omega L \\ 0 & 1 \end{bmatrix} \begin{bmatrix} 1 & 0 \\ j\omega C & 1 \end{bmatrix} \begin{bmatrix} 1 & j\omega L \\ 0 & 1 \end{bmatrix} \\ &= \begin{bmatrix} 1 - \omega^2 LC & j\omega L\,(2 - \omega^2 LC) \\ j\omega C & 1 - \omega^2 LC \end{bmatrix} \end{aligned} \tag{9.212}$$

Equating the corresponding elements of both matrices yields the simple equations to determine the corresponding parameters of the lumped $T$-network through the transmission-line parameters as

$$C_{\pm\delta} = \frac{\sin(\theta \pm \delta)}{\omega Z_0} \tag{9.213}$$

$$L_{\pm\delta} = \frac{Z_0}{\omega}\tan\frac{\theta \pm \delta}{2} \tag{9.214}$$

where $Z_0$ is the transmission-line characteristic impedance.

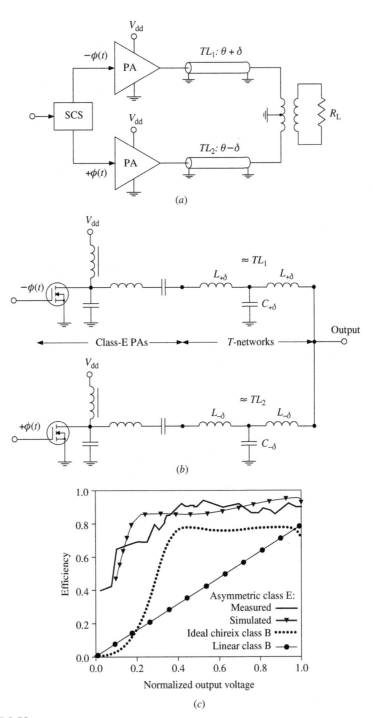

**FIGURE 9.58**

Asymmetric Chireix outphasing systems with Class-E power amplifiers.

Figure 9.58(*c*) shows the simulated and measured results of a Class-E outphasing efficiency versus normalized output voltage for $\theta = 147.5°$ and $\delta = 17.4°$, from which it follows that a drain efficiency of 85% or better for amplitude range of 15 dB from 0.8 W to full output of 27.5 W in simulation and for amplitude range of 10 dB in measurement is maintained. In contrast, the efficiency for outphasing with ideal Class-B power amplifiers is no better than 78.5%. The value of the difference $\delta$ in a transmission-line length (or its equivalent *LC* T-network) does not appear to be critical and high efficiency is obtained for $\delta$ equals to 10°, 18.4°, 22°, and 48°.

The schematic of a CMOS outphasing system with a transformer-based power combiner intended for broadband operation is shown in Fig. 9.59(*a*) [68]. Here,

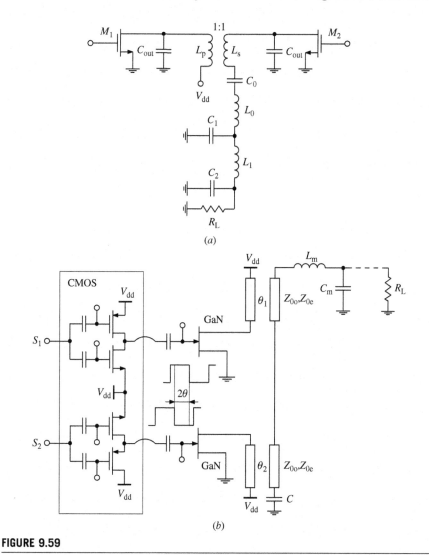

FIGURE 9.59

CMOS Chireix outphasing systems with Class-E power amplifiers.

the series resonators of both Class-E power amplifiers are shifted to the secondary side of the combiner, in series with the transformer leakage inductance defined as $L_{\text{leak}} = L_p(1 - k^2)/k^2$, where $k$ is the transformer coupling coefficient. The leakage inductance is merged with the resonator inductor and inductor of the wideband two-section output matching circuit. To achieve a transformation ratio equal to 1, the primary inductance $L_p$ and secondary inductance $L_s$ must relate to each other as $L_p/L_s = 1/k$. The compensation susceptance of the transistor $M_2$ is created by changing the value of the dc-feed inductance to a larger value. The primary transformer winding serves as a dc-feed inductance for the transistor $M_1$ and a compensation susceptance. The resonator at the secondary side has a low loaded quality factor to enable a high-bandwidth operation. For a larger bandwidth, the inductance $L_p$ must be small and coupling coefficient $k$ must be close to 1. As a result, such a Chireix outphasing system with a combining circuit implemented on a three-layer PCB and two 65-nm CMOS Class-E power amplifiers achieved a peak power of more than 30 dBm and a drain efficiency of greater than 46.4% at 6-dB power backoff over a frequency range of 600–800 MHz.

Since a lumped-element combining circuit is difficult to implement for high powers at microwave frequencies, coupled lines can be used to combine the outputs like in a Marchand balun. Figure 9.59(b) shows the circuit schematic of a two-stage Chireix outphasing system where the push-pull drivers are fabricated in standard 65-nm CMOS technology and the Class-E power amplifiers are based on a 28-V GaN HEMT technology [69]. The CMOS drivers are ac-coupled to overcome the negative bias voltage required for GaN HEMT devices. The electrical lengths of the coupled lines with odd-mode characteristic impedance $Z_{0o} = 1/Y_{0o}$ and even-mode characteristic impedance $Z_{0e} = 1/Y_{0e}$ are unequal and the terminating capacitance is defined as

$$C = \frac{Y_{0o} + Y_{0e}}{2\omega} \cot \theta_2 \qquad (9.215)$$

that tunes out the leakage inductance of the coupled-line transformer. A fourth-order Butterworth matching filter transforms the 50-$\Omega$ antenna impedance to the required Class-E load and sets the bandwidth property. As a result, a drain efficiency of 65.1% and a total system efficiency of 51.6% with an output power of 19 W were measured for a single-carrier WCDMA signal with $PAR$ of 7.5 dB at 1.95 GHz. Besides, a peak drain efficiency of more than 60% can be achieved in a frequency range of 1800–2050 MHz at 6-dB power backoff.

The basic advantage of a Chireix combining technique is the fact that the combiner is ideally lossless, and that the real components of the effective load admittances seen by the individual power amplifiers vary with outphasing so as to minimize power amplifier losses when output power reduces. However, the reactive portions of the effective load admittances are only zero for at most two outphasing angles, and become large outside of a limited power range. This limits efficiency, due both to loss associated with added reactive currents and to degradation of power amplifier performance with reactive loading. In this case,

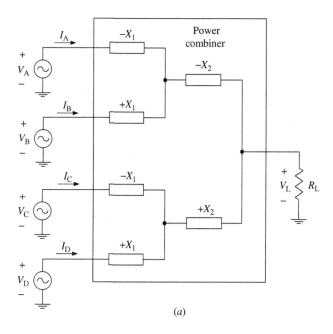

(a)

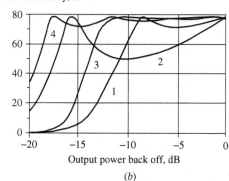

(b)

**FIGURE 9.60**

Four-stage outphasing architecture and instantaneous efficiencies.

multistage techniques applied to a conventional Chireix outphasing system can overcome the loss and reactive loading problems, providing an ideally lossless power combining with nearly resistive loading of the individual power amplifiers over a very wide output power range when high average efficiencies are achieved even for large peak-to-average power ratios [70].

Figure 9.60(a) shows the simplified four-stage outphasing architecture, where the power amplifiers are modeled as ideal voltage sources. The lossless power combiner has four input ports and one output port connecting to the load, and it comprises reactive elements having specified reactances at the operating frequency. The system behavior can be described based on the relationships between the source voltages and input currents of the network of Fig. 9.60(a) according to

$$
\begin{bmatrix} I_A \\ I_B \\ I_C \\ I_D \end{bmatrix} = \frac{1}{X_1} \begin{bmatrix} \gamma + j(1-\beta) & -\gamma + j\beta & \gamma & -\gamma \\ -\gamma + j\beta & \gamma - j(\beta+1) & -\gamma & \gamma \\ \gamma & -\gamma & \gamma + j(\beta+1) & -\gamma - j\beta \\ -\gamma & \gamma & -\gamma - j\beta & \gamma + j(\beta-1) \end{bmatrix} \begin{bmatrix} V_A \\ V_B \\ V_C \\ V_D \end{bmatrix}
$$

$$(9.216)$$

where $\gamma = R_L/X_1$, $\beta = X_2/X_1$, and voltage amplitude are equal having different control angles used for outphasing. In this case, $+X_1$ and $+X_2$ represent inductive reactances due to inductances and $-X_1$ and $-X_2$ represent capacitive reactances due to capacitances. The effective admittance at a combiner input port is the complex ratio of current to voltage at the port with all sources active. The effective admittances represent the admittances seen by the sources when they are operating under outphasing control. A key advantage of this outphasing system is that the susceptive portion of admittance loading the power amplifiers is substantially smaller than with a conventional Chireix combining.

The drain efficiency of a multistage outphasing system using ideal saturated Class-B power amplifiers can be obtained by

$$
\eta = \frac{\pi}{4} \frac{\sum\limits_{j=1}^{N} \mathrm{Re}(Y_{\mathrm{eff},j})}{\sum\limits_{j=1}^{N} |Y_{\mathrm{eff},j}|}
$$

$$(9.217)$$

where $Y_{\mathrm{eff},j}$ is the effective admittance at a corresponding combiner input port seen by the $j$th source and $N$ is the number of ports. For example, for the output probability density function (PDF) representing a WLAN signal with a *PAR* of 9.01 dB, the average drain efficiency remains almost constant for a much larger range of backoff output powers for a four-stage outphasing architecture (curves 3 and 4) compared with a conventional Chireix outphasing architecture (curves 1 and 2), as shown in Fig. 9.60(b), with average drain efficiencies of 38% and 46.1% corresponding to curves 1 and 2, and average drain efficiencies of 56.9% and 69%, corresponding to curves 3 and 4, respectively [70].

Figure 9.61(a) shows the implementation of a Chireix outphasing system, where each of the two outphased sources is constructed of a pair of the power amplifiers designed to operate in a Class-E/F$_{\mathrm{odd}}$ mode at 27.12 MHz with output powers up to 750 W using ARF521 power MOSFETs. The two pairs of power

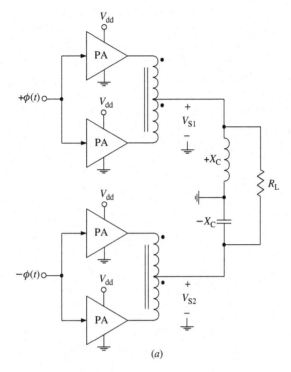

(a)

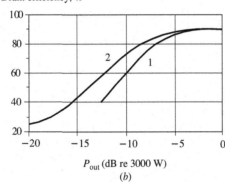

$P_{out}$ (dB re 3000 W)

(b)

**FIGURE 9.61**

Chireix outphasing power amplifier system and instantaneous efficiencies.

amplifiers are outphased using a Chireix combiner having $X_c = 13.6 \ \Omega$ and supply a load resistance $R_L = 13 \ \Omega$. For the four-stage outphasing system shown in Fig. 9.60(a), the power combiner is used with a load resistance of 50 $\Omega$ and reactance values $X_1 = 35 \ \Omega$ and $X_2 = 48.78 \ \Omega$. The drain efficiency *versus* output power for both the four-stage combining system and conventional Chireix system is shown in Fig. 9.61(*b*), where the four-stage outphasing architecture (curve 2) demonstrates superior performance at low power levels compared with a conventional Chireix system (curve 1), with more than 12% higher efficiency at 300-W output at 10-dB power backoff and more than 20% higher efficiency at 12.5-dB power backoff. The difference in efficiency arises both due to the higher reactive currents in the conventional Chireix architecture and because the greater reactive loading on the Chireix power amplifiers at low power levels causes them to lose zero-voltage switching, resulting in substantial capacitive discharge loss. At the same time, the four-stage outphasing system is advantageous in that it maintains desirable switching conditions down to far lower power levels than is possible in a conventional Chireix system.

## REFERENCES

1. Zhukov SA, Kozyrev VB. Push—pull switching-mode tuned power amplifier without commutation losses (in Russian). *Poluprovodnikovye Pribory v Tekhnike Elektrosvyazi.* 1975;15:95—106.
2. El-Hamamsy S-A. Design of high-efficiency RF Class-D power amplifier. *IEEE Trans. Power Electronics.* May 1994;PE-9:297—308.
3. Koizumi H, Suetsugu T, Fujii M, Shinoda K, Mori S, Ikeda K. Class DE high-efficiency tuned power amplifier. *IEEE Trans. Circuits and Systems — I: Fundamental Theory Appl.* January 1996;CAS-I-43:51—60.
4. Albulet M. An exact analysis of Class DE amplifier at any output *Q. IEEE Trans. Circuits and Systems — I: Fundamental Theory Appl.* October 1999;CAS-I-46:1228—1239.
5. Alipov A, Kozyrev V. Push/pull Class-DE switching power amplifier. *2002 IEEE MTT-S Int. Microwave Symp. Dig.* 3:1635—1638.
6. Grebennikov A. *RF and Microwave Power Amplifier Design.* New York: McGraw-Hill; 2004.
7. Sekiya H, Watanabe T, Suetsugu T, Kazimierczuk MK. Analysis and design of Class DE amplifier with nonlinear shunt capacitances. *IEEE Trans. Circuits and Systems — I: Regular Papers.* October 2009;CAS-I-56:2362—2371.
8. Sekiya H, Sagawa N, Kazimierczuk MK. Analysis of Class DE amplifier with nonlinear shunt capacitances at any grading coefficient for high *Q* and 25% duty ratio. *IEEE Trans. Power Electron.* April 2010;PE-25:924—932.
9. Sekiya H, Sagawa N, Kazimierczuk MK. Analysis of Class-DE amplifier with linear and nonlinear shunt capacitances at 25% duty ratio. *IEEE Trans. Circuits and Systems — I: Regular Papers.* September 2010;CAS-I-57:2334—2342.
10. de Vries I. High power and high frequency Class-DE inverters. PhD thesis, Department of Electrical Engineering, University of Cape Town, South Africa, August 1999.
11. Phinney JW, Perreault DJ, Lang JH. Radio-frequency inverters with transmission-line input networks. *IEEE Trans. Power Electronics.* July 2007;PE-22:1154—1161.
12. Grebennikov A. High-efficiency Class FE tuned power amplifiers. *IEEE Trans. Circuits and Systems — I: Regular Papers.* November 2008;CAS-I-55:3284—3292.

13. Kessler DJ, Kazimierczuk MK. Power losses and efficiency of Class-E power amplifier at any duty ratio. *IEEE Trans. Circuits and Systems − I: Regular Papers*. September 2004; CAS-I-51:1675−1689.

14. Thian M, Fusco VF. Analysis and design of Class-$E_3F$ and transmission-line Class-$E_3F_2$ power amplifiers. *IEEE Trans. Circuits and Systems − I: Regular Papers*. May 2011;CAS-I-58:902−912.

15. Kaczmarczyk Z. High-efficiency Class E, $EF_2$, and E/$F_3$ inverters. *IEEE Trans. Industrial Electronics*. October 2006;IE-53:1584−1593.

16. Rivas JM, Han Y, Leitermann O, Sagneri AD, Perreault DJ. A high-frequency resonant inverter topology with low-voltage stress. *IEEE Trans. Power Electronics*. July 2008;PE-23:1759−1771.

17. Kee SD, Aoki I, Hajimiri A, Rutledge D. The Class-E/F family of ZVS switching amplifiers. *IEEE Trans. Microwave Theory Tech*. June 2003;MTT-51:1677−1690.

18. Kee SD, Aoki I, Rutledge D. 7-MHz, 1.1-kW demonstration of the new E/$F_{2,odd}$ switching amplifier class. *2001 IEEE MTT-S Int. Microwave Symp. Dig*. 3:1505−1508.

19. Iwadare M, Mori S, Ikeda K. Even harmonic resonant Class E tuned power amplifier without RF choke. *Electronics and Communications in Japan*. January 1996;79:23−30.

20. Bohn F, Kee S, Hajimiri A. Demonstration of the new E/$F_{odd}$ dual-band power amplifier. *2002 IEEE MTT-S Int. Microwave Symp. Dig*. 3:1631−1634.

21. Grebennikov A. High-efficiency Class E/F lumped and transmission-line power amplifiers. *IEEE Trans. Microwave Theory Tech*. June 2011;MTT-59:1579−1588.

22. Molnar B. Basic limitations on waveforms achievable in single-ended switching-mode tuned (Class E) power amplifiers. *IEEE J. Solid-State Circuits*. February 1984;SC-19:144−146.

23. Telegdy A, Molnar B, Sokal NO. Class-$E_M$ switching-mode tuned power amplifier − high efficiency with slow-switching transistor. *IEEE Trans. Microwave Theory Tech*. June 2003;MTT-51:1662−1676.

24. Miyahara R, Sekiya H, Kazimierczuk MK. Novel design procedure for Class-$E_M$ power amplifiers. *IEEE Trans. Microwave Theory Tech*. December 2010;MTT-58:3607−3616.

25. Kazimierczuk M. Class E tuned power amplifier with shunt inductor. *IEEE J. Solid-State Circuits*. February 1981;SC-16:2−7.

26. Avratoglou CP, Voulgaris NC. A Class E tuned amplifier configuration with finite DC-Feed inductance and no capacitance in parallel with switch. *IEEE Trans. Circuits and Systems*. April 1988;CAS-35:416−422.

27. Brabetz T, Fusco VF. Voltage-driven Class E amplifier and applications. *IEE Proc. Microwaves Antennas Propag*. October 2005;152:373−377.

28. Thian M, Fusco VF. Series-L/parallel-tuned versus Shunt-C/series-tuned Class-E power amplifier comparison. *IEE Proc. Circuits Devices Syst*. December 2005;152:709−717.

29. Herman KJ, Zulinski RE. The infeasibility of a zero-current switching Class E amplifier. *IEEE Trans. Circuits and Systems*. January 1990;CAS-37:152−154.

30. Thian M, Fusco VF. Inverse Class-E amplifier with transmission-line harmonic suppression. *IEEE Trans. Circuits and Systems − I: Regular Papers*. July 2007;CAS-I-54:1555−1561.

31. Thian M, Xiao M, Gardner P. Digital baseband predistortion based linearized broadband inverse Class-E power amplifier. *IEEE Trans. Microwave Theory Tech*. February 2009;MTT-57:323−328.

32. Mazumder SR, Azizi A, Gardiol FE. Improvement of a Class-C transistor power amplifier by second-harmonic tuning. *IEEE Trans. Microwave Theory Tech.* May 1979;MTT-27:430–433.

33. Nojima T, Nishiki S. High efficiency microwave harmonic reaction amplifier. *1988 IEEE MTT-S Int. Microwave Symp. Dig.* 2:1007–1010.

34. Khatibzadeh MA, Tserng HQ. Harmonic tuning of power FETs at X-Band. *1990 IEEE MTT-S Int. Microwave Symp. Dig.* 3:989–992.

35. Kopp B, Heston DD. High-efficiency 5-Watt power amplifier with harmonic tuning. *1988 IEEE MTT-S Int. Microwave Symp. Dig.* 2:839–842.

36. Colantonio P, Giannini F, Giofre R, et al. A C-Band high-efficiency second-harmonic-tuned hybrid power amplifier in GaN technology. *IEEE Trans. Microwave Theory Tech.* June 2006;MTT-54:2713–2722.

37. Yamasaki T, Kittaka Y, Minamide H, et al. A 68% efficiency, C-Band 100 W GaN power amplifier for space applications. *2010 IEEE MTT-S Int. Microwave Symp. Dig.* 1384–1387.

38. Kim J, Mkadem F, Boumaiza S. A high-efficiency and multi-band/multi-mode power amplifier using a distributed second harmonic termination. *Proc. 40th Europ. Microwave Conf.* 2010;1662–1665.

39. Roberg M, Popovic Z. Analysis of high-efficiency power amplifiers with arbitrary output harmonic terminations. *IEEE Trans. Microwave Theory Tech.* August 2011; MTT-59:2037–2048.

40. Cusak JM, Perlow SM, Perlman BS. Automatic load contour mapping for microwave power transistors. *IEEE Trans. Microwave Theory Tech.* December 1974;MTT-22:1146–1152.

41. Takayama Y. A new load-pull characterization method for microwave power transistors. *1976 IEEE MTT-S Int. Microwave Symp. Dig.* 218–220.

42. Berini P, Desgagne M, Ghannouchi FM, Bosisio RG. An experimental study of the effects of harmonic loading on microwave MESFET oscillators and amplifiers. *IEEE Trans. Microwave Theory Tech.* June 1994;MTT-42:943–950.

43. Bensmida S, Bergeault E, Abib GI, Huyart B. Power amplifier characterization: an active load-pull system based on six-port reflectometer using complex modulated carrier. *IEEE Trans. Microwave Theory Tech.* June 2006;MTT-54:2707–2712.

44. Watanabe S, Takatsuka S, Takagi K, Kuroda H, Oda Y. Effect of source harmonic tuning on linearity of power GaAs FET under Class AB operation. *IEICE Trans. Electron.* May 1996;E79-C:611–616.

45. Hall LC, Trew RJ. Maximum efficiency tuning of microwave amplifiers. *1991 IEEE MTT-S Int. Microwave Symp. Dig.* 1:123–126.

46. Cai Q, Herber J, Peng S. A systematic scheme for power amplifier design using a multi-harmonic loadpull simulation technique. *1998 IEEE MTT-S Int. Microwave Symp. Dig.* 161–165.

47. Staudinger J. Multiharmonic load termination effects on GaAs MESFET power amplifiers. *Microwave J.* April 1996;39:60–77.

48. Colantonio P, Giannini F, Leuzzi G, Limiti E. Multiharmonic manipulation for highly efficient microwave power amplifiers. *Int. J. RF and Microwave Computer-Aided Eng.* November 2001;11:366–384.

49. Colantonio P, Giannini F, Limiti E, Teppati V. An approach to harmonic load- and source-pull measurements for high-efficiency PA design. *IEEE Trans. Microwave Theory Tech.* January 2005;MTT-52:191–198.

50. Tang OSA, Liu SMJ, Chao PC, et al. Design and fabrication of a wideband 56- to 63-GHz monolithic power amplifier with very high power-added efficiency. *IEEE J. Solid-State Circuits*. September 2000;SC-35:1298−1306.

51. Simpson G. A comparison of harmonic tuning methods for load pull systems. *Maury Microwave Corp.* 2004.

52. Application Note 58. Comparing harmonic load pull techniques with regards to power-added efficiency (PAE). *Focus Microwaves Inc.*, 2007.

53. Chireix H. High power outphasing modulation. *Proc. IRE*. November 1935;23:1370−1392.

54. Cox DC. Linear amplification with nonlinear components. *IEEE Trans. Commun.* December 1974;COM-22:1942−1945.

55. Raab FH. Average efficiency of outphasing power-amplifier systems. *IEEE Trans. Commun.* October 1985;COM-33:1094−1099.

56. Hetzel SA, Bateman A, McGeehan JP. A LINC transmitter. *Proc. 41st IEEE Vehicular Technol. Conf.* 1991;133−137.

57. Casadevall F, Olmos JJ. On the behavior of the LINC transmitter. *Proc. 40th IEEE Vehicular Technol. Conf.* 1990;29−34.

58. Birafane A, Kouki AB. On the linearity and efficiency of outphasing microwave amplifiers. *IEEE Trans. Microwave Theory Tech.* July 2004;MTT-52:1702−1708.

59. Zhang X, Larson LE, Asbeck PM, Nanawa P. Gain/phase imbalance-minimization techniques for LINC transmitters. *IEEE Trans. Microwave Theory Tech.* December 2001;MTT-49:2507−2515.

60. Hakala I, Choi DK, Gharavi L, Kajakine N, Koskela J, Kaunisto R. A 2.14-GHz chireix outphasing transmitter. *IEEE Trans. Microwave Theory Tech.* June 2005; MTT-53:2129−2138.

61. Hung T-P, Choi DK, Larson LE, Asbeck PM. CMOS outphasing Class-D amplifier with chireix combiner. *IEEE Microwave and Wireless Lett.* August 2007;17:619−621.

62. Lee S, Nam S. A CMOS outphasing power amplifier with integrated single-ended chireix combiner. *IEEE Trans. Circuits and Systems − II: Express Briefs*. June 2010; CAS-II-57:411−415.

63. Huttunen A, Kaunisto R. A 20-W Chireix outphasing transmitter for WCDMA base stations. *IEEE Trans. Microwave Theory Tech.* December 2007;MTT-55:2709−2718.

64. Helaoui M, Boumaiza S, Ghannouchi FM. On the outphasing power amplifier nonlinearity analysis and correction using digital predistortion technique. *Proc. 2008 IEEE Radio and Wireless Symp.* 751−754.

65. Gerhard W, Knoechel R. Improvement of power amplifier efficiency by reactive chireix combining, power back-off and differential phase adjustment. *2006 IEEE MTT-S Int. Microwave Symp. Dig.* 1887−1890.

66. Beltran R, Raab FH, Velazquez A. HF outphasing transmitter using Class-E power amplifiers. *2009 IEEE MTT-S Int. Microwave Symp. Dig.* 757−760.

67. Beltran R, Raab FH, Velazquez A. An outphasing transmitter using Class-E PAs and asymmetric combining. *High Frequency Electronics*. May 2011;10:18−26:April, pp. 34−46

68. van Schie MCA, van der Heijden MP, Acar M, de Graauw AJM, de Vreede LCN. Analysis and design of a wideband high efficiency CMOS outphasing amplifier. *2010 IEEE RFIC Symp. Dig.* 399−402.

69. van der Heijden MP, Acar M, Vromans JS, Calvillo-Cortes DA. A 19 W high-efficiency wide-band CMOS-GaN Class-E Chireix RF outphasing power amplifier. *2011 IEEE MTT-S Int. Microwave Symp. Dig.* 1−4.

70. Perreault DJ. A new power combining and outphasing modulation system for high-efficiency power amplification. *IEEE Trans. Circuits and Systems − I: Regular Papers.* October 2011;CAS-I-58:1713−1726.

# High-Efficiency Doherty Power Amplifiers

# 10

## INTRODUCTION

This chapter describes the historical aspect of the Doherty approach to the power amplifier design and modern trends in Doherty amplifier design techniques using multistage and asymmetric multi-way architectures. To increase efficiency over the power-backoff range, the switchmode Class-E, conventional Class-F, or inverse Class-F operation mode by controlling the second and third harmonics can be used in the load network. The Doherty amplifier with a series connected load, inverted, balanced, push–pull, and dual Doherty architectures are also described and discussed. Finally, examples of the lumped Doherty amplifier implemented in monolithic microwave integrated circuits, digitally driven Doherty technique, and broadband capability of the two-stage Doherty amplifier are given.

## 10.1 Historical aspects and conventional Doherty architecture

A new power amplifier technique for amplitude-modulated (AM) radio-frequency signals was introduced by William H. Doherty in broadcasting in the middle of the 1930s as a more efficient alternative to both conventional amplitude-modulation techniques and Chireix outphasing [1,2]. This new technique achieves plate circuit efficiencies of up to 60–65% independent of modulation by means of a combined action of the variation of load distribution of the vacuum tubes, and the variation of the circuit impedance over the modulation cycle. When Doherty joined the radio development department of the Bell Telephone Laboratories in

June 1929, he was engaged in the development of high-power radio transmitters for transoceanic radiotelephony and broadcasting. As a result, in 1936 he invented a means to greatly improve the efficiency of radio-frequency power amplifiers, quickly termed the "Doherty amplifier". It was first used in a 50-kW transmitter with audio-frequency feedback providing a resulting distortion level of less than 1% at lower frequencies to a few per cent at high audio frequencies. In this case, the power amplifier was operated at an efficiency of 60% representing a reduction of nearly one half in all-day power consumption as compared with the power required in the conventional type of linear power amplifier operating at 33% efficiency [3]. The IRE Morris Liebmann Memorial Award was voted to Doherty in May 1937 for his improvement in the efficiency of radio-frequency power amplifiers [4]. By 1940, the Doherty amplifiers were incorporated in 35 commercial radio stations worldwide, at powers of up to 50 kW.

In subsequent years, Doherty amplifiers continued to be used in a number of medium- and high-power low-frequency (LF) and medium-frequency (MF) vacuum-tube AM transmitters [5,6]. In August 1953, a one megawatt vacuum-tube transmitter operating in the long-wave band began regular operations in Europe where the outputs of the two 500-kW Doherty amplifiers were joined in a bridge-type combiner. The Doherty amplifiers had also been considered for use in solid-state MF and high-frequency (HF) systems, as well as in high-power ultra-high-frequency (UHF) transmitters [7–9]. The practical implementation of a classical triode-based Doherty scheme was restricted by its substantial nonlinearity for both linear amplification of AM signals and grid-type signal modulation that required complicated envelope correction and feedback linearization circuits. At the same time, the Doherty amplifiers employing tetrode transmitting tubes could improve their overall performance when the modulation was applied to the screen grids of both the carrier and peaking tubes, while the control grids of both tubes are fed by an essentially constant level of RF excitation [10]. This resulted in the peaking tube being modulated upward during the positive half of the modulating cycle and the carrier tube being modulated downward during the negative half of the modulating cycle.

Generally, a Doherty amplifier system combines the outputs of two (or more) linear RF power amplifiers (PAs) through an impedance-inverting network composed of the lumped elements or represented by a quarterwave transmission line. The two fundamental forms of the Doherty amplifier are shown in Fig. 10.1, with a shunt connected load in Fig. 10.1($a$), and with a series connected load in Fig. 10.1($b$) [2]. In the former case, the load impedance used is $R/2$, which is the same as would be employed if the tubes were to be connected in parallel in the conventional type of power amplifier. In the latter case, as long as the right-hand or peaking tube does not conduct, the impedance-inverting network provides zero impedance being terminated as open circuit, and the left-hand or carrier tube operates into an impedance of $2R$. However, when the peaking tube is permitted to conduct, each tube is operating into the impedance $R$ at the peak of modulation and delivering twice the carrier power, so that the total instantaneous output is the

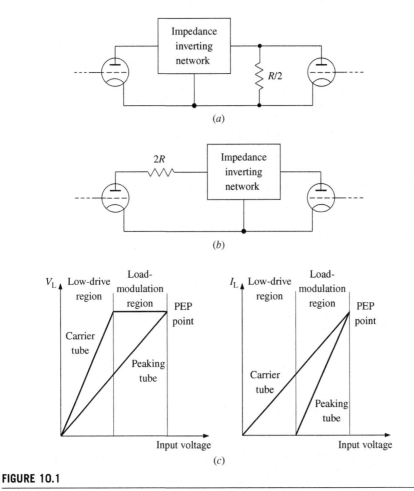

**FIGURE 10.1**

Doherty fundamental load-network structures and their ideal voltage and current behavior.

required value of four times the carrier power. The shunt connection appears to be more advantageous for most practical applications because the load circuit is grounded, while the load is neither grounded nor balanced to ground in the series arrangement. The ideal load voltage ($V_L$) and current ($I_L$) behavior in the carrier and peaking tubes as the amplitude of the grid excitation is varied is shown in Fig. 10.1(c). Here, for the classical Doherty amplifier with equal-power tubes, the transition voltage is half the peak-envelope point (PEP), and the total output power of the amplifier comes from the carrier tube for input amplitudes less or equal to the transition point. The region between the transition point and PEP values represents the load modulation region and the voltage on the carrier tube remains constant at the PEP level. At the same time, the voltage across the peaking tube continues to rise linearly, with its current commencing and rising twice as fast as the current in the carrier tube in order to reach its PEP value at maximum output power.

Thus, at low output power levels, the carrier amplifier operates linearly, reaching saturation that corresponds to maximum efficiency at some transition voltage below the system peak-output voltage. However, at higher output power levels, the carrier amplifier remains saturated while the peaking amplifier operates linearly. The average output power for the modulated signal is equal to

$$P_{avr} = P_0 \left( 1 + \frac{m^2}{2} \right) \tag{10.1}$$

where $m$ is called the modulation factor and denotes the ratio of the variation of the modulated carrier amplitude to the unmodulated carrier amplitude and $P_0$ is the unmodulated carrier output power. As a result, the average efficiency through the modulation cycle is found to be

$$\eta_{avr} = \eta_0 \frac{1 + \frac{m^2}{2}}{1 + \frac{2mq}{\pi}} \tag{10.2}$$

where $\eta_0$ is the efficiency for zero modulation and $q$ is the variable factor which ranges from about 0.7 for zero modulation to 0.93 for full modulation [1].

Figure 10.2(a) shows the Doherty amplifier schematic where the plates of the carrier and peaking tubes are connected together by a $\pi$-type 90° lumped network which introduces a lagging phase shift of 90° from the plate of the carrier tube to the plate of the peaking tube [2,10]. Such a 90° lumped network is used due to its impedance-inverting characteristics. This means that, if the terminating impedance at the point in the network where the peaking tube is located is reduced, the impedance seen by the carrier tube will increase. In this case, since the load network of the high-power transmitter was approximately 35 Ω, the 90° lumped network was set to provide an impedance at the carrier tube of about 140 Ω. As a result, the series inductance and two shunt capacitors are each selected to have a reactance equal to $\sqrt{35\,\Omega \times 140\,\Omega} = 70\,\Omega$. To compensate for the output 90° phase shift, the input to the grid of the peaking tube is delayed by 90° by similar means. If the input excitation is applied to the grid of the peaking tube, then a $\pi$-type lumped network consisting of the series capacitor and two shunt inductances is added between the grids of the carrier and peaking tubes to compensate for the output phase shift of 90°, as shown in Fig. 10.2(b).

The simplified two-stage transmission-line Doherty power-amplifier architecture shown in Fig. 10.3(a) incorporates the carrier and peaking power amplifiers, separated by a quarterwave transmission line in the carrier amplifier path [11,12]. Such a section of line, known as a quarterwave transformer, has the ability to invert impedances according to

$$Z_0 = \sqrt{Z_{in} Z_{out}} \tag{10.3}$$

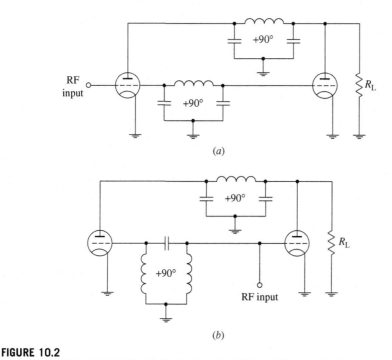

**FIGURE 10.2**

Doherty amplifier basic schematics with lumped elements.

where $Z_0$ is the characteristic impedance of the transmission line. This property can be seen more clearly by rewriting Eq. (3) as

$$Z_{out} = \frac{Z_0^2}{Z_{in}} \tag{10.4}$$

from which it follows that the output impedance $Z_{out}$ increases inversely with the input impedance $Z_{in}$ for constant $Z_0$. The quarterwave transmission line at the input of the peaking amplifier is required to compensate for the 90° phase shift caused by the quarterwave transmission line at the output of the carrier amplifier. The output quarterwave line with $Z_0 = \sqrt{25\ \Omega \times 50\ \Omega} = 35\ \Omega$ is required to match the standard load impedance of 50 $\Omega$ when both carrier and peaking amplifiers deliver maximum power, each of which designed in a 50-$\Omega$ environment.

An input drive controller is used to turn on the peaking amplifier (bias control) when the carrier amplifier starts to saturate since it is assumed that the carrier and peaking amplifiers are biased in Class B for idealized system analysis. However, in practice the carrier amplifier is biased in the Class-B mode, while the peaking amplifier is biased in the Class-C mode. At a backoff power level of −6 dB, the saturated output power of the carrier amplifier is four times lower than the peak output power $P_{PEP}$. This indicates that its collector (or drain) efficiency when

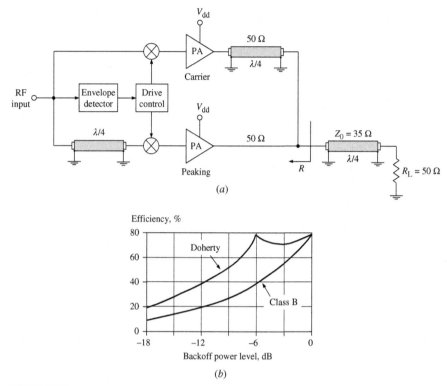

**FIGURE 10.3**

Doherty amplifier architecture with quarterwave lines and collector efficiencies.

operated in an ideal Class-B mode is twice than that of a conventional Class-B power amplifier, achieving a maximum efficiency of 78.5%, as shown in Fig. 10.3(b).

The basic operation principle of the conventional Doherty power-amplifier architecture shown in Fig. 10.4(a) can be analyzed for low, medium, and peak output power regions separately [12]. Figure 10.4(b) shows the current and voltage behavior for ideal transistors and lossless matching circuits, where $V_L$ is the load voltage and $I_L$ is the load current. The condition of power conservation for a lossless output transmission line results in

$$I_3 = I_1 \sqrt{\frac{R_1}{R_3}}$$

(10.5)

while the current division ratio $\beta$ is defined by

$$\beta = \frac{I_3}{I_2 + I_3}.$$

(10.6)

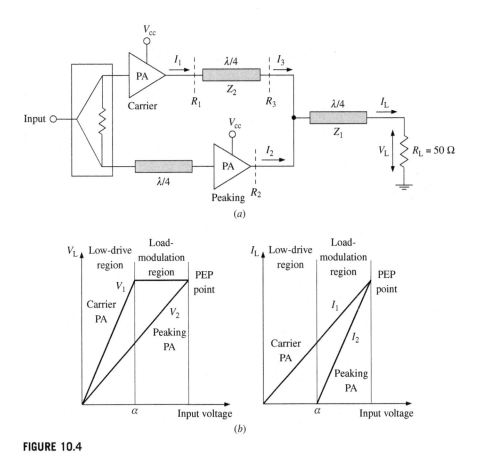

**FIGURE 10.4**

Basic two-stage Doherty amplifier architecture and its operation principle.

As a result, the overall output power $P_{out}$ is the sum of the carrier (main) amplifier output power $P_1 = \beta P_{out}$ and peaking (auxiliary) amplifier output power $P_2 = (1 - \beta)P_{out}$. The impedance seen at the output of the transmission line in the carrier-amplifier path is

$$R_3 = \frac{I_2 + I_3}{I_3} \frac{Z_1^2}{R_L} = \frac{Z_1^2}{\beta R_L} \qquad (10.7)$$

and the impedance seen by the peaking power amplifier is

$$R_2 = \frac{I_2 + I_3}{I_2} \frac{Z_1^2}{R_L} = \frac{Z_1^2}{(1 - \beta) R_L}. \qquad (10.8)$$

At peak output power $P_{PEP}$ when both carrier and peaking amplifiers are saturated, the resultant collector efficiency is equal to the maximum achievable efficiency $\eta = \pi/4 \approx 78.5\%$ for an ideal Class-B operation. For the classical Doherty power-amplifier architecture shown in Fig. 4(a) with the current and power division ratios $\beta = \alpha = 0.5$ when both carrier and peaking amplifiers produce equal output powers, their load impedances are equal to $R_1 = R_3 = R_2 = Z_2 = 2Z_1^2/R_L$. If the characteristic impedance of the output transmission line is chosen to be equal to $Z_1 = 35\ \Omega$, then $R_1 = R_3 = R_2 = Z_2 = R_L = 50\ \Omega$.

At lower power levels in a low-drive region, the peaking amplifier is turned off because the instantaneous amplitude of the input signal is insufficient to overcome the negative Class-C bias and appears as an open circuit, whereas the carrier amplifier operates in the active region. In this case, the load impedance seen by the carrier amplifier is

$$R_1 = \left(\frac{Z_2}{Z_1}\right)^2 R_L \qquad (10.9)$$

resulting in $R_1 = 2R_L = 100\ \Omega$ when $Z_1 = 35\ \Omega$ and $Z_2 = R_L = 50\ \Omega$. Because the output power of the carrier amplifier in saturation is four times less than the peak output power $P_{PEP}$, the collector efficiency of the carrier amplifier in an ideal Class-B mode will be twice than that of a conventional Class-B power amplifier, achieving a maximum of 78.5% at backoff power level of $-6$ dB, as shown in Fig. 10.3(b).

At medium power levels in a load-modulation region, the carrier amplifier is saturated, whereas the peaking amplifier is turned on and operates in the active region. Since the output voltage of the carrier amplifier $V_1 = I_1 R_1$ is constant under saturation conditions, from Eq. (10.5) it follows that the current $I_3$ is constant in the medium power region as well. The collector efficiency of the carrier amplifier remains at its maximum value, whereas the collector efficiency of the peaking amplifier increases up to its maximum value for Class-B operation at peak output power $P_{PEP}$. As a result, the Doherty amplifier architecture achieves maximum efficiency at both the transition $-6$ dB backoff point and the peak output power, and remains relatively high in between, as seen from Fig. 10.3(b).

In the mid 1990s, it was found that by using existing microwave design techniques and by implementing a few modifications, a microwave version of a Doherty amplifier could be realized [13]. In this case, an efficiency of 61% was achieved at 1-dB gain compression point and this level of efficiency was maintained through a 5.5-dB reduction in output power at an operating frequency of 1.37 GHz. A few years later, the possibility of achieving high-efficiency performance of a microwave monolithic Doherty amplifier was demonstrated in the *Ku*-band and *K*-band frequencies using pHEMT and InP DHBT technologies [14,15]. The first fully integrated Doherty amplifier MMIC with a chip size of 2 mm$^2$ operating in a frequency range of 38$-$46 GHz was developed using a 0.15-$\mu$m GaAs HEMT process [16]. By using modern high-voltage HBT and GaN HEMT technologies, a high average efficiency of more than 50% can be achieved

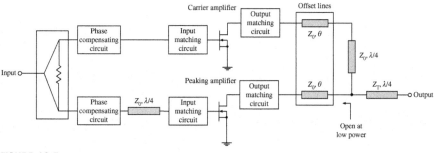

**FIGURE 10.5**

Block diagram of a microstrip LDMOSFET Doherty amplifier.

for multicarrier WCDMA signals using a high-power two-way symmetrical Doherty amplifier [17,18]. At high frequencies, it is necessary to take into account that the transistor input impedance is varied with bias voltage and the transistor output reactance should be compensated to provide the required open-circuit condition by using so-called offset lines with optimized electrical lengths. Figure 10.5 shows the schematic diagram of a fully matched microwave Doherty amplifier with offset lines in the output circuit and phase-compensating circuits in the input circuit required to reduce AM/PM variations [19,20]. To closer approximate ideal Doherty amplifier performance, an adaptive power-dependent input power distribution between the carrier and peaking amplifier can be provided so as to deliver more power to the carrier amplifier in the low-power region and to the carrier amplifier in the high-power region, which can result in a linearity improvement of 5−7 dB over a wide range of output powers and an increased efficiency up to 5% for WCDMA signals at 2.4 GHz [21].

Generally, the non-ideal power gain and phase performance in a high-power nonlinear region can cause a significant linearity problem for transmitting signals with a nonconstant envelope in wireless communication transmitters. To solve this problem, an improved Doherty power-amplifier architecture can be used by employing an envelope tracking technique to control the gate bias voltage of the peaking amplifier in accordance with the input signal envelope. Such an approach can also provide a higher efficiency with a lower bias voltage for higher output powers. This ensures both high efficiency and good linearity requirements over a wide range of output powers. Figure 10.6(*a*) shows the block diagram of a 2.14-GHz LDMOSFET microstrip Doherty power amplifier for WCDMA applications with adaptive gate bias control [22]. For the same average output powers of 32.7 dBm, such a two-stage Doherty power amplifier demonstrates an improvement in a *PAE* of 15.2% at an *ACLR* of −30 dBc compared to its Class AB counterpart. This is because, in a Doherty power amplifier, the quiescent current is maintained at a constant level only for the carrier amplifier, whereas the bias

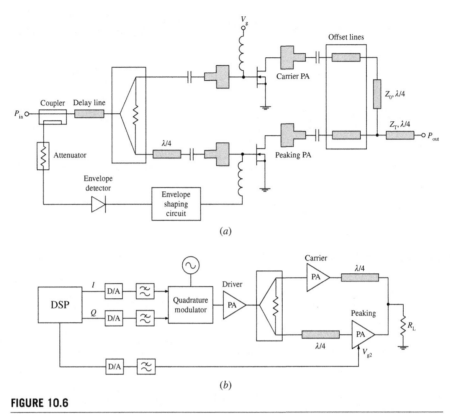

**FIGURE 10.6**

Block diagrams of Doherty amplifier architectures with adaptive control.

point of the peaking amplifier is varied according to the input signal envelope. However, it should be noted that the wider the modulation bandwidth of the transmitting signal, the more problematic it is to implement such a technique in practice to achieve significant linearity improvement.

The linearity of the power amplifier can be improved by using digital signal processing to provide more accurate gate-bias control alongside the digital predistortion needed for the simultaneous correction of the gain and phase characteristics in a high-power region. Figure 10.6(*b*) shows the block schematic of an 840-MHz MESFET Doherty power amplifier for CDMA applications with the digital signal processor (DSP) implemented externally on a board controlled by a personal computer [23]. In this case, the DSP generates both the baseband in-phase (*I*) and quadrature (*Q*) signals, which are upconverted to form an RF signal using the quadrature modulator. The DSP unit also generates the voltage signal $V_{g2}$, which is applied to the peaking amplifier as the gate bias. This results in an

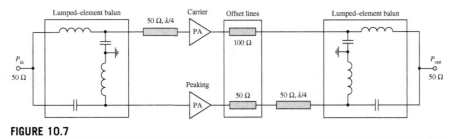

**FIGURE 10.7**

Block diagram of a Doherty amplifier with series connected load.

efficiency improvement by the dynamic gate biasing of the peaking amplifier according to the instantaneous envelope of the input signal. At the same time, the phase performance is corrected by the phase predistortion at baseband level based on the dynamic gate bias-voltage values from the gain correction, thus resulting in a linearity improvement. An overall improvement of *PAE* from 3% to 5% and an *ACPR* of about 10 dB at an average output power of 23 dBm can be achieved by utilizing such a DSP technique. Besides, a simple bias-switching technique can be introduced to Doherty-type amplifiers, so that they can satisfy the linearity requirements of the power amplifiers for CDMA handset applications over the entire dynamic power range [24]. In addition to the bias-switching technique, a dual-mode matching approach can be used to optimally design a dual-mode Doherty amplifier operated simultaneously in HPSK (hybrid phase shift keying) mode and OFDM (orthogonal frequency-division multiplexing) 64-QAM (quadrature amplitude modulation) mode for mobile terminals [25].

In comparison with the shunt connected load type, the series connected load combines the output powers of the carrier and peaking amplifiers in a manner similar to that of the push−pull amplifiers, having the same capability to suppress even-order harmonic components in the output signal spectrum. Figure 10.7 shows the block diagram of a Doherty amplifier with a series-connected load where the input and output baluns are implemented with the lumped inductances and capacitors [26]. The lumped-element balun can be designed for an arbitrary unbalanced-load value by properly selecting its element values. In this case, an unbalanced load of 50 Ω and a balanced port load of 100 Ω were chosen. At low power levels when the peaking amplifier is turned off, one balanced port of the balun connected to the open-circuited quarterwave transmission line is shorted; that results in a load of 100 Ω for the carrier amplifier. At high power levels when the peaking amplifier becomes active, both the carrier and the peaking amplifiers are operated in near push−pull mode. For such a GaN HEMT Doherty amplifier operating at 1.8 GHz, high efficiencies of 31% and 56% were achieved at 24- and 31-dBm saturated output powers, respectively.

## 10.2 Carrier and peaking amplifiers with harmonic control

Further efficiency improvement of the vacuum-tube Doherty amplifiers were achieved by creating biharmonic modes at the input and load networks of the carrier and peaking amplifiers. Specifically, additional parallel resonant circuits tuned to the third harmonic were used that have resulted in an efficiency of over 80% [27]. Ideally, an infinite number of odd-harmonic resonators results in an idealized Class-F mode with a square voltage waveform and a half-sinusoidal current waveform, whereas an infinite number of even-harmonic resonators results in an idealized inverse Class-F mode with a half-sinusoidal voltage waveform and a square current waveform at the device output terminal. However, in practice at microwave frequencies, it is enough to control the second- and third-harmonic components to provide a high operational efficiency of the power amplifier, and it is preferable to use short-circuit and open-circuit stubs instead of lumped capacitors in the load network.

Figure 10.8(a) shows the block schematic of a fully matched microwave Doherty amplifier including the harmonic-control circuits employed in front of the output matching circuits and offset lines. To approximate a Class-F operation mode, the harmonic-control circuit includes both arm shunt stubs for a better harmonic trap and a series tuning line to compensate for the output device parasitic elements, as shown in Fig. 10.8(b). Despite the fact that the Doherty amplifier with harmonic control provides worse linearity than the conventional Doherty amplifier in terms of *AM/AM* and *AM/PM* performance, applying a digital feedback predistortion linearizer allows the *ACLR* to be significantly improved. For example, the *ACLR* of a 2.14-GHz Class-F two-stage GaN HEMT Doherty amplifier with an average output power of 36 dBm and a drain efficiency of 52.4% was improved by more than 20 dB [28]. The inverse Class-F two-stage LDMOSFET Doherty amplifier using the harmonic-control circuits shown in Fig. 10.8(c) is capable of providing a drain efficiency of 54.7% at an average output power of 32 dBm for a 1-GHz forward-link WCDMA signal [29].

In some cases, a Class-F design strategy can be applied to the carrier amplifier only, while the peaking amplifier operates in a conventional Class-C mode [30]. Figure 10.9 shows the simplified circuit schematic of a Doherty amplifier, in which the drain bias supplies are connected to the corresponding device drain terminals through the short-circuited quarter-wavelength transmission lines providing inherent even-harmonic suppression [31,32]. In this case, the drain current of the Class-B biased carrier FET contains the dc, fundamental-frequency, and even-harmonic components. However, the drain current of the Class-C biased peaking FET is ideally purely sinusoidal since its odd-harmonic components are shunted by the short-circuited bias-feed quarterwave line connected to the carrier FET, thus resulting in no harmonics appearing at the peaking FET output. However, the effect of output parasitics of the carrier FET reduces the quality of the odd-harmonic short-circuit

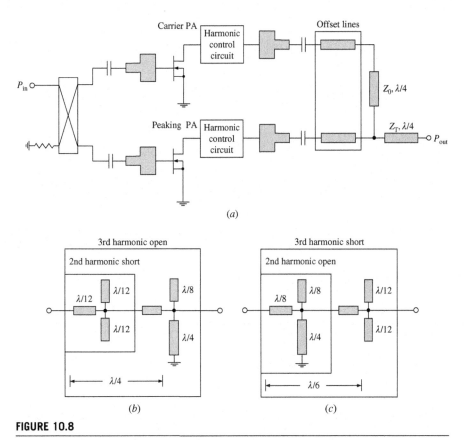

**FIGURE 10.8**

Block diagram of a Doherty amplifier with harmonic control circuits.

termination at the peaking FET, resulting in power-amplifier performance degradation when measured efficiency was slightly above 60% in saturation at an operating frequency of 770 MHz for a supply voltage of 3.5 V.

The simulation setup of a highly efficient 2.14-GHz, 2.5-W transmission-line two-stage GaN HEMT Doherty amplifier with an input branch-line coupler operating in an inverse Class-F mode and implemented in a 30-mil RO4350 substrate is shown in Fig. 10.10. The simple output load network of the both carrier and peaking amplifiers provides the load matching at the fundamental frequency and the corresponding second- and third-harmonic control. Here, high impedance at the second harmonic and low impedance at the third harmonic are created at the output of each device using a short-length series microstrip line short-circuited at the second and third harmonics by a quarterwave short-circuit stub and an open-circuit stub with an electrical length of 30°, respectively [33]. The exact electrical

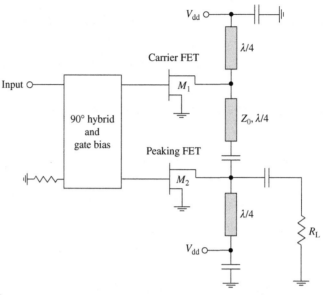

**FIGURE 10.9**

Schematic of a Class-F/Class-C Doherty amplifier.

lengths of the microstrip lines depend on the values of the device output shunt capacitance and series inductance. Efficiency enhancement of a Doherty amplifier can also be provided with a combination of Class-F and inverse Class-F schemes for the carrier and peaking amplifiers, respectively, and an efficiency of 45% can be achieved at 10-dB backoff [34].

A high efficiency of the Doherty amplifier can also be achieved using a Class-E mode when the device output shunt capacitance can be considered as an internal element of the Class-E load network [35]. If the device shunt capacitance is larger than the nominal Class-E capacitance, an additional compensation circuit with a series capacitor and a shunt inductance needs to be added to the device output. For example, this configuration was used in the design of a 2.14-GHz Class-E Doherty power amplifier using 25-W GaN HEMT devices in the carrier and peaking amplifiers, resulting in a *PAE* of 44.8% at an average output power of 37 dBm for a single-carrier WCDMA signal [36]. The symmetrical Doherty configuration using packaged devices in the carrier and peaking amplifiers can also be designed in a Class-E mode when the device output capacitance is used to compose a $\pi$-section Class-E load network [37]. In this case, the ideal high-$Q$ series resonant circuit required for the Class-E mode is replaced by a low-pass matching circuit with the series inductance, implemented as a microstrip line on a ceramic substrate, and a shunt ceramic chip capacitor.

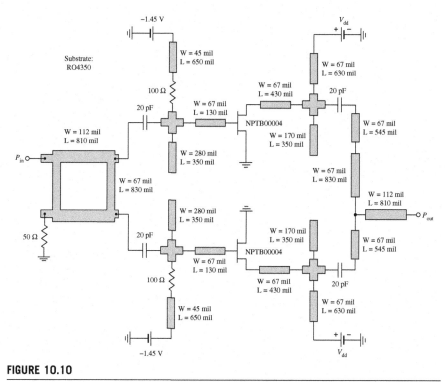

**FIGURE 10.10**

Circuit schematic of an inverse Class-F GaN HEMT Doherty amplifier.

## 10.3 Balanced, push–pull, and dual Doherty amplifiers

In order to increase the overall output power of the Doherty amplifier system, two individual two-stage conventional Doherty amplifiers can be connected in a push–pull configuration using the coaxial-line baluns at both input and output ports to provide unbalanced-to-balanced impedance transformation by combining the outputs of the Doherty amplifiers [38].

Figure 10.11 shows the balanced configuration using 90° hybrid branch-line couplers to split input power and to combine output powers from two identical conventional Doherty amplifiers [39]. In this case, to achieve higher efficiency up to 6-dB backoff output power, a combined design technique based on an uneven power divider and asymmetrical devices with optimized size ratio can be used [40]. However, better linearity and higher power capability can be achieved with a balanced structure which combines two identical Doherty amplifiers. To match each individual Doherty structure with the output coupler having 50-$\Omega$ inputs, it is necessary to choose the characteristic impedances of the series transforming quarter-wavelength transmission lines equal to $50/\sqrt{2} = 35\ \Omega$. The balanced

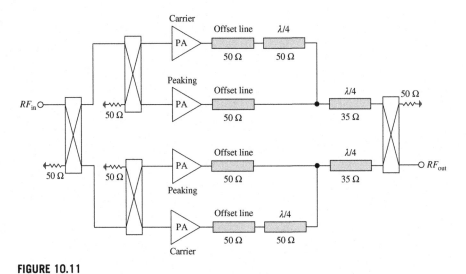

**FIGURE 10.11**

Balanced Doherty amplifier architecture.

Doherty amplifier where the peaking amplifier of the top two-stage Doherty structure and the carrier amplifier of the bottom two-stage Doherty structure are interchanged since they have equal 90° phase shifts at their outputs can provide better
frequency flatness and circuit stability [41].

The distributed Doherty amplifier uses two individual dual-fed distributed
amplifiers where these two distributed amplifiers are coupled in the conventional
Doherty amplifier configuration with a 90° hybrid at the input and a quarterwave
transmission line at the output [42]. By interleaving one peaking amplifier and
one carrier amplifier, the input hybrid can be replaced with a quarterwave transmission line, thus resulting in an interleaved four-way Doherty amplifier where
two carrier amplifiers and two peaking amplifiers are separated by the input and
output quarterwave transmission lines that reduce the overall system size [42,43].
A distributed Doherty architecture can also represent a single-ended dual-fed distributed structure without the need for a two-way input divider and output combiner when the individual identical two-stage conventional Doherty amplifiers are
separated by a $\lambda/2$ at the operating frequency using half-wavelength 50-$\Omega$ microstrip lines [44]. However, in this case, both individual Doherty amplifiers should
be matched to a 100-$\Omega$ output or an additional output impedance transformer
must be used.

Figure 10.12($a$) shows the dual Doherty amplifier architecture where it is
enough to use the quarterwave 50-$\Omega$ transmission lines only, one in the input circuit to provide a 90° phase shift and two in the output circuit to provide a parallel
in-phase connection of two individual Doherty amplifiers. This is because the

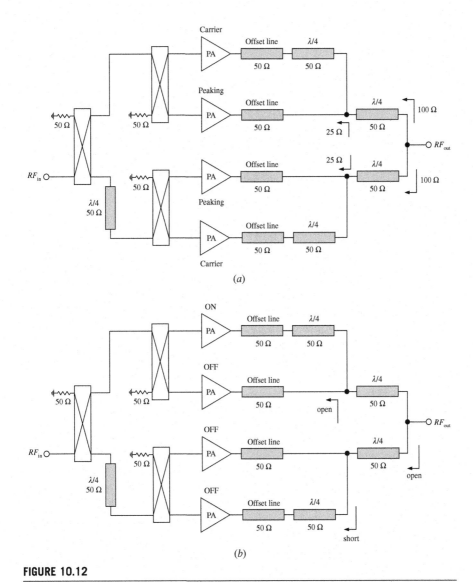

**FIGURE 10.12**

Dual Doherty amplifier architecture.

input of each output quarterwave transmission line sees 25 Ω, which is then transformed to 100 Ω by a quarterwave 50-Ω transmission line. Finally, a parallel connection of two 100-Ω output impedances results in the required 50-Ω output. To provide a high impedance of the peaking amplifiers when they are turned off and to compensate for the effect of their output matching elements and phase delays,

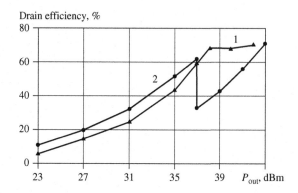

**FIGURE 10.13**

Simulated drain efficiencies versus output power.

offset lines are used in each carrier and peaking amplifying path. In addition, such a parallel dual Doherty architecture can provide high efficiencies at very low power levels since three of the four identical amplifiers can be switched off providing high impedance at the corresponding ends of the output 50-$\Omega$ lines, as shown in Fig. 10.12(*b*). As a result, the remaining single power amplifier operates in a 50-$\Omega$ environment, providing an efficient and linear operating condition at low output power levels.

Figure 10.13 shows the simulated continuous-wave (CW) drain efficiency *versus* output power for the dual Doherty amplifier shown in Fig. 10.12(*a*) (curve 1) and for the switched-path power amplifier shown in Fig. 10.12(*b*) (curve 2). In this case, for the switched-path GaN HEMT power amplifier based on four 5-W NPTB00004 devices and designed in an inverse Class-F mode when only the top transistor is turned on, the drain efficiency of 28.5% at an output power of 30 dBm can be achieved at an operating frequency of 2.14 GHz. This efficiency is 7.5% higher than that for a dual Doherty amplifier (curve 1). This means that the power amplifier configuration shown in Fig. 10.12(*b*) can operate as a dual Doherty amplifier at high output power levels and as a single power amplifier at low output power levels, thus providing a significant dc-power saving capability.

## 10.4 Asymmetric Doherty amplifiers

There is a possibility to extend the region of high efficiency over a wider range of output powers if the carrier and peaking amplifiers are designed to operate with different output powers—smaller for the carrier amplifier and larger for the peaking amplifier. For instance, for a power-division ratio $\alpha = P_{\text{carrier}}/P_{\text{peaking}} = 0.25$, the transition point with maximum drain efficiency corresponds to the backoff power level of $-12\,\text{dB}$ from peak output power [12]. At peak output power when the carrier and peaking amplifiers shown in Fig. 10.4(a) are saturated, it follows from consideration of their output powers that $R_1 = Z_2 = R_3 = 3R_2$. As a result, $I_2 = 3I_3$ and $\beta = 0.25$. The output impedances $R_2$ and $R_3$ as functions of the load resistance $R_L$ and characteristic impedance $Z_1$ can be obtained from Eqs (10.7) and (10.8). For example, if one can choose the characteristic impedance of the output transmission line and load resistance equal to $Z_1 = 15\,\Omega$ and $R_L = 50\,\Omega$, respectively, then the characteristic impedance of the quarterwave transformer and load impedance for the carrier amplifier are $Z_2 = R_1 = 18\,\Omega$, whereas the output impedance of the peaking amplifier is equal to $R_2 = R_1/3 = 6\,\Omega$.

Since from Eqs (10.7) and (10.8) it follows that

$$R_1 = \frac{Z_2^2}{\beta R_3},\tag{10.10}$$

hence, at lower power levels when the peaking amplifier is turned off, the output impedance $R_1$ is four times higher than that at peak output power where $R_1 = Z_2 = R_3$. The scaling ratio of 4:1 was used for GaAs HBT devices with total emitter areas of 3360 and 840 $\mu\text{m}^2$ for the peaking and carrier amplifiers, respectively, to implement the extended Doherty technique into the monolithic power amplifier developed for CDMA handset applications. As a result, the power-added efficiencies of 45% and 23% were measured at the highest output power of 25 dBm and at 10-dB backoff level, respectively [45]. Note that the conventional Class-AB power amplifiers designed for the same application normally have the power-added efficiency of about four times lower at this backoff power.

For the packaged devices when it is difficult to choose the proper power ratio between the devices, it is convenient to use the identical power amplifiers which can compose ideally the $N$-way Doherty architecture where one carrier power amplifier is in parallel with $(N-1)$ numbers of the peaking amplifiers. This is the simplest hybrid approach to acquire an $(N-1)$ times larger-sized peaking amplifier compared with the carrier amplifier for an asymmetric two-way Doherty amplifier configuration [46]. Figure 10.14(a) shows the schematic diagram of an $N$-way Doherty amplifier with a parallel connection of one carrier amplifier and $(N-1)$ numbers of the identical peaking amplifiers. The ideal drain efficiencies of the $N$-way Doherty amplifier (DA) architectures with peak values at $-6$, $-9.5$, $-12$, and $-14\,\text{dB}$ power backoff points, according to $P_{\text{backoff}} = 20\log_{10} N$ for the two-, three-, four-, and five-ways structures, respectively, and conventional Class-B power amplifier are shown in Fig. 10.14(b).

Generally, the $N$-way Doherty amplifier is composed of the $N$-way power splitter, identical fully matched carrier and $(N-1)$ peaking amplifiers, $N$ offset lines, and an output combiner representing a quarterwave impedance transformer. The characteristic impedance of a quarterwave transmission line for converting the load of the carrier amplifier is $Z_0 = R_0/\sqrt{\alpha}$ and the common load resistance is

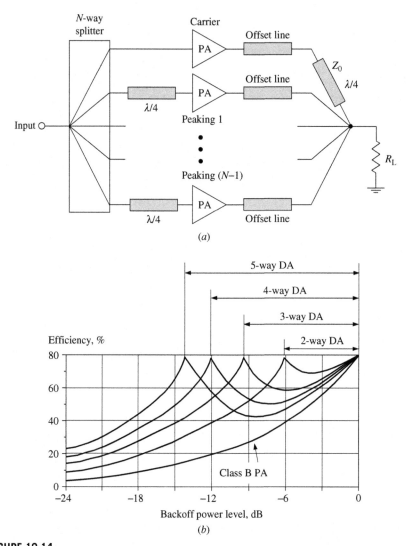

**FIGURE 10.14**

Asymmetric *N*-way Doherty amplifier and efficiencies.

$R_L = R_0/(\alpha + 1)$, where $R_0$ is the matched load for both carrier and peaking amplifiers, usually equal to 50 Ω [47]. For asymmetric two-way and three-way Doherty amplifiers with optimized individual bias conditions and load matching for the carrier and peaking amplifiers, applying an uneven drive results in more linear

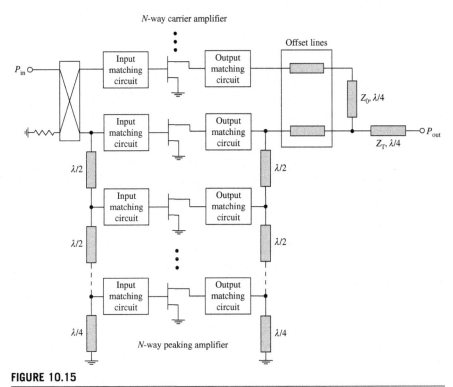

**FIGURE 10.15**

Distributed N-way Doherty amplifier.

operation and produces more power than an even drive [48,49]. For example, a three-way Doherty amplifier based on Class-F load networks and fabricated using 10-W GaN HEMT devices achieved a *PAE* of 45.9% for a single-carrier WCDMA signal with a *PAR* of 10 dB at 2.14 GHz with an *ACLR* of −49.2 dBc using a digital feedback predistortion technique [50].

In spite of the efficiency improvements offered by the asymmetric N-way Doherty amplifier over its symmetric two-way Doherty amplifier counterpart, its total power gain, which significantly depends on the power gain of the carrier amplifier, will be reduced due to the corresponding insertion loss in the required input N-way power splitter. This issue is circumvented by using distributed amplification. In particular, distributed amplification is a technique whereby power combining is performed directly at the transistor level without the need for an N-way power combiner. Figure 10.15 shows the simplified schematic of a distributed Doherty amplifier where the powers of both N-carrier and N-peaking amplifiers are combined using half-wave and quarterwave microstrip lines [51,52].

The desired location of peak efficiency points of such a distributed $N$-way Doherty amplifier can be given in decibels by

$$P_{\text{backoff}} = 20 \log_{10} \left( \frac{K}{M} + 1 \right) \tag{10.11}$$

where $K$ and $M$ are the numbers of the carrier and peaking amplifiers, respectively. Practically, in order to design a high-efficiency three-way distributed Doherty amplifier, the two peaking amplifiers can be combined using a dual-fed distributed structure. The measured results of a three-way distributed 2.14-GHz Doherty amplifier using three 45-W LDMOSFET devices indicate that a *PAE* of 39.5% with a power gain of 11 dB was achieved at 9.5-dB backoff.

## 10.5 Multistage Doherty amplifiers

An asymmetric Doherty architecture exhibits a significant drop in efficiency in the region between the efficiency peaking points, especially for large power ratios between the carrier and peaking amplifiers. However, it is possible to use more than two power amplifiers to prevent significant deterioration of efficiency at backoff output power levels. This can be provided by the so-called multistage Doherty amplifiers, the operation of which is analogous to that of the conventional two-stage Doherty amplifier [12].

The basic multistage Doherty power amplifier architecture shown in Fig. 10.16(*a*) comprises more than one peaking amplifier, with the quarterwave transmission lines to combine their output powers [53]. The characteristic impedances of each output quarterwave transmission line depend on the levels of backoff power and can be calculated from

$$Z_{0i} = R_L \prod_{j=1}^{i} \gamma_j \tag{10.12}$$

$$\prod_{j=k}^{(i+k)/2} \gamma_{(2j-k)} = 10^{(B_i/20)} \tag{10.13}$$

where $i = 1, 2, \ldots, N-1$, $k = 1$ (for odd $i$) or 2 (for even $i$), $N$ is the total number of amplifier stages, and $B_i$ is the backoff level (positive value in decibels) from the maximum output power of the system, at which the efficiency peaks. The maximum level of backoff $B_{N-1}$ is set by the carrier amplifier, while the number of efficiency peaking points is directly proportional to the number of amplifier stages used in the design.

Figure 10.17 shows the instantaneous drain efficiencies of the multistage Doherty amplifier (DA) architectures for the two, three, and four stages, having maximum efficiencies at the transition points of $-6$, $-12$, and $-18$ dB backoff output power levels, respectively. From Fig. 10.17, it follows that the multistage

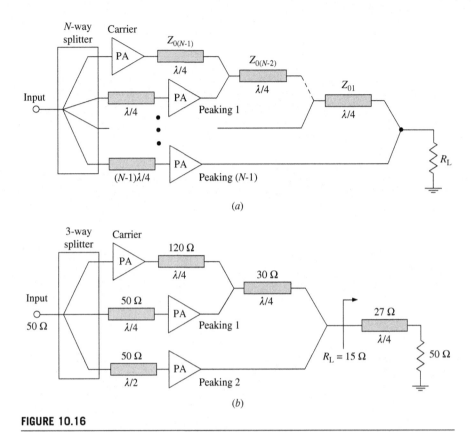

**FIGURE 10.16**

Multistage Doherty amplifier architectures.

architecture provides higher efficiencies at backoff levels between the efficiency peaking points compared with an asymmetric Doherty architecture and significantly higher efficiency at all backoff output power levels compared with the conventional Class-B power amplifiers. For the most practical case of a three-stage Doherty amplifier whose block schematic is shown in Fig. 10.16(b), the characteristic impedances of each output quarterwave transmission line can be obtained from Eqs (10.12) and (10.13) to be

$$Z_{01} = \gamma_1 R_L \tag{10.14}$$

$$Z_{02} = \gamma_1 \gamma_2 R_L \tag{10.15}$$

where

$$\gamma_1 = 10^{(B_1/20)} \tag{10.16}$$

$$\gamma_2 = 10^{(B_2/20)} \tag{10.17}$$

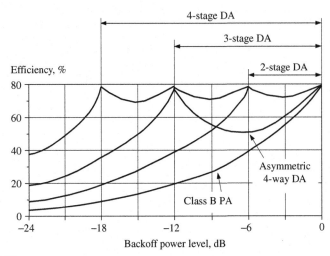

**FIGURE 10.17**

Efficiencies of the different Doherty amplifier architectures.

where $B_1 = 6$ and $B_2 = 12$ for peak efficiencies at $-6$ and $-12$ dB backoff points, respectively, resulting in $Z_{01} = 30\,\Omega$ and $Z_{02} = 120\,\Omega$ for $R_L = 15\,\Omega$.

For a 1.95-GHz WCDMA application, a three-stage Doherty amplifier structure using GaAs MESFET devices with the device periphery ratio of 1:2:4 and microstrip power combining elements provides a *PAE* of 48.5% and a power gain of 12 dB at $P_{1dB} = 33$ dBm. The peak power-added efficiencies of 42% and 27% were measured at the $-6$ and $-12$ dB backoff levels [53]. Efficiencies at backoff points can be increased by optimizing the input drive conditions for the peaking amplifiers [54]. Moreover, further efficiency improvement of the three-stage Doherty amplifier at the maximum output power and backoff points can be achieved by using highly effective GaN HEMT devices and applying digital predistortion technique for linearization. In this case, the drain efficiency at the $-12$ dB output power backoff point can be increased to be higher than 60% [55].

A typical problem associated with the conventional three-stage Doherty amplifier is that the load-line modulation of the carrier stage stops at a certain power level, leaving the carrier amplifier in deep saturation and leading, consequently, to a significant degradation of its linear performance. In addition, when the carrier and peaking amplifiers have equal configurations with the same device periphery sizes, similar performance is obtained with regards to the symmetrical two-stage Doherty amplifier, with the efficiency peaking points at $-3.5$ and $-6$ dB backoff output powers. These problems can be partially solved by using a modified three-stageDoherty amplifier architecture with a parallel combination of one

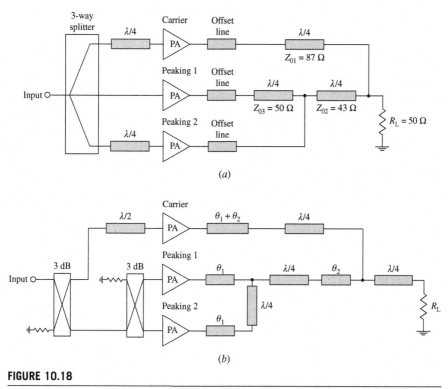

**FIGURE 10.18**

Modified three-stage Doherty amplifier architectures.

carrier and one Doherty amplifier used as a peaking amplifier, as shown in Fig. 10.18(*a*) [56]. In this case, a novel way of combining enables high instantaneous efficiencies at −6 and −9.5 dB backoff output powers with a single device size. The characteristic impedances of the transforming quarterwave transmission lines are calculated as $Z_{01} = \sqrt{3}R_L$, $Z_{02} = (\sqrt{3}/2)R_L$, and $Z_{03} = R_L$, where $R_L$ is the load resistance [57,58]. Figure 10.18(*b*) shows the other modified three-stage Doherty architecture operating in a 2.14-GHz WCDMA system based on three 10-W GaN HEMT transistors, where both peaking amplifiers represent in turn the conventional two-way Doherty configuration [59]. In this case, only the carrier amplifier is turned on at the low-power region, the carrier amplifier is saturated and the first peaking amplifier is turned on at the medium-power region, and the carrier and first peaking amplifier are both in saturation and the second peaking amplifier is turned on at the high-power region. The optimum electrical lengths of the required 50-Ω offset lines are $\theta_1 = 0.28\lambda$, $\theta_2 = 0.36\lambda$, and $\theta_1 + \theta_2 = 0.64\lambda$, respectively. As a result, the carrier amplifiers are turned on at near −9 dB and −6 dB backoff output power levels. The efficiency and linearity can be optimized by using

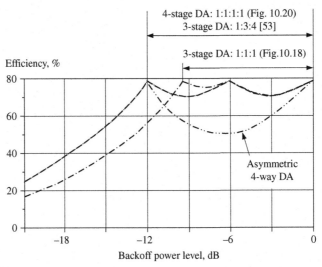

**FIGURE 10.19**

Efficiencies of different Doherty amplifier architectures.

the two driving amplifiers connected to the output ports of the input 3-dB coupler and biased in Class B and Class C, respectively.

Figure 10.19 shows the theoretical instantaneous drain efficiencies of the multistage (three and four stages) and four-way asymmetric Doherty amplifier (DA) architectures for different power (or device size) ratios, with peak efficiencies ranging from the −12 dB output power backoff levels. From Fig. 10.19, it follows that the four-way or any asymmetric multi-way Doherty architecture provides significantly lower efficiency between the corresponding efficiency peaking points. However, for a multistage Doherty configuration, the efficiency peaking points at lower output power backoff levels can be achieved using an optimum device size ratio. For example, a peak efficiency at the lowest backoff point of −12 dB is achieved for a device periphery ratio of 1:3:4 in a three-stage Doherty amplifier, while the lowest backoff of about −9.5 dB corresponds to the peak efficiency for an equal device periphery size of 1:1:1 in the modified three-stage Doherty amplifiers shown in Fig. 10.18(*a*) and 10.18(*b*) [53,58,59].

In a classical four-stage Doherty power amplifier with corresponding peak efficiencies at −6, −12, and −18 dB backoff output power points, the maximum ratio between the characteristic impedances of the quarterwave transmission lines is equal to 16 [53]. For example, for $R_L = 6 \, \Omega$, the characteristic impedances of the consecutive quarterwave transmission lines are $Z_{01} = 12 \, \Omega$, $Z_{02} = 48 \, \Omega$, and $Z_{03} = 192 \, \Omega$, respectively. These values are difficult to correctly implement using microstrip lines on a single substrate with a fixed thickness and dielectric

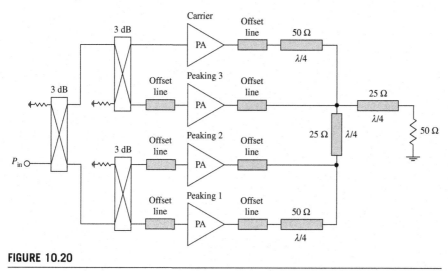

**FIGURE 10.20**

Modified four-stage Doherty amplifier architecture.

permittivity. In a modified four-stage Doherty configuration with the device size ratio of 1:1:1:1 shown in Fig. 10.20, where the two conventional two-stage Doherty amplifiers are combined in a final four-stage Doherty configuration (Doherty in Doherty), the maximum ratio between the transmission-line character-istic impedances is equal to 50 $\Omega$/25 $\Omega$ = 2 only [60]. Figure 10.19 shows the three efficiency peaking points provided by the modified four-stage Doherty amplifier with equal gate bias voltages for the second and third peaking ampli-fiers. Furthermore, the optimization of their gate bias voltages can change the efficiency profile between the peak power and −6 dB backoff points and contrib-ute to linearity improvement. Because of the device input and output parasitics such as the gate-source and drain-source capacitances, additional input offset lines are implemented at the input of the peaking amplifiers and identical output offset lines which introduce the compensating inductive reactances are connected in series to each output circuit. This is a very practical version of a four-stage Doherty amplifier, capable of achieving high output powers with high drain effi-ciency and having three 90° hybrid couplers at the input and four quarterwave microstrip lines at the output.

Figure 10.21 shows the simplified practical topology of a modified four-stage GaN HEMT Doherty power-amplifier architecture based on four 25-W CGH40025F devices and fabricated using a 30-mil RO4350 substrate [60]. The input dividing network includes three commercial 90° hybrid couplers, while a 30-dB directional coupler required to sampling output power for linearization loop need to be connected to the output port. In a CW operation mode when all transistors are biased with the same gate voltage of −3.4 V, an output power of

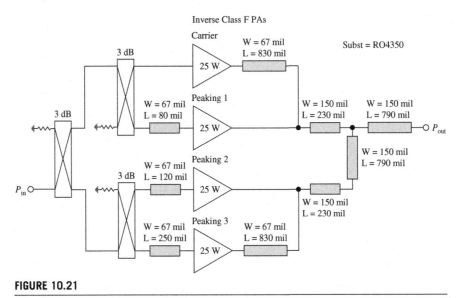

**FIGURE 10.21**

Simplified topology of 2.14 GHz 100-W four-stage Doherty GaN HEMT amplifier.

50 dBm (100 W) and a drain efficiency of 77% were achieved at a supply voltage of 34 V. In a single-carrier 2.14-GHz WCDMA operation mode with a *PAR* of 6.5 dB, a drain efficiency of 61% was achieved at an average output power of 43 dBm (20 W), corresponding to a 7-dB backoff from the saturated output power. In this case, an adjacent channel leakage ratio ($ACLR_1$) was measured at −31 dBc level, with an alternate channel leakage ratio ($ACLR_2$) of −38.5 dBc.

## 10.6 Inverted Doherty amplifiers

Figure 10.22 shows a schematic diagram of an inverted Doherty amplifier configuration with a quarterwave-line transformer connected to the output of the peaking amplifier. The quarterwave-line transformer can be implemented in a more compact form than in the conventional Doherty amplifier, which is more suitable for use in mobile applications [61]. In this case, at low-power levels, a quarterwave line is used to transform very low output impedance after the offset line to high impedance seen from the load junction. In particular, by taking into account the device package parasitic elements of the peaking amplifier, an optimized output matching circuit and a proper offset line are designed to provide the maximum output power from the device and to rotate the output impedance near to 0 Ω at the input of the quarterwave line [62]. At a high power level, for the matched phase difference between identical carrier and peaking amplifiers, the

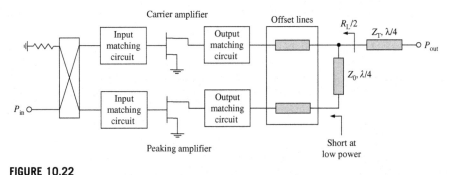

Carrier amplifier

**FIGURE 10.22**

Schematic diagram of an inverted Doherty amplifier.

load impedance seen from each amplifier after the offset lines is equal to $R_L$. The combination of internal package components, output matching circuit, and offset line for the carrier amplifier transforms the load impedance from half ($R_L/2$) to twice of the optimum internal load impedance ($2R_L$). Using a four-carrier downlink WCDMA signal, a *PAE* of 32% with an *ACLR* of −30 dBc at an average output power level as high as 46.3 dBm was achieved for an inverted 2.14-GHz LDMOSFET Doherty amplifier. This provides a 9.5% improvement in efficiency and 1-dB improvement in the output power under the same *ACLR* conditions as for the balanced Class-AB operation using the same devices [63].

In a Doherty configuration, both the Class-AB carrier amplifier and Class-C peaking amplifier are not fully isolated from each other. This results in a serious problem to robustly design the optimum load impedance shift presented to both transistors for high efficiency and low distortion [64]. From the load-pull measurement for a unit-cell 28-V GaAs HJFET device, it was observed that, in order to obtain high efficiency and low distortion, the carrier amplifier load impedance should change from the maximum efficiency point to the maximum output power point at Class AB, while the peaking amplifier load impedance should vary from the small-signal gain point to the maximum output power point in Class C [65]. In this case, the load impedance corresponding to the maximum efficiency point is lower than the load impedance corresponding to the maximum output power point. An inverted Doherty architecture has proved very suitable to realize the carrier amplifier load impedance variation from lower impedance to higher impedance in accordance with the increase of the input power level. The external input and output matching circuits are necessary to optimize the load impedance shift presented to both carrier and peaking amplifiers as a function of the input power level. As a result, a drain efficiency of 42% at an output power of 49 dBm around the 6-dB backoff level was achieved for a two-carrier WCDMA signal of 2.135 GHz and 2.145 GHz with a third-order intermodulation distortion ($IM_3$) of −37 dBc.

Figure 10.23 shows the three-stage inverted Doherty amplifier configuration where the quarter-wavelength transmission lines are added in the outputs of the carrier and peaking amplifiers to provide a proper load modulation ratio [66]. The half-wave transmission line in the input path of the carrier amplifier is used to compensate for the delay provided by the output load network. The characteristic impedances of the quarterwave transmission lines are optimized to provide a high efficiency over wide output power backoff range. If the device size ratio of the carrier, first peaking, and second peaking amplifiers is $1:m_1:m_2$, respectively, the characteristic impedances of the quarterwave transmission lines at the full power loading condition can be calculated from

$$Z_T = Z_1 \sqrt{\frac{1}{1 + m_1 + m_2}} \tag{10.18}$$

$$Z_4 = Z_1 \frac{Z_3}{Z_2} \sqrt{\frac{1}{m_1}} \tag{10.19}$$

$$Z_5 = Z_1 \frac{Z_3}{Z_2} \sqrt{\frac{1}{m_2}} \tag{10.20}$$

assuming the same 50-$\Omega$ load conditions for the standard load and the carrier and peaking amplifiers at full loading conditions. As a result, for the same device sizes for carrier and peaking amplifiers when $m_1 = m_2 = 1$, $Z_2 = Z_3 = 50\,\Omega$, and $Z_1 = 70\,\Omega$, from Eqs (10.18) to (10.20) it follows that $Z_T = 40.4\,\Omega$ and $Z_4 = Z_5 = 50\,\Omega$, respectively. In this case, the drain efficiency for a single-carrier 2.14-GHz WCDMA signal with a *PAR* of 10.5 dB can be improved by 5% over a wide range of output powers.

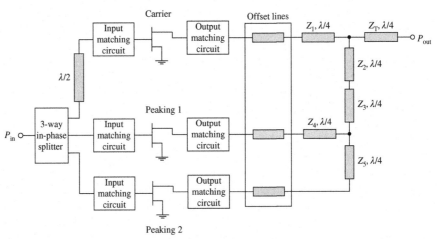

**FIGURE 10.23**

Schematic diagram of three-stage inverted Doherty amplifier.

## 10.7 Integration

The transmission-line two-stage Doherty amplifier can easily be implemented into the monolithic microwave integrated circuit (MMIC) by using a pHEMT or CMOS process. For example, a fully integrated $Ku$-band MMIC Doherty amplifier using a 0.25-μm pHEMT technology achieved a two-tone $PAE$ of 40% with a corresponding $IM_3$ of −24 dBc at 17 GHz, while a single-tone $PAE$ of 38.5% at 1-dB compression point was measured for a 20-GHz MMIC Doherty amplifier implemented in a 0.15-μm pHEMT process for use in digital satellite communication (DSC) systems [67,68]. Furthermore, by using a 0.13-μm RF CMOS technology, a transmission-line MMIC Doherty amplifier with a cascode configuration of the carrier and peaking amplifiers achieved a saturation output power of 7.8 dBm from a supply voltage of 1.6 V at an operating frequency of 60 GHz for use in wireless personal area network (WPAN) transceivers [69]. However, the efforts to directly apply the Doherty technique to the design of power-amplifier integrated circuits with a high level of integration at lower frequencies face difficulties, since the physical size of the quarterwave transmission lines is too large in this case. For example, for an FR4 substrate with effective dielectric permittivity of $\varepsilon_r = 3.48$, the geometrical lengths of the quarterwave transmission lines are 48, 19, and 8.7 mm at the operating frequencies of 900 MHz, 2.4 GHz, and 5.2 GHz, respectively. Therefore, one of the acceptable solutions for the fabrication of small-size Doherty amplifier MMICs intended to operate in WLAN or WiMAX transmitter systems is to replace each quarterwave line in the input combining circuit and output impedance transformer by its low-pass $\pi$-type lumped-distributed equivalent with a short-length series transmission line and two shunt capacitors connected to its both ends [70,71]. Additionally, simple and small-size second-harmonic termination circuits can be realized with integrated MIM capacitors and bondwires at the end of the carrier and peaking amplifier collectors [70].

In order to minimize the inherently high substrate loss and increase the level of integration for the implementation of the Doherty amplifier in CMOS process, the branch-line coupler and quarterwave transformer in the amplifier input and output circuits are fully substituted by their lumped equivalents [72]. By considering the transmission $ABCD$-matrices for a quarterwave transmission line shown in Fig. 10.24($a$) and a $\pi$-type low-pass lumped circuit consisting of a series inductance and two shunt capacitors, as shown in Fig. 10.24($b$), and equating the corresponding elements of both matrices, the ratio between the circuit elements can be written as

$$Z_0 \omega C = \frac{Z_0}{\omega L} = 1 \tag{10.21}$$

where $Z_0$ is the characteristic impedance of the quarterwave line. A high-power Doherty amplifier MMIC can be integrated with lumped elements in a standard discrete package where the compensation series circuits (each consisting of an inductance and a capacitor) are connected to the drain terminals of the carrier and

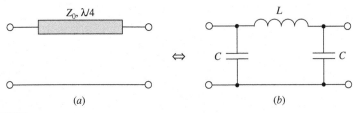

**FIGURE 10.24**

Quarterwave transmission line and its single-frequency lumped equivalent.

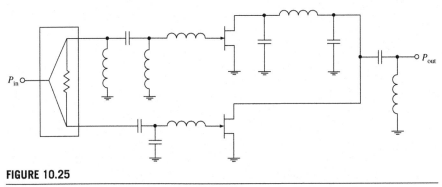

**FIGURE 10.25**

Schematic of an 900-MHz Doherty amplifier implemented with lumped elements.

peaking transistors to compensate for their output capacitances [73]. For example, an integrated solution based on four 10-W MMIC Doherty amplifier cells combined in parallel achieves a drain efficiency of 39.8% at an average output power of 7.5 W with an *ACLR* of −50 dBc using a digital predistortion technique for a two-carrier 2.14-GHz WCDMA signal with a *PAR* of 7.6 dB [74].

Similarly, the input in-phase Wilkinson divider can be replaced by a $\pi$-type low-pass lumped circuit in its each branch. Figure 10.25 shows the simplified schematic of a two-stage 900-MHz GaAs MESFET lumped Doherty amplifier architecture where the output quarterwave transmission line connected to the carrier amplifier output is replaced by a $\pi$-type low-pass lumped circuit, whereas the input phase-shifting three-quarterwave transmission line connected to the carrier amplifier input is replaced by a $\pi$-type high-pass lumped circuit. In addition, the output quarterwave transformer is replaced by an *L*-type high-pass matching circuit, whereas two *L*-type low-pass matching circuits are used to provide the input matching of the carrier and peaking amplifiers. At the carrier amplifier input path, the right-hand shunt inductance as a part of the equivalent three-quarterwave phase shifter and the shunt capacitor as a part of the input *L*-type low-pass

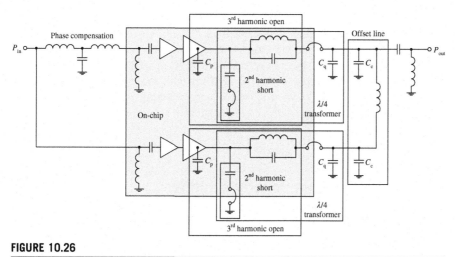

**FIGURE 10.26**

Schematic of a Doherty amplifier for handset applications.

matching circuit can be combined into a single shunt inductance. For such a lumped two-stage Doherty amplifier, a *PAE* of about 52% is achieved at maximum output power, while the power usage efficiency according to a probability density function (PDF) for the CDMA signal is 14.1%, which is more than three times higher compared to the conventional Class-AB power amplifier with a *PAE* of 4.4% [75].

As the Doherty amplifier for handset applications should be compact, a direct input-dividing circuit considering the impedance variations of the carrier and peaking amplifiers can be used instead of the Wilkinson power combiner. In this case, since the input impedance of the carrier amplifier remains almost constant, while that of the peaking amplifier changes significantly because of the Class-C bias, this effect can be utilized for the uneven input dividing [48,76]. As a result, more power is delivered to the carrier amplifier at the low-power region, and the power gain at the low-power region becomes much higher than that at the high-power region, deteriorating the gain flatness and linearity of the Doherty amplifier. Figure 10.26 shows the full circuit schematic of the Doherty amplifier with two-stage carrier and peaking amplifiers [77]. Here, the output matching circuit takes a role of a quarterwave transformer including parasitics, with the phase compensation network employed at the input of the carrier path, and the offset line employed at the output of the peaking path. The second and third harmonics are properly controlled to enhance the efficiency of both carrier and peaking amplifiers. Moreover, the second- and third-harmonic control circuits are also utilized for the quarterwave transformer by connecting the capacitor $C_q$, forming a $\pi$-network where the device output capacitance $C_p$ and second-harmonic control circuit are considered as one capacitor and the parallel resonant *LC*-circuit is inductive at the fundamental. The capacitors $C_c$ of the offset line are combined with capacitors $C_q$ to reduce the number of components. As a result, the Doherty amplifier MMIC implemented in a 2-μm InGaP/GaAs HBT process presents a *PAE* of 40.2% at the output power of 26 dBm with the error vector magnitude (EVM) of 3% for a 16-QAM m-WiMAX signal having a 9.54-dBc crest factor and 8.75-MHz bandwidth.

## 10.8 Digitally driven Doherty amplifier

In a digitally driven dual-input Doherty amplifier architecture, the input signal of each branch is digitally preprocessed and supplied separately to each branch of the Doherty amplifier to optimize its overall performance. In this case, digital signal processing is applied to reduce the performance degradation due to phase impairment in the Doherty amplifier branches achieved by adaptively aligning the phases of the carrier and peaking paths for all power levels after the peaking amplifier is turned on. Generally, a digital signal processor (DSP) includes a digital predistortion (DPD) system to improve linearity which can be configured to provide a carrier signal component along a carrier amplifier path and a peaking signal component along a peaking amplifier path from a digital input signal [78]. In this case, the carrier and peaking amplifiers can amplify the signal components according to the programmable proportions of the split input signal, and not based on a saturation condition of the carrier amplifier, thus resulting in a higher efficiency. Because the signals are isolated prior to being input to the Doherty amplifier, the Doherty amplifier need not include an asymmetric splitting with input phase-matching delay and input impedance-matching circuitry is simplified. The DPD system also performs phase and gain adjustments to each of the signal components. To further improve efficiency performance of a two-stage Doherty amplifier, the separated amplitude and phase modulated signals, produced by the DSP, drive through the corresponding quadrature upconverters both the carrier and peaking amplifiers, each operated in a Class-E mode [79].

Figure 10.27(*a*) shows the block diagram of a dual-input digitally driven Doherty amplifier with digital signal processing and a dual channel upconverter [80]. In this case, direct access and software control of the individual inputs can bring an improvement in efficiency of a Doherty amplifier between the two efficiency peaking points at peak and 6-dB backoff powers, as shown in Fig. 10.27(*b*). The Doherty amplifier design is performed by deriving the offset line with electrical length $\theta_p$ to be inserted at the output of the peaking branch to ensure a quasi-open circuit condition and prevent leakage from the carrier amplifier to the output of the peaking amplifier at the low-power region. The offset line with electrical length $\theta_c$ was optimized to maximize efficiency around the turn-on point of the Doherty amplifier. Because of the different bias conditions for the carrier

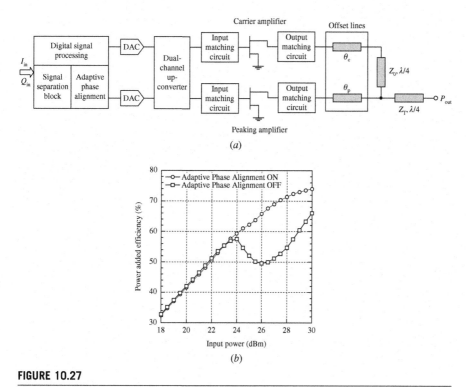

**FIGURE 10.27**

Block diagram and simulated performance of a dual-input digital Doherty amplifier.

amplifier (Class AB) and the peaking amplifier (Class C), the degradation in output power due to phase imbalance condition can be as high as 40% after the peaking amplifier is fully turned on, which directly translates into a significant deterioration in drain efficiency of the Doherty amplifier. However, the dual-input digitally driven Doherty architecture, allowing for the adoption and implementation of a power adaptive phase-alignment mechanism, can minimize the adverse effects of phase imbalance between the carrier and peaking branches. The power-dependent phase offset is adjusted using a power-indexed lookup table (LUT) to correct for the phase disparity at all power levels, where both the carrier and peaking amplifiers contribute to the total output power of the Doherty amplifier. As a result, the phase difference between the carrier and peaking branches is reduced to 0° over the input power range spanning from the turn on of the peaking amplifier until the saturation of the Doherty amplifier. The phase-aligned Doherty amplifier based on two 10-W GaN HEMT transistors demonstrates a *PAE* higher than 50% over an 8-dB output-power backoff range and a *PAE* of 57% at an average output power of 37.8 dBm for a single-carrier WiMAX signal with a *PAR* of 7 dB, which corresponds to an improvement of 7% in *PAE* and

1 dB in average output power with similar linearity performance corresponding to *ACPR* of −22 dBc compared to the fully analog Doherty amplifier [80].

Efficiency enhancement in a digital Doherty amplifier over wide power range can also be achieved by using a digitally controlled dynamic input power distribution scheme to minimize the drive power waste into the peaking branch at backoff power levels [81]. In this case, the carrier amplifier should get significantly more input power in comparison to the peaking amplifier at low-power drive, while the carrier amplifier should get slightly less input power in comparison to the peaking amplifier after turn-on point. As a result, the efficiency can be improved by 7% compared to the conventional fully analog symmetrical Doherty amplifier based on two 10-W GaN HEMT devices and operating at 2.14 GHz for a single-carrier WiMAX signal with a 9-dB *PAR* and 10-MHz bandwidth.

## 10.9 Multiband and broadband capability

A multiband capability of the conventional two-stage Doherty amplifier, whose block schematic is shown in Fig. 10.28(*a*), can be achieved when all of its components are designed to provide their corresponding characteristics over the required bandwidth of operation. In this case, the carrier and peaking amplifiers should provide broadband performance when, for example, their input and interstage matching circuits are designed as broadband and the load network can generally represent a low-pass structure with two or three sections tuned to the required frequencies. In a broadband Class-E mode, the load network can be composed of the consecutive series and parallel resonant circuits using lumped or transmission-line elements according to reactance compensation technique. For a multiband operation with the center frequency ratio at each of the frequency bands of 2 or greater, the input divider can be configured by a multisection Wilkinson power divider or coupled-line directional coupler. In a dual-band operation mode, an input power splitter can represent a $\pi$-shape or *T*-shape stub tapped branch-line coupler composed of four dual-band quarter-wavelength transmission lines, and an impedance inverter network introducing a 90° phase shift can be based on a $\pi$-type or *T*-type transmission-line impedance-inverting section with proper selected transmission-line characteristic impedances and electrical lengths, where shunt elements are provided by the open-circuit stubs [82,83].

The delay transmission line at the input of the peaking amplifier can be constructed in a similar way as the multiband impedance transformer at the output of the carrier amplifier by allowing the phase of the signal transmitted through the carrier amplifier path to match the phase of the signal in the peaking amplifier path. However, it should be noted that it is not easy to design a single multiband impedance transformer which should provide adequately two separate matching options simultaneously: firstly, to operate in a 50-$\Omega$ environment without affecting power-amplifier performance in a high-power region; and secondly, to

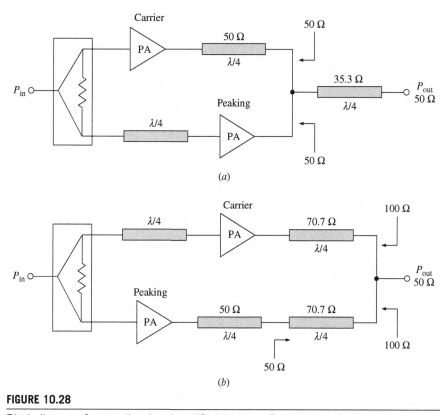

**FIGURE 10.28**

Block diagram of conventional and modified two-stage Doherty amplifiers.

provide an impedance matching from 25 to 100 $\Omega$ in a low-power region. In this case, the possible simple solution is to switch between two quarterwave impedance transformers in a dual-band operation when each of transformers is tuned to the corresponding center bandwidth frequency. However, it may not be so simple in practical implementation because of a load-network complexity and additional power losses. The multiband output combiner required to combine the output powers from the carrier and peaking amplifiers and match the resulting 25-$\Omega$ impedance to the standard load impedance of 50 $\Omega$ can be realized using the two quarterwave transmission lines where the characteristic impedance of the first transmission line can be equal to $\sqrt{25\ \Omega \times 35\ \Omega} = 29.58\ \Omega$ and the characteristic impedance of the second transmission line can be equal to $\sqrt{35\ \Omega \times 50\ \Omega} = 41.83\ \Omega$ for an intermediate impedance of 35 $\Omega$. Generally, a simple two-stepped transmission-line impedance transformer can provide a two-pole response with a different characteristic impedance ratio and different lengths of electrical lines of

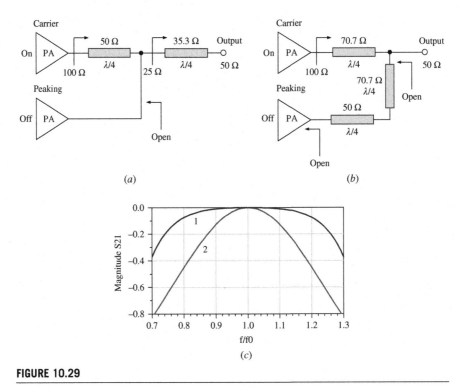

**FIGURE 10.29**

Load-network schematics and broadband properties.

the transmission-line sections [84]. In a monolithic integrated CMOS process, the multiband impedance transformer and combiner can be designed with lumped elements in the form of *LC* ladder low-pass networks [85].

The classical two-stage Doherty amplifier has limited bandwidth capability in a low-power region since it is necessary to provide an impedance transformation from 25 to 100 $\Omega$ when the peaking amplifier is turned off, as shown in Fig. 10.29(a), thus resulting in a loaded quality factor $Q_L = \sqrt{100\,\Omega/25\,\Omega - 1} = 1.73$ at 3-dB output-power reduction level which is sufficiently high for broadband operation. The parallel architecture of a two-stage Doherty amplifier with modified modulated load network, whose block schematic is shown in Fig. 10.28(b), can improve bandwidth properties in a low-power region by twice reducing the impedance transformation ratio [86]. In this case, the load network for the carrier amplifier consists of a single quarterwave transmission line required for impedance transformation, the load network for the peaking amplifier consists of a 50-$\Omega$ quarterwave transmission line followed by another quarterwave transmission line required for impedance transformation, and the quarterwave transmission line at the input of the carrier amplifier is necessary for phase compensation. Both impedance-transforming quarterwave transmission lines, having a characteristic impedance of 70.7 $\Omega$ each, provide a parallel connection of the carrier and peaking amplifiers in a high-power region by parallel combining of the two 100-$\Omega$ impedances at their output into a 50-$\Omega$ load, with 50-$\Omega$ impedances at their inputs seen by each amplifier output. In a low-power region below −6 dB output-power backoff point, when the peaking amplifier is turned off, the required impedance of 100 $\Omega$ seen by the carrier-amplifier output is achieved by using a single quarterwave transmission line with a characteristic impedance of 70.7 $\Omega$ to match with a 50-$\Omega$ load, as shown in Fig. 10.29(*b*). This provides a loaded quality factor $Q_L = \sqrt{100\,\Omega/50\,\Omega - 1} = 1$, resulting in a 1.73 times wider frequency bandwidth, as shown in Fig. 10.29(*c*) by curve 1 compared with a conventional case (curve 2). Since the load network of the peaking amplifier contains two quarterwave transmission lines connected in series, this provides an overall half-wavelength transmission line, and an open circuit at the peaking-amplifier output directly transforms to the load providing a significant isolation of the peaking-amplifier path from the carrier-amplifier path in a wide frequency range. The input in-phase divider and phase-compensating transmission line can be replaced by a coupled-line 90° hybrid coupler.

Since a frequency-dependent quarterwave transformer generally cannot provide a broadband operation of a conventional Doherty amplifier, a quarterwave transmission line as an additional matching element can be added at the output of the peaking amplifier in series with the offset line to minimize the loaded quality factor for broader operation [87]. In a handset monolithic application, the transmission-line quarterwave impedance transformer and offset line are implemented with lumped elements, representing the equivalent $\pi$-type low-pass and $\pi$-type high-pass *LC* networks, respectively, as shown in Fig. 10.30(*a*). In this case, the network parameters are optimized to provide an open-circuit condition at the output of the peaking branch over broadband frequency range when the peaking amplifier is turned off. Figure 10.30(*b*) shows the circuit schematic of a 2-μm GaAs HBT Doherty amplifier where all of the components are fully integrated on a chip. In this case, the inductors are implemented using bondwires and slab inductors, the input dividing circuits are broadband based on low-*Q* matching networks, and the second- and third-harmonic impedances are controlled for high efficiency across the wide bandwidth. The open-circuit conditions are achieved by optimizing all load-network elements including drain bondwires and device output capacitances. For a mobile 8.75-MHz 16-QAM m-WiMAX application with a 9.6-dB crest factor, such a lumped Doherty amplifier exhibits a *PAE* of over 27% and an output power of over 23.6 dBm across 2.2−2.8 GHz using a DPD technique. A similar Doherty amplifier with broadband lumped networks can provide a *PAE* over 30% and an output power of over 28 dBm across 1.6−2.1 GHz for a 10-MHz LTE signal with a *PAR* of 7.5 dB [88].

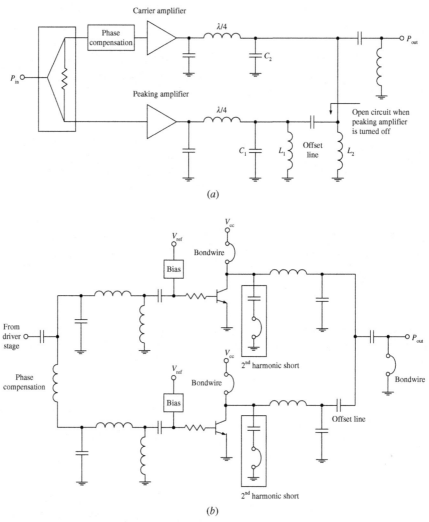

**FIGURE 10.30**

Schematics of broadband Doherty amplifiers for handset applications.

## REFERENCES

1. Doherty WH. Amplifier. U.S. Patent 2,210,028, August 1940 (filed April 1936).
2. Doherty WH. A new high efficiency power amplifier for modulated waves. *Proc. IRE*. September 1936;24:1163−1182.
3. Doherty WH, Towner OW. A 50-kilowatt broadcast station utilizing the Doherty amplifier and designed for expansion to 500 kilowatts. *Proc. IRE*. September 1939; 27:531−534.

4. Doherty WH. *Proc. IRE*. August 1937;25:922.

5. Smith CE, Hall JR, Weldon JO. Very high power long-wave broadcast station. *Proc. IRE*. August 1954;42:1222–1235.

6. Sainton JB. A 500 kilowatt medium frequency standard broadcast transmitter. *Cathode Press* (Machlett Company). 1965;22(4):22–29.

7. Rozov VM, Kuzmin VF. Use of the Doherty circuit in SSB transmitters. *Telecommunications and Radio Eng*. 1970–1971.

8. Vinogradov PY, Vorobyev NI, Sokolov EP, Fuzik NS. Amplification of a modulated signal by the Doherty method in a transistorized power amplifier. *Telecommunications and Radio Eng*. October 1977;31(part 1):38–41.

9. Development of circuitry for multikilowatt transmitter for space communications satellites. In: *Report No. CRI19803*, (NASA N71-29212). General Electric Company, Space Systems Division; February 1971.

10. Mina A, Parry F. Broadcasting with megawatts of power: the modern era of efficient powerful transmitters. *IRE Trans. Broadcast*. June 1989;BC-35:121–130.

11. Clark G. A comparison of current broadcast amplitude-modulation techniques. *IEEE Trans. Broadcast*. June 1975;BC-21:25–31.

12. Raab FH. Efficiency of Doherty RF power-amplifier systems. *IEEE Trans. Broadcast*. September 1987;BC-33:77–83.

13. McMorrow RJ, Upton DM, Maloney PR. The microwave Doherty amplifier. *1994 IEEE MTT-S Int. Microwave Symp. Dig*. 1653–1656.

14. Campbell CF. A full integrated Ku-Band Doherty amplifier MMIC. *IEEE Microwave Guided Wave Lett*. March 1999;9:114–116.

15. Kobayashi KW, Oki AK, Gutierrez-Aitken A, et al. An 18–21 GHz InP DHBT linear microwave Doherty amplifier. *2000 IEEE RFIC Symp. Dig*. 179–182.

16. Tsai J-H, Huang T-W. A 38–46 GHz MMIC Doherty amplifier using post-distortion linearization. *IEEE Microwave Wireless Compon. Lett*. May 2007;17:388–390.

17. Steinberser C, Landon T, Suckling C, et al. 250 W HVHBT Doherty with 57% WCDMA efficiency linearized to −55 dBc for 2c11 6.5 dB PAR. *IEEE J. Solid-State Circuits*. October 2008;SC-43:2218–2228.

18. Deguchi H, Ui N, Ebihara K, Inoue K, Yoshimura N, Takahashi H. A 33 W GaN HEMT Doherty amplifier with 55% drain efficiency for 2.6 GHz base stations. *2009 IEEE MTT-S Int. Microwave Symp. Dig*. 1273–1276.

19. Yang Y, Yi J, Woo YY, Kim B. Optimum design for linearity and efficiency of a microwave Doherty amplifier using a new load matching technique. *Microwave J*. December 2001;44:20–36.

20. Wong GK, Shah TR, Titizer K. Doherty power amplifier with phase compensation. U.S. Patent 7,295,074, November 2007 (filed March 2005).

21. Nick M, Mortazawi A. Adaptive input-power distribution in Doherty power amplifiers for linearity and efficiency enhancement. *IEEE Trans. Microwave Theory Tech*. November 2010;MTT-58:2764–2771.

22. Yang Y, Cha J, Shin B, Kim B. A microwave Doherty amplifier employing envelope tracking technique for high efficiency and linearity. *IEEE Microwave Wireless Compon. Lett*. September 2003;13:370–372.

23. Zhao Y, Iwamoto M, Larson LE, Asbeck PM. Doherty amplifier with DSP control to improve performance in CDMA operation. *2003 IEEE MTT-S Int. Microwave Symp. Dig*. 2:687–690.

24. Bae S, Kim J, Nam I, Kwon Y. Bias-Switching Quasi-Doherty-Type amplifier for CDMA handset applications. *2003 IEEE RFIC Symp. Dig.* 137−140.

25. Kato T, Yamaguchi K, Kuriyama Y, Yoshida H. An HPSK/OFDM 64-QAM Dual-Mode Doherty power amplifier module for mobile terminals. *IEICE Trans. Electron.* September 2007;E90-C:1678−1684.

26. Kawai S, Takayama Y, Ishikawa R, Honjo K. A GaN HEMT Doherty amplifier with a series connected load. *Proc. 2009 Asia-Pacific Microwave Conf.* 325−328.

27. Bowers DF. HEAD − a high efficiency amplitude-modulation system for broadcasting transmitters. *Communication and Broadcasting*. February 1982;7:15−23.

28. Kim J, Moon J, Woo YY, et al. Analysis of a fully matched saturated Doherty amplifier with excellent efficiency. *IEEE Trans. Microwave Theory Tech*. February 2008; MTT-56:328−338.

29. Kim J, Kim B, Woo YY. Advanced design of linear Doherty amplifier for high efficiency using saturation amplifier. *2007 IEEE MTT-S Int. Microwave Symp. Dig.* 1573−1576.

30. Colantonio P, Giannini F, Giofre R, Piazzon L. Theory and experimental results of a class F AB-C Doherty power amplifier. *IEEE Trans. Microwave Theory Tech*. August 2009;MTT-57:1936−1947.

31. Eccleston KW, Smith KJI, Gough PT, Mann SI. Harmonic load modulation in Doherty amplifiers. *Electronics Lett*. January 2008;44:128−129.

32. Eccleston KW, Smith KJI, Gough PT. A Compact Class-F/Class-C Doherty amplifier. *Microwave and Optical Technology Lett*. July 2011;53:1606−1610.

33. Grebennikov A. High-Efficiency Transmission-Line GaN HEMT Inverse Class F power amplifier for active antenna arrays. *Proc. 2009 Asia-Pacific Microwave Conf*. 317−320.

34. Goto S, Kunii T, Inoue A, Izawa K, Ishikawa T, Matsuda Y. Efficiency enhancement of Doherty amplifier with combination of Class-F and Inverse Class-F schemes for S-Band base station application. *2004 IEEE MTT-S Int. Microwave Symp. Dig.* 2:839−842.

35. Choi GW, Kim HJ, Hwang WJ, Shin SW, Choi JJ, Ha SJ. High efficiency Class-E Tuned Doherty amplifier using GaN HEMT. *2009 IEEE MTT-S Int. Microwave Symp. Dig.* 925−928.

36. Lee Y-S, Lee M-W, Jeong Y-H. Highly efficient Doherty amplifier based on Class-E topology for WCDMA applications. *IEEE Microwave Wireless Compon. Lett*. September 2008;18:608−610.

37. Takahashi E, Ishikawa T, Kashimura K, Adachi N. High-efficiency four-stage Class-E Doherty amplifier for W-CDMA base stations. *Proc. 38th Europ. Microwave Conf*. 2008:234−237.

38. Shiikuma K. Circuit for parallel operation of Doherty amplifiers. U.S. Patent 7,262,656, August 2007 (filed August 2005).

39. Cho K-J, Kim W-J, Kim J-H, Stapleton SP. Linearity optimization of a high power Doherty amplifier based on post-distortion compensation. *IEEE Microwave Wireless Compon. Lett*. November 2005;15:748−750.

40. Markos AZ, Bathich K, Al Tanany A, Gruner D, Boeck G. Design of a 120 W balanced GaN Doherty power amplifier. *Proc. 6th German Microwave Conf*. 2011:1−4.

41. Mobbs CI. Doherty amplifier. U.S. Patent 7,301,395, November 2007 (filed March 2005).

42. Eccleston KW. Analysis of a multi-transistor interleaved Doherty amplifier. *Proc. 2009 Asia-Pacific Microwave Conf*. 1581−1584.

43. Eccleston KW. Four-Transistor interleaved Doherty amplifier. *Electronics Lett.* July 2009;45:792−794.

44. Lee Y-S, Lee M-W, Kam S-H, Jeong Y-H. A new wideband distributed Doherty amplifier for WCDMA repeater applications. *IEEE Microwave Wireless Compon. Lett.* October 2009;19:668−670.

45. Iwamoto M, Williams A, Chen P-F, Metzger AG, Larsson LE, Asbeck PM. An extended Doherty amplifier with high efficiency over a wide power range. *IEEE Trans. Microwave Theory Tech.* December 2001;MTT-49:2472−2479.

46. Yang Y, Cha J, Shin B, Kim B. A fully matched N-Way Doherty amplifier with optimized linearity. *IEEE Trans. Microwave Theory Tech.* March 2003;MTT-51:986−993.

47. Takayama Y, Harada T, Fujita T, Maenaka K. Design method of microwave Doherty power amplifiers and its application to Si Power MOSFET amplifiers. *Electronics and Communications in Japan.* April 2005;88(part 2):9−17.

48. Kim J, Cha J, Kim I, Kim B. Optimum operation of asymmetrical-cells-based linear Doherty power amplifiers − uneven power drive and power matching. *IEEE Trans. Microwave Theory Tech.* May 2005;MTT-53:1802−1809.

49. Kim I, Cha J, Hong S, et al. Highly linear three-way Doherty amplifier with uneven power drive for repeater system. *IEEE Microwave Wireless Compon. Lett.* April 2006;16:176−178.

50. Moon J, Kim Ja, Kim I, Kim Ju, Kim B. Highly efficient three-way saturated Doherty amplifier with digital feedback predistortion. *IEEE Microwave Wireless Compon. Lett.* August 2008;18:539−541.

51. Cho KJ, Kim WJ, Stapleton SP, et al. Design of N-way distributed Doherty amplifier for WCDMA and OFDM applications. *Electronics Lett.* May 2007;43:577−578.

52. Kim WJ, Cho KJ, Stapleton SP, Kim JH. N-way Doherty distributed power amplifier. U.S. Patent 7,688,135, March 2010 (filed April 2008).

53. Srirattana N, Raghavan A, Heo D, Allen PE, Laskar J. Analysis and design of a high-efficiency multistage Doherty power amplifier for wireless communications. *IEEE Trans. Microwave Theory Tech.* March 2005;MTT-53:852−860.

54. Neo WCE, Qureshi J, Pelk MJ, Gajadharsing JR, de Vreede LCN. A mixed-signal approach towards linear and efficient N-way Doherty amplifiers. *IEEE Trans. Microwave Theory Tech.* May 2007;MTT-55:866−879.

55. Pelk MJ, Neo WCE, Gajadharsing JR, Pengelly RS, de Vreede LCN. A high-efficiency 100-W GaN three-way Doherty amplifier for base-station applications. *IEEE Trans. Microwave Theory Tech.* July 2008;MTT-56:1582−1591.

56. Gajadharsing J, Neo WCE, Pelk M, de Vreede LCN, Zhao J. 3-way Doherty amplifier with minimum output network. U.S. Patent 8,022,760, September 2011 (filed December 2008).

57. Kim B, Kim I, Moon J. Advanced Doherty architecture. *IEEE Microwave Mag.* August 2010;11:72−86.

58. Kim I, Moon J, Jee S, Kim B. Optimized design of a highly efficient three-stage Doherty PA using gate adaptation. *IEEE Trans. Microwave Theory Tech.* October 2010;MTT-58:2562−2574.

59. Lee Y-S, Lee M-W, Kam S-H, Jeong Y-H. Advanced design of a double Doherty power amplifier with a flat efficiency range. *2010 IEEE MTT-S Int. Microwave Symp. Dig.* 1500−1503.

60. Grebennikov A. A high-efficiency 100-W Four-Stage Doherty GaN HEMT power amplifier module for WCDMA systems. *2011 IEEE MTT-S Int. Microwave Symp. Dig.* 1−4.

61. Stengel RF, Gu W-CA, Leizerovich GD, Cygan LF. High efficiency power amplifier having reduced output matching networks for use in portable devices. U.S. Patent 6,262,629, July 2001 (filed July 1999).

62. Ahn G, Kim M, Park H, et al. Design of a high-efficiency and high-power inverted Doherty amplifier. *IEEE Trans. Microwave Theory Tech.* June 2007;MTT-55:1105−1111.

63. Kwon S, Kim M, Jung S, et al. Inverted-load network for high-power Doherty amplifier. *IEEE Microwave Mag.* February 2009;10:93−98.

64. Sirois J, Boumaiza S, Helaoui M, Brassard G, Ghannouchi FM. A robust modeling and design approach for dynamically loaded and digitally linearized Doherty amplifiers. *IEEE Trans. Microwave Theory Tech.* September 2005; MTT-53:2875−2883.

65. Takenaka I, Ishikura K, Takahashi H, et al. A distortion-cancelled Doherty high-power amplifier using 28-V GaAs heterojunction FETs for W-CDMA base stations. *IEEE Trans. Microwave Theory Tech.* December 2006;MTT-54:4513−4521.

66. Lee M-W, Kam S-H, Lee Y-S, Jeong Y-H. Design of highly efficient three-stage inverted Doherty power amplifier. *IEEE Microwave Wireless Compon. Lett.* July 2011;21:383−385.

67. Campbell CF. A fully integrated *Ku*-Band Doherty amplifier MMIC. *IEEE Microwave Guided Wave Lett.* March 1999;9:114−116.

68. McCarroll CP, Alley GD, Yates S, Matreci R. A 20 GHz Doherty power amplifier MMIC with high efficiency and low distortion designed for broad band digital communication systems. *2000 IEEE MTT-S Int. Microwave Symp. Dig.* 6:537−540.

69. Yu D, Kim Y, Han K, Shin J, Kim B. A 60-GHz fully integrated Doherty power amplifier based on 0.13-μm CMOS process. *2008 IEEE RFIC Symp. Dig.* 69−72.

70. Yu D, Kim Y, Han K, Shin J, Kim B. Fully integrated Doherty power amplifiers for 5 GHz Wireless-LANs. *2006 IEEE RFIC Symp. Dig.* 177−180.

71. Elmala M, Paramesh J, Soumyanath K. A 90-nm CMOS Doherty power amplifier with minimum AM−PM distortion. *IEEE J. Solid-State Circuits.* June 2006; SC-41:1323−1332.

72. Tongchoi C, Chongcheawchamnan M, Worapishet A. Lumped element based Doherty power amplifier topology in CMOS process. *2003 IEEE Int. Circuits and Systems Symp. Dig.* 1:I-445−I-448.

73. Blednov II. High power Doherty amplifier. U.S. Patent 7,078,976, July 2006 (filed October 2005).

74. Blednov II, van der Zanden J. High power LDMOS integrated Doherty amplifier for W-CDMA. *2006 IEEE RFIC Symp. Dig.* 1−4.

75. Zhao Y, Iwamoto M, Kimball D, Larsson LE, Asbeck PM. A 900 MHz Doherty amplifier implemented with lumped elements. *IEEE Topical Workshop on Power Amplifiers for Wireless Commun.* September 2003.

76. Nick M, Mortazawi A. A Doherty power amplifier with extended resonance power divider for linearity improvement. *2008 IEEE MTT-S Int. Microwave Symp. Dig.* 423−426.

77. Kang D, Choi J, Kim D, Kim B. Design of Doherty power amplifiers for handset applications. *IEEE Trans. Microwave Theory Tech.* August 2010;MTT-58:2134−2142.

78. Sperlich R, Copeland GC, Hoppenstein R. Hybrid Doherty amplifier system and method. U.S. Patent Appl. 2008/0111622, May 2008 (filed November 2007).

79. Pehlke DR. Class E Doherty amplifier topology for high efficiency signal transmitters. U.S. Patent 6,396,341, May 2002 (filed December 2000).

80. Darraji R, Ghannouchi FM, Hammi H. A dual-input digitally driven Doherty amplifier for performance enhancement of Doherty transmitters. *IEEE Trans. Microwave Theory Tech.* May 2011;MTT-59:1284–1293.

81. Darraji R, Ghannouchi FM. Digital Doherty amplifier with enhanced efficiency and extended range. *IEEE Trans. Microwave Theory Tech.* November 2011;MTT-59:2898–2909.

82. Zhang H, Chen KJ. A stub tapped branch-line coupler for dual-band operations. *IEEE Microwave Wireless Compon. Lett.* February 2007;17:106–108.

83. Chen W, Bassam SA, Li X, et al. Design and linearization of concurrent dual-band Doherty power amplifier with frequency-dependent power ranges. *IEEE Trans. Microwave Theory Tech.* October 2011;MTT-59:2537–2546.

84. Monzon C. A small dual-frequency transformer in two sections. *IEEE Trans. Microwave Theory Tech.* April 2003;MTT-51:1157–1161.

85. Hamedi-Hagh S, Salama CAT. Wideband CMOS integrated RF combiner for LINC transmitters. *2003 IEEE MTT-S Int. Microwave Symp. Dig.* 1: pp. A41–A44.

86. Grebennikov A, Wong J. Parallel Doherty amplifier. Patent pending (filed February 2012).

87. Kang D, Kim D, Kim B. Broadband HBT Doherty power amplifiers for handset applications. *IEEE Trans. Microwave Theory Tech.* December 2010;MTT-58:4031–4039.

88. Kang D, Kim D, Cho Y, Park B, Kim J, Kim B. Design of bandwidth-enhanced Doherty power amplifiers for handset applications. *IEEE Trans. Microwave Theory Tech.* December 2011;MTT-59:3474–3483.

# Predistortion Linearization Techniques

## INTRODUCTION

Wireless devices have become an ubiquitous part of daily life for billions of people. The projections of sales for these devices, particularly those tailored for mobile applications, do not seem to show any signs of slowing down. Since the radio spectrum is a finite natural resource, one method to accommodate an ever-increasing number of users is to utilize highly efficient modulation schemes that can convey as much information as possible within the narrowest bandwidth. Like most things in nature, there is a trade-off with this approach. Highly spectrum-efficient modulation schemes utilize multi-level combinations of amplitude and phase modulation, which results in a non-constant envelope. A very low level of distortion will be required for a receiver to be able to recover the information conveyed during the modulation process at the transmitter. Most modern communication systems make use of error-correcting codes to maximize the throughput of information in spite of channel distortions. However, the RF power amplifier remains the major contributor of distortion in the communication channel.

Several techniques can be used to minimize the distortion of an RF power amplifier. Among the simplest is to operate the amplifier within its linear region, also known as "back-off" operation [1]. Unfortunately, the efficiency of the amplifier is degraded by this approach. A lower efficiency translates into more energy consumption and higher heat dissipation. Systematic methods to reduce power amplifier distortion, or linearization, which allow for an increase in power efficiency, have been extensively developed. Among the most common linearization techniques, predistortion is the most popular due to its ease of use and possibility to be implemented utilizing analog or digital techniques.

This chapter treats the power amplifier from a system perspective, introducing a simple black-box model. It continues describing the fundamentals of predistortion linearization, including the sensitivity of the method to various parameters. Finally, it presents practical predistorters implemented using analog and digital techniques.

## 11.1 Modeling of RF power amplifiers with memory

Distortion in radio frequency (RF) power amplifiers is a major problem in modern communication systems. Complex modulation schemes with non-constant envelopes require highly linear processing. Power amplifiers play an important role because they are a main contributor to the overall distortion of a communication system. If the dynamic range of an amplifier is fully utilized, high-amplitude signals will be subject to significant nonlinear distortion.

Fig. 11.1 shows a plot of the measured gain of a typical solid state power amplifier as a function of its input power when driven by a constant-frequency single tone. Gain compression and phase shift are observed at large signal levels. These effects are also know as AM/AM and AM/PM, respectively, since they denote a variation in amplitude (AM) or phase (PM) at the output with respect to a variation in amplitude at the input.

When gain compression occurs, the input/output transfer characteristic of a power amplifier may look like Fig. 11.2. The output signal $y$ increases linearly with respect to the input signal $x$ for small drive levels. Consequently, this is the linear region of the amplifier transfer characteristic, and its gain is constant. However, when the input signal is further increased, the transfer characteristic

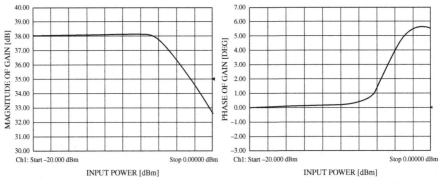

**FIGURE 11.1**

Measured magnitude and phase of the gain of a typical solid state RF power amplifier as a function of its input power.

becomes nonlinear and a decrease in the rate at which the output power increases with respect to the input power is observed. This phenomenon is equivalent to a decrease in gain. All practical power amplifiers will reach a saturation point where there will be no further increase in output power in spite of an increase in input power.

The memoryless transfer characteristic $y(t) = T[x(t)]$ shown in Fig. 11.2 can be approximated with a polynomial of degree $n$. Complex coefficients are required since the power amplifier can exhibit a phase shift as a function of input level,

$$y(t) \approx \sum_{i=0}^{n} (a_i + jb_i)\, x^i(t). \tag{11.1}$$

Since the saturation of a real power amplifier is unavoidable due to physical limitations, the best possible transfer characteristic for a power amplifier in terms of linearity is shown in Fig. 11.3. This type of characteristic is often called *ideal limiter*. The gain of the amplifier is constant until saturation is reached; above this point the gain decreases at the same rate the input power increases. The phase shift must be zero for all input power levels below saturation.

The transfer characteristics shown so far occupy two quadrants and represent input/output voltage relations. This allows for the representation of even-order distortion, which can only occur when there is asymmetry in the transfer characteristic, i.e. the first and third quadrants of the transfer characteristic are different from each other. However, quite often the transfer characteristic of a power amplifier is shown for the first quadrant only. This allows for the use of the input/output power relation instead of the voltage, but it does not permit the representation of any even-order distortion. The transfer characteristic of a power amplifier can be approximated by Eq. (11.1) over a very narrow frequency range. In order to extend the model over a wider spectrum, memory could be incorporated by

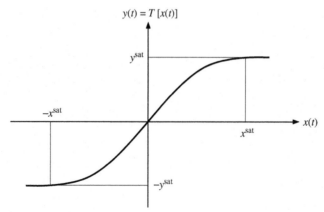

**FIGURE 11.2**

Transfer characteristic of a memoryless RF power amplifier.

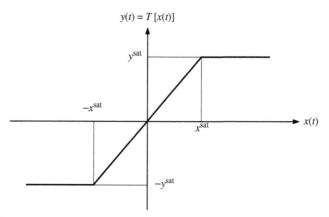

**FIGURE 11.3**

Transfer characteristic of a memoryless RF power amplifier for optimum linearity up to the saturation point.

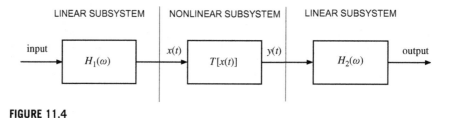

**FIGURE 11.4**

Power amplifier model.

using, for example, a Volterra series based model [2]. However, alternative simpler models can be successfully used over moderate bandwidths, depending on the particular characteristics of the amplifier.

Frequency dependence can be added to a memoryless nonlinearity by inserting a linear filter in front of it [3]. This technique is based on the work by Wiener. The insertion loss of the filter at each frequency will determine the signal level that reaches the input of the memoryless nonlinearity, which will result in a different nonlinear behavior as a function of frequency. An alternative approach, developed by Hammerstein [4,5], involved the insertion of a linear filter after the nonlinearity. In both cases, the use of only one filter adds memory effects to the model. It is possible to utilize two filters for enhanced accuracy, one before and one after the nonlinearity. This is known as the Weiner-Hammerstein model [6,7]. Figure 11.4 shows such a system, in which the memoryless nonlinearity $T[x(t)]$ of Fig. 11.2 is preceded by the linear system $H_1$ and followed by the linear system $H_2$.

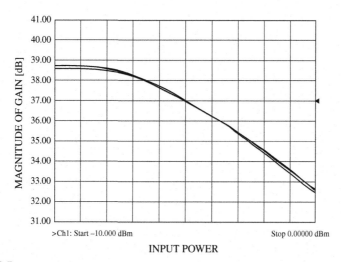

**FIGURE 11.5**

Measured magnitude of the gain as a function of input power for a solid state power amplifier at 1.0 GHz, 1.1 GHz and 1.2 GHz.

The small signal transfer function $H_{ss}(\omega)$ of the system of Fig. 11.4 can be written as

$$H_{ss}(\omega) = H_1(\omega)T_{ss}H_2(\omega) \tag{11.2}$$

where $T_{ss}$ is the small signal gain of the nonlinear subsystem defined as

$$T_{ss} = \frac{|y(t)|}{|x(t)|} \bigg|_{|y(t)| \ll y^{sat}} \tag{11.3}$$

From Eq. (11.2), the transfer function $H_1(\omega)$ of the first filter can then be written as

$$H_1(\omega) = \frac{H_{ss}(\omega)}{T_{ss}H_2(\omega)} \tag{11.4}$$

The transfer function of the second filter $H_2(\omega)$ is given by

$$H_2(\omega) = H_{sat}(\omega)\, T_{sat} \tag{11.5}$$

where $H_{sat}(\omega)$ is the transfer function of the amplifier at the saturation point and $T_{sat}$ is the gain of the nonlinear subsystem at saturation, defined as

$$T_{sat} = \frac{|y(t)|}{|x(t)|} \bigg|_{|y(t)| \geq y^{sat}} \tag{11.6}$$

Both $H_{sat}(\omega)$ and $H_{ss}(\omega)$ are measurable parameters in a power amplifier.

Figure 11.5 shows the measured magnitude of the gain of a power amplifier at 1.0, 1.1 and 1.2 GHz for an input power level ranging from −10 dBm

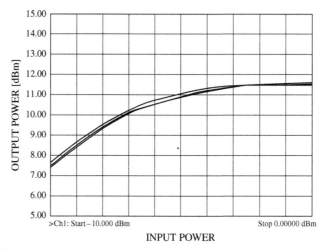

**FIGURE 11.6**

Measured output power as a function of input power for a solid state power amplifier at 1.0 GHz, 1.1 GHz and 1.2 GHz.

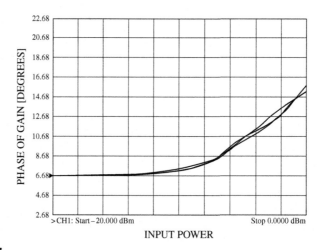

**FIGURE 11.7**

Measured phase of the gain as a function of input power for a solid state power amplifier at 1.0 GHz, 1.1 GHz and 1.2 GHz.

to 0 dBm, respectively, while Fig. 11.6 shows the measured output power versus the input power under the same conditions. The measured phase of the gain at the same three frequencies for an input power of −20 dBm to 0 dBm is shown in Fig. 11.7. The phase measurements were normalized to the same

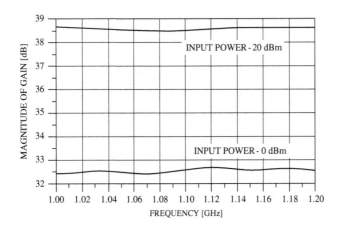

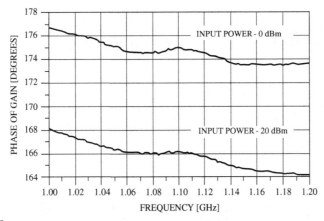

**FIGURE 11.8**

Measured magnitude and phase of the gain of a solid state amplifier as a function of frequency.

phase value at the minimum input power for an easier comparison. The measurements show that the response of the amplifier is very similar across the evaluated frequency range, except for some small displacements in the vertical and horizontal axis.

Figure 11.8 depicts the measured complex gain of the amplifier from 1.0 to 1.2 GHz. This data was used in conjunction with Eqs (11.4) and (11.5) to calculate the frequency response of the linear subsystems $H_1$ and $H_2$, as shown in Fig. 11.9. The results are only presented in magnitude for simplicity.

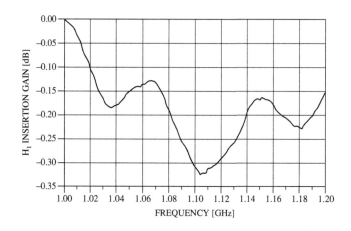

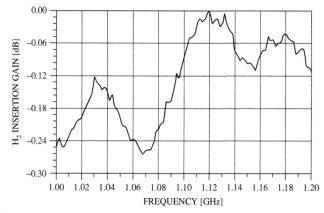

**FIGURE 11.9**

Calculated frequency response (magnitude) of the linear subsystems $H_1$ and $H_2$.

## 11.2 Predistortion linearization

### 11.2.1 Introduction

Linearization is any systematic method of nonlinear distortion reduction. Predistortion linearization is performed by a predistorter, which is a device that always precedes the RF power amplifier. The magnitude of the predistorter gain increases when the magnitude of the power amplifier gain decreases and the phase of the predistorter gain is the negative of the phase of the power amplifier gain. The net result is that the magnitude and phase of the gain of the two devices in cascade becomes approximately a constant until the power amplifier reaches

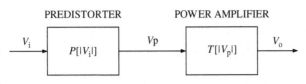

**FIGURE 11.10**

Block diagram of a predistorter-power amplifier system.

saturation. The block diagram of a predistorter-power amplifier cascaded system is shown in Fig. 11.10.

When the frequency of operation of the amplifier covers less than an octave, only the odd-order intermodulation products need to be canceled by the predistorter since even-order intermodulation distortion and all harmonics fall outside the amplifier's passband and can be attenuated by a filter. Consequently, the following analysis of the predistorter will be restricted to the first zone.

If the input signal to the predistorter $V_i$ is given by

$$V_i = r(t)\, e^{j\theta(t)} \tag{11.7}$$

where $r(t)$ is the magnitude and $\theta(t)$ is the phase of $V_i$, the output of the predistorter will be

$$V_p = P_m[r(t)]\, e^{j(\theta(t)+P_p[r(t)])} \tag{11.8}$$

where $P_m[.]$ and $P_p[.]$ are the magnitude and the phase, respectively, of the gain $P$ of the predistorter. Both $P_m[.]$ and $P_p[.]$ are a function of the magnitude of $V_i$, i.e. $r(t)$.

The output of the power amplifier that follows the predistorter is

$$V_o = T_m\{P_m[r(t)]\}\ e^{j\{\theta(t)+P_p[r(t)]+T_p(P_p[r(t)])\}} \tag{11.9}$$

where $T_m[.]$ and $T_p[.]$ are the magnitude and the phase of the gain of the power amplifier, also known as AM/AM and AM/PM, respectively.

The goal of the predistorter is to achieve a constant magnitude of the predistorter-power amplifier system gain $k_m$

$$T_m\{P_m[r(t)]\} = k_m r(t) \tag{11.10}$$

and a constant phase $k_{ph}$

$$P_p[r(t)] + T_p\{P[r(t)]\} = k_{ph.} \tag{11.11}$$

Figure 11.11 shows the magnitude of the gain and the transfer characteristic of an ideal predistorter for the correction of odd order intermodulation distortion of a power amplifier. The gain of the power amplifier and the predistorter were normalized to unity at saturation.

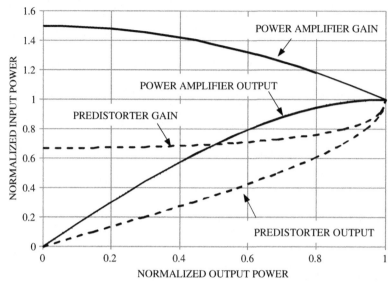

**FIGURE 11.11**

Normalized magnitude of the gain and transfer characteristic of an ideal predistorter for the correction of the odd-order intermodulation distortion of a power amplifier.

### 11.2.2 Memoryless predistorter for octave-bandwidth amplifiers

At first glance, it could be presumed that the predistorter gain could be calculated as the reciprocal of the power amplifier gain. Unfortunately, every time the predistorter increases its gain to compensate for the power amplifier gain compression, the predistorter output signal level also increases. This forces the gain of the power amplifier (at this new larger input signal) to compress even more. This effect can be overcome if the predistorter transfer characteristic is calculated as a function of *the output of the power amplifier instead of the input to the predistorter*, after the gain and output signal of the power amplifier are both normalized to unity at saturation. From Fig. 11.10, it follows that

$$P[|V_o|] = \frac{1}{T[|V_i|]} \tag{11.12}$$

where $P$ is the transfer characteristic of the predistorter and $T$ is the transfer characteristic of the power amplifier, both normalized to unity at saturation. It must be noted that an iterative solution to Eq. (11.12) is required.

For a better comprehension of the meaning of Eq. (11.12), the memoryless transfer characteristics of the power amplifier and the predistorter are shown in

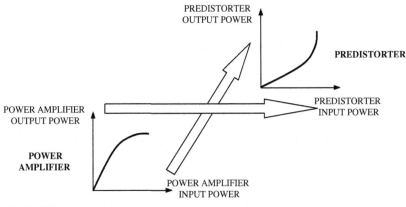

**FIGURE 11.12**

The transfer characteristic of the predistorter can be obtained by swapping the axes of the amplifier transfer characteristic after its normalization to unity at saturation.

Fig. 11.12. Since the power amplifier transfer characteristic had been normalized to unity at saturation, the process to obtain the predistorter transfer characteristic simply involves the swapping of the axes.

The following conclusions can be drawn at this point:

- The slope of the predistorter transfer characteristic at the saturation point is infinite, which could only be achieved by infinite predistorter gain. Therefore, it is not possible to effectively predistort once the amplifier reaches the saturation point.
- The gain of the power amplifier cannot be zero in any part of its domain since it would require infinite gain from the predistorter. This could be the case of a class-C power amplifier in which the conduction angle is less than 180°.
- The transfer characteristic must be a monotonically increasing function to avoid ambiguity in the calculation of the predistorter transfer characteristic.

The last two cases are illustrated in Fig. 11.13.

The predistorter transfer characteristic can also be calculated by inverting the power series that represents the power amplifier transfer characteristic. If only the odd intermodulation distortion is considered, the transfer characteristic of the power amplifier can be written as

$$y(t) = \sum_{n=0}^{\infty} k_{(2n+1)} x^{(2n+1)}(t) \tag{11.13}$$

where $n$ is any positive integer, $y(t)$ and $x(t)$ are the output and input signals of the power amplifier, respectively, and $k_{(2n+1)}$ are complex coefficients.

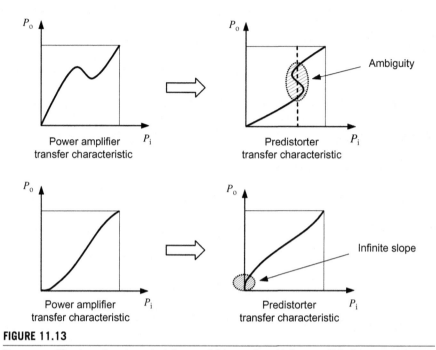

**FIGURE 11.13**

Power amplifier transfer characteristic special cases for which a predistorter cannot be implemented. $P_o$ and $P_i$ are the output and input powers, respectively.

For the analysis, it will be supposed that the power amplifier only has third order intermodulation distortion by truncating Eq. (11.13) to its third power, which results in

$$y(t) = k_1 x(t) + k_3 x^3(t). \tag{11.14}$$

If $y(t)$ and $x(t)$ are normalized to unity at saturation by manipulating the coefficients $k_1$ and $k_3$, then the transfer characteristic of the predistorter can be obtained by inverting Eq. (11.14),

$$x'(t) = \sum_{n=0}^{\infty} K_{(2n+1)} y'^{(2n+1)}(t) \tag{11.15}$$

where $x'$ is now the output of the predistorter (the same as the input to the power amplifier), $y'$ is the input to the predistorter (the same as the output of the power amplifier since its output at saturation is unity) and $n$ is any positive integer. As an example, the coefficients $K$ up to order 7 can be calculated as follows [8]:

$$K_1 = k_1^{-1} \tag{11.16}$$

$$K_3 = -k_1^{-4} k_3 \tag{11.17}$$

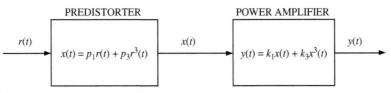

**FIGURE 11.14**

Third-order predistorter with coefficients $p_1$ and $p_3$ followed by a third-order power amplifier with coefficients $k_1$ and $k_3$.

$$K_5 = 3k_1^{-7}k_3^2 \tag{11.18}$$

$$K_7 = -12k_1^{-10}k_3^3. \tag{11.19}$$

It is important to observe in Eq. (11.15) that the predistorter transfer characteristic contains an infinite number of terms even when the power amplifier transfer characteristic was truncated to the third degree. A predistorter transfer characteristic of this type cannot be realized in practice since it would require an infinite bandwidth. As a result, only a partial cancellation of the intermodulation distortion products produced by the power amplifier can be achieved.

Bandwidth restrictions in the predistorter can also be understood by analyzing Fig. 11.14, in which a third-order predistorter is placed in front of a power amplifier also with a third-order transfer characteristic. A two-tone signal $r(t)$ of the form

$$r(t) = \cos \omega_1 t + \cos \omega_2 t \tag{11.20}$$

is applied to the input of the predistorter. The tones of angular frequency $\omega_1$ and $\omega_2$ are of unity amplitude and zero phase in order to simplify the analysis. The output of the predistorter will be

$$x(t) = p_1(\cos \omega_1 t + \cos \omega_2 t) + p_3(\cos \omega_1 t + \cos \omega_2 t)^3. \tag{11.21}$$

By operating and expanding the cubic term, the following can be obtained

$$x(t) = p_1 \cos \omega_1 t + p_1 \cos \omega_2 t$$
$$+ p_3(\cos^3 \omega_1 t + 3 \cos^2 \omega_1 t \cos \omega_2 t + 3 \cos \omega_1 t \cos^2 \omega_2 t + \cos^3 \omega_2 t). \tag{11.22}$$

After grouping the terms of equal frequency and eliminating the harmonics,

$$x(t) = \left[ p_1 + \frac{9}{4}p_3 \right] \cos \omega_1 t$$
$$+ \left[ p_1 + \frac{9}{4}p_3 \right] \cos \omega_2 t + \frac{3}{4}p_3[\cos(2\omega_1 - \omega_2)t + \cos(2\omega_2 - \omega_1)t]. \tag{11.23}$$

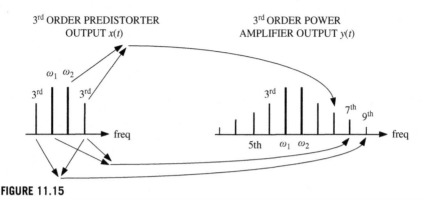

**FIGURE 11.15**

Graphical representation of the output spectrum of the predistorter and power amplifier of Fig. 11.14.

From Eq. (11.23), it follows that the output of the predistorter only contains third-order intermodulation distortion (IMD), which will effectively cancel out the power amplifier's third-order IMD products. However, the predistorter's output contains *four tones* ($\omega_1$, $\omega_2$, $2\omega_1 - \omega_2$ and $2\omega_2 - \omega_1$), two more than the original signal $r(t)$. When these four tones are applied to the third-order power amplifier, a large number of IMD products will occur. Since the mathematical resolution of this problem is rather complex, Fig. 11.15 shows this process graphically, where only the upper intermodulation distortion products are shown for clarity. The third-order IMD products are mixed with the fundamental tones to produce fifth- and seventh-order IMD products, while the third-order tones provide ninth-order IMD products by themselves. The only way the predistorter could be able to cancel the fifth-, seventh- and ninth-order IMD products is by also generating IMD products of that same order, which requires an increase in bandwidth. Unfortunately, the process would repeat infinitely, which means that the predistorter bandwidth must be unlimited for the perfect cancelation of the intermodulation distortion of a power amplifier with just a third-order transfer characteristic.

The effects of bandwidth restrictions can also be analyzed by evaluating the nonlinear gain characteristics of the predistorter. Figure 11.16 shows the power gain and output power of a third-, fifth-, seventh- and infinite-bandwidth predistorter followed by a third-order power amplifier. The output power is shown for the predistorter and the power amplifier individually, as well as for the cascaded connection of both. The band-limited predistorters are not capable of effectively linearizing the power amplifier to saturation.

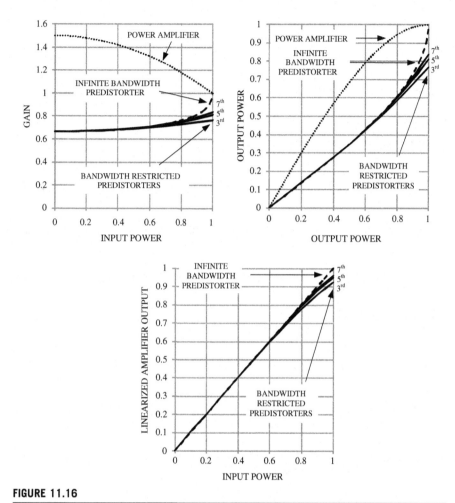

**FIGURE 11.16**

Effects of bandwidth restrictions in predistortion.

## 11.2.3 **Predistorter with memory for octave-bandwidth amplifiers**

Figure 11.4 shows the power amplifier model that contains memory introduced by two linear subsystems located before and after the memoryless nonlinearity. This memory must be removed before calculating the predistorter transfer characteristic using Eq. (11.12), which can be accomplished by placing two more linear subsystems, one before and one after the predistorter, as shown in Fig. 11.17. The transfer functions of the linear subsystems $H_4$ and $H_3$ are calculated so they cancel out the memory introduced by $H_1$ and $H_2$. The net result is that the gain of the

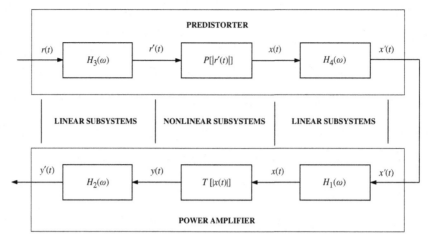

**FIGURE 11.17**

System block diagram used for the calculation of the predistorter function $P[|r'(t)|]$ when the power amplifier has memory.

cascaded systems $H_1-H_4$ and $H_2-H_3$ becomes unity with zero phase shift across the passband of the amplifier.

The linear subsystems $H_3$ and $H_4$ affect the predistorter-amplifier system frequency response, forcing it to be flat across the passband of the power amplifier. In order to recover the original frequency response of the power amplifier, the predistorter-amplifier system must look like the one in Fig. 11.18, where the memory between the predistorter and the power amplifier is eliminated, allowing for memoryless predistortion, but the original frequency response of the power amplifier is preserved by replacing $H_3$ with $H_1$.

## 11.2.4 Postdistortion

The use of a predistortion linearizer can reduce the intermodulation distortion of a nonlinear system, but it can also affect its performance. For example, the noise figure of an amplifier can be degraded by the insertion of a predistorter in front of it, [9] or the use of a predistorter may be simply not feasible, as in the case of an optical receiver [10].

In cases like those described above, the linearizer can be inserted after the nonlinear device, in which case it is called a *postdistorter*. Its transfer characteristic is identical to that of the predistorter given in Eq. (11.12) since the predistorter transfer characteristic was actually calculated as a postdistorter when the output of the power amplifier was used as the input to the predistorter after the normalization to unity at saturation, as described in Section 11.2.2 while analyzing a memoryless predistorter for octave-bandwidth amplifiers.

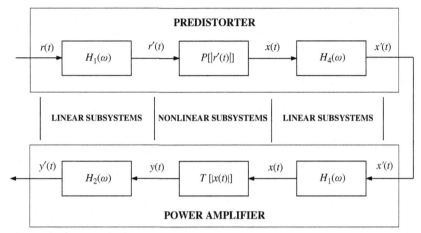

**FIGURE 11.18**

System block diagram of the predistorter-power amplifier with memory. $H_3$ is replaced by $H_1$ from Fig. 11.17 to preserve the power amplifier frequency response.

## 11.3 Analog predistortion implementation

### 11.3.1 Introduction

It is certainly possible to implement the transfer characteristic of a predistorter using analog components. The predistorter can directly be implemented at the operating frequency of the RF power amplifier. However, in cases where there is an up-conversion in the early stages of the signal generation, the predistorter can be inserted at an intermediate frequency (IF). In all cases, bandwidth restrictions must be taken into account to determine the effectiveness of the predistorter. This is particularly important with IF implementations since the signal is usually tightly filtered to remove spurious caused by the up-mixing process.

Increasing the gain of a device and changing its insertion phase as a function of its input signal level is not a trivial task if such performance will effectively predistort a power amplifier. Moreover, achieving this goal over a wide bandwidth is even more complex. The following sections will show the principles of the most common analog predistortion techniques available.

### 11.3.2 Reflective predistorters

The most common reflective predistorter consists of a branch-line coupler (quadrature hybrid), an open circuited transmission line, and a nonlinearity implemented with two diodes in anti-parallel configuration, also known as limiter [11].

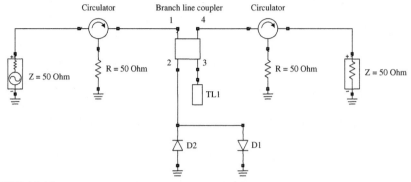

**FIGURE 11.19**

Simplified schematic of a reflective predistortion linearizer utilizing a branch-line coupler and a diode limiter.

Figure 11.19 shows a simplified schematic of such a predistorter. An alternative implementation can include nonlinearities connected to ports 2 and 3 of the branch-line coupler.

The branch-line coupler can be represented by the following S-parameter matrix,

$$S_{90} = -\frac{1}{\sqrt{2}} \begin{vmatrix} 0 & j & 1 & 0 \\ j & 0 & 0 & 1 \\ 1 & 0 & 0 & j \\ 0 & 1 & j & 0 \end{vmatrix}. \tag{11.24}$$

The coupler equally splits the input signal applied to port 1 into two components shifted 90° from each other, available at ports 2 and 3. Since these two ports are terminated into a mismatch, the majority of the energy is reflected back into the coupler. The energy reflected back into port 2 is now equally split between ports 1 and 4, while the energy reflected back into port 3 is equally split between ports 4 and 1.

The voltage $V_3$ at port 3 is given by

$$V_3 = S_{31} V_1 \Gamma_3 = -\frac{V_1}{\sqrt{2}} \frac{Z_{\text{in}} - Z_0}{Z_{\text{in}} + Z_0} \tag{11.25}$$

where $V_1$ is the voltage at port 1, $\Gamma_3$ is the reflection coefficient at port 3 and $Z_{\text{in}}$ is the input impedance of the open circuited transmission line $TL1$ of electrical length $\theta$, which can be written as

$$Z_{\text{in}} = -jZ_0 \cot \theta. \tag{11.26}$$

The voltage $V_2$ at port 2 is given by

$$V_2 = S_{21} V_1 \Gamma_2 = -\frac{jV_1}{\sqrt{2}} \frac{Z_d - Z_0}{Z_d + Z_0} \tag{11.27}$$

where $\Gamma_2$ is the reflection coefficient at port 2, and $Z_d$ is the impedance of the diode anti-parallel, which is an inverse function of the voltage applied across the diodes.

The output voltage of the predistorter $V_4$ can be calculated from Eqs (11.24), (11.25) and (11.27), as

$$V_4 = S_{42} V_2 + S_{43} V_3 = -\frac{V_2}{\sqrt{2}} - \frac{jV_3}{\sqrt{2}} \tag{11.28}$$

$$V_4 = \frac{jV_1}{2} \frac{Z_d - Z_0}{Z_d + Z_0} + \frac{jV_1}{2} \frac{Z_{in} - Z_0}{Z_{in} + Z_0}. \tag{11.29}$$

The gain of the reflective predistorter $G_{RP}$ can be derived from Eqs (11.26) and (11.29) as

$$G_{RP} = \frac{V_4}{V_1} = \frac{j}{2} \left( \frac{Z_d - Z_0}{Z_d + Z_0} - \frac{Z_0 + jZ_0 \cot \theta}{Z_0 - jZ_0 \cot \theta} \right). \tag{11.30}$$

By carefully choosing $\theta$, it is possible to obtain gain expansion (lower insertion loss) when $Z_d$ decreases after the application of a large input signal level at port 1. Since the input impedance of the reflective predistorter varies as a function of the input signal level, an isolator is necessary at the input of the device to ensure proper performance. An additional isolator at the output of the predistorter must be added if a good output impedance match is desired.

Figure 11.20 provides simulation results of the reflective predistorter of Fig. 11.19.

### 11.3.3 Transmissive predistorters

The basic transmissive predistorter is depicted in Fig. 11.21. It consists of a linear signal path in parallel with a nonlinear signal path. The nonlinear signal path gain is reduced by the nonlinearity at large input signal levels. By subtracting the nonlinear path from the linear path, gain expansion and phase shift can be achieved.

As in the reflective predistorter, the nonlinearity can be implemented with a diode limiter. A phase shifter and an attenuator can be used to adjust the nonlinear characteristics of the predistorter. The predistorter input impedance varies as a function of the input signal level, for which an isolator is needed at the input for proper performance. An optional isolator at the output can improve the output impedance match as well.

Figure 11.22 shows the simulated magnitude and phase of the gain of the transmissive predistorter of Fig. 11.21.

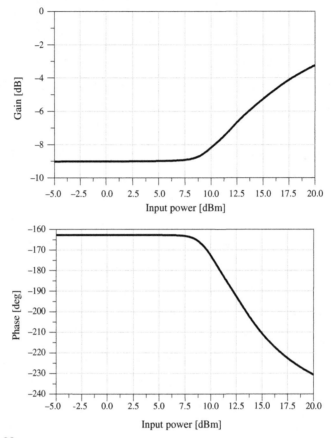

**FIGURE 11.20**

Simulated magnitude and phase of the gain of the reflective predistortion linearizer shown in Fig. 11.19.

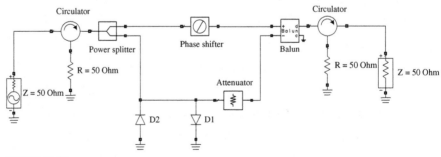

**FIGURE 11.21**

Simplified schematic of a transmissive predistortion linearizer utilizing parallel paths and a diode limiter.

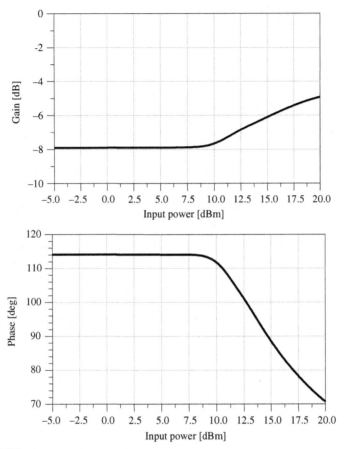

**FIGURE 11.22**

Simulated magnitude and phase of the gain of the transmissive predistortion linearizer shown in Fig. 11.21.

It is possible to achieve gain expansion and phase shift as a function of signal level without resorting to the subtraction of parallel paths. A typical example is the series diode predistorter [12] shown in Fig. 11.23.

The current through a forward biased diode $I_D$ can be represented by Schokley's equation,

$$I_D = I_S \left( e^{\frac{V_D}{V_T}} - 1 \right) \tag{11.31}$$

where $I_S$ is the reverse saturation current, $V_D$ is the voltage across the diode, and $V_T$ is the thermal voltage (26 mV at room temperature). The diode is assumed to be ideal.

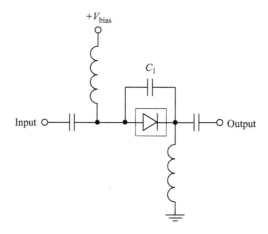

**FIGURE 11.23**

Simplified schematic of a transmissive predistortion linearizer utilizing a series diode [12].

The dynamic conductance of the diode can be obtained by differentiating Eq. (11.31) with respect to $V_D$ as,

$$g_d = \frac{\partial I_D}{\partial V_D} = \frac{I_S}{V_T} \exp\left(\frac{V_D}{V_T}\right). \tag{11.32}$$

Expressing Eq. (11.32) in terms of the dynamic resistance of the diode $r_d = 1/g_d$ results in

$$r_d = \frac{V_T}{I_S \exp\left(\frac{V_D}{V_T}\right)}. \tag{11.33}$$

Equation (11.33) is significantly nonlinear for small values of $V_D$. If the pre-distorter input signal is superimposed to a low dc voltage, the resulting nonlinearity will produce a larger decrease of the dynamic resistance of the diode during the positive half-cycle of the input signal compared to the increase that occurs during the negative half-cycle of the input signal. The net result of this process is an average decrease in the diode dynamic resistance $r_d$ for large input signals to the predistorter. Since the diode in the predistorter is in series with the signal path, any $r_d$ decrease at high input signal levels will result in a reduction in the predistorter insertion loss, which can also be interpreted as a gain expansion.

The predistorter's gain expansion can be used to compensate for the effects of gain compression in the power amplifier (AM/AM). In order to also obtain a nonlinear phase shift in the predistorter to compensate for the amplifier's AM/PM, an impedance can be added in parallel with the diode. This impedance,

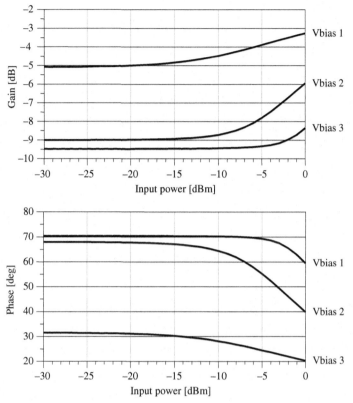

**FIGURE 11.24**

Simulated magnitude and phase of the gain of the diode predistorter of Fig. 11.23 as a function of input power. Values for three different bias voltages.

represented by $C_1$ in Fig. 11.23 in conjunction with the dynamic resistance of the diode, will introduce a phase shift that is a function of the input signal level to the predistorter.

The predistorter gain can be controlled by adjusting the diode bias voltage and the value of the capacitor $C_1$. Figure 11.24 shows the magnitude and phase of the gain of the predistorter for three different bias voltages with a capacitor placed in parallel with the diode. The capacitor introduces a negative AM/PM behavior.

Similarly to the other predistorters described earlier, the series diode predistorter requires isolators at the input and the output for good performance since the input and output impedances are a function of the input signal level.

Other transmissive predistorter implementations include the parallel diode approach [13], and field effect transistors [14,15].

## 11.4 Digital predistortion implementation

### 11.4.1 Introduction

Most modern communication systems utilize digital techniques for signal generation, which simplifies the implementation of predistortion linearization at baseband or at an intermediate frequency while the signal is still in the digital domain, i.e. in the form of discrete time samples instead of a continuous function of time. The analog predistortion concepts described earlier in the chapter also apply to digital predistortion.

Digital predistortion has the advantage of being able to generate predistortion transfer characteristics that would be difficult to implement with discrete components and analog signals. The fabrication cost of a predistorter can be relatively low since the digital processing power is already present for the signal generation, and the predistortion function may consist of additional lines of code. However, the signal instantaneous bandwidth, which includes its intermodulation distortion products, is limited by the sampling rate of the digital system. As a result, wideband digital predistortion is much more challenging than its analog counterpart.

### 11.4.2 Principles of memoryless digital predistortion

In today's designs, it is quite common to generate the baseband signal in digital form. Usually, the Cartesian, in-phase $I$ and quadrature $Q$ components are used for most types of modulation schemes. A possible predistorter implementation consists of a device with its outputs $I'$ and $Q'$ mapped to its inputs $I$ and $Q$. The mapping is defined by $P[I,Q]$ where $P$ is the gain of the predistorter. Note that $P$ is a two-dimensional function, and it consists of $P_1[I,Q]$ and $P_2[I,Q]$. The output of the mapping device can be written as

$$I' = P_1[I, Q] \tag{11.34}$$

$$Q' = P_2[I, Q]. \tag{11.35}$$

A block diagram of this type of linearizer, called *mapping predistorter* is shown in Fig. 11.25. The mapping can be implemented by means of a look-up table stored in memory [16,17]. Because every possible input complex signal amplitude must have its correspondent complex output, the amount of memory needed can be quite large. The total number of memory entries $M$ as a function of the quantization level $n$ is

$$M = 2 (2^{2n}). \tag{11.36}$$

The memory size requirements can be reduced with a *gain based predistorter* [18], as shown in Fig. 11.26. The idea behind this predistorter consists of using interpolation to find intermediate values not included in the predistorter gain table, and to use the magnitude squared of the complex input signal as the table index, which restricts the nonlinear distortion cancellation to odd-order intermodulation

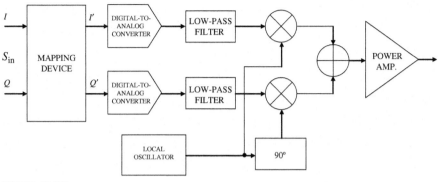

**FIGURE 11.25**

Digital mapping predistorter block diagram.

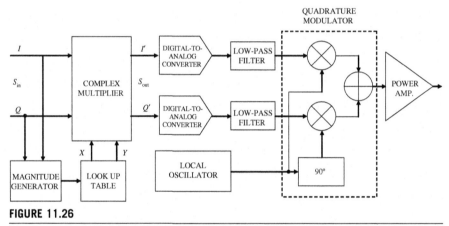

**FIGURE 11.26**

Gain based predistorter simplified block diagram.

distortion products. Individual tables for the $I$ and $Q$ signals are required. It is possible to make the input to the tables a non-uniform distribution to maximize performance. The power amplifier nonlinearity is stronger at high input signal levels, so a greater number of table entries can be stored for large input signals, leaving a smaller number for weak signals, for which the power amplifier is almost linear. Only 64 entries per table were found to be satisfactory for good performance [18], which compares very favorably to the mapping predistorter (Eq. (11.36)).

The input signal to the predistorter $S_{in}$, with real and imaginary components $I$ and $Q$, can be written as

$$S_{in} = I + jQ. \tag{11.37}$$

The complex predistortion coefficient $P$, with real and imaginary parts $X$ and $Y$, is given by

$$P[|S_{in}|^2] = X + jY. \tag{11.38}$$

The output of the predistorter $S_{out}$ will be

$$S_{out} = P[|S_{in}|^2] \, S_{in} = (X + jY)(I + jQ) \tag{11.39}$$

$$S_{out} = \begin{cases} I' = IX - QY \\ Q' = IY + QX \end{cases} \tag{11.40}$$

$$S_{out} = I' + jQ'. \tag{11.41}$$

Section 11.2.2 showed that the predistorter increases the bandwidth of the input signal to include its own distortion products intended to cancel the intermodulation distortion of the power amplifier. Therefore, the predistorter of Fig. 11.26 must operate at a sampling rate that is high enough to include all the distortion produced by the predistortion function. The simplest way to accomplish this task is to upsample the input signal to the predistorter to accommodate the wider bandwidth [19]. The upsampler inserts $U-1$ zero-amplitude samples between existing time samples, increasing the sample rate by $U$ times, where $U$ is the upsampling rate. The output $X_{out}$ of the upsampler as a function of the sample index or discrete time variable $n$ can be written as

$$X_{out}[n] = \begin{cases} X_{in}[n/U] & n = kU \\ 0 & \text{otherwise} \end{cases} \tag{11.42}$$

where $X_{in}$ is the input to the upsampler, and $k$ is any integer.

The insertion of zero samples introduces changes in the spectrum of the signal. Therefore, a low pass filter is required to remove undesired frequency components. From a time domain perspective, the filter performs a linear interpolation between the samples for the reconstruction of the original signal. The cut-off frequency $f_c$ of this lowpass filter is given by

$$f_c = \frac{f_s}{2U} \tag{11.43}$$

where $f_s$ is the original sampling frequency.

Figure 11.27 shows a block diagram of the digital signal processing section of the memoryless gain based predistorter.

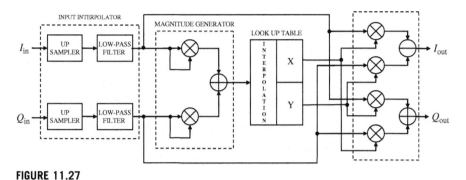

**FIGURE 11.27**

Digital signal processing section of the memoryless gain-based predistorter.

### 11.4.3 Digital predistortion adaptation

The analog predistorters presented in Sections 11.3.2 and 11.3.3 are open-loop devices, i.e. they are designed to perform a specific predistortion function without the possibility of adapting to changes in the transfer characteristic of the power amplifier over time. The dynamic adaptation of a predistorter is a difficult task that must be carefully implemented in order to guarantee system stability and actual distortion reduction under a wide range of operating parameters.

The situation is quite different for a digital predistorter. The predistortion function is not produced by analog components but by coefficients stored in a look-up table. It is, therefore, much simpler to implement an adaptation algorithm through the straightforward manipulation of the look-up table coefficients compared to trying to alter the nonlinear characteristics of an analog predistorter by varying, for instance, a bias voltage.

A typical adaptation algorithm performs the comparison between the output signal of the power amplifier and the input signal to the predistorter. The difference between these two signals becomes the error function to minimize by operating on the values stored in the look-up table, with the goal of obtaining a constant magnitude and phase of the gain of the predistorter-power amplifier system as a function of the input power, as described by Eqs (11.10) and (11.11).

Several convergence algorithms can be used for the minimization of the error function [20,21]. The look-up table values can be dynamically modified after the initial convergence is achieved in order to maintain optimum performance under variations of the operating parameters. The mapping predistorter described earlier will be slower to converge than the gain-based predistorter because every possible entry of its large look-up table must be updated in order to complete the adaptation process.

Figure 11.28 shows the simplified block diagram of an adaptive digital predistorter. The output of the power amplifier is sampled and downconverted to baseband using a quadrature demodulator that shares the same local oscillator used by the quadrature modulator. A time delay must be introduced to align the feedback and forward signals in time. The adaptation algorithm performs the calculation of the look-up table coefficients. A signal that exercises the full dynamic range of the power amplifier is necessary to allow for the calculation of the look up table coefficients. A two-tone signal or a single-tone ramp can be used, provided that the peak output power of the amplifier reaches the saturation point to ensure that the look up table values include the entire dynamic range of

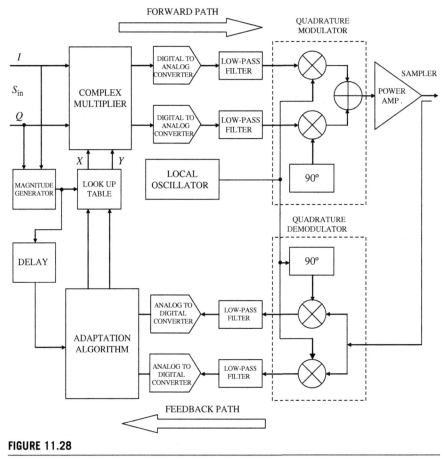

**FIGURE 11.28**

Simplified block diagram of an adaptive digital predistorter.

the power amplifier. Other modulated signals that follow the above criteria can be used, but their probability density function can affect the look up table accuracy since the signal may only spend brief periods of time at certain power levels.

### 11.4.4 **Digital predistorter performance**

Figure 11.29 shows the spectrum of a solid-state power amplifier after the application of digital predistortion. The test signal consists of two identical tones with a separation of 2 MHz. The power amplifier was operated at 3 dB output backoff, which coincides with the peak to average ratio of the two-tone signal. The digital predistorter is capable of reducing the third-order intermodulation distortion by more than 10 dB.

Figure 11.30 shows the spectrum of a digitally predistorted traveling wave tube amplifier. In this case, eleven equal-amplitude carriers are used as the test signal. The carriers are all in phase, for which the peak to average ratio is 10.4 dB. The spectrum on the left was obtained by operating the amplifier at 10 dB output backoff, while the spectrum on the right was measured at 11 dB output back-off. A significant improvement in distortion cancelation can be achieved when the output power level is such that the peak amplitude of the signal does not reach the saturation level of the amplifier.

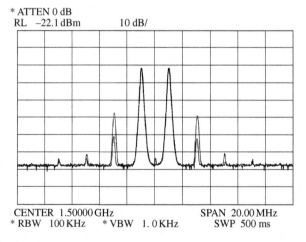

**FIGURE 11.29**

Digital predistortion of a solid-state power amplifier using a two-tone signal.

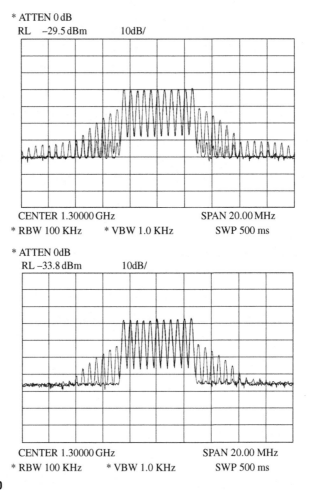

**FIGURE 11.30**

Digital predistortion of a traveling wave tube amplifier using 11 equal-amplitude, coherent carriers. Top: 10 dB output back-off; bottom: 11 dB output back-off.

# REFERENCES

1. Raab FH, Asbeck P, Cripps S, et al. RF and microwave power amplifier and transmitter technologies — part 1. *High Frequency Electronics.* May 2003;2(3):22−36.
2. Zhu A, Wren M, Brazil T. An efficient volterra-based behavioral model for wide-band RF power amplifiers. *IEEE Microwave & Wireless Components Letters.* December 2004;14(12).
3. Silveira D, Gadringer M, Arthaber H, et al. RF-Power amplifier characteristics determination using parallel cascade wiener models and pseudo-inverse techniques. *2005 Asia-Pacific Microwave Conference Proceedings.* vol. 1.

4. Moon J, Kim B. Enhanced hammerstein behavioral model for broadband wireless transmitters. *IEEE Transactions on Microwave Theory and Techniques*. 2007;59 (4, Part 1):924–933.

5. Liu T, Boumaiza S, Ghannouchi F. Augmented hammerstein predistorter for linearization of broad-band wireless transmitters. *IEEE Transactions on Microwave Theory and Techniques*. 2006;54(4 part 1):1340–1349.

6. Taringou F, Srinivasan B, Malhame R, et al. Hammerstein-wiener model for wideband RF transmitters using base-band data. *2007 Asia-Pacific Microwave Conference*. 1–4.

7. Gilabert P, Montoro G, Bertran E, On the wiener and hammerstein models for power amplifier predistortion. *2005 Asia-Pacific Microwave Conference Proceedings*. 2:4.

8. Abramowitz M, Stegun I. *Handbook of Mathematical Functions*. National Bureau of Standards; 1970.

9. Kim N, Aparin V, Barnett K, Persico C. A cellular-band CDMA 0.25-$\mu$m CMOS LNA linearized using active post-distortion. *IEEE Journal of Solid State Circuits*. July 2006;41(7):1275–1278.

10. Basak J, Jalali B. Photodetector linearization using adaptive electronic post-distortion. *2005 Optical Fiber Communication Conference*. 6–11 March 2005;4:3.

11. Jeong H, Park S, Ryu N, Jeong Y, Yom I, Kim Y. A Design of K-band Predistortion Linearizer using Reflective Schottky Diode for Satellite TWTAs. *2005 European Microwave Conference*. 4–6 October 2005;4.

12. Yamauchi K, Mori K, Nakayama M, Itoh Y, Mitsui Y, Ishida O. A novel series diode linearizer for mobile radio power amplifiers. *1996 IEEE International Microwave Symposium Digest*. 17–21 June 1996;2:831–834 [San Francisco, CA].

13. Yamauchi K, Mori K, Nakayama M, Mitsui Y, Takagi T. A microwave miniaturized linearizer using parallel diode. *1997 IEEE International Microwave Symposium Digest*. 8–13 June 1997;3:1199–1202 [Denver, CO].

14. Hau G, Nishimura T, Iwata N. A highly efficient linearized wide-band CDMA handset power amplifier based on predistortion under various bias conditions. *IEEE Transactions on Microwave Theory and Techniques*. June 2001;49(6):1194–1201.

15. Kim J, Jean M, Lee J, Kwon Y. A new active predistorter with high gain and programmable gain and phase characteristics using cascode-FET structures. *IEEE Transactions on Microwave Theory and Techniques*. November 2002;50(11):2459–2466.

16. Nagata Y. Linear amplification technique for digital mobile communications. *Proceedings of the 39th IEEE Vehicular Technology Conference*. 1989;1:159–164.

17. Ren Q, Wolff I. Improvement of digital mapping predistorters for linearising transmitters. *1997 IEEE International Microwave Symposium Digest*. 08–13 June 1997;3:1691–1694 [Denver, CO].

18. Cavers J. Amplifier linearization using a digital predistorter with fast adaptation and low memory requirements. *IEEE Transactions on Vehicular Technology*. November 1990;39(4).

19. Oppenheim A, Schafer R, Buck J. *Discrete-Time Signal Processing*. 2nd ed. Upper Saddle River, New Jersey: Prentice Hall; 172–176, ISBN 0-13-754920-2.

20. Lee K, Gardner P. "Comparison of Different Adaptation Algorithms for Adaptive Digital Predistortion based on EDGE Standard". *2001 IEE International Microwave Symposium Digest*. 20–25 May 2001;2:1353 [Phoenix, AZ].

21. Levy Y, Karam G, Sari H. Adaptation of a digital predistortion technique based on intersymbol interpolation. *1996 IEEE Global Telecommunications Conference Digest*. 13–17 November 1995;1:145–150 [Singapore].

# Computer-Aided Design of Switchmode Power Amplifiers

## INTRODUCTION

Nonlinear circuit simulation in the frequency and time domains is a very important tool for analysis, design, and optimization of high-efficiency switchmode power amplifiers of Classes D, DE, and E. The advantages are significantly reduced development time and final product cost, better understanding of the circuit behavior, and faster obtaining of the optimum design. It is especially important at very high frequencies, including microwaves, and for MMIC development, where the transistor and circuit parasitics can significantly affect the overall power-amplifier performance. Therefore, it is very important to incorporate into the simulator as accurate a transistor model as possible, to approximate correctly the device behavior, not only at the fundamental frequency but also at the second- and higher-order harmonics of the operating frequency. In this chapter, the different CAD programs are described to analyze the time-domain and frequency-domain behavior of the switchmode high-efficiency power amplifiers in different frequency ranges from high frequencies to microwaves, including HB-PLUS and SPICE CAD tools for Class-D and-DE circuits and HEPA-PLUS, SPICE, and ADS CAD tools for Class-E circuits. The detailed simulation example of a high-efficiency two-stage 1.75-GHz HBT MMIC power amplifier that uses microstrip transmission lines in the supply and load networks will be presented and discussed.

## 12.1 HB-PLUS program for half-bridge and full-bridge direct-coupled voltage-switching Class-D and Class-DE circuits

### 12.1.1 Program capabilities

The HB-PLUS CAD program can simulate and automatically optimize resonant and non-resonant half-bridge or full-bridge circuits with a load that receives the output power, driven through a load network [1]. Several versions of the circuit are described in Chapter 2; the simplest tuned version is shown in Fig. 2.6. The switching power transistors can be operated in the hard-switchmode (Class-D; both switches operating at 50% duty ratio, and the transistor switch is turned on with the full dc supply voltage across the transistor), or the soft-switchmode (Class-DE; each transistor is turned on with nearly zero voltage across the transistor, achieved by using smaller than 50% duty ratio of each transistor). The circuit can be used as an RF or microwave power amplifier, as a dc−dc converter, as a dc−ac inverter, or as an ac (including RF and microwave) power generator. The HB-PLUS program performs the following functions:

- Simulates very rapidly the steady-state periodic response of the circuit, 100−1000 times as fast as with SPICE and about 10 times as fast as with Agilent ADS. The outputs are time-domain graphical plots and frequency domain spectra of all important circuit voltage and current waveforms, and a tabulation of input power from the dc supply, RF output power, efficiency, inefficiency, and power dissipation in each circuit element (including the parasitic power losses in each inductor and capacitor). The spectrum is computed for frequencies from dc through the 31st harmonic of the operating frequency.
- Simulates the cycle-by-cycle transient response, from any computed or user-specified starting condition, to any combination of time-varying circuit parameters, such as switching frequency, switch duty ratio, dc supply voltage, and load impedance. The available outputs are the same as for the steady-state periodic response.
- Sweeps up to 28 circuit parameters and plots one or two output variables *versus* the swept parameter.
- Optimizes the design automatically, according to tradeoffs that can be specified among input dc power, output RF power, efficiency, inefficiency, and each component of power loss.
- Computes transfer functions: any output variable *versus* any circuit parameter.

The steady-state periodic response of the circuit to the periodic input drive is the circuit operation after the start-up transient has died away, which is the operation, usually observed in experimental results. HB-PLUS simulates the voltage-switching Class-D or Class-DE circuit very quickly and the simulation requires about 8 ms on a Pentium III/667-MHz computer.

Using the cycle-by-cycle transient response, it is possible to evaluate transient stresses:

- occurring during equipment turn-on and turn-off
- resulting from transient changes of load (including load faults of shorted or open load, or any arbitrary impedance magnitude and angle)
- resulting from application or removal of input drive with the dc supply voltage present
- resulting from transient changes of dc supply voltage
- resulting from failure of any circuit element

or simulate the transient response to modulation of any combination of time-varying circuit parameters. This text refers to the half-bridge circuit; the *HB-PLUS User Manual* also explains how to simulate and optimize the full-bridge circuit.

### 12.1.2 Circuit topologies

Figure 12.1(*a*) shows the general model of a half-bridge voltage-switching Class-D or Class-DE circuit used in HB-PLUS including a dc power supply $V_{cc}$, a load

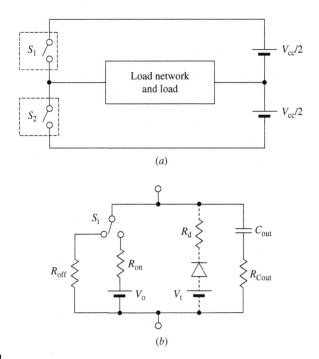

(*a*)

(*b*)

**FIGURE 12.1**

General model of a half-bridge voltage-switching Class-D power amplifier: (*a*) basic circuit; (*b*) equivalent switch model.

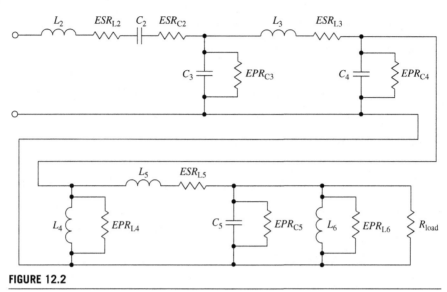

**FIGURE 12.2**

Load network and load in HB-PLUS.

network and load, and two switches. The switch-enhanced macro-model, which can accurately represent any type of input-controlled solid-state switches based on different types of bipolar junction transistors (BJTs), MOSFETs or MESFETs is shown in Fig. 12.1(b) where $V_o$ is the saturation offset voltage, $V_t$ is the turn-on threshold voltage of internal or external anti-parallel diode (the parasitic substrate diode of a silicon MOSFET or an external diode intentionally connected from emitter to collector of a BJT), $R_d$ is the ohmic resistance of the anti-parallel diode, $R_{on}$ is the on-resistance ($R_{ds(on)}$ for a MOSFET or $R_{ce(sat)}$ for a BJT), and $C_{out}$ is the equivalent transistor output capacitance characterized by its parasitic resistance $R_{Cout}$ [2].

The load-network configuration shown in Fig. 12.2 includes a general-purpose multiple-inductor/multiple-capacitor network. It can be used to represent a non-resonant network or a resonant network that is series-loaded, parallel-loaded, or split-reactor-loaded, where either loss resistances or quality factors for all inductors and capacitors can be specified. The extended configuration of the load network provides considerable flexibility for representing a particular circuit arrangement, where $R_{load}$ is a resistor to which the RF output power is delivered. The inductor and capacitor that are adjacent to $R_{load}$ allow representing the load as generic $RLC$ impedance. Alternatively, $R_{load}$ and its adjacent $L$ and $C$ can represent the input-port impedance of an external two-port network that performs the functions of coupling, tuning, and impedance-transformation of a load located at the output port. Which $L$ and $C$ components will be adjacent to $R_{load}$ depends on which reactive components the user includes in the load network; the user can remove unwanted components. The two-port network's input/output transmission function can include attenuation, by using the *ESR* and *EPR* components that are associated with the *L* and *C* that are adjacent to $R_{load}$. For example, the two-port network's input/output transmission function has an attenuation ($1/EPR_{C5}$ + $1/EPR_{L6})/(1/EPR_{C5}$ + $1/EPR_{L6}$ + $1/R_{load}$). The full high-frequency *T*-model of a transformer includes mutual inductance and core loss ($L_4$ and $EPR_{L4}$), primary and secondary leakage inductances ($L_3$ and $L_5$), winding resistances ($ESR_{L3}$ and $ESR_{L5}$), and winding capacitances ($C_3$ and $C_5$). The capacitor $C_4$ represents a distributed model of winding capacitance. The inductor $L_2$ can be the inductor of a resonant load network or the commuting inductor of a non-resonant soft-switching (zero-voltage turn-on) load network. The capacitor $C_2$ can be the resonance capacitor of a resonant load network or the dc-blocking capacitor of a non-resonant network. If the user is not using the two-port macro-modeling technique discussed previously, $R_{load}$, $L_6$, $C_5$, $EPR_{L6}$, and $EPR_{C5}$ can represent a general-purpose *RLC* network that models any chosen four parameters of any arbitrary load impedance, e.g. magnitude and phase at the operating frequency, magnitude or phase at the third-harmonic frequency (the even-harmonic voltages and currents are usually negligibly small in bridge-type power converters), and fractional power loss (attenuation) at all frequencies, as discussed earlier. Alternatively, $C_4 L_4$ can represent a parallel resonator, $C_2 C_3$ can represent a split resonance capacitor, $L_3 L_4$ (or $L_5 L_6$) can represent a split resonance inductor and so on. Using a two-port macro-model for the external two-port network yields accurate results with faster computation than would be obtained by computing separately the voltages and currents of all of the individual components that collectively provide the functions modeled by the two-port macro-model.

### 12.1.3 Class-D *versus* Class-DE

P. J. Baxandall published the first description of a Class-D circuit, in a paper discussing voltage-switching and current-switching self-excited oscillators [3]. Later publications by other authors discussed the use of the circuit in tuned and untuned amplifiers. In those early publications, the transistor switches were operated with 'on' duty ratio of 50%, resulting in 'hard turn-on switching' of the transistors: a turning-on upper transistor ($S_1$ in Fig. 12.1(*a*)) would charge the switching-node capacitance from ground up to the positive-supply rail, and a turning-on lower transistor ($S_2$ in Fig. 12.1(*a*)) would discharge that capacitance from the positive-rail voltage down to ground. The switching-node capacitance is the output capacitances of the two transistors, plus circuit wiring capacitance. The charging and discharging power dissipations resulting from hard-switching were ($CV^2 f)/2$ at each turn-on transition (twice per cycle of the switching frequency), where *C* is the node capacitance, *f* is the switching frequency, and *V* is the dc supply voltage (labeled '$2V_{cc}$' in Figs. 2.6 and 2.7, and '$V_{cc}$' in Fig. 12.1(*a*)). When the circuit was being used at audio frequencies, that power dissipation was not a problem,

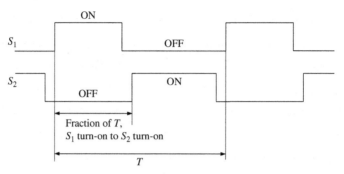

**FIGURE 12.3**

Switch timing for a Class-DE.

but as users applied the technique at frequencies higher than a few-hundred kHz, the capacitor-discharge power dissipation became objectionable.

SA Zhukov and VB Kozyrev published a technique for eliminating the power dissipation associated with charging and discharging the node capacitance [4]. The duty ratio of each transistor switch is reduced from 50% to a lower value, leaving a "conduction gap" during which both switches are "off", as shown in Fig. 12.3. During that conduction gap, the current in the inductor of the load network can charge the node capacitance up to the positive rail, and discharge that capacitance down to ground. They suggested choosing the circuit parameters to yield a switch-voltage waveform that has zero voltage and zero-voltage slope, at the time the switch will be turned "on". Those conditions are the same as in the nominal waveform of a Class-E amplifier described in Chapter 5. Because the circuit operates like Class-D, modified to have turn-on transitions like Class-E, later authors referred to that variant of Class-D as "Class-DE". DC Hamill updated this publication and made it easily available to the English-speaking part of the world [5].

If the load-network inductance value is intentionally made larger than the value needed for resonance with the network capacitance at the switching frequency, zero-voltage turn-on switching can be achieved with a larger switch duty ratio (narrower conduction gap) than would be needed if the inductance resonated with the capacitance at the switching frequency. That increases the output power available at a given level of peak current in the transistor, or it decreases the peak current needed for a specified output power. A unique feature of the HB-PLUS program is the ability to automatically adjust the values of the circuit elements (capacitors, inductors, and resistors) and circuit parameters (switching frequency and switch conduction duty ratio) to achieve optimum performance. The meaning of "optimum" is defined by an *objective function* which comprises a series of terms, each of which pertains to one aspect of circuit performance. For each performance parameter of interest, a *target* value and a *handicap* (tradeoff) value are entered. The target value is the value which is required to achieve representing, for example, the number of watts of output power. The handicap value represents the relative number of units of that performance parameter that corresponds in importance to the specified handicap values of each of the other performance parameters. For example, if the handicap for efficiency is 2.0 and the handicap for output power is 5.0, the program will understand that an increase of efficiency of 2.0 per cent points has the same importance as an increase of output power of 5.0 W. A handicap value of zero instructs the program to ignore that parameter in computing the objective function.

## 12.2 **HEPA-PLUS CAD program for Class-E**

### 12.2.1 **Program capabilities**

The HEPA-PLUS CAD program can be used to simulate a variety of topologies of tuned and untuned single-ended and push—pull power amplifiers operating in Classes AB, B, and C in output-power saturation and in Class-E and some types of Class-F [6]. The program performs the following design functions:

- Evaluates *a priori* the performance achievable with a candidate transistor in a nominal-waveform Class-E circuit using inductors and capacitors with specified quality factors. The output is plots of efficiency *versus* frequency at four values of output power for specified ranges of frequency and output power.
- Automatically makes a preliminary circuit design to meet the specified design goals using inductors and capacitors with specified quality factors. The output is a complete set of circuit-element values, which can be a starting point for a circuit optimization.
- Simulates very rapidly the steady-state periodic response of the circuit 100 to 1000 times as fast as with SPICE and about 10 times as fast as with ADS. The outputs are time-domain graphical plots and frequency-domain spectra of all circuit voltage and current waveforms and a tabulation of input power from the dc supply, output power, efficiency, inefficiency, power dissipation in each element (including the parasitic power losses in each inductor and capacitor), and transistor peak voltage and current in the normal and inverse directions.
- Simulates the cycle-by-cycle transient response, from any starting condition, to a transient change of any combination of input drive, dc supply voltage, and circuit-parameter values. The available outputs are the same as for the steady-state periodic response.
- Sweeps any of the 25 circuit parameters and plots any one or two output variables *versus* the swept parameter.
- Optimizes the design automatically according to the specified trade-offs, among seven evaluation factors that include output power, efficiency, and transistor stresses.

- Computes required control functions: value of any circuit parameter *versus* any other circuit parameter to maintain a specified output voltage, current, or power, e.g. transistor conduction angle (duty ratio) *versus* required output-voltage amplitude, in a linear-amplifier or high-level modulator application.

### 12.2.2 Steady-state periodic response

The steady-state periodic response of the circuit to the periodic input drive is the circuit operation after the start-up transient has died away, which is the operation usually observed in experimental measurements. HEPA-PLUS simulates the switchmode Class-E circuit very quickly. For example, the simulation requires about 8 ms on a 667-MHz computer making a 256-points frequency sweep from 13.5 to 14 MHz of the built-in default circuit.

### 12.2.3 Transient response

Using the cycle-by-cycle transient response, it is possible to evaluate transient stresses:

- occurring during equipment turn-on and turn-off
- resulting from transient changes of load (including load faults of shorted or open load, or any arbitrary impedance magnitude and angle)
- resulting from application or removal of input drive with the dc supply voltage present
- resulting from transient changes of dc supply voltage
- resulting from failure of any circuit component

or simulate the transient response to modulation of any combination of time-varying circuit parameters such as dc supply voltage, input-drive frequency, transistor duty ratio (conduction angle), or any circuit-element value.

### 12.2.4 Circuit topology

Figure 12.4 shows the generic single-ended topology used in HEPA-PLUS including both a switch and a load network. The switch model comprises a saturation offset voltage $V_o$, a turn-on threshold voltage $V_t$ of internal or external anti-parallel diode (the parasitic substrate diode of a silicon MOSFET or an external diode intentionally connected from emitter to collector of a BJT), an ohmic resistance of the anti-parallel diode $R_d$, an on-resistance $R_{on}$ ($R_{ds(on)}$ for a MOSFET or $R_{ce(sat)}$ for a BJT), an equivalent nonlinear transistor output capacitance $C_{out}$ characterized by its parasitic resistance $R_{Cout}$, and a series parasitic inductance $L_q$ modeling the inductance of the emitter or source bondwires. The load network includes five resonators (main series resonator $L_2C_2$ with parasitic loss resistances $ESR_{L2}$ and $ESR_{C2}$, main parallel resonator $L_3C_3$ with parasitic loss resistances $EPR_{L3}$ and $EPR_{C3}$, two auxiliary series resonators $L_QC_{out}$ and $L_{C1}C_1$ and auxiliary parallel

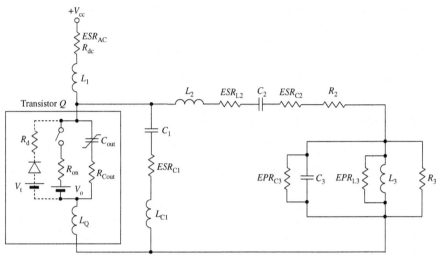

**FIGURE 12.4**

Generic single-ended circuit topology (two such circuits are used for push—pull amplifiers; three types of combiners are available for combining the "push" and "pull" halves of the push—pull circuit).

resonator $C_{out}L_{C1}$), an arbitrary $RLC$ load impedance that can be placed at the $R_2$ location in series with the resonator $L_2C_2$ or at the $R_3$ location in parallel with resonator $C_3L_3$, and parasitic loss resistances for all circuit components. The resonators can each be made resonant at any chosen frequency, not necessarily at the operating frequency and not necessarily all at the same frequency. Components not used in a particular circuit can be deleted from the generic circuit topology by giving them negligibly small or negligibly large values, or removing them.

## 12.2.5 Optimization

The HEPA-PLUS CAD program optimizes all components of the load network shown in Fig. 12.4, including simultaneous optimization of the load-network resistance and the impedance transformation ("matching") network that transforms the external load impedance (connected at the matching-network's output port) to the optimum value of load-network resistance (presented at the matching-network's input port). That is, the external load $R_3$ is shunted by $C_3$, and the equivalent series $C$ and lower-resistance $R$ are tuned to purely resistive by adding an inductance in series with the equivalent $C$ and $R$. The optimum values of $C_3$ and the added inductance are found automatically by the optimizer. That needed inductance is automatically added to the value of $L_2$ that would have been used as a resonator with $C_2$ and a purely resistive load. The external load impedance can be any arbitrary parallel *RLC* circuit (e.g., $R_3$, $L_3$, and $C_3$), not necessarily the commonly used purely resistive 50 Ω. The *HEPA-PLUS User Manual* gives full information on this useful feature.

## 12.3 Effect of Class-E load-network parameter variations

Mistuning of the Class-E load network parameters can have a significant effect on the output power and efficiency. In this case, it is convenient to define this effect visually based on the collector current and voltage waveforms to tune the load-network parameters to their optimum values quicker. The effect of a change in the load angle $\psi = \tan^{-1}(\omega L/R)$ or normalized series inductance $L$ on collector voltage and current waveforms of a Class-E power amplifier with shunt capacitance is shown in Fig. 12.5 [7]. For values of $\psi$ slightly less inductive than optimum of 49.05°, the collector voltage at the turn-on instant attains a certain positive value, as shown in Fig. 12.5(a). However, for values of $\psi$ slightly more inductive than optimum, a small negative voltage swing appears, as shown in Fig. 12.5(b). For capacitive loading and larger inductive loading, the collector voltage becomes large at the turn-on time, causing low efficiency. For extremely reactive loads, the collector voltage tends toward the ramp shape, which it would have with no load at all. The negative current capability is needed whenever the

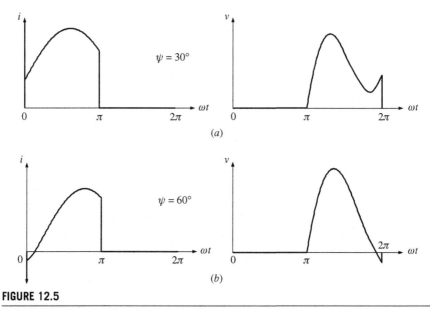

**FIGURE 12.5**

Collector waveforms as functions of a load angle.

load is more inductive than the ideal 49.05°. It should be mentioned that the efficiency remains close to 100% for load angles between about 40° and 70°.

Figure 12.6 illustrates the effect in the collector current and voltage waveforms of a change in the normalized shunt susceptance $\xi = \omega CR$ in a Class-E power amplifier with shunt capacitance of otherwise-nominal design [7]. In this case, low values of shunt susceptance produce a collector voltage that swings to both large positive and large negative values, as shown in Fig. 12.6(a). While the voltage at turn-on time is never negative, the negative current is required for some values of $\xi$, as shown in Fig. 12.6(b). As the shunt susceptance becomes very large for the same load resistance, it tends to dominate, producing a ramp voltage waveform and drawing such current as is necessary to charge the capacitor. Under these conditions, the output power becomes constant but is small. It should be noted that the power amplifier is quite tolerant of variations of shunt susceptance and maintains high efficiency for values $\xi$ between 0.06 and 0.3. The effect caused by deviation of the load resistance $R$ from its optimum value shows that efficiency varies gradually, remaining at 95% or more for variations in the load resistance of +55% to −37%.

Figure 12.7(a) shows the low-order Class-E load network with shunt capacitance and the typical mistuned collector voltage waveform demonstrating the transistor turn-on, turn-off, and waveform "trough". The inductor $L_1$ is considered an RF choke with high reactance to isolate dc and RF paths between each other. The reactance of the series circuit composing of $L_2$ and $C_2$ is inductive. At the turn-on time of the transistor, the collector voltage waveform has zero slope and zero

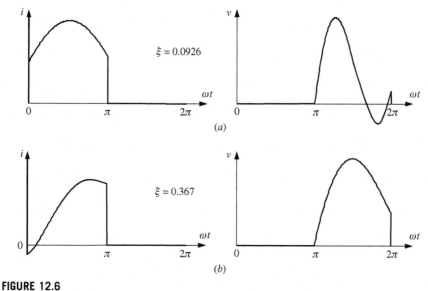

**FIGURE 12.6**

Collector waveforms as functions of normalized shunt capacitance.

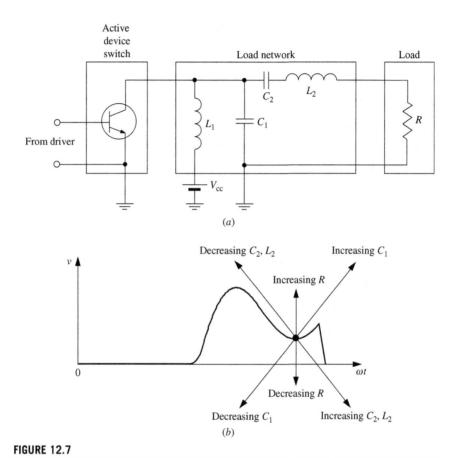

**FIGURE 12.7**

Load network and effect of adjusting its components.

voltage. An actual load network can approximate the idealized optimum (or nominal) conditions by adjusting its parameters $C_1$, $C_2$, and/or $L_2$. If $R$ is not already the desired value for the desired output power, it may need adjustment as well. The circuit will operate with the nominal Class-E waveform, while delivering the specified output power at the specified frequency, if the chosen parameter values are installed in the actual hardware. The possible need for tuning results from tolerances on the component values (normally not a problem, because Class-E has low sensitivity to component tolerances) and the possibility of unknown-value reactances in series with $R$ (therefore, in series with $L_2$ and $C_2$) after the load resistance has been transformed to the chosen value of $R$. Those series reactances require that the reactances of $L_2$ and $C_2$ be reduced by the amounts of the unknown inserted inductive and capacitive series reactances, but how can we do that when those inserted reactances are unknown?

If we know how changes in $L_2$ and $C_2$ will affect the collector voltage waveform, we can adjust $L_2$ and $C_2$ to meet two criteria at the operating frequency: achieve close to the nominal collector voltage waveform and deliver the specified value of output power [8]. Figure 12.7(*b*) shows how $L_2$ and $C_2$ affect the collector voltage waveform. We know also that increasing $L_2$ reduces the output power and *vice versa*. In practice, with an oscilloscope displaying the voltage waveform and a directional power meter indicating the power delivered to the load, we can adjust $L_2$ and $C_2$ to simultaneously fulfill the two desired conditions (nominal waveform and desired output power) even if the reactances in series with $R$ are unknown.

If $C_1$ (comprised of the transistor output capacitance and the external capacitor connected in parallel with it) is within about 10% of the intended value, $C_1$ will normally not need adjustment. When there is a large deviation from the design value, $C_1$ can be adjusted to achieve the nominal collector voltage waveform, using the information in Fig. 12.7(*b*) about the effects of $C_1$ on this waveform. In that case, the three components $C_1$, $C_2$, and $L_2$ can be adjusted to achieve three conditions simultaneously at the operating frequency: desired output power, transistor saturation voltage just before transistor turn-on, and zero slope of the voltage waveform just before turn-on.

Changes in the values of the load network components affect the collector voltage waveform as follows:

- Increasing $C_1$ moves the trough of the waveform upwards and to the right.
- Increasing $C_2$ moves the trough of the waveform downwards and to the right.
- Increasing $L_2$ moves the trough of the waveform downwards and to the right.
- Increasing $R$ moves the trough of the waveform upwards ($R$ is not normally an adjustable circuit element).

Knowing these effects, it is easy to adjust the load network for a nominal Class-E operation by observing the collector voltage waveform. The adjustment procedure is:

- Set $R$ to the desired value or accept what exists.
- Set $L_2$ for the desired $Q_L = \omega L_2/R$ or accept what exists.
- Set the frequency as desired.
- Adjust $C_1$ and/or $C_2$ as shown in Fig. 12.7(*b*).

## 12.4 HB-PLUS CAD examples for Class-D and Class-DE

Section 12.1 introduces the basic features of the HB-PLUS program. Now we will demonstrate the use of HB-PLUS to simulate and analyze a 3-kW, 3-MHz half-bridge (voltage-switching Class-D) tuned dc–ac power inverter, ac power amplifier, or RF power generator (different specialists use those different terms for what is basically the same thing). The circuit uses a pair of International Rectifier IRFP460 TO-247 MOSFETs, operating from a 450-V dc supply.

After that, we reduce the duty ratio of both switches to generate a Class-DE version of the circuit in the simplest possible way, using the same transistors, inductors, and capacitors as in the Class-D circuit. For this simplest possible demonstration of the advantages of Class-DE operation, we omit the full optimization that could be achieved by having the optimizer adjust the values of all of the important parameters, avoiding the need for a more arcane explanation of subtle details. In this example, we demonstrate the benefit of using soft switching (i.e. having the switch turn on at nearly zero switch voltage, and therefore not discharging, through the turning-on switch, a node capacitance that is charged to the full supply voltage). To obtain soft switching, a conduction gap (dead time) is provided between the turn-off of one switch and the turn-on of the other switch shown in Fig. 12.3, and the load network can also be made slightly inductive at the switching frequency. The inductive characteristic can shift the phase of the sinusoidal current, so that the current at switch turn-off can be larger (higher up on the falling part of the sine wave). That higher current will discharge the node capacitance more quickly. During the switch conduction gap, the inductive load current charges the node capacitance from the former voltage level to the new voltage level, or to nearly that level. Consequently, the turning-on transistor turns on at near-zero voltage, rather than with the node capacitance charged to the full power supply voltage, thus avoiding the power dissipation that would occur if the turning-on transistor would charge or discharge the node capacitance. This example shows the general case, illustrating both waveform features: (*a*) the node capacitance is charged by the inductive load current to nearly the new voltage level, and (*b*) the turning-on switch charges the capacitance the rest of the way. In a full optimization, that combination might result in the smallest total power dissipation at the specified output power and frequency, depending on the set of numerical values of all of the circuit parameters.

### 12.4.1 **Class-D with hard switching**

As noted previously, the example Class-D circuit is a 3-kW, 3-MHz power amplifier with a duty ratio of 50% for each of the two switches. Hence, this circuit does not yet have the conduction gap (dead time, as shown in Fig. 12.3) that provides soft switching; we will add that later, after we compute the efficiency of the amplifier output stage and the power losses at different places in the circuit, without yet providing the soft-switching operation. The "Enter Circuit Parameters" screens of Figs. 12.8 and 12.9 show the numerical values of all of the circuit parameters. The "Efficiency and Powers" screen of Fig. 12.10 shows the drain efficiency and inefficiency, and each of the components of power dissipation. Summarizing the results for Class-D operation: 3213.4 W output at 82.733% drain efficiency, with 612.94 W power dissipation in the two transistors (19.1% of the output power). That amount of power dissipation is higher than desired, and can be reduced significantly by reducing the duty ratio of the two switches to provide soft switching in the Class-DE mode of operation. A major portion of the power

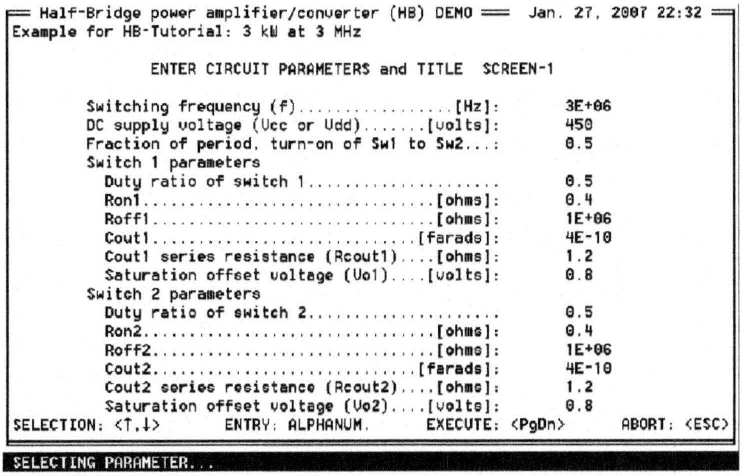

```
== Half-Bridge power amplifier/converter (HB) DEMO ==  Jan. 27, 2007 22:32 ==
Example for HB-Tutorial: 3 kW at 3 MHz

              ENTER CIRCUIT PARAMETERS and TITLE  SCREEN-1

        Switching frequency (f)................[Hz]:      3E+06
        DC supply voltage (Ucc or Udd).......[volts]:     450
        Fraction of period, turn-on of Sw1 to Sw2...:     0.5
        Switch 1 parameters
          Duty ratio of switch 1....................      0.5
          Ron1.............................[ohms]:         0.4
          Roff1............................[ohms]:         1E+06
          Cout1............................[farads]:       4E-10
          Cout1 series resistance (Rcout1)....[ohms]:      1.2
          Saturation offset voltage (Uo1)....[volts]:      0.8
        Switch 2 parameters
          Duty ratio of switch 2....................      0.5
          Ron2.............................[ohms]:         0.4
          Roff2............................[ohms]:         1E+06
          Cout2............................[farads]:       4E-10
          Cout2 series resistance (Rcout2)....[ohms]:      1.2
          Saturation offset voltage (Uo2)....[volts]:      0.8
SELECTION: <↑,↓>      ENTRY: ALPHANUM.      EXECUTE: <PgDn>      ABORT: <ESC>

SELECTING PARAMETER...
```

**FIGURE 12.8**

'Enter Circuit Parameters' – Screen 1 for Class-D.

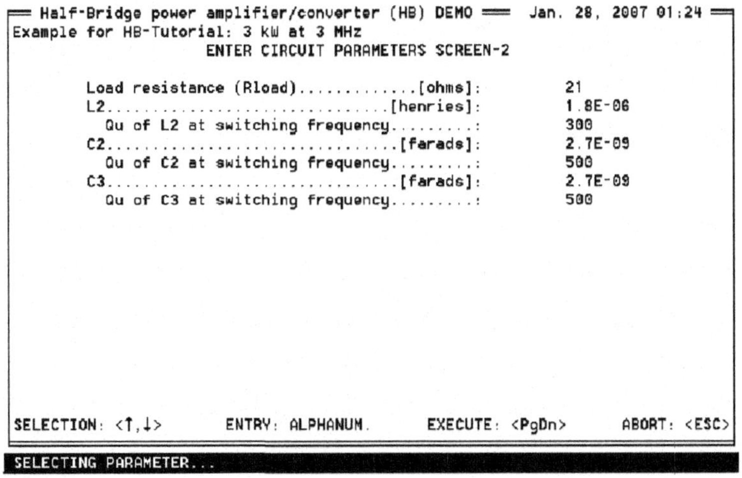

```
== Half-Bridge power amplifier/converter (HB) DEMO ==  Jan. 28, 2007 01:24 ==
Example for HB-Tutorial: 3 kW at 3 MHz
                ENTER CIRCUIT PARAMETERS SCREEN-2

        Load resistance (Rload)............[ohms]:       21
        L2.............................[henries]:        1.8E-06
          Qu of L2 at switching frequency.........:     300
        C2.............................[farads]:         2.7E-09
          Qu of C2 at switching frequency.........:     500
        C3.............................[farads]:         2.7E-09
          Qu of C3 at switching frequency.........:     500













SELECTION: <↑,↓>      ENTRY: ALPHANUM.      EXECUTE: <PgDn>      ABORT: <ESC>

SELECTING PARAMETER...
```

**FIGURE 12.9**

'Enter Circuit Parameters' – Screen 2 for Class-D.

loss in each transistor occurs during turn-on switching; that portion will be essentially eliminated by providing soft-switching operation.

Figure 12.11 shows graph plots of the waveforms $V_{sw1}$ (the voltage across switch $S_1$) and $I_{sw1(actv)}$ (the current in the active portion of switch $S_1$, i.e. not including the current in the output capacitance of the switch). Note the large spike

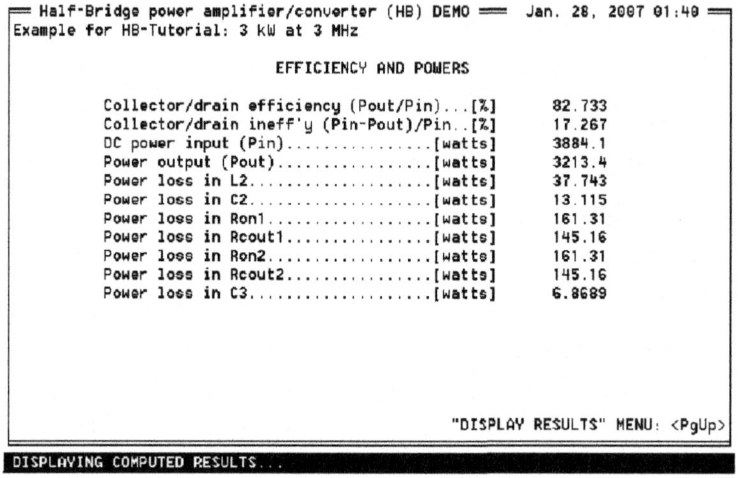

Half-Bridge power amplifier/converter (HB) DEMO Jan. 28, 2007 01:40
Example for HB-Tutorial: 3 kW at 3 MHz
EFFICIENCY AND POWERS
Collector/drain efficiency (Pout/Pin)...[%] 82.733
Collector/drain ineff'y (Pin-Pout)/Pin..[%] 17.267
DC power input (Pin)...............[watts] 3884.1
Power output (Pout)................[watts] 3213.4
Power loss in L2...................[watts] 37.743
Power loss in C2...................[watts] 13.115
Power loss in Ron1.................[watts] 161.31
Power loss in Rcout1...............[watts] 145.16
Power loss in Ron2.................[watts] 161.31
Power loss in Rcout2...............[watts] 145.16
Power loss in C3...................[watts] 6.8689
"DISPLAY RESULTS" MENU: <PgUp>
DISPLAYING COMPUTED RESULTS...

**FIGURE 12.10**

"Efficiency and Powers" screen for Class-D.

Class-D example, 3 kW at 3 MHz, D = 0.50 for both switches
Jan. 25, 2007 20:07
Vswl (V)
0
100
200
300
400
500
5.00E–08 1.00E–07 1.50E–07 2.00E–07 2.50E–07 3.00E–07 3.50E–07
Time (seconds)
Iswl (actv)
0
100
200
300
400
500
5.00E–08 1.00E–07 1.50E–07 2.00E–07 2.50E–07 3.00E–07 3.50E–07
Time (seconds)

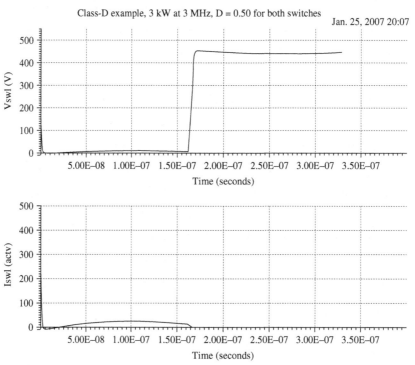

**FIGURE 12.11**

Voltage and current waveforms for Class-D.

of switch current at transistor turn-on, while the node capacitance is being discharged from +450 V down to ground; that current spike is the source of much of the transistors' power dissipation, in the on-resistance ($R_{on}$) of the turning-on transistor and in the series resistances ($R_{Cout}$) of the output capacitances of both transistors.

### 12.4.2 Class-DE with soft switching

To obtain soft-switching operation in a simple way, as explained in Section 12.1, we change only the duty ratio of the two transistors, from the original 50% in Class-D operation down to 40%. Repeating the simulation with only that one change, we obtain the following results: Fig. 12.12 shows the voltage and current waveforms for $S_1$. Note that the current spike previously existing during the Class-D turn-on switching shown in Fig. 12.11 has almost completely disappeared because the conduction gap allows the inductive load current to pull almost all of

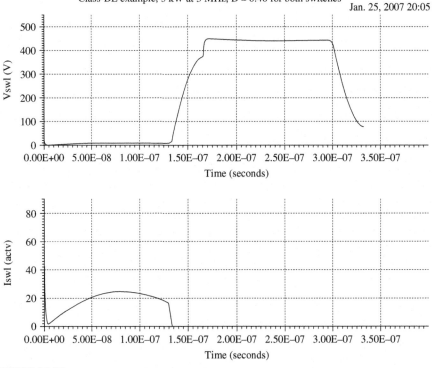

**FIGURE 12.12**

Voltage and current waveforms for Class-DE.

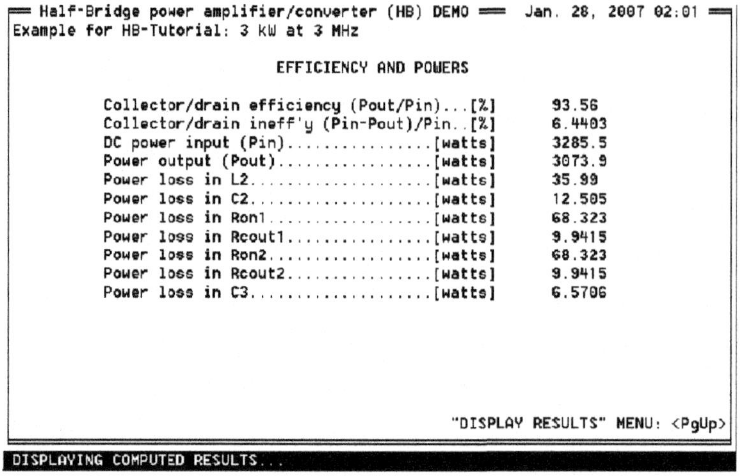

```
== Half-Bridge power amplifier/converter (HB) DEMO ==  Jan. 28, 2007 02:01 ==
Example for HB-Tutorial: 3 kW at 3 MHz

                          EFFICIENCY AND POWERS

         Collector/drain efficiency (Pout/Pin)...[%]      93.56
         Collector/drain ineff'y (Pin-Pout)/Pin..[%]       6.4403
         DC power input (Pin)..............[watts]       3285.5
         Power output (Pout)...............[watts]       3073.9
         Power loss in L2..................[watts]         35.99
         Power loss in C2..................[watts]         12.505
         Power loss in Ron1................[watts]         68.323
         Power loss in Rcout1..............[watts]          9.9415
         Power loss in Ron2................[watts]         68.323
         Power loss in Rcout2..............[watts]          9.9415
         Power loss in C3..................[watts]          6.5706



                                     "DISPLAY RESULTS" MENU: <PgUp>

DISPLAYING COMPUTED RESULTS...
```

**FIGURE 12.13**

'Efficiency and Powers' screen for Class-DE.

the stored energy from the node capacitance before the turning-on switch is commanded to turn on (i.e. the node voltage swings most of the way toward the new voltage level before the switch turns on). When the switch turns on, it pulls the node voltage the rest of the way to the new voltage level, in approximately a step change of voltage. Figure 12.13 shows the "Efficiency and Powers" screen for a Class-DE operation that can be compared with Fig. 12.10 for the Class-D circuit. Now the drain efficiency is 93.56%, 10.8 percentage points higher than the 82.733% obtained with switch duty ratios of 0.5. The output power is 3073.9 W, 95.7% of the 3213.4 W output of the Class-D circuit. But the transistors' power dissipation has been reduced from 612.9 W in the two transistors in Class-D operation to only 156.5 W in Class-DE operation, a reduction by a factor of 3.92. This reduction in power dissipation and increase in drain efficiency are the result of the soft-switching circuit operation. In this simple example, about 95% of the voltage swing is achieved by lossless charging of the node capacitance by the inductive load-network current; the remaining approximately 5% of the voltage swing is accomplished by the turning-on transistor charging the capacitance. The combination of load-network parameter values determines what fraction of the voltage swing is accomplished by lossless charging of the capacitance; under some conditions, it can be the full voltage swing.

We can sweep the duty ratio $D$ of both switches from 0.38 to 0.50 to see continuous functions of the efficiency and output power *versus* duty ratio to verify that the chosen duty ratio is a good choice. The drop in efficiency between the almost-zero-voltage-switching operation at $D = 0.4$ and the hard-switching operation at $D = 0.5$ should be approximately a quadratic function of the increase in

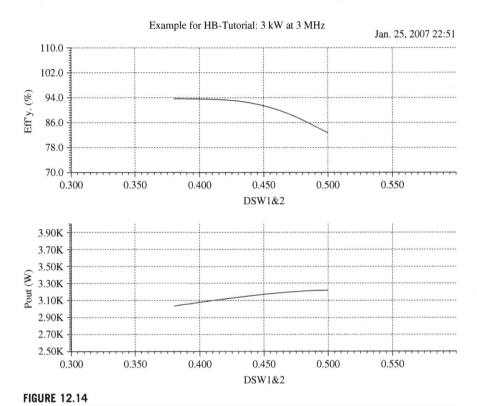

**FIGURE 12.14**

Efficiency and output power *versus* duty ratio for Class-DE.

duty ratio beyond the value at the borderline of soft switching. This is because (*a*) the power dissipation when the transistors charge the node capacitance is proportional to the square of the voltage change through which the transistors charge the capacitance, and (*b*) the voltage change is approximately proportional to the shortage of the conduction gap below the value that just provides soft switching. We can confirm this prediction by sweeping the duty ratio of both switches from 0.40 to 0.50; the result is shown in Fig. 12.14. Sweeping the duty ratio and observing the change in efficiency in the duty ratio also demonstrates that the sensitivity of efficiency to changes in the duty ratio is quite small in the vicinity of the chosen operating point of $D = 0.4$.

The program displays the numerical results for each point in the sweep as that point is computed. In about a second, the program completes the sweep and plots the two requested variables (efficiency and output power) *versus* the swept variable (duty ratio of switches $S_1$ and $S_2$). Note that the shape of the curve of efficiency *versus* duty ratio confirms the predictions in the preceding paragraph.

## 12.5 HEPA-PLUS CAD example for Class-E

Section 12.2 introduces the basic features of HEPA-PLUS program. Now, we will demonstrate a typical design procedure using HEPA-PLUS to design a 13.56-MHz, 100-W Class-E RF power amplifier with a bandwidth of about 5 MHz.

### 12.5.1 Evaluate a candidate transistor

As a first step, it is necessary to evaluate a candidate transistor. In this case, the user examines the amplifier efficiency that can be achieved with one or more candidate transistors at the desired output power of 100 W at 13.56 MHz. For this example, we use the transistor parameters for the International Rectifier IRF540N silicon MOSFET, and quality factors of 150 for inductors and 1000 for capacitors. We assume that the transistor internal temperature is 95°C, obtained as a 60°C rise above an ambient air temperature of 35°C. We use the vendor-specified maximum $R_{ds(on)}$ of 0.044 $\Omega$ at 25°C, and increase that value by a factor of 1.72 to obtain the increased value at 95°C. $R_{Cout}$ is never specified by the vendors of silicon-gate transistors; it is typically five times as large as $R_{ds(on)}$ [2]; we used that factor.

To begin, the user goes from the "Main Menu" shown in Fig. 12.15 to "Transistor Evaluation" shown in Fig. 12.16. The user enters the transistor parameters, the desired frequency range, the amplifier relative bandwidth (the ratio of the frequencies at the upper and lower edges of the desired operating frequency band), and the desired range of output power to be accessed. The desired output power is 100 W; the user brackets that value by specifying the power range to be

**FIGURE 12.15**

"Main Menu".

evaluated as 50−200 W, to be able to see quantitatively the trade-off between efficiency and output power (lower output power yields higher efficiency). The user presses < Page Down > to execute the transistor evaluation. After a few seconds, a family of curves is displayed with frequency on the X-axis, efficiency on the Y-axis, and output power as a parameter, as shown in Fig. 12.17. Observe that for a 13.56-MHz Class-E circuit with an output power of 100 W, the expected efficiency of a nominal-waveforms Class-E circuit is about 87.5%. For the present design example, we will assume that this efficiency is satisfactory. In an actual design, if there were other candidate transistors, the user would check the expected efficiency of each, in turn, by modifying the transistor characteristics entered on the "Transistor Evaluation" screen.

### 12.5.2 Use the automatic preliminary design module to obtain a nominal-waveform Class-E design

The user goes to the "Automatic Preliminary Design" screen by pressing A at the "Main Menu". The Automatic Preliminary Design module designs a nominal-waveforms Class-E circuit (transistor voltage and voltage slope both zero at transistor turn-on time [8]), and calculates the expected amplifier efficiency, using parameters:

- desired frequency band
- two of the following: output power, dc supply voltage and ac load resistance
- transistor parameters (saved from "Transistor Evaluation")
- quality factors of inductors and capacitors (saved from "Transistor Evaluation").

```
══ High-Efficiency Power Amplifier CAD (HEPA-PLUS/WB)   Feb. 01, 2007 12:45 ══
13.56 MHz Class E, International Rectifier IRF540N, Tj=95 C
                        TRANSISTOR EVALUATION
                         Single transistor
     Transistor application
         Frequency range.......[Hz]:   Min.:  6.7802E+06 Max.:  2.7121E+07
         Power range........[watts]:   Min.:  50          Max.:  200
         Amplifier relative bandwidth [f(upper)/f(lower)]:  1.5
     Transistor characteristics
         Peak voltage at nominal tuning..........[volts]:   66
         Turn-off and turn-on times, each..........[sec]:   7.37E-09
         "On" resistance (Ron)....................[ohms]:   0.07568
         Output capacitance (Cout)..............[farads]:   2.7376E-10
         Series resistance of output capacitance...[ohms]:  0.3784
         Transistor saturation offset voltage (Vo)[volts]: 0
     Passive components characteristics
         Unloaded Q of L1 at operating frequency.........   150
         Unloaded Q of L2 at operating frequency.........   150
         Unloaded Q of L3 at operating frequency.........   150
         Unloaded Q of C1 at operating frequency.........   1000
         Unloaded Q of C2 at operating frequency.........   1000
         Unloaded Q of C3 at operating frequency.........   1000
SELECTION: <↑,↓,←,→>   ENTRY: ALPHANUM.       EXECUTE: <PgDn>     ABORT: <ESC>

  SELECTING PARAMETER...
```

**FIGURE 12.16**

"Transistor Evaluation".

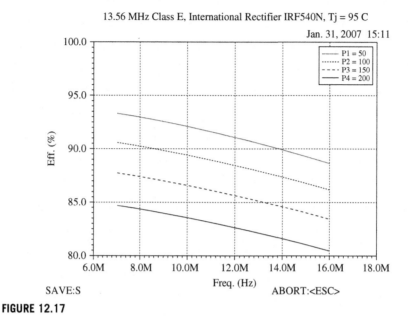

**FIGURE 12.17**

Transistor evaluation: efficiency *versus* frequency and output power.

To begin, the user specifies a 1.5:1 frequency band with a geometric mean of about 13.56 MHz: 11.296–16.296 MHz (a bandwidth of 5 MHz centered at 13.56 MHz). The user specifies a desired output power of 100 W and a dc drain-supply voltage $V_{dd}$ of 22 V to have the nominal $V_{ds(pk)}$ (estimated as approximately 3.4$V_{dd}$ after optimization) of about 75% of the maximum rated $V_{dd}$ of 100 V, to provide a 30% safety factor to allow for off-nominal loads and component tolerances. The user enters "Transistor turn-off and turn-on times" as (in this example) 10% of the RF period, hence 7.37 ns (5–10% of the RF period is usually quite satisfactory). The user presses <Page Down> to execute the Automatic Preliminary Design. Almost immediately, the program displays (without making a simulation) a predicted drain efficiency, for this nominal-waveforms design of 87.8%, and displays the load resistance to be provided at the $R_3$ location as 1.923 $\Omega$. The impedance-transformation ("matching") network (not yet designed) will transform the external load resistance (e.g., 50 $\Omega$) to this computed value. The results so far are deemed to be acceptable, so there is no need to evaluate other candidate transistors. The user transfers the computed circuit design to the "Enter Circuit Parameters" screen by pressing <Page Up>, after which the user will simulate the circuit to verify its correctness. If it is deemed correct, it will be the starting point for an optimization. The result is shown in the screen "Enter Circuit Parameters and/or Title", as shown in Fig. 12.18.

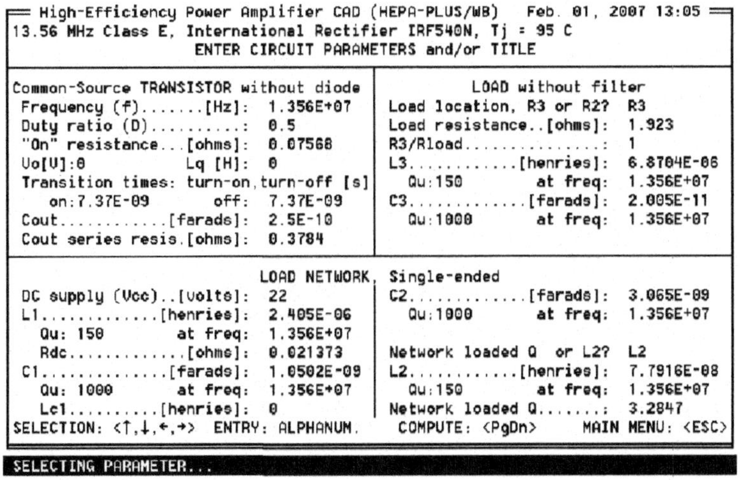

```
== High-Efficiency Power Amplifier CAD (HEPA-PLUS/WB)   Feb. 01, 2007 13:05 ==
   13.56 MHz Class E, International Rectifier IRF540N, Tj = 95 C
                     ENTER CIRCUIT PARAMETERS and/or TITLE
   ┌─────────────────────────────────────┬──────────────────────────────────────┐
   │ Common-Source TRANSISTOR without diode │        LOAD without filter           │
   │  Frequency (f).......[Hz]: 1.356E+07 │ Load location, R3 or R2?   R3          │
   │  Duty ratio (D)..........:  0.5      │ Load resistance..[ohms]:   1.923      │
   │  "On" resistance...[ohms]: 0.07568   │ R3/Rload................:   1         │
   │  Uo[U]:0        Lq [H]:   0          │ L3...........[henries]:    6.8704E-06  │
   │  Transition times: turn-on,turn-off [s] │  Qu:150       at freq:   1.356E+07 │
   │    on:7.37E-09        off:  7.37E-09 │ C3...........[farads]:     2.005E-11   │
   │  Cout...........[farads]:  2.5E-10   │  Qu:1000      at freq:    1.356E+07    │
   │  Cout series resis.[ohms]: 0.3784    │                                       │
   ├─────────────────────────────────────┼──────────────────────────────────────┤
   │             LOAD NETWORK, Single-ended                                       │
   │ DC supply (Vcc)..[volts]:  22        │ C2...........[farads]:     3.065E-09   │
   │ L1............[henries]:   2.405E-06 │  Qu:1000      at freq:    1.356E+07    │
   │  Qu: 150      at freq: 1.356E+07     │                                       │
   │  Rdc...........[ohms]:  0.021373     │ Network loaded Q  or L2?   L2          │
   │ C1............[farads]: 1.0502E-09   │ L2...........[henries]:    7.7916E-08  │
   │  Qu: 1000     at freq: 1.356E+07     │  Qu:150       at freq:    1.356E+07    │
   │  Lc1.........[henries]:  0           │ Network loaded Q........:  3.2847      │
   │ SELECTION: <↑,↓,←,→> ENTRY: ALPHANUM.  COMPUTE: <PgDn>   MAIN MENU: <ESC>    │
   └─────────────────────────────────────────────────────────────────────────────┘
   ▌SELECTING PARAMETER...
```

**FIGURE 12.18**

'Enter Circuit Parameters and/or Title.'

### 12.5.3 Simulate the nominal-waveforms circuit

Pressing < Page Down > at the "Enter Circuit Parameters" screen simulates the circuit in about eight ms on a Pentium III/667 MHz, and faster on faster computers. The result is shown in the screen "Efficiency, Powers, and Stresses," as shown in Fig. 12.19. The drain efficiency ($P_{out}/P_{in,dc}$) is 87.686% and the output power is 96.848 W. The peak transistor voltage and current are 75.346 V and 13.827 A, well within the transistor's maximum ratings of 100 V and 110 A. The Automatic Preliminary Design module, using explicit design equations and no simulation, was designed for 100 W output and achieved within 3.2% of that target value, and had predicted 87.8% efficiency, within 0.11 percentage points of the results of the accurate simulation. Figure 12.20 shows the transistor voltage and current waveforms of this nominal-waveforms Class-E power amplifier. They are seen to be the same as the theoretical waveforms in Figs. 5.5 and 6.8 except that the simulation includes the nonzero switching times, not included in those theoretical waveforms. Note that *high voltage and high current do not exist simultaneously*, even if the transistor turn-on and turn-off transition times are appreciable fractions of the RF cycle. That is the fundamental characteristic of a high-efficiency power amplifier. In this 13.56-MHz design, the transistor turns on at the left edge of the plot at $t = 0$, and begins to turn off near the middle of the plot at $t = 33.2$ ns.

### 12.5.4 RF output spectrum

A spectrum of the RF output power delivered to $R_3$ is available at the screen "Display Spectrum of Plot Variable VR3/L3/C3", shown for the optimized circuit

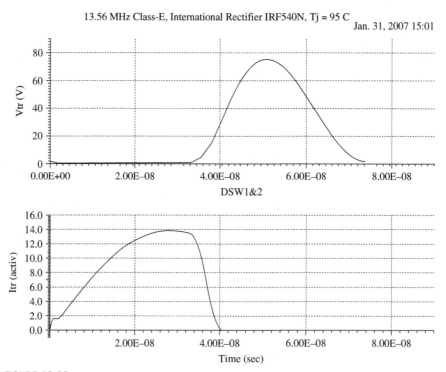

```
 == High-Efficiency Power Amplifier CAD (HEPA-PLUS/WB)   Feb. 01, 2007 13:07 ==
 13.56 MHz Class E, International Rectifier IRF540N, Tj : 95 C
                    EFFICIENCY, POWERS, and STRESSES
                           Single-ended
          Collector/drain efficiency..........[Pout/Pin]      87.686%
          Collector/drain inefficiency...[(Pin-Pout)/Pin]     12.314%
          Overall efficiency............[Pout/(Pin+Pid)]      84.993%
          DC power input (Pin)...................[watts]      110.45
          Input-drive power (Pid)...............[watts]         3.5
          Power output (Pout)...................[watts]      96.848
          Power loss in L1......................[watts]      0.55901
          Power loss in L2......................[watts]       2.229
          Power loss in C2......................[watts]      0.19286
          Resistive power loss of transistor & C1 [watts]    9.3761
          Turn-off power loss of transistor......[watts]      1.1435
          Turn-on power loss of transistor.......[watts]     0.097623
          Power loss in L3......................[watts]      0.002121
          Power loss in C3......................[watts]     0.00031816

     Output voltage, current (at Rload)  [V,A]:     13.647       7.0967
  Transistor peak voltages....[volts]:  normal     75.346   inverse   None
  Transistor peak currents..[amperes]:  normal     13.827   inverse  -0.060441
                                               "DISPLAY RESULTS" MENU: <PgUp>

 DISPLAYING COMPUTED POWERS...
```

**FIGURE 12.19**

"Efficiency, Powers, and Stresses".

**FIGURE 12.20**

Transistor voltage and current waveforms of nominal-waveforms for a Class-E power amplifier.

```
== High-Efficiency Power Amplifier CAD (HEPA-PLUS/WB)   Feb. 01, 2007 13:09 ==
13.56 MHz Class E, International Rectifier IRF540N, Tj : 95 C
            DISPLAY SPECTRUM OF SELECTED VARIABLE:   UR3/L3/C3
   Index   Real [U|A]   Imaginary [U|A]   Magnitude [U|A]/[dB]   Phase [DEG]
     0    -0.002672           0           0.002672 / -74.06        180.00
     1     8.458            10.51          13.49 /   0.00           51.17
     2    -1.408             1.43           2.007 / -16.55         134.56
     3    -0.02661          -0.2968         0.298 / -33.11         264.88
     4    -0.02972           0.09503        0.09957 / -42.64       107.37
     5    -0.004858         -0.03743        0.03774 / -51.06       262.60
     6    -0.001651          0.01417        0.01427 / -59.51        96.64
     7     0.0003398        -0.006672       0.006725 / -66.05      -82.83
     8     0.001435          0.003905       0.004161 / -70.22       69.82
     9    -0.0008718         0.0003009      0.0009222 / -83.30     160.96
    10    -0.0002246        -0.0004733      0.0005239 / -88.21     244.61
    11     5.648E-05         0.0004415      0.0004451 / -89.63      82.71
    12     7.456E-06        -0.0004093      0.0004093 / -90.36     -88.96
    13     0.0002482        -0.0004129      0.0004817 / -88.94     -58.99
    14     0.0009398        -8.411E-05      0.0009436 / -83.10      -5.11
    15     0.0004091         0.0001675      0.0004421 / -89.69      22.27
    16     0.0002795         0.0004242      0.000508 / -88.48       56.62
    17    -0.0002444        -0.0002604      0.0003571 / -91.54     226.82
SCROLL: <PgUp,PgDn,↑,↓>                        "DISPLAY RESULTS" MENU: <Enter>
```

```
DISPLAYING COMPUTED SPECTRUM...
```

**FIGURE 12.21**

'Display Spectrum of Plot Variable VR3/L3/C3.'

in Fig. 12.21. Other voltages and currents can be chosen also as the variable to be displayed. For example, the harmonic content of $I_{L1}$, the dc current drawn from the dc power supply, gives an indication of how much bypass-capacitance filtering is needed on the $V_{dd}$ line to maintain the conducted EMI current below a desired maximum allowed value. The first spectrum screen shown in Fig. 12.21 lists the dBc values from dc (actually zero output, listed as −74.06 dBc, an approximation to −∞ dBc) through the 17th harmonic. A similar second screen (not shown here) lists the values for the 18th through 31st harmonics. At each harmonic, the difference between the listed value and the maximum allowed harmonic output is the number of dB of suppression needed in the post-amplifier harmonic-suppression filter. For example, if all harmonics must be below −65 dBc, the 7th and higher harmonics need no suppression, $2f$ (− 16.55 dBc) needs 65−16.55 = 48.45 dB of suppression, $3f$ (− 33.11 dBc) needs 65−33.11 = 31.89 dB of suppression, and so on.

## 12.5.5 Optimize the design, using the nominal-waveforms design as a starting-point

The Optimizer module is entered by pressing O at the Main Menu or by highlighting that line and pressing the < Enter > key. That takes the user to the screen "Inputs to Optimizer, Screen 1—Objectives of Optimization", shown as the upper screen in Fig. 12.22. That upper screen shows the first four of the seven performance characteristics that can be used to define the meaning of "optimum" for this application; the next three characteristics are on the continuation Screen 2, shown below Screen 1 in Fig. 12.22. For this example, we will give the

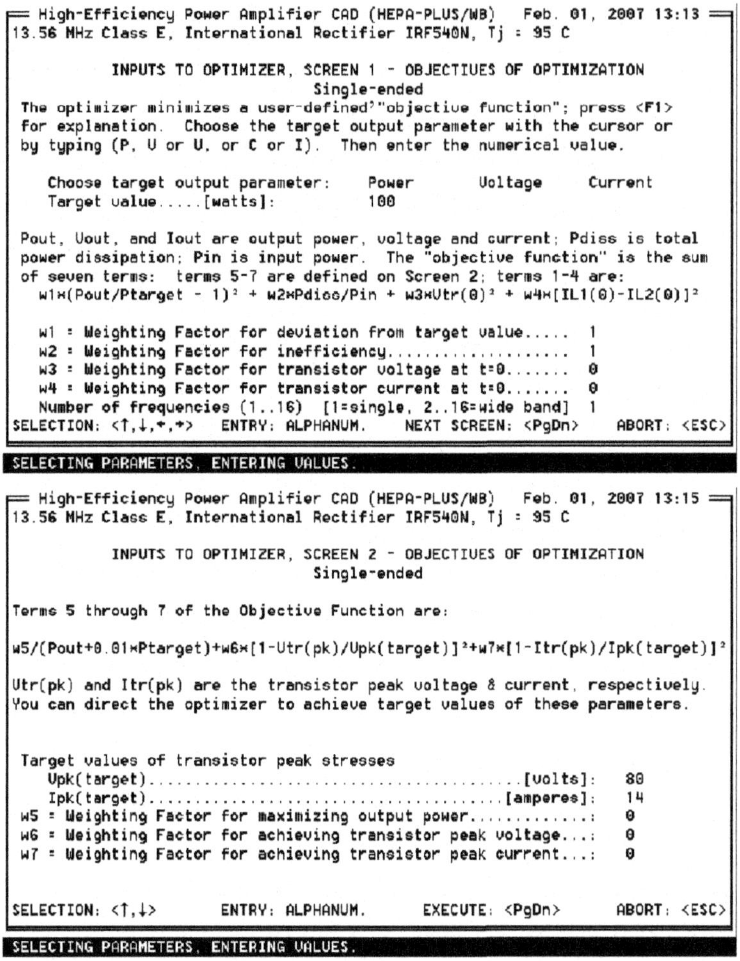

```
== High-Efficiency Power Amplifier CAD (HEPA-PLUS/WB)  Feb. 01, 2007 13:13 ==
13.56 MHz Class E, International Rectifier IRF540N, Tj : 95 C

        INPUTS TO OPTIMIZER, SCREEN 1 - OBJECTIVES OF OPTIMIZATION
                          Single-ended
The optimizer minimizes a user-defined "objective function"; press <F1>
for explanation.  Choose the target output parameter with the cursor or
by typing (P, V or U, or C or I).  Then enter the numerical value.

    Choose target output parameter:    Power      Voltage     Current
    Target value.....[watts]:          100

Pout, Vout, and Iout are output power, voltage and current; Pdiss is total
power dissipation; Pin is input power.  The "objective function" is the sum
of seven terms:  terms 5-7 are defined on Screen 2; terms 1-4 are:
   w1*(Pout/Ptarget - 1)² + w2*Pdiss/Pin + w3*Utr(0)² + w4*[IL1(0)-IL2(0)]²

    w1 : Weighting Factor for deviation from target value.....  1
    w2 : Weighting Factor for inefficiency...................  1
    w3 : Weighting Factor for transistor voltage at t=0.......  0
    w4 : Weighting Factor for transistor current at t=0.......  0
    Number of frequencies (1..16) [1=single, 2..16=wide band]  1
SELECTION: <↑,↓,←,→>   ENTRY: ALPHANUM.    NEXT SCREEN: <PgDn>    ABORT: <ESC>
```
**SELECTING PARAMETERS, ENTERING VALUES.**
```
== High-Efficiency Power Amplifier CAD (HEPA-PLUS/WB)  Feb. 01, 2007 13:15 ==
13.56 MHz Class E, International Rectifier IRF540N, Tj : 95 C

        INPUTS TO OPTIMIZER, SCREEN 2 - OBJECTIVES OF OPTIMIZATION
                          Single-ended

Terms 5 through 7 of the Objective Function are:

w5/(Pout+0.01*Ptarget)+w6*[1-Utr(pk)/Upk(target)]²+w7*[1-Itr(pk)/Ipk(target)]²

Utr(pk) and Itr(pk) are the transistor peak voltage & current, respectively.
You can direct the optimizer to achieve target values of these parameters.


    Target values of transistor peak stresses
      Upk(target)......................................[volts]:   80
      Ipk(target)......................................[amperes]: 14
    w5 : Weighting Factor for maximizing output power.............  0
    w6 : Weighting Factor for achieving transistor peak voltage...  0
    w7 : Weighting Factor for achieving transistor peak current...  0


SELECTION: <↑,↓>      ENTRY: ALPHANUM.     EXECUTE: <PgDn>     ABORT: <ESC>
```
**SELECTING PARAMETERS, ENTERING VALUES.**

**FIGURE 12.22**

"Inputs to Optimizer—Objectives of Optimization".

weighting factors $w_1$ (deviation from the target value of output power) and $w_2$ (inefficiency) equal weights of 1, and leave the remaining five weighting factors at their default values of zero. Note that all mathematicians except the Russians optimize by *minimizing* the "badness" of the design (here, $w_1$ is for the *deviation* from the target value of output power and $w_2$ is for the *in*efficiency); the Russians optimize by *maximizing* the "goodness" of the design. Both methods work well and are basically equivalent.

The optimization can be done at one frequency or at up to 16 frequencies simultaneously, useful for wideband designs. For simplicity, we will illustrate the optimization at one frequency of 13.56 MHz. The user proceeds to the screen

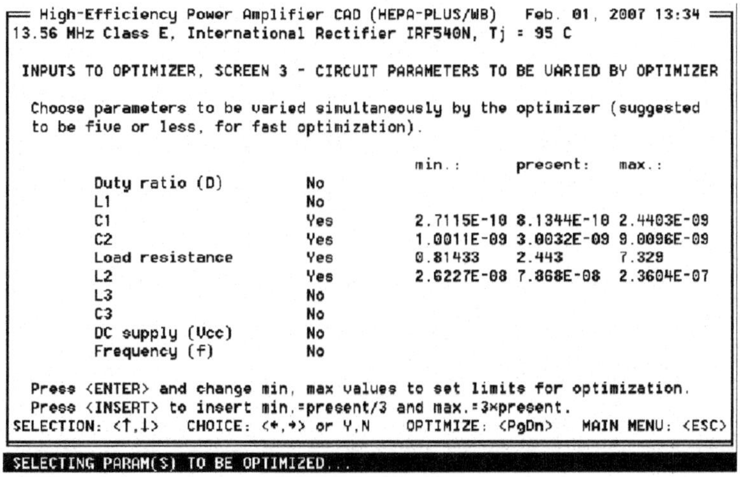

```
═ High-Efficiency Power Amplifier CAD (HEPA-PLUS/WB)   Feb. 01, 2007 13:34 ═
13.56 MHz Class E, International Rectifier IRF540N, Tj = 95 C

 INPUTS TO OPTIMIZER, SCREEN 3 - CIRCUIT PARAMETERS TO BE VARIED BY OPTIMIZER

 Choose parameters to be varied simultaneously by the optimizer (suggested
 to be five or less, for fast optimization).

                                          min.:      present:    max.:
           Duty ratio (D)         No
           L1                     No
           C1                     Yes     2.7115E-10 8.1344E-10 2.4483E-09
           C2                     Yes     1.0011E-09 3.0032E-09 9.0096E-09
           Load resistance       Yes     0.81433    2.443      7.328
           L2                     Yes     2.6227E-08 7.868E-08  2.3604E-07
           L3                     No
           C3                     No
           DC supply (Vcc)        No
           Frequency (f)          No

  Press <ENTER> and change min, max values to set limits for optimization.
  Press <INSERT> to insert min.=present/3 and max.=3×present.
 SELECTION: <↑,↓>   CHOICE: <◄,►> or Y,N    OPTIMIZE: <PgDn>   MAIN MENU: <ESC>

 SELECTING PARAM($) TO BE OPTIMIZED...
```

**FIGURE 12.23**

"Inputs to Optimizer, Screen 3—Circuit Parameters to Be Varied by the Optimizer".

"Inputs to Optimizer, Screen 3—Circuit Parameters to be Varied by the Optimizer", shown in Fig. 12.23. Here, the user can choose which parameters the Optimizer is allowed to vary, and the minimum and maximum allowed values of those parameters, as constraints on the Optimizer. We will choose $C_1$, $C_2$, $L_2$, and Load Resistance as the parameters to be varied in the search for the optimum combination. Parameters not chosen to be varied:

1. Duty ratio is set to 50%, already known to be a reasonable compromise among conflicting trade-off factors.
2. $L_1$ is effectively "infinite"; higher inductance will have essentially no effect, and lower inductance is not needed.
3. $L_3$ and $C_3$ are tuned to the operating frequency, but are given high enough reactance values that they are essentially removed from the circuit. They are available if a tuned load is desired, but that is not needed in this case. For harmonic suppression, a better way to use an inductor and a capacitor (such as $L_3$ and $C_3$) would be to place them in a low-pass $LC$ filter ($-6$ dB per octave above the operating frequency), rather than as a parallel resonator at the load ($-3$ dB per octave above the operating frequency).
4. The dc supply voltage was already chosen as 22 V to provide sufficient margin against excessive peak transistor voltage.
5. The operating frequency was already chosen as 13.56 MHz, according to the requirements placed on the design.

The user presses the < Page Down > key to begin the optimization. In a few seconds, the result is shown at the screen "Results of Optimization, Completed at Power Evaluation No. nnn", where *nnn* is the number of evaluations made on the

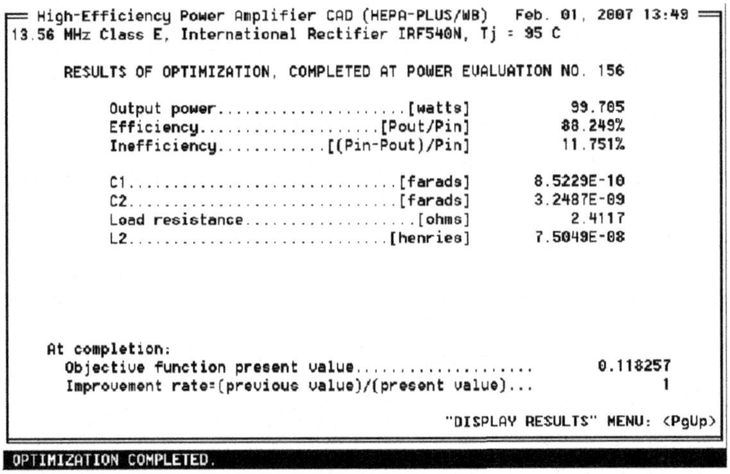

```
== High-Efficiency Power Amplifier CAD (HEPA-PLUS/WB)   Feb. 01, 2007 13:49 ==
13.56 MHz Class E, International Rectifier IRF540N, Tj = 95 C

     RESULTS OF OPTIMIZATION, COMPLETED AT POWER EVALUATION NO. 156

       Output power...................[watts]          99.785
       Efficiency.................[Pout/Pin]          88.249%
       Inefficiency...........[(Pin-Pout)/Pin]        11.751%

       C1..........................[farads]         8.5229E-10
       C2..........................[farads]         3.2487E-09
       Load resistance.............[ohms]             2.4117
       L2..........................[henries]        7.5049E-08




    At completion:
       Objective function present value...............     0.118257
       Improvement rate=(previous value)/(present value)...    1

                            "DISPLAY RESULTS" MENU: <PgUp>

 OPTIMIZATION COMPLETED.
```

**FIGURE 12.24**

"Results of Optimization, Completed at Power Evaluation No. 156."

way to finding the optimum (156 in this example). Figure 12.24 shows that screen for the case computed here. The initial value of efficiency of the starting circuit (87.686% for the nominal-waveforms design shown in Fig. 12.19, already quite high) was improved by 0.563 percentage points to the listed value of 88.249% shown in Fig. 12.24. The optimizer usually reduces the power dissipation by about 30% of its value for the starting circuit, usually increasing the drain efficiency by about 5 percentage points. But if the starting circuit already has high efficiency, the increase in efficiency from optimizing can be small. To illustrate the larger improvement that can be obtained when the nominal-waveforms circuit does not *already* have high efficiency, the authors increased the required output power from 100 to 300 W, and reduced the $Q$ values of $L$ and $C$ from 150 and 1000 to 50 and 200, respectively; both changes increase the resistive power losses. As a result, for nominal-waveforms circuit: 300 W at 75.207% efficiency, with waveforms similar to Fig. 12.20; for optimized circuit: 302 W at 80.585% efficiency, efficiency increased by 5.378 percentage points, with waveforms similar to Fig. 12.25.

The transistor voltage and current waveforms in the first optimized circuit are shown in Fig. 12.25; they are off-nominal as compared with the non-ideal nominal waveforms shown in Fig. 12.20 or the idealized nominal waveforms shown in Figs. 5.5 and 6.8. The user can read out $X$ and $Y$ numerical values from the plot by pressing the left/right cursor keys to move a vertical read-out cursor to the point of interest. The numerical values are displayed at the upper-left and lower-left corners of the plots. The read-out cursor is positioned at the peak of the transistor-voltage waveform; the peak voltage is seen to be 74.79 V, the same as the value that would be listed at the bottom of the screen "Efficiency, Powers, and Stresses" for the optimized circuit.

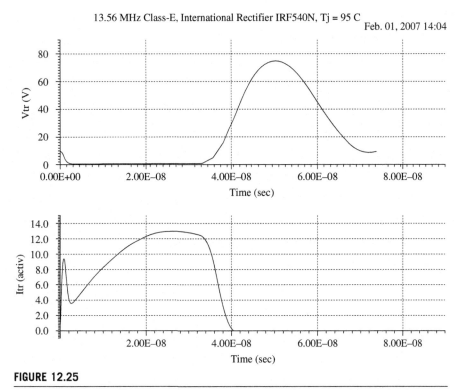

13.56 MHz Class-E, International Rectifier IRF540N, Tj = 95 C
Feb. 01, 2007 14:04

**FIGURE 12.25**

Transistor optimized off-nominal voltage and current waveforms in optimized circuit.

### 12.5.6 Use the SWEEP function

The program can sweep any of 25 circuit parameters and display graph-plots of one or two of 18 available output variables. At the Main Menu, the user presses S for Sweep. The program presents the screen "Sweep Menu" shown in Fig. 12.26. For this demonstration, we will sweep Frequency, chosen from a menu of 20 available parameters. The user enters values to compute the results at 13 points from 11 to 17 MHz in an algebraic sweep (equal steps of the swept variable between adjacent points). Also available is Geometric sweep, that has a given multiplication factor on the swept variable between adjacent points. The program determines the step value, given the user-specified range, number of points, and type of sweep.

The user chooses to plot the default output variables: Output Power and Efficiency. The user presses < Page Down > to execute the sweep. During the sweep calculations, the latest values of the swept parameter and the variables to be plotted are displayed in a table at the right side of the screen; the user can watch that table change during the sweep. At the completion of the computation, the program displays graphs of Output Power and Efficiency *versus* frequency, as shown in Fig. 12.27.

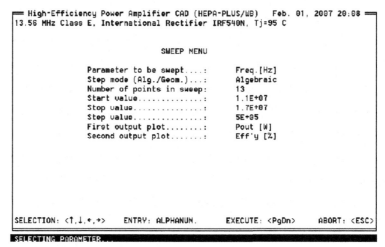

**FIGURE 12.26**

"Sweep Menu".

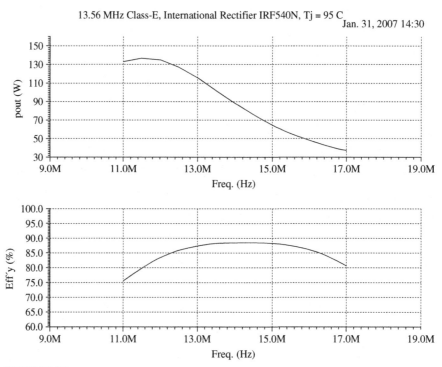

**FIGURE 12.27**

Output power and efficiency *versus* swept frequency.

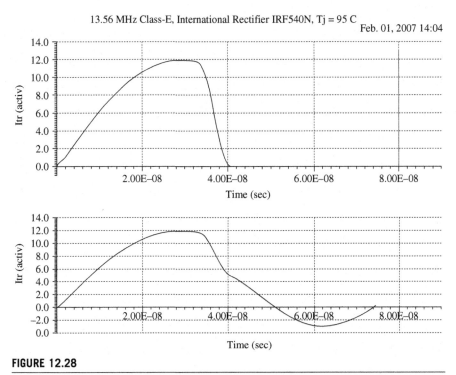

**FIGURE 12.28**

Waveforms of current in the active part of a transistor and the total transistor current.

Two more interesting waveforms to display are the currents in the power transistor. We can examine both $I_{tr(activ)}$ (the current in *the active part* of the transistor) and $I_{tr(total)}$ (*the terminal current*, which is the sum of the currents in the active part and in the transistor's output capacitance), and we can compare the two current waveforms. If we wish, we can also display separately the current in the transistor's output capacitance. In order to begin, return to the screen "Variable(s) to Be Plotted" by pressing <Enter>, <Page Up>, and P. Then press <6> to open a window showing a list of variables that can be plotted. Move the cursor to $I_{tr(activ)}$ and press <Enter>. Then press <6> to place the cursor on $I_{tr(total)}$ and press <Enter>. Press <Page Down> to display the two waveforms, as shown in Fig. 12.28.

In the laboratory, using an oscilloscope, you can observe only $I_{tr(total)}$, and that only up to about 50 MHz. Unfortunately, this does not give a clear picture of the details of circuit operation. With the program, however, you can observe clearly the circuit operation with $I_{tr(activ)}$, and you can also see $I_{tr(total)}$ at the same time. By means of the latter, you can compare experimental observations with the computer simulation, or you can *predict* what you can *expect* to see in a laboratory test.

Similarly, you can observe two versions of the transistor voltage waveform when the transistor internal wiring inductance ($L_Q$ in Fig. 12.4) is nonzero: the transistor *internal* voltage at the transistor die, termed $V_{tr(inter)}$, and the transistor *terminal* voltage $V_{tr(term)}$, which includes also the voltage across $L_Q$. Experimentally, you can observe only $V_{tr(term)}$, but $V_{tr(inter)}$, which you cannot observe experimentally, gives you much more information about the circuit operation. You can easily observe both voltages with the program if you have set $L_Q$ to a nonzero value at the screen "Enter Circuit Parameters." If $L_Q$ is zero, the internal and external voltages are identical, so the transistor voltage is called simply $V_{tr}$.

## 12.6 Class-E power amplifier design using SPICE

The high-voltage MOSFET devices are preferable to provide a very high output power at high frequencies operating in the switchmode Class-E. Figure 12.29 shows the simulated circuit schematic of a 13.56-MHz, 400-W MOSFET Class-E power amplifier implemented in SPICE [9]. The MOSFET device IRFP450LC is modeled as a switch with a linear capacitor $C_t$ and a nonlinear voltage source $E_{ct}$ to model the square-root behavior of the drain-source capacitance. To simulate an accurate frequency-domain behavior, it is important to include the parasitic inductance $L_{cd}$ of the drain-source capacitor $C_d$ and the parasitic capacitor $C_r$ due to the

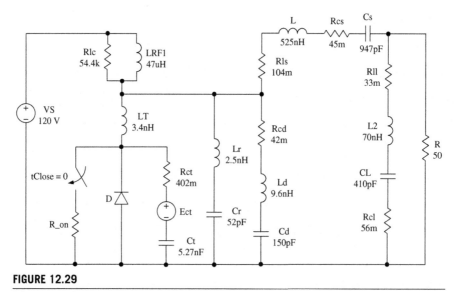

**FIGURE 12.29**

13.56-MHz Class-E power amplifier SPICE model.

heatsink, insulating pad, and MOSFET assembly. The sum of the inductances $L_T$ and $L_r$ along with a capacitance $C_r$ predicts the resonant frequency at the 22nd harmonic component or 298 MHz. The final SPICE simulation involved the calculation of component losses using the measured values of the resistances for the inductors and capacitors. The accuracy of the drain efficiency and output power simulations was of about 1% compared to the experiment. The practical circuit schematic of this high-power Class-E MOSFET power amplifier is shown in Fig. 5.32(a) in Chapter 5.

High-voltage MOSFETs are available in mirror image pairs and the heat spreader of the plastic package is connected to the source. They can operate up to supply voltage of 300 V and at frequencies up to 100 MHz. The device used for this amplifier is ARF448A (or its more rugged version ARF460A) having a breakdown voltage of 450 V and drain junction thermal resistance of 0.55°C/W. The power gain in Class C is more than 25 dB at 27 MHz. The drain-source capacitance $C_{oss}$ is 125 pF at a supply voltage $V_{dd}$ of 125 V, and the saturation resistance $R_{ds(on)}$ is 0.4 Ω. These data-sheet parameters were entered in the HEPA-PLUS Class-E design program and it generated starting design values for the Class-E load network. The circuit schematic of a 27.12-MHz, 500-W MOSFET power amplifier common to most Class-E amplifiers is shown in Fig. 12.30 [10]. In Class-E amplifiers, there is a capacitor shunted across the drain to source. There is none used here because the output capacitance of the device is slightly larger than the optimum value for that component required for Class-E. This defines the upper-frequency limit for efficient Class-E operation of a particular device. The parasitic series resistance of 3 Ω of the lossy drain-source capacitance $C_{oss}$ is one of the primary loss mechanisms in the circuit. However, $C_{oss}$ exhibits little of the parasitic inductance. The output circuit values were adjusted slightly for maximum efficiency at an output power of 490 W.

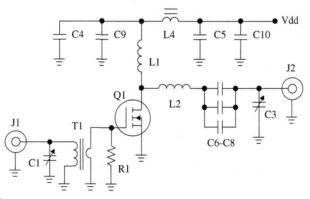

**FIGURE 12.30**

Circuit schematic of 27.12-MHz MOSFET Class-E power amplifier.

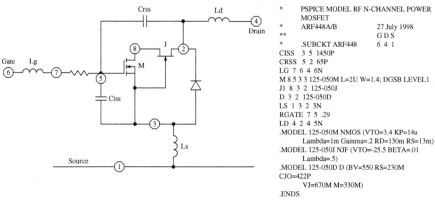

```
*        PSPICE MODEL RF N-CHANNEL POWER
         MOSFET
*        ARF448A/B                 27 July 1998
**                                 G D S
*        .SUBCKT ARF448            6 4 1
CISS  3 5 1450P
CRSS  5 2 65P
LG 7 6 4 6N
M 8 5 3 3 125-050M L=2U W=1.4; DGSB LEVEL1
J1 8 3 2 125-050J
D 3 2 125-050D
LS 1 3 2 3N
RGATE 7 5 .29
LD 4 2 4 5N
.MODEL 125-050M NMOS (VTO=3.4 KP=14u
        Lambda=1m Gamma=.2 RD=130m RS=13m)
.MODEL 125-050J NJF (VTO=-25.5 BETA=.01
        Lambda=.5)
.MODEL 125-050D D (BV=550 RS=230M
CJO=422P
        VJ=670M M=330M)
.ENDS
```

**FIGURE 12.31**

SPICE model for ARF448 MOSFET.

While the designing of the output circuit is straightforward enough, it was not reduced to practice very easily. The main problem was providing enough RF voltage to the gate to drive the drain into voltage saturation. The input impedance of the gate at 27 MHz is $(0.1 - j2.7)$ $\Omega$ and the gate-source capacitance $C_{iss}$ is 1400 pF. If 10 V of peak gate drive is needed, a reasonable match between the drive source and gate is required. There is approximately 9 nH of parasitic series gate inductance. This is enough inductance to make it impossible to observe the actual voltage applied to the gate with an oscilloscope. A SPICE macro model for the device is shown in Fig. 12.31 [10]. The goal was to design a network using SPICE to match the gate impedance sufficiently to permit sine-wave drive.

The input circuit using a 4:1 transformer and a low-pass $L$-network was designed using Smith chart software program. To stabilize the power amplifier operation, a 25-$\Omega$, 5-W padding resistance was placed across the gate representing the parallel equivalent of the capacitance 2200 pF in parallel with the resistance 210 $\Omega$. This raises the effective input impedance to $(0.38 - j2.6)$ $\Omega$, lowers the network quality factor, and makes it much easier to match and drive properly. The input transformer is made from a two-hole ferrite "binocular" bead balun. The secondary winding consists of two 7/8-inch pieces of 3/16-inch diameter brass tubing connected with copper shim stock. The four-turn primary is wound inside the tubes for maximum coupling and minimum leakage which measured 19 nH referenced to the secondary side. High-quality passive components are required in the output network. Most important of these is the series inductor $L_2$ with a quality factor of 375 and a calculated dissipation of 4.2 W, which was wound from bare copper wire, as described in Table 12.1. This coil is not capable of continuous duty operation unless it is attached with high-temperature solder and/or separate mechanical termination support is used. It was necessary to parallel three 10-nF ceramic coupling capacitors (type BX or Y5V) to carry the RF

---

**Table 12.1** 27.12-MHz Class-E Power Amplifier Component Values

| | |
|---|---|
| C1, C3 | 75–380 pF mica trimmer, ARCO 465 |
| C4–C8 | 0.01 μF 1 kV disc ceramic |
| C9, C10 | 0.01 μF 500 V disc ceramic |
| L1 | 0.6 μH. 25t #24 ga.enam. 0.5" diameter |
| L2 | 210 nH. 4t #8 ga. .75" id, 1" long |
| L4 | 2t #20 PTFE on 0.5" ferrite bead μ = 850 |
| Q1 | APT ARF448A |
| R1 | 25 Ω 5 W non-inductive |
| T1 | Pri: 4t #20 PTFE, Sec: 1t brass tube on two-hole balun bead Fair-Rite #2843010302 μ = 850 |

current. Table 12.1 shows the components values of a 27.12-MHz, 500-W Class-E power amplifier [10].

A simple SPICE model of the power amplifier was compared with HEPA simulation and measured results. Overall, the agreement was good, especially between HEPA and the ideal circuit SPICE model. Attempts to insert the SPICE macro model of the transistor into the amplifier were not successful. A much more sophisticated model is needed to adequately simulate the effects of the nonlinear capacitances of the MOSFET. However, the SPICE model was very useful for understanding the gate-drive problem mentioned earlier. HEPA assumes that the device is being driven into voltage saturation; the only input parameter it considers is input drive power used for the overall efficiency calculation. The gate drive was the biggest problem in the design because a large RF-capable power MOSFET has very small input impedance. One of the best tests for reliability of an amplifier is mismatch load testing. The ARF448 has about 175 W of available dissipation in the test amplifier with its air cooled heat sink. The performance at eight points around a 2:1 *VSWR* mismatch circle was calculated using HEPA. It was found that there is a region with the high inductive impedance that should be avoided. Without protection, the output power soared to 750 W and the drain current was almost twice normal. The efficiency stayed quite constant, never losing more than 11% at any load angle. The measured gate and drain voltage waveforms are shown in Fig. 12.32 [10]. The peak drain voltage measures 430 V. The highest efficiency of 83% was measured at 490 W, which is 5.5% below the simulated due to the lossy nonlinear drain-source capacitance and longer switching time.

Figure 12.33 shows the PSPICE model describing the transistor switchmode operation with the output voltage-dependent capacitance [11]. The five basic elements are included in this model: ideal switch SW, transistor on-resistance $R_{on}$ which normally can be found from data sheet, saturation offset voltage $V_o$ which

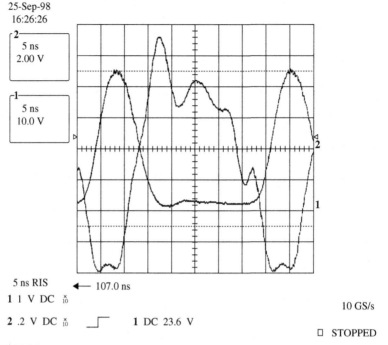

**FIGURE 12.32**

Measured gate and drain voltage waveforms.

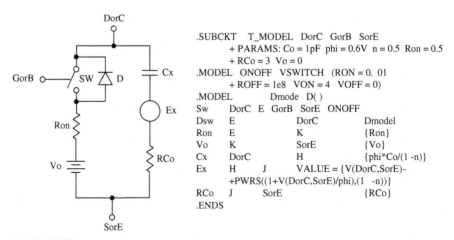

**FIGURE 12.33**

PSPICE model of transistor switchmode.

is equal to zero for FET devices, nonlinear equivalent for the output capacitance ($C_X$ in series to $E_X$), and equivalent series resistance $R_{Co}$ due to lossy output capacitance.

To implement the nonlinear voltage dependence of the output capacitance (the SPICE standard capacitor supports only linear capacitance), its nonlinear behavior can be represented by the junction voltage dependence in the form of

$$C(v) = \frac{C_0}{\left(1 + \frac{v}{\phi}\right)^n} \qquad (12.1)$$

where $C_0$ is the capacitance for $v = 0$ and $n$ is the junction profile coefficient. During optimization procedure, the parameters $C_0$, $n$, and $\phi$ are considered the fitting parameters that can be extracted from the data-sheet curve for the output MOSFET capacitance $C_{oss}$.

The nonlinear voltage dependence can be modeled with a fixed capacitor $C_X$ and a series voltage-controlled voltage source $E_X$, which can be found from the expression for the total voltage $v$ given by

$$v = v_C + E_X(v) \qquad (12.2)$$

where $v_C$ is the voltage across the capacitor $C_X$ and $E_X$ is defined as

$$E_X(v) = v - \left(1 + \frac{v}{\phi}\right)^{1-n}. \qquad (12.3)$$

By differentiating both parts of Eq. (12.2) and taking into account that $i/C = dv/dt$ and $i/C_X = dv_C/dt$, the fixed capacitor can be determined using Eq. (12.1) as

$$C_X = C_0 \frac{\phi}{1-n}. \qquad (12.4)$$

The model parameters $R_{on}$ and $R_{Co}$ can be measured or estimated from the device data sheet. The sub-circuit shown in Fig. 12.33 must be connected to nodes DorC (drain or collector), SorE (source or emitter), and GorB (gate or base). A VSWITCH element should be used for the ideal switch SW, with two resistance values (low and high) following the control voltage applied to the gate terminal. If this voltage is equal or greater than VON (default value 4 V), the switch turns on to low resistance RON (default value 0.01 $\Omega$). Otherwise, the switch turns off to high resistance ROFF (default value 1E8 $\Omega$). The diode D parameters can be adjusted through its .MODEL description. The parameters $C_0$, $n$, $\phi$, $R_{Co}$, and $R_{on}$ can be more accurately modeled using the optimization procedure based on a discrete-time technique for the steady-state analysis of the nonlinear switched circuits with inconsistent initial conditions [12]. Using NGSPICE which is a latest version of Berkeley SPICE, the optimization procedure for the entire Class-E power amplifier (including input circuit) based on the time-domain shooting method for finding the steady-state responses can be fully implemented [13].

## 12.7 ADS circuit simulator and its applicability to switchmode Class-E

The Agilent Advanced Design System (ADS) circuit simulator representing a comprehensive simulator of linear and nonlinear circuits in the frequency and time domains. It can be used directly to model and simulate the performance of the switchmode Class-E power amplifiers. This can be done using the transient, envelope, and harmonic balance simulation engines.

Figure 12.34 shows the simulation setup for ideal parallel-circuit Class-E operation in the time domain. The active device is represented by a voltage-controlled switch with off-resistance of 1 MΩ and small finite on-resistance, the value of which can generally be varied. The input source represents a voltage source with pulse train defined at discrete time steps used in envelope and transient simulators. The use of discrete time pulse source, as opposed to a standard pulse source, can guarantee that there is no timing jitter in the pulse edge due to the waveform being sampled asynchronously by a fixed-time interval simulation. The simulation time is significantly faster than the period of a square wave.

To provide the circuit simulation in time domain, the transient simulator is added to the simulation template. The stop time of 20 s is chosen for a normalized frequency of 1 Hz which is sufficient to reach a steady-state mode for a simulated operating frequency normalized to unity, as shown on the example of the waveforms of the (a) switch voltage and (b) load current shown in Fig. 12.35. The inductors and capacitors are lossless and the loaded quality factor $Q_L$ of the series resonant circuit is chosen to be as high as 20. The measure equations MeasEqn include the conditions when the switch voltage V_sw and its voltage derivative V_sw_der must take zero values at the instant just before the switch is turned on. The efficiency is calculated in the 19th + 20th period since the products of instantaneous current and voltage are integrated over these two periods and divided by two. The function "integrate" automatically deals with the nonconstant time steps in the transient simulation results. The term "switch_index" is the number (index) of the simulation points for 19 s, the instant when the switch is turned on, while the term "switch_index-1" is therefore the simulation point just before the switch is turned on.

After the transient simulation has settled to a steady-state mode, the simulation results for the optimum parallel-circuit Class-E load-network parameters calculated from Eqs (6.79) to (6.81) in Chapter 6 demonstrate the' ideal Class-E voltage and current waveforms. The optimization simulator added to the simulation template shown in Fig. 12.34 is necessary to optimize the load-network parameters by varying their factors for a non-ideal switch with finite on-resistance. The optimization is performed to minimize the switch voltage and voltage-derivative values to zero.

Figure 12.36 shows a set of the switch (a) voltage and (b) current waveforms with a duty cycle (or ratio) of 0.5 obtained for zero voltage and voltage-derivative

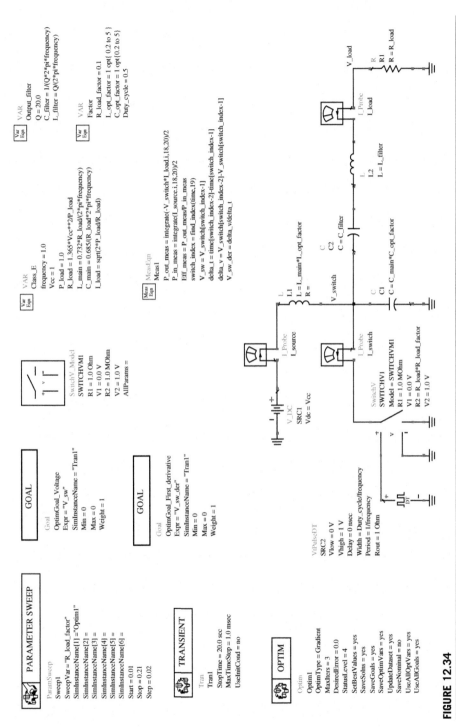

**FIGURE 12.34**

Simulation setup to maintain a Class-E mode in time domain.

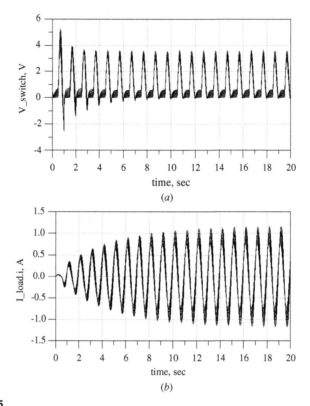

**FIGURE 12.35**

Transient response of switch voltage and load current.

conditions by sweeping the switch resistance load factor from 0.01 to 0.21 with a step of 0.02. The total simulation time for a 1.6-GHz processor is 1.2 h. In this case, the peak voltage and current values are the smallest for maximum values of the switch on-resistance, and the saturation voltage becomes significant resulting in a lower output power and efficiency. The output power and efficiency drop by approximately 45% and 39% respectively, when a ratio $r_{sat}/R$ reaches the value of 0.15, as shown in Fig. 12.37($a$). This is achieved for an increased capacitance by 29% and a reduced inductance by 29%, as shown in Fig. 12.37($b$). When $r_{sat}/R = 0.1$, the efficiency is equal to 73.4%.

However, the Class-E zero voltage and zero voltage-derivative conditions become non-optimum for finite values of the on-resistance. This means that higher efficiency can be achieved when these Class-E conditions are nonzero. Consequently, some voltage is present at the capacitor before the switch is turned on. By maintaining the switch transient time to almost zero and the optimum parameters of the load network, one can observe the discharge of this voltage in the shape of a current spike. Figure 12.38 shows the switch ($a$) voltage and

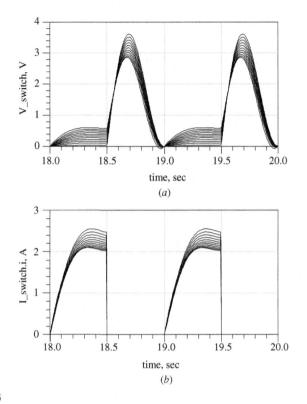

**FIGURE 12.36**

Parallel-circuit Class-E optimum waveforms with finite on-resistance.

(b) current waveforms as a function of the normalized switch saturation resistance $r_{sat}/R_L$ varying from 0.05 to 0.3 with a step of 0.05. Here, higher spikes correspond to lower values of $r_{sat}/R_L$, and then decrease with larger values of $r_{sat}/R_L$. As a result, for $r_{sat}/R = 0.1$, the efficiency is equal to 75.7%, which is by 2.3% greater than in a nominal case; for $r_{sat}/R = 0.15$, the efficiency is equal to 67.2%, which is by 6.2% greater than in a nominal case. This means that, for the normalized saturation resistance $r_{sat}/R$ equal or smaller than 0.1, it makes sense to use the nominal values of a parallel-circuit Class-E load network, since it will significantly simplify the entire design procedure (no need optimization) and the efficiency will be close to theoretically achievable maximum.

Figure 12.39 shows the simulation setup for ideal parallel-circuit Class-E operation in the frequency domain. Using the frequency domain enables the overall simulation procedure to be much faster than that in the time domain and can take a few seconds. However, because the number of harmonic components is not infinite, the simulation waveforms and numerical results for the optimum load-network parameters are not so accurate. In this case, the input source is changed

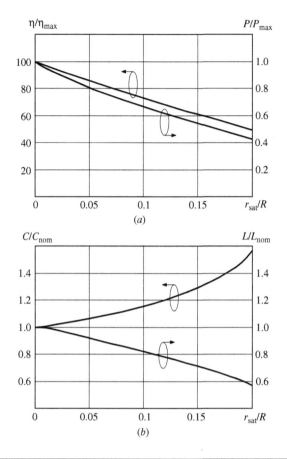

**FIGURE 12.37**

Optimum parameters *versus* on-resistance.

and represents a voltage source with Fourier-series expansion of period square wave used in a harmonic-balance simulator. The harmonic order is chosen to 100. The optimization procedure can be applied with respect to the efficiency as an optimization parameter. Since the simulation time is very short, a number of iterations can significantly be increased for more accuracy. Figure 12.40 shows the switch (*a*) voltage and (*b*) current waveforms obtained for the optimum parameters of the parallel-circuit Class-E load network. Unlike the time-domain simulations, there are smoother transitions between the positions when the switch is turned on and the switch is turned off and *vice versa*. Nevertheless, for $r_{sat}/R = 0.01$, the efficiency is equal to 96.9%, which is only by about 0.1% smaller than that using the time-domain simulation.

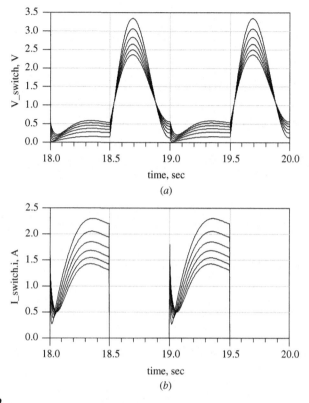

**FIGURE 12.38**

Parallel-circuit Class-E waveforms with finite on-resistance.

## 12.8 ADS CAD design example: high-efficiency two-stage 1.75-GHz MMIC HBT power amplifier

Since the ADC harmonic-balance circuit simulator gives a real possibility of describing the nonlinear circuit behavior in the frequency domain accurately enough, including active device and passive circuits, it can be very useful to simulate the entire high-efficiency power amplifier in a switchmode Class-E approximation as well. Generally, the high-efficiency power amplifier design procedure can include the following several basic steps:

- The proper selection of the active device, accurate measurements of its small-signal $S$-parameters under different bias conditions in a wide frequency range, transformation of the $S$-parameters to the impedance $Z$- and admittance $Y$-parameters to describe the device electrical behavior through the nonlinear

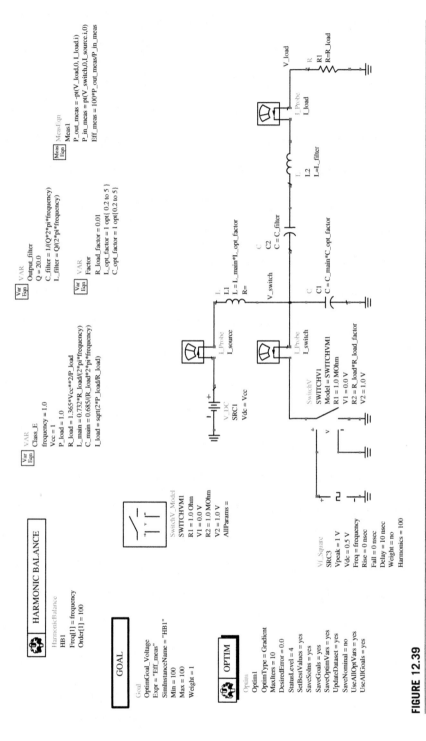

**FIGURE 12.39**

Simulation setup to maintain a Class-E mode in frequency domain.

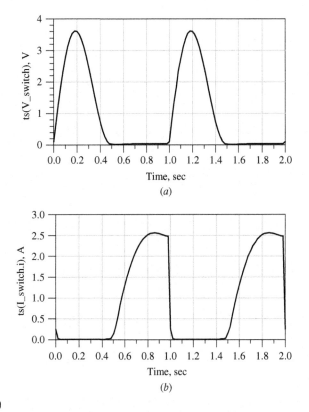

**FIGURE 12.40**

Parallel-circuit Class-E nominal switch waveforms.

model in the form of its equivalent circuit with generally nonlinear parameters.

- The analytical calculation of the optimum parameters of the proper load network in a high-efficiency mode for given output power, supply voltage, and device output capacitance to provide the maximum collector efficiency and required harmonic suppression.
- The choice of a proper bias circuit to optimize the quiescent current and minimize the reference current and variations over temperature for maximum power gain and power-added efficiency.
- The design of the input and interstage matching circuits to provide the minimum input return loss, maximum power gain and power-added efficiency, and stable operating conditions.
- The final circuit parameter optimization to maximize power-added efficiency.

As an example, our objective is to design a high-efficiency monolithic Class-E bipolar power amplifier operating at 1.75 GHz with an output

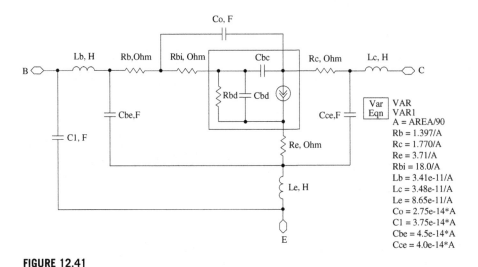

**FIGURE 12.41**

Nonlinear high-frequency InGaP/GaAs HBT equivalent circuit.

fundamental-frequency power $P_{out} = 33$ dBm, a transducer power gain $G_T = 27$ dB, and a supply voltage $V_{cc} = 5$ V. To satisfy these requirements, our decision is to choose a two-stage topology and HBT devices with a minimum saturation voltage at low supply voltage and a transition frequency $f_T > 25$ GHz. Figure 12.41 shows the nonlinear equivalent circuit of an InGaP/GaAs HBT transistor with the circuit parameters corresponding to the device emitter area of $3 \times 30 \ \mu m^2$ (when A = 1). The most nonlinear parameters such as $R_{bd}$ and $C_{bd}$ representing the input diode forward-biased junction can be found by S-parameter simulation as a result of the S-parameter transformation to the input impedance $Z_{in}$ or input admittance $Y_{in}$. Since, in a high-efficiency mode, the voltage swing is very high, the junction base-collector (or collector) capacitance $C_{bc}$ can be chosen at some middle bias point, between the base and collector dc voltages. In our case, $C_{bc} = 4.25e-14*A$ in farads.

Ideally, in a switchmode operation, the active device should act as an ideal switch, driven to be turned on or off by the input RF signal. Assuming the collector efficiency of the second-stage transistor of 80%, the dc power is equal to $P_0 = 2$ W/ $0.8 = 2.5$ W with the dc supply current $I_0 = P_0/V_{cc} = 2.5$ W/5V = 500 mA. Since, for a selected HBT device, the recommended dc current density for a reliable operation over wide temperature range should not exceed 15 mA per 90 $\mu m^2$, the overall emitter area was chosen with some margin to be equal to 5400 $\mu m^2$. The dc output characteristics $I_{ce}(V_{ce})$ of such an InGaP/GaAs HBT transistor for different base bias voltage $V_{be}$ varying from 1.2 to 1.6 V with a voltage step of 30 mV are shown in Fig. 12.42. In this case, the peak collector voltage for a parallel-circuit Class-E mode according to Eq. (6.84) given in Chapter 6 is equal to $I_{max} = 2.647I_0 = 2.15$ A. The idealized Class-E load line (with instant transition between pinch-off and

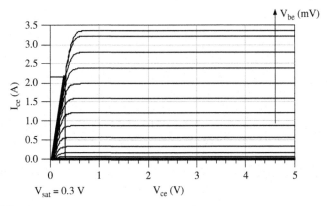

**FIGURE 12.42**

Device output current–voltage characteristics and load line.

saturation regions) is shown as a heavy line in two sections: horizontal at zero current (transistor is pinched off) and slanted upwards to the right (transistor conducts current). It can be seen that the operating point moves along the horizontal $V_{ce}$ axis (pinch-off region) and then along the collector current saturated line (voltage saturation region) until $I_{ce} = 2.15$ A. At this final point, the saturation voltage can be found as equal to $V_{sat} = 0.3$ V. This means that the power loss due to the finite device saturation resistance can be calculated as $P_{sat} = I_0 V_{sat} = 150$ mW with degradation in the collector efficiency of $P_{sat}/P_0 = 0.06$ or 6%.

First, we will design the second stage for a maximum collector efficiency with some intermediate value of the source resistance much less than 50 Ω. The optimum parameters of a parallel-circuit Class-E load network can be calculated using Eqs (6.79) to (6.81) given in Chapter 6 (including the saturation voltage $V_{sat}$) as

$$R = 1.365 \frac{(V_{cc} - V_{sat})^2}{P_{out}} = 15 \ \Omega$$

$$L = 0.732 \frac{R}{\omega} = 1.0 \text{ nH}$$

$$C = \frac{0.685}{\omega R} = 3.9 \text{ pF}.$$

It should be noted that, in view of a saturation voltage, the calculated parameters of the Class-E load network generally cannot be considered optimum unlike the ideal case of zero saturation voltage. However, they can be considered as a sufficiently accurate initial guess for final design and optimization when efficiency is sufficiently high. In addition, the collector capacitance can be larger than required, especially at higher frequencies and for the transistor with large emitter area required for high output power. In our case, the total output capacitance can be estimated as $C_{out} = C_{ce} + C_0 + C_{bc} = 6.6$ pF for an emitter area of 5400 $\mu m^2$, which is 1.5 times greater than required for nominal parallel-circuit Class-E mode. The excessive output capacitance of $(6.6 - 3.9) = 2.7$ pF is responsible for additional switching losses which occur during the transitions from the saturation mode to the pinch-off mode of the device operation. To compensate for this capacitance at the fundamental frequency, it is necessary to connect the inductance in parallel, which value is equal to 3 nH. Hence, the parallel connection of the two inductors with values of 1 nH and 3 nH results in a final value of $(1 \times 3)/(1 + 3) = 0.75$ nH.

Assuming the quality factor of the series filter of $Q_L = 10$, the series capacitance $C_0$ and inductance $L_0$ are calculated as

$$C_0 = \frac{1}{\omega R Q_L} = 0.6 \text{ pF}$$

$$L_0 = \frac{1}{\omega^2 C_0} = 13.8 \text{ nH}.$$

Since the input device impedance is sufficiently low, the intermediate source resistance is chosen to be equal to 5 $\Omega$. In this case, it will be enough to use only one input matching section. The input matching circuit is composed in the form of a high-pass $L$-type section with a shunt inductor and a series capacitor. The parameters of this matching circuit can be simulated in a large-signal mode to minimize input return loss as a criterion by using the Smith chart tuning procedure.

Figure 12.43 shows the simulation setup of the second power-amplifier stage designed to operate in a parallel-circuit Class-E mode. To simulate the electrical characteristics and waveforms, the corresponding current probes and voltage wire labels are incorporated into the circuit. As a current-controlled device, the bipolar transistor requires the dc base driving current, the value of which depends on the output power and device parameters. Because technologically the bipolar device represents a parallel connection of the basic cells, the important issue is to use the ballasting series resistors to avoid the current imbalance and possible device collapse at higher current density levels. Generally, different types of the current-mirror bias circuits can be effectively used to bias a transistor. In our case, the emitter-follower bias circuit that provides the temperature compensation and minimizes the reference current is used [14]. It is very important to provide the proper ratio between the ballasting resistors $R_{34}$ and $R_{33}$, which is equal to the ratio of the corresponding device areas of $5400/270 = 20$. The emitter-follower bias circuit normally requires only several tens of microamperes of reference current.

Figure 12.44 shows the measurement equations required to plot the small- and large-signal power-amplifier characteristics. The small-signal frequency dependence of the stability factor $K$ is shown in Fig. 12.45(a) demonstrating the unconditional stable conditions over entire frequency range with $K > 1$. However, it is not enough just simulate the small-signal performance. It is also important to

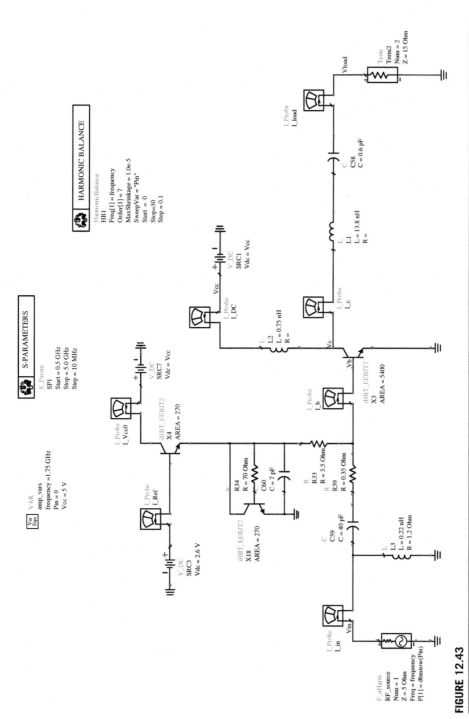

**FIGURE 12.43**

Simulation setup for a Class-E second stage.

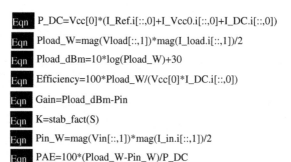

**FIGURE 12.44**

Measurement equations.

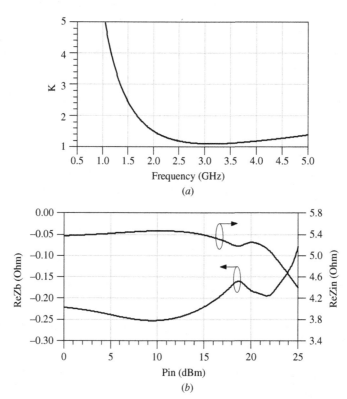

**FIGURE 12.45**

Input impedance and stability.

verify the potential instability which can occur in the form of an injection-locking effect at a large-signal mode near the operating frequency. The dependences of the real part of the device impedance at the base terminal $ReZ_b$ and the real part of the input impedance $ReZ_{in}$ on the input power $P_{in}$ seen by the source are shown in Fig. 12.45($b$). It is seen that the real part of the base impedance is slightly negative. In order to compensate for this negative resistance, it is necessary to connect the resistance of 0.35 $\Omega$ in series to the base compromising the stable operation and sufficient power gain. The real part of the input impedance seen by the source is close to the required 5 $\Omega$ at nominal large-signal operation.

As a result, the second power-amplifier stage exhibits a linear power gain of about 13 dB and a saturated output power of 33.1 dBm, whose dependences are shown in Fig. 12.46($a$), with a maximum collector efficiency of 85.3% and a maximum *PAE* of 60.5%, whose dependences on the input power $P_{in}$ are shown in Fig. 12.46($b$). It is seen that the collector efficiency improvement is achieved at the expense of the significant power gain reduction. However, the maximum

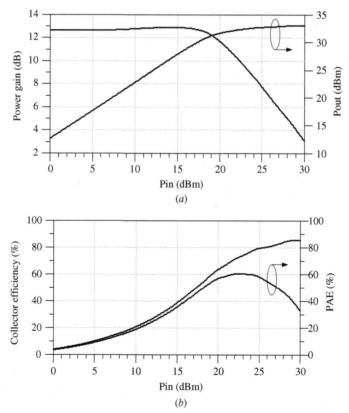

**FIGURE 12.46**

Power gain, output power, and collector efficiency.

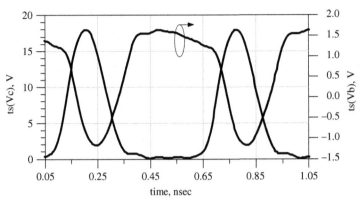

**FIGURE 12.47**

Collector and base voltage waveforms.

*PAE* occurs when the power gain is reduced by about 3 dB, which is a result of the contribution of the driving power.

Figure 12.47 shows the collector and base voltage waveforms corresponding to a Class-E approximation. The collector voltage waveform with a peak factor of about 18 V/5 V = 3.6 was achieved for a maximum collector efficiency with the base voltage waveform having a flattened top part and causing the device to operate in a switchmode with minimum switching loss.

To match the Class-E load resistance $R = 15\ \Omega$ with a standard load resistance $R_L = 50\ \Omega$, it is necessary to use a matching circuit with the series inductor as a first element to provide high impedance at the second- and higher-order harmonics. Since the ratio of impedances is not so high, we can use the low-pass $L$-type matching section with design equations given by Eqs (5.40) to (5.42) in Chapter 5. As a result,

$$Q_L = \sqrt{\frac{50}{15} - 1} \cong 1.5$$

$$L = \frac{Q_L R}{\omega} = 2.1\ \text{nH}$$

$$C = \frac{Q_L}{\omega R} = 2.9\ \text{pF}.$$

Figure 12.48 shows the simulation setup of a Class-E second stage with output matching circuit. Since the loaded quality factor $Q_L$ is sufficiently small, the conditions at the fundamental-frequency and other harmonic components will be slightly different compared to the ideal case. In this case, the easiest way to maximize collector efficiency is to manually tune the parameters of the basic elements such as the parallel inductor, which is shown inside the dashed circle in

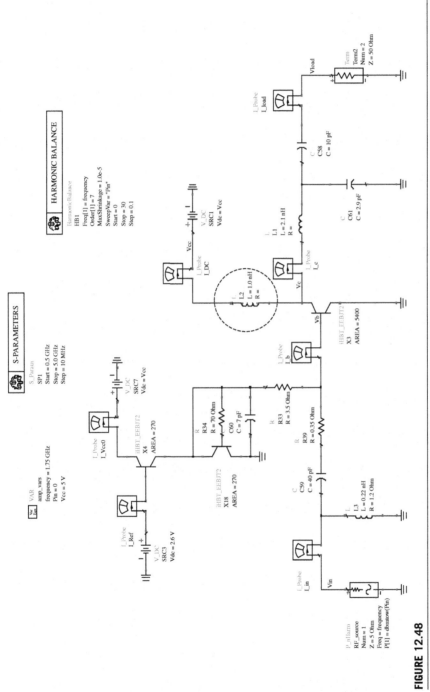

**FIGURE 12.48**

Simulation setup for a Class-E second stage with an output matching circuit.

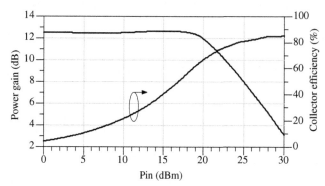

**FIGURE 12.49**

Power gain and collector efficiency.

Fig. 12.48. The maximum collector efficiency of 84.8% and power gain of about 13 dB were achieved, as shown in Fig. 12.49, by tuning the value of the inductance from 0.75 to 1.0 nH. However, a single low-pass *L*-type matching section can provide a second-harmonic suppression of about 20 dB. Consequently, to improve the spectral performance, it is necessary to use the two or more low-pass sections in succession, with equal quality factors to better provide the wider frequency bandwidth [14].

Figure 12.50 shows the simulation setup of the first power-amplifier stage designed to operate in a Class-AB mode. Since it is sufficient to provide an output power from the first stage of not more than 200 mW, the device emitter area size was chosen to be 720 $\mu m^2$ and the equivalent output resistance is assumed to be 50 $\Omega$, which is closed to the calculated value of $R_{out} = (V_{cc} - V_{sat})^2/2P_{out} = (4.7 \times 4.7)/0.4 = 55.2\ \Omega$. Then, to conjugately match the intermediate load resistance of 5 $\Omega$ with the chosen output resistance $R_{out} = 50\ \Omega$, it is convenient to use a high-pass *L*-type matching section with the shunt inductor as a first element followed by the series capacitor. In this case, the shunt inductor with a bypass capacitor at its end in practical implementation can also serve as a dc power supply path. The parameters of the output matching circuit are calculated as

$$Q_L = \sqrt{\frac{50}{5} - 1} = 3$$

$$L = \frac{R_{out}}{Q_L \omega} = 1.5\ \text{nH}$$

$$C = \frac{1}{5 Q_L \omega} = 6.1\ \text{pF}.$$

To obtain a final value of the shunt inductor, it is necessary to take into account the device collector capacitance. In our case, the total output capacitance

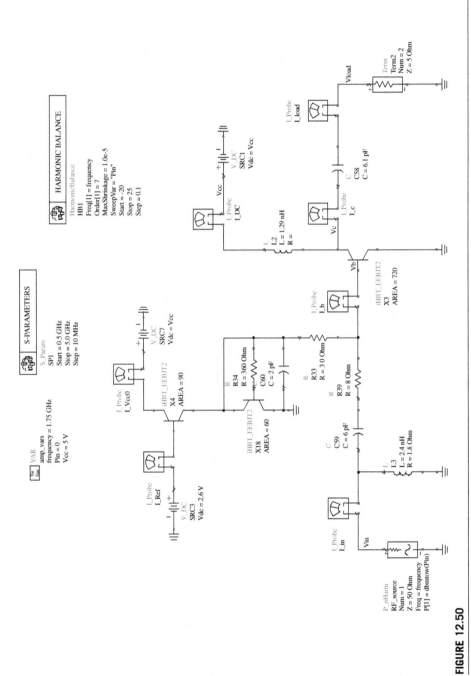

**FIGURE 12.50**

Simulation setup for Class-AB first stage.

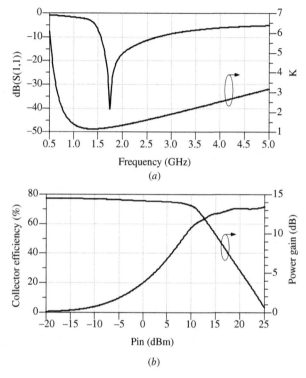

**FIGURE 12.51**

Electrical performance of the first power-amplifier stage.

can be estimated as $C_{out} = 6.6$ pF $\times (720 \; \mu m^2 / 5400 \; \mu m^2) = 0.88$ pF for an emitter area of $720 \; \mu m^2$. To compensate for this capacitance at the fundamental frequency, it is necessary to connect the inductor of 9.4 nH in parallel. Hence, the parallel connection of the two inductors with values of 9.4 and 1.5 nH results in a final value of 1.29 nH. The input matching circuit is composed in the form of a high-pass $L$-type section, the parameters of whose elements can be simulated in a large-signal mode to minimize input return loss as a criterion by using Smith chart tuning procedure. To improve stability factor and simplify the input matching, the resistor of 8 Ω, connected in series to the transistor base terminal, was included.

The small-signal frequency dependence of the stability factor $K$ and $S_{11}$ are shown in Fig. 12.51(a) demonstrating the unconditional stable conditions over entire frequency range with $K > 1$ and input return loss better than 30 dB at the operating frequency. Under a large-signal operation, the first power-amplifier stage exhibits a linear power gain of about 15 dB and a maximum collector efficiency of about 72%, the dependences of which on the input power $P_{in}$ are shown in Fig. 12.51(b). It is seen that the collector efficiency improvement is achieved at the expense of the significant power gain reduction. However, the sufficiently high efficiency of about 65% can be achieved at a 3-dB compression point when the power gain is reduced by 3 dB from its linear value.

The next step is to combine the first power-amplifier stage with the load impedance of 5 Ω and the second power-amplifier stage with the source impedance of 5 Ω in a two-stage power amplifier. Figure 12.52 shows the simulation setup of the two-stage Class-E power amplifier with lumped elements. The main attention should be paid to the interstage matching circuit consisting of the two high-pass matching sections since the impedances are not exactly the same, having also some certain values at the harmonics. In this case, it is necessary to provide some tuning of the elements around their initial values shown in a dashed circle. It can be done sufficiently fast even with manual tuning.

Figure 12.53 shows the measurement equations required to plot the small- and large-signal power-amplifier characteristics including all dc power sources. The small-signal frequency dependences of the stability factor $K$ and $S_{11}$ are shown in Fig. 12.54(a) demonstrating the unconditional stable conditions over entire frequency range with $K > 2$ and input return loss better than 13 dB at the operating frequency. Under large-signal operation, the two-stage parallel-circuit Class-E power amplifier exhibits a linear power gain of more than 29 dB and a maximum *PAE* of about 68%, the dependences of which on the input power $P_{in}$ are shown in Fig. 12.54(b). The sufficiently high efficiency of about 57.5% can be achieved at a 1-dB compression point when the power gain is reduced by 1 dB from its small-signal linear value.

Finally, the lumped inductors must be replaced by the microstrip lines using standard FR4 substrate. The lumped capacitors in the output matching circuit can be implemented as the MIM capacitors using a separate die to provide their high quality factors. To calculate the parameters of the microstrip lines, let us represent the input impedance of a transmission line with arbitrary load impedance as

$$Z_{in} = Z_0 \frac{Z_L + jZ_0 \ \tan \theta}{Z_0 + jZ_L \ \tan \theta} \qquad (12.5)$$

where $\theta$ is the electrical length, $Z_0$ is the characteristic impedance, and $Z_L$ is the load impedance [14]. For a short-circuited condition of $Z_L = 0$ or when $Z_0 \gg Z_L$, we can write a simple equation

$$Z_{in} = jZ_0 \ \tan \theta. \qquad (12.6)$$

Since the input impedance of a lumped inductance $L$ is written as $Z_{in} = j\omega L$, the electrical length $\theta$ can be written as

$$\theta = \tan^{-1} \frac{\omega L}{Z_0}. \qquad (12.7)$$

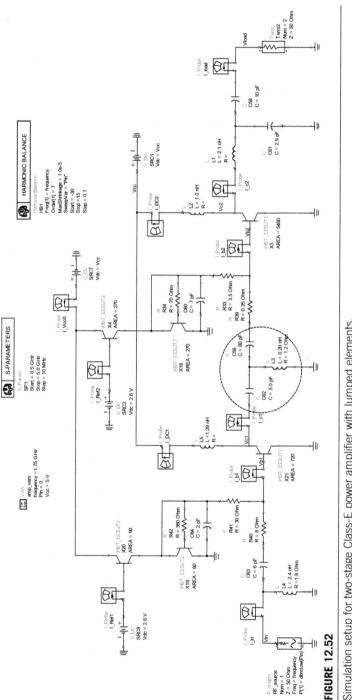

**FIGURE 12.52**

Simulation setup for two-stage Class-E power amplifier with lumped elements.

Eqn Pin_W = mag(Vin[::,1])*mag(I_in.i[::,1])/2

Eqn Pload_W = mag(Vload[::,1])*mag(I_load.i[::,1])/2

Eqn Pload_dBm = 10*log(Pload_W)+30

Eqn Efficiency = 100*Pload_W/(Vcc[0]*I_DC2.i[::,0])

Eqn Gain = Pload_dBm-Pin

Eqn K = stab_fact(S)

Eqn I_DC = I_Ref+I_Vcc0.i[::,0]+I_DC1.i[::,0]+I_DC2.i[::,0]

Eqn PAE = 100*(Pload_W-Pin_W)/(Vcc[0]*I_DC)

Eqn I_Ref = I_Ref1.i[::,0]+I_Ref2.i[::,0]

**FIGURE 12.53**

Measurement equations.

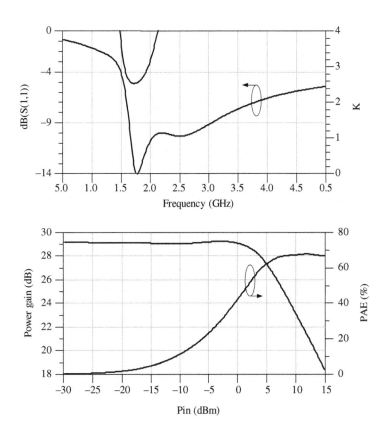

**FIGURE 12.54**

Electrical performance of a two-stage power amplifier with lumped elements.

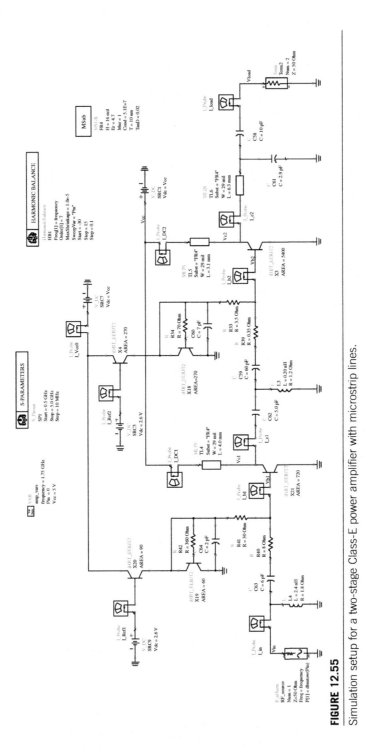

**FIGURE 12.55**

Simulation setup for a two-stage Class-E power amplifier with microstrip lines.

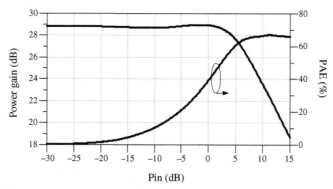

**FIGURE 12.56**

Electrical performance of a two-stage Class-E power amplifier with microstrip lines.

Then, the electrical lengths of the first-stage collector microstrip line, second-stage collector microstrip line, and series microstrip line in an output matching circuit can be calculated, respectively, as

$$\theta_1 = \tan^{-1}(0.284) = 15.85°$$

$$\theta_2 = \tan^{-1}(0.220) = 12.41°$$

$$\theta_3 = \tan^{-1}(0.462) = 24.79°.$$

Using the LineCalc program, available in Tools displayed by the project menu, results in the following geometrical length of the microstrip lines with a 29-mil width (characteristic impedance of approximately 50 $\Omega$) implemented into the FR4 substrate with the parameters given in Fig. 12.55:

$$l_1 = 4.0 \text{ mm}$$

$$l_2 = 3.1 \text{ mm}$$

$$l_3 = 6.3 \text{ mm}.$$

Figure 12.55 shows the simulation setup of the two-stage Class-E power amplifier with microstrip lines. In principle, there is no need to tune any circuit element. In this case, the maximum *PAE* of 67.1% and power gain of 22.5 dB with the output power of 33.6 dBm can be achieved, as shown in Fig. 12.56. At the same time, the output power of 33 dBm and power gain of 27 dB are provided with a *PAE* as high as 62.7%. Some small additional tuning of circuit parameters around their nominal values can probably slightly improve the performance. Hence, based on the example described above, it can be seen how effective a simple analytical approach with quick manual tuning (which can provide high performance and can significantly speed up the entire high-efficiency power-amplifier design procedure) is.

# REFERENCES

1. Tutorial guide to the demonstration version of the HB-PLUS computer program for simulating and optimizing resonant and non-resonant voltage-switching (half-bridge and full-bridge) Classes D and DE DC−DC converters, DC−AC inverters, and AC (including RF) power amplifiers and power generators. Design Automation, Suite 221, 130-D Seminary Ave., MA 02466-2660, U.S.A., January 2007.
2. Sokal NO, Redl R. Power transistor output port model. *RF Design*. June 1987;10:45−53.
3. Baxandall PJ. Transistor sine wave LC oscillators, some general considerations and new developments. *IEE Proc. Electric Power Appl*. May 1959;106:748−758.
4. Zhukov SA, Kozyrev VB. Push−pull switching-mode tuned power amplifier without commutation losses (in Russian). *Poluprovodnikovye Pribory v Tekhnike Elektrosvyazi*. 1975;15:95−106.
5. Hamill DC. Class DE inverters and rectifiers for dc−dc conversion. *Proc. 27th IEEE Power Electronics Spec. Conf.* 1996;I:854−860.
6. Tutorial guide to the demonstration version of the HEPA-PLUS program for design, simulation, and optimization of high-efficiency RF/microwave power amplifiers. Design Automation, Suite 221, 130-D Seminary Ave., MA 02466-2660, U.S.A., September 2006.
7. Raab FH. Effects of circuit variations on the Class E tuned power amplifier. *IEEE J. Solid-State Circuits*. April 1978;SC-13:239−246.
8. Sokal NO. Class−E high−efficiency RF/microwave power amplifiers: principles of operation, design procedure, and experimental verification. In: Huijsing JH, Steyaert M, van Roermund A, eds. *Analog Circuit Design: Scalable Analog Circuit Design, High Speed D/A Converters, RF Power Amplifiers*. The Netherlands: Kluwer Academic Publishers; 2002:269−301.
9. Davis JF, Rutledge DB. A low-cost Class-E power amplifier with sine-wave drive. *1998 IEEE MTT-S Int. Microwave Symp. Dig.* 2:1113−1116.
10. Frey R. 500 W, Class E 27.12 MHz amplifier using a single plastic MOSFET. *Application Note APT9903*, Advanced Power Technology, June 1999.
11. Mediano A, Molina P, Navarro J. An easy to use PSPICE model for power transistors in RF switching operation and including voltage dependent output capacitance. *53rd Automatic RF Techniques Group (ARFTG) Conf Dig.* 1999;25−30.
12. Molina P, del Aguila F, Pala P, Navarro J. Simple nonlinear large signal MOSFET model parameter extraction for Class E amplifiers. *Proc. 9th Int. Conf Electronics, Circuits and Systems.* 2002;1:269−272.
13. Tanji Y, Sekiya H, Asai H. Optimization procedure of Class E amplifiers using SPICE. *Proc. 2005 Europ. Conf. Circuit Theory and Design* 3:133−136.
14. Grebennikov A. *RF and Microwave Power Amplifier Design*. New York: McGraw-Hill; 2004.

# Index

*Note*: Page numbers followed by "*f*" and "*t*" refer to figures and tables, respectively

## A

Active device models, 20
  GaAs MESFETs, 24–29
  GaN HEMTs, 24–29
  LDMOSFETs, 20–24
  low- and high-voltage HBTs, 29–32
Adaptive digital predistorter, 602–603, 602*f*
Advanced Design System (ADS) circuit
    simulator
  switchmode Class-E approximation
    applicability to, 644–649
  high-efficiency power amplifier in,
    649–667
Alternative and mixed-mode high-efficiency
    power amplifiers, 429
  biharmonic Class–$E_M$ power amplifier,
    488–495
  Chireix outphasing power amplifiers,
    512–524, 514*f*, 518*f*
    CMOS outphasing system with Class-E
      power amplifiers, 519*f*
    and instantaneous efficiencies, 523*f*
  Class-DE power amplifier, 430–444, 431*f*,
    432*f*
    with nonlinear shunt capacitances, 440*f*
    with parasitic inductors, 437*f*
    with rectangular drive, 442*f*
    with saturation resistances, 439*f*
    with sinusoidal drive, 443*f*
  Class-E/F power amplifiers, 462
    harmonic-tuned, 463*f*
    single-ended Class-E/$F_3$ mode, 471–488
    symmetrical push–pull configurations,
      465–471
  Class-FE power amplifiers, 444–462
    equivalent circuits of, 448*f*
    with series quarterwave transmission line,
      457*f*
  harmonic tuning using load-pull techniques,
    503–512
  inverse Class-E power amplifiers, 495–502,
    499*f*
Amplifier stability, 54–62
Amplitude distortion, 74

## B

Analog predistortion techniques, 591
  reflective predistorters, 591–592
  transmissive predistorters, 593–597
Angelov model, 25, 27
Asymmetric Doherty amplifiers, 546–550
Asymmetric *N*-way Doherty amplifier, 548*f*, 549

Balanced Doherty amplifiers, 543–546, 544*f*
Balanced transistor, 45–46, 45*f*
Baxandall, P. J., 611
Bias circuits, 67–72
Biharmonic and polyharmonic operation modes
  Class-F power amplifiers, 129–139
  with even-harmonic resonant tanks, 132*f*
  inverse Class-F, 195–202
  with low-pass filters, 134*f*
Biharmonic bipolar-transistor power amplifier,
  135
Biharmonic Class-$E_M$ power amplifier, 488–495
  optimum operation, 491
Biharmonic power amplifier
  with cathode harmonic control, 139
  with input harmonic control, 135, 201
Bipolar current-switching Class-D power
  amplifier, 119–120, 120*f*
Bipolar high power amplifier, 228*f*
Bipolar junction transistors (BJTs), 610
Bondwire inductance effect, 332–333
Broadband capability, 564–568
  of Class-F power amplifiers, 181–183
  of two-stage Doherty amplifier, 564–568
Broadband Class-E power amplifier, 387
  broadband GaN HEMT power amplifier, 407
  circuit, conductance and susceptance of, 402*f*
  CMOS power amplifiers, 424–426
  high-efficiency microstrip LDMOSFET power
    amplifier, 415*f*
  high-power RF power amplifiers, 416–419
  load network with bandpass filter and
    impedance, 403, 404*f*
  microwave monolithic power amplifiers,
    419–424

Broadband Class-E power amplifier (*Continued*)
parallel-circuit power amplifier, 409–416
reactance compensation circuit with lumped
elements and transmission line, 405*f*
reactance compensation technique,
387–400
with shunt capacitance, 400–408
simulated broadband LDMOSFET power
amplifier, 411*f*
single- and double-susceptance compensation
circuits, 409*f*
two-stage InGaP/GaAs HBT power amplifier,
415–416
*See also* Class-E power amplifiers

## C

Capacitive load impedance, 288
Cascaded tuner method, 512
Cascode Class-E power amplifier
with compensating inductor, 350
with finite dc-feed inductance, 348
with output transformer, 349
*See also* Class-E power amplifiers
CDMA2000, 231
CGH40010, 155, 236, 237, 379, 384, 482, 488
CGH40010P, 237
Chireix combining technique, 520
Chireix outphasing power amplifiers, 512–524,
514*f*, 518*f*
CMOS outphasing system with Class-E power
amplifiers, 519*f*
and instantaneous efficiencies, 523*f*
Circuit theory, 9–10
Class Φ, 445
Class-A power amplifier, 7–13
voltage and current waveforms in, 8*f*
Class-AB power amplifier, 7–13, 661
collector current waveforms in, 14*f*
with source-harmonic tuning, 508*f*
Class-B power amplifier, 7–13
with harmonic tuning, 505*f*
voltage and current waveforms in, 11*f*
Class C harmonic reactance, 19
Class-C power amplifier, 7–13
collector current waveforms in, 14*f*
Class-D power amplifier, 83
versus Class-DE circuit, 611–613
complementary voltage-switching
configuration, 92–97
drive and transition time, 111–118
"Efficiency and Powers" for, 622*f*
"Enter Circuit Parameters" for, 621*f*

for digital pulse-modulation transmitters,
123–127
with hard switching, 620–623
implementation, 118–122
switchmode power amplifiers with resistive
load, 83–92
symmetrical current-switching configuration,
103–107
transformer-coupled current-switching
configuration, 99–103
transformer-coupled voltage-switching
configuration, 97–99
voltage and current waveforms for, 622*f*
voltage-switching configuration with reactive
load, 107–111
Class-DE power amplifier, 429, 430–444,
431*f*, 432*f*
"Efficiency and Powers" for, 624*f*
with nonlinear shunt capacitances, 440*f*
with parasitic inductors, 437*f*
with rectangular drive, 442*f*
with saturation resistances, 439*f*
with sinusoidal drive, 443*f*
with soft switching, 623–625
voltage and current waveforms for, 623*f*
Class-E Doherty power amplifier, 542
Class-E/F impedances at fundamental and
harmonics, 464*t*
Class-E/F power amplifiers, 429, 462
single-ended Class-E/F₃ mode, 471–488
symmetrical push–pull configurations,
465–471
Class-E harmonic reactance, 19
Class-E LDMOSFET power amplifier,
broadband performance of, 412*f*
Class-E power amplifiers, 490, 495, 496, 501
"Automatic Preliminary Design" screen,
627–628
in frequency domain, 650*f*
load-network parameter variations, 616–619
with lumped elements, 664*f*
with microstrip lines, 666*f*
power amplifier design using SPICE,
638–643
second stage, 655*f*
in time domain, 645*f*
"Transistor Evaluation", 627*f*
*See also* Broadband Class-E power amplifier;
Cascode Class-E power amplifier
Class-E power amplifiers with shunt
capacitance, 281
basic circuits of, 251

component resistance and predicted power loss, 263

detuned resonant circuit effect, 245–250

driving signal and finite switching time, 263–270

high-power high-frequency Class-E MOSFET power amplifiers, 293

high-power VHF Class-E power amplifier, 294

with impedance-matching circuits, 257

load networks with transmission lines, 281–291

load network with shunt capacitor and series filter, 250–256

lumped-elements schematics, 258, 259

nonlinear shunt capacitance effect, 270–272

nonsinusoidal output voltage, basic circuits of switchmode power amplifier with, 247

optimum, nominal, and off-nominal Class-E operation, 272–277

practical Class-E power amplifiers and applications, 291–300

push–pull operation mode, 277–281

saturation resistance effect, 260–263

standard load, matching with, 256–260

Class-E power amplifier with finite dc-feed inductance

bondwire inductance effect, 332–333

Class-E with one capacitor and one inductor, 305–313

CMOS technology, 348–354

even-harmonic Class-E power amplifier, 330–332

generalization of, 313–320, 321, 322

load network with transmission lines, 333–340

operation beyond maximum Class-E frequency, 340–345

parallel-circuit Class-E power amplifier, 324–330

power gain, 345–348

subharmonic Class-E power amplifier, 320–324

Class-E power amplifier with shunt inductance, 495f

Class-E with quarterwave transmission line, 357

10-W, 2.14-GHz Class-E GaN HEMT power amplifier, 378–385

current waveforms of, 370f

design example, 378–385

equivalent circuit of, 359f

load network with parallel quarterwave line, 357–364

load network with parasitic shunt capacitance and bondwire inductance, 358f

load network with series quarterwave line and shunt filter, 376–377

load network with zero series reactance, 367–371

matching circuit with lumped elements, 372–373

matching circuit with transmission lines, 373–376

optimum impedances at fundamental and harmonics for, 371t

optimum load-network parameters, 364–367

voltage and current waveforms of, 369f

Class-$E_3$F and Class-$E_3F_2$ power amplifiers, 458f

Class-$E_M$ and Class-E power amplifiers efficiency versus switching time for, 494f

Class-$E_M$ MOSFET power amplifier, 490

Class-$EF_2$ power amplifier, 460f

Class-$E/F_{2,odd}$ power amplifier and collector waveforms, 468f with third-harmonic trap, 470f

Class-$E/F_3$ power amplifier, 476f, 478, 482 frequency response of lumped network for, 483f

GaN HEMT power amplifier, 484f simulated results of, 485f simulation setup for, 484f transmission-line, 489f

in time domain simulation setup for, 479f transmission-line, 486f

Class-$E/F_n$ load network at fundamental frequency, 474f

Class-$E/F_{odd}$ power amplifier circuit implementation and collector waveforms, 466f dual-band, 471f

Class-$E/F_{x,odd}$ circuit implementation, 469

Class-F harmonic reactance, 19

Class-F power amplifiers, 129, 195, 510 biharmonic and polyharmonic operation modes, 129–139 broadband capability of, 181–183 effect of saturation resistance and shunt capacitance, 157–162 Fourier analysis, 131, 132f idealized Class-F mode, 139–143 LDMOSFET power amplifier design examples, 176–181

Class-F power amplifiers (*Continued*)
load networks with lumped elements, 162–169
load networks with transmission lines,
169–176
with maximally flat waveforms, 143–151
practical Class-F power amplifiers and
applications, 183–190
with quarterwave transmission line,
151–157, 186*f*
GaN HEMT power amplifier, 156*f*, 157*f*
*See also* Inverse Class-F power amplifiers
Class-FE load-network current waveforms, 456*f*
Class-FE power amplifiers, 429, 444–462
equivalent circuits of, 448*f*
in frequency domain, 454*f*
with lumped resonant circuits, 461*f*
with series quarterwave transmission line, 457*f*
Class-FE switch voltage and current waveforms,
455*f*
CMOS, 227*f*
CMOS Class-E power amplifiers, 348–354,
424–426
CMOS outphasing system, 519
CMOS voltage-switching Class-D power
amplifiers, 124
Coding efficiency, 125
Collector capacitance, 161
Collector efficiency, evaluation of, 159
Complementary voltage-switching Class-D
amplifier, 92–97
Computer-aided design of switchmode power
amplifiers, 607
ADS circuit simulator, its applicability to
switchmode Class-E, 644–649
Class-E load-network parameter variations,
616–619
Class-E power amplifier design using SPICE,
638–643
HB-PLUS CAD program, 608–613
capabilities, 608–609
circuit topologies, 609–611
Class-D versus Class-DE circuit, 611–613
examples, 619–625
HEPA-PLUS CAD program for Class-E
circuit topology, 614–615
examples, 626–638
optimization, 615–616
program capabilities, 613–614
steady-state periodic response, 614
transient response, 614
high-efficiency two-stage power amplifier,
649–667

Conduction angle, 4*f*
Critical mode, 8
Current-switching push–pull power amplifier,
87–88

**D**
Delay distortion, 74
Delta-sigma modulator (DSM), 123, 124, 125
Design principles, of power amplifier
active device models, 20
GaAs MESFETs, 24–29
GaN HEMTs, 24–29
LDMOSFETs, 20–24
low- and high-voltage HBTs, 29–32
amplifier stability, 54–62
bias circuits, 67–72
Classes of operation
Class-A, 7–13
Class-AB, 7–13
Class-B, 7–13
Class-C, 7–13
finite number of harmonics, 17–20
distortion fundamentals, 72
distortion, types of, 74
distortion of electrical signals, 73–74
linearity, 72–73
memory, 73
nonlinear distortion analysis for sinusoidal
signals, 75
time variance, 73
high-frequency conduction angle, 32–38
load line and output impedance, 13–17
load–pull characterization, 52–54
nonlinear effect of collector capacitance,
38–42
parametric oscillations, 62–66
power gain and impedance matching, 47–51
push–pull power amplifiers, 42–47
spectral-domain analysis, 1–7
Detuned resonant circuit effect, 245–250
Device under test (DUT), 53, 507, 511*f*
Digitally driven Doherty amplifier, 562–564
Digital predistortion (DPD) techniques, 562, 598
adaptation, 601–603
memoryless digital predistortion, 598–600
performance, 603–604
Digital pulse-modulation transmitters, Class-D
power amplifiers for, 123–127
Digital signal processor (DSP), 299, 514, 562
Distortion, 72, 575, 576
of electrical signals, 73–74
linearity, 72–73

memory, 73
nonlinear distortion analysis for sinusoidal
    signals, 75
time variance, 73
types, 74
Distributed N-way Doherty amplifier, 549*f*, 550
Doherty, William H., 529
Doherty power amplifiers, high-efficiency,
    536, 537
    asymmetric Doherty amplifiers, 546–550
    balanced Doherty amplifiers, 543–546
    broadband capability, 564–568
    digitally driven Doherty amplifier, 562–564
    dual Doherty amplifiers, 543–546
    for handset applications, 568*f*
    with harmonic control circuits, 541*f*
    historical aspects and conventional, 529
    integration, 559–562
    inverted Doherty amplifiers, 556–558
    multiband capability, 564–568
    multistage Doherty amplifiers, 550–556
    push–pull Doherty amplifiers, 543–546
Double-reactance compensation circuit and
    impedances, 393*f*, 394, 395*f*
Double-susceptance compensation circuit, 390,
    393, 409*f*, 410
Drain-source capacitance, 17, 49, 63, 165, 189,
    208, 221, 379, 411, 414, 470, 471, 482,
    502, 638, 639, 641
    of MOSFET device, 440
Dual-band Class-E/F$_{odd}$ power amplifier, 471*f*
Dual Doherty amplifiers, 543–546
    architecture, 545*f*
Dual-input digital Doherty amplifier, 563*f*

**E**
"Efficiency and Powers", 620
    for Class-D, 622*f*
    for Class-DE, 624*f*
Electrical length of transmission line, 488
"Enter Circuit Parameters", for Class-D
    operation, 621*f*, 628, 629
Envelope elimination and restoration (EER)
    approach, 299
    Kahn approach, 189–190
Even-harmonic Class-E power amplifier, 330–332

**F**
Filter-diplexer, 90–92
Finite number of harmonics, 1, 17–20
Four-stage Doherty amplifier architecture, 555*f*
Four-stage outphasing architecture, 521, 521*f*

**G**
GaAs MESFETs, 24–29, 138, 185, 202, 421,
    502, 509, 552, 560
Gain based predistorter, 598, 599*f*, 601
GaN HEMT, 24–29
    device, 501
    inverse Class-F power amplifier, 221
    power amplifier, 237*f*
    for WCDMA systems, 231–242
Giacoletto model, for bipolar transistor, 33*f*

**H**
Half-bridge voltage-switching Class-D power
    amplifier, 123, 609*f*
Harmonic active load-pull measurement system,
    507*f*
Harmonic-reaction amplifier, 504*f*
Harmonic-tuned Class-E/F power amplifier, 463*f*
Harmonic tuning
    Class-B power amplifier with, 505*f*
    using load-pull techniques, 503–512
HB-PLUS CAD program, 608–613
    capabilities, 608–609
    circuit topologies, 609–611
    Class-D versus Class-DE circuit, 611–613
    for Class-E power amplifier
        automatic preliminary design module use,
            627–629
        candidate transistor, 626–627
        circuit topology, 614–615
        design optimization, 631–635
        example, 626–638
        nominal-waveforms circuit, 629
        optimization, 615–616
        program capabilities, 613–614
        RF output spectrum, 629–631
        steady-state periodic response, 614
        SWEEP function, 635–638
        transient response, 614
    examples, 619–625
    load network and load in, 610*f*
High-efficiency Doherty power amplifiers *See*
    Doherty power amplifiers, high-efficiency
High-efficiency two-stage transmission-line
    power amplifier, 230*f*
High-frequency conduction angle, 32–38
High-power RF Class-E power amplifiers, 416–419
High-voltage HBTs, 29–32

**I**
Idealized transmission-line, 213*f*, 219*f*
Ideal limiter, 577

InGaP/GaAs HBT equivalent circuit, 652*f*
Input impedance and stability, 656*f*
Intermodulation distortion (IMD), 75, 588
Inverse Class C harmonic reactance, 19
Inverse Class-E power amplifiers, 495–502, 499*f*
    transmission-line, 502*f*
    with series inductance, 501*f*
Inverse Class-F power amplifiers, 195
    basic circuits, 203*f*
    biharmonic operation modes, 195–202
    GaN HEMT Doherty amplifier, 543*f*
    GaN HEMT power amplifiers for WCDMA
        systems, 231–242
    harmonic reactance, 19
    idealized mode, 202–205
    ideal waveforms of, 204*f*
    LDMOSFET power amplifier design
        examples, 222–226
    load networks with lumped elements,
        208–211
    mode, 510
    polyharmonic operation modes, 195–202
    practical implementation, 226–231
    with quarterwave transmission line, 205–208
    resonant circuits, 203*f*
    *See also* Class-F power amplifiers
Inverted Doherty amplifiers, 556–558

**K**

Kirchhoff's equations, 2, 114, 448

**L**

LDMOSFET, 20–24
    current-switching Class-D power amplifier, 122
    inverse Class-F power amplifier, 221
    power amplifier design examples, 222–226
    power amplifier design examples
        Class-F power amplifiers, 176–181
        inverse Class-F, 222–226
LINC (linear amplification using nonlinear
        components), 231, 378, 512, 516
Linearization, 575, 582
Linear mode, 7–8
Load line, 14
    and output impedance, 13–17
Load network, 233*f*
    Class-E power amplifiers
        with finite dc-feed inductance, 333–340
        with shunt capacitance, 281–291, 401*f*
    with distributed second-harmonic control, 506*f*
    impedance, 483, 488
    input impedance, 162
    with lumped elements
        broadband Class-E, 388–394
        Class-F power amplifiers, 162–169
        inverse Class-F, 208–211
    second-harmonic control, 215*f*
    seen by a device output at harmonics, 487*f*
    with shunt capacitor and series filter, 250–256
    third-harmonic control, 215*f*
    with transmission lines
        broadband Class-E, 394–400
        Class-E with shunt capacitance, 281–291
        Class-F power amplifiers, 169–176
        inverse Class-F, 212–222
Load–pull characterization, 52–54
Load resistance, 501
Low-voltage HBTs, 29–32
Lumped resonant circuits, Class-FE
    approximation with, 461*f*

**M**

Mapping predistorter, 598, 599*f*
Memoryless digital predistortion, 598–600
Memoryless predistorter for octave-bandwidth
    amplifiers, 584–588
Memoryless RF power amplifier, transfer
    characteristic of, 577*f*, 578*f*
MESFET, 27, 31, 152, 162, 185, 189, 264, 285,
    295, 300, 417, 538, 560, 610
    device, 138
    power amplifier, 503, 508
Microwave current-switching Class-D power
    amplifiers, 106
    with multiharmonic transformation networks, 121
Microwave monolithic Class-E power amplifiers,
    419–424
Mixed-mode high-efficiency power amplifiers
    *See* Alternative and mixed-mode
    high-efficiency power amplifiers
Monolithic microwave integrated circuit
    (MMIC), 334, 529, 559
MOSFET, 1, 17, 20, 23, 24, 59, 67, 85, 92, 107,
    114, 135, 138, 226, 262, 264, 278, 291,
    418, 419, 443, 470, 471, 610, 614, 626,
    638–639
    Class-E power amplifier, 639*f*
    device, drain-source capacitance of, 440
    and MESFET transistors, 263–264
    power amplifier, 470
    voltage-switching Class-D power amplifier,
    118–119
Multiband capability
    of two-stage Doherty amplifier, 564–568

Multiharmonic load-pull design, 511*f*
  procedure, circuit topologies for, 509*f*
Multistage Doherty amplifiers, 550—556
  architectures, 551*f*
Multistage outphasing system, 522

## N

Network impedance, 337, 403, 462, 480
Nitronex RF power transistor, 238
Nonlinear distortion, 3, 74, 75
Nonlinear effect of collector capacitance, 38—42
Nonlinear mode, 8, 12, 22, 346
Nonlinear shunt capacitance effect,
  270—272
*n*th-harmonic peaking
  current waveforms for, 147*f*, 149
  voltage waveforms for, 145*f*, 146

## O

Octave-bandwidth amplifiers
  memoryless predistorter for, 584—588
  predistorter with memory for, 589—590, 591*f*
OFDM (orthogonal frequency-division
  multiplexing), 231
Optimum, nominal, and off-nominal Class-E
  operation, 272—277
Optimum Class-DE operation mode, waveforms
  for, 436*f*
Optimum Class-FE operation mode, waveforms
  for, 450*f*
Optimum load resistance, 500
Optimum phase angle, 500
Optimum series inductance, 501
Outphasing power amplifier system, 513*f*

## P

Parallel-circuit Class-E power amplifier,
  324—330
Parallel-circuit Class-E waveforms, 649*f*
  nominal switch waveforms, 651*f*
Parametric oscillations, 62—66
Parasitic bipolar oscillators, 57—58
Phase distortion, 74
Phase-locked loop (PLL), 299—300
PHEMT device, 138, 297, 501—502
Piecewise-linear approximation
  technique, 2, 3*f*
Polar transmitter architecture, 299
Postdistorter, 590
Power-added efficiency (PAE), 12
Power gain, 47, 345—348
  and impedance matching, 47—51

Predistortion linearization techniques, 575,
  582—583
  analog predistortion implementation, 591
    reflective predistorters, 591—592
    transmissive predistorters, 593—597
  digital predistortion implementation, 598
    digital predistorter performance,
      603—604
    digital predistortion adaptation, 601—603
    memoryless digital predistortion, principles
      of, 598—600
  octave-bandwidth amplifiers
    memoryless predistorter for, 584—588
    predistorter with memory for, 589—590,
      591*f*
  postdistortion, 590
  RF power amplifiers modeling with memory,
    576—582
Probability density function (PDF), 522, 561
Push—pull Class-E power amplifier, 277—281
Push—pull Doherty amplifiers, 543—546
Push—pull power amplifiers, 42—47

## Q

Quarter-wavelength transmission-line
  transformer, 394
Quarterwave transmission line
  1.78-GHz power amplifier with, 229*f*
  ABCD-matrix for, 398
  Class-E power amplifier with, 357
  Class-F power amplifier with, 186*f*
  ideal current waveform in, 155
  inverse Class-F with, 205—208
  *See also* Transmission line
Quasi-complementary MOSFET voltage-
  switching Class-D power amplifier,
  118—119

## R

Radio frequency (RF) power amplifiers with
  memory, 576—582
Reactance compensation technique, 387
  double-reactance compensation circuits, 390,
    393*f*
  load networks
    with lumped elements, 388—394
    with transmission lines, 394—400
  single-reactance compensation circuits, 388*f*
    and impedances, 392*f*
  single-susceptance compensation circuit and
    admittances, 391*f*
Reflective predistorters, 591—592

## S

Saturation resistance, effect of, 260–263
Second-harmonic resonator
   1-GHz power amplifier with, 229f
Second-harmonic-tuned power-amplifier design
      strategy, 504
Shunt capacitance, 501
   Class-E power amplifiers with *See* Class-E
         power amplifiers with shunt capacitance
Signal component separator (SCS), 512
Silicon LDMOSFET devices, 294
Simulated 500-MHz single-stage microstrip
      power amplifier, 222f, 224f
Single-ended circuit topology, 615f
Single-ended Class-E/F$_n$ power amplifier, 472f
Single-ended X-band MESFET power amplifier,
      504
Single-susceptance compensation circuit and
      admittances, 391f
Sinusoidal 180° out-of-phase driving waveforms,
      441f
Solid-state power amplifier, digital predistortion
      of, 603f
Source- and load-pull simulation setup, 511f
Source-harmonic tuning, Class-AB power
      amplifier with, 508f
Spectral-domain analysis, 1–7
SPICE model, 641
   for ARF448 MOSFET, 640f
   of transistor switching mode, 642f
Stub resonator method, 512
Subharmonic Class-E power amplifier, 320–324
Superposition, 72
Susceptance compensation load network,
      390, 391f
Switchmode power amplifiers with resistive
      load, 83–92
Switch voltage and current waveforms
   in frequency domain, 481f
   in time domain, 480f
Symmetrical current-switching Class-D power
      amplifier, 103–107

## T

T-type transmission-line impedance transformer,
      185
Terminating capacitance, 520
Three-stage Doherty amplifier architecture, 553f
Three-stage inverted Doherty amplifier, 558f
Total harmonic distortion (THD), 75
Transformer-coupled current-switching Class-D
      power amplifier, 99–103

Transformer-coupled current-switching
      push–pull power amplifier, 89–90
Transformer-coupled voltage-switching Class-D
      power amplifier, 97–99
Transistor saturation resistance, Class-E load
      network with, 260–263
Transmission line, 214f, 217f, 220f, 221f
   5-W inverse Class-F GaN HEMT power
         amplifier, 240f
   50-W inverse Class-F GaN HEMT power
         amplifier, 241f
   electrical length of, 488
   *See also* Quarterwave transmission line
Transmission line, load networks with
   Class-E power amplifiers
      with finite dc-feed inductance, 333–340
      with shunt capacitance, 281–291
Transmission-line 10-W inverse Class-F GaN
      HEMT power amplifier, 236f
Transmission-line broadband Class-E load
      network, 412–413, 414f
Transmission-line Class-E GaN HEMT power
      amplifier, 292f
Transmission-line Class-E load network, 283f,
      287–288, 290, 337
   with susceptance compensation, 414f
Transmission-line Class-E power amplifier, 373,
      374, 374f
Transmission-line Class-E/F$_3$ GaN HEMT power
      amplifier, 489f
Transmission-line Class-E/F$_3$ power amplifier,
      486f
Transmission-line Class-F load network, 170,
      173f, 175f
Transmission-line inverse Class-E power
      amplifier, 502
Transmission-line inverse Class-F power
      amplifier, 214f, 217f, 220f, 221f, 232f,
      237–238
   5-W inverse Class-F GaN HEMT power
         amplifier, 240f
   50-W inverse Class-F GaN HEMT power
         amplifier, 241f
Transmission-line modeling (TLM) technique,
      334
Transmission-line reactance compensation
      circuits and design equations, 397t
Transmission-line single frequency equivalence
      technique, 400f
Transmission-line susceptance compensation
      circuit, 396
Transmissive predistorters, 593–597

Traveling wave tube amplifier, digital
  predistortion of, 603, 604*f*
Triplexer tuning method, 511
Two-stage injection-locked Class-E power
  amplifier, 353

## V
VDMOSFET, 20
VHF high-efficiency bipolar power amplifiers, 184
Virtual grounding, 45
Voltage standing wave ratio (VSWR), 62, 276
Voltage-switching Class-D and Class-F power
  amplifiers, 446*f*

Voltage-switching Class-DE and Class-FE power
  amplifiers, 447*f*
Voltage-switching Class-D power amplifier, 123
  with reactive load, 107−111
Voltage-switching push−pull power amplifier,
  85−87

## W
WCDMA systems, 231−241

## Z
Z-parameters, 28*f*

Printed in the United States
By Bookmasters